Handbook of Measurement Science

Volume 2 Practical Fundamentals

Handbook of Measurement Science

Volume 2　Practical Fundamentals

Edited by
P. H. Sydenham
School of Electronic Engineering
South Australian Institute of Technology

A Wiley-Interscience Publication

JOHN WILEY & SONS
Chichester · New York · Brisbane · Toronto · Singapore

05853495

Library of Congress Cataloging in Publication Data (Revised):
Main entry under title:

Handbook of measurement science.
  'A Wiley–Interscience publication.'
  Bibliography: v. 1, p.
  Vol. 2. includes index.
  Contents: v. 1. Theoretical fundamentals—
v. 2. Practical fundamentals.
  1. Mensuration. I. Sydenham, P. H.
T50.H26   620'.0044   81-14628

ISBN 0 471 10037 4 (v. 1)   AACR2

British Library Cataloguing in Publication Data:
Handbook of measurement science.
  Vol. 2: Practical fundamentals
  1. Measuring instruments
  I. Sydenham, P. H.
  620'.0044 ✓   QC100.5

ISBN 0 471 10493 0

Filmset and printed in Northern Ireland at The Universities Press (Belfast) Ltd.
Bound at the Pitman Press Ltd., Bath, Avon.

D

620 · 0044

HAN

To R. V. Jones
for expanding my vision of instrumentation and
providing guidance and understanding

Contributing Authors

K. A. BLAKE

Barr and Stroud Ltd, Anniesland, Glasgow, UK

E. BONOLLO

Department of Design, Royal Melbourne Institute of Technology, Australia

D. BOSMAN

Department of Electrical Engineering, Twente University of Technology, The Netherlands

T. J. S. BRAIN

National Engineering Laboratory, East Kilbride, Glasgow, UK

L. FINKELSTEIN

Department of Physics, The City University, London, UK

T. P. FLANAGAN

Sira Institute Limited, Chislehurst, Kent, UK

J. A. GILMOUR

National Association of Testing Authorities, Chatswood, New South Wales, Australia

J. W. HOBSON

Munitions Supply Division, Department of Industry and Commerce, Victoria, Australia

E. K. LASKARIS

Formerly Instrument Evaluation Department, Sira Institute Limited, Chislehurst, Kent, UK

G. MITRA

Optics Division, Central Scientific Instruments Organization, Chandigarh, India

F. G. Peuscher   *Department of Electrical Engineering, Twente University of Technology, The Netherlands*

J. Prasad   *Optics Division, Central Scientific Instruments Organization, Chandigarh, India*

L. Schnell   *Department of Measurement and Instrument Engineering, Technical University, Budapest, Hungary*

P. H. Sydenham   *School of Electronic Engineering, South Australian Institute of Technology, Australia*

W. Trylinski   *Formerly Faculty of Fine Mechanics, Warsaw Technical University, Poland*

R. D. Watts   *Department of Systems Science, The City University, London, UK*

R. S. Watts   *School of Chemical Technology, South Australian Institute of Technology, Australia*

Contents

19. Human factors in display design

D. Bosman

797

20. Measurement of electrical signals and quantities

L. Schnell

823

21. Mechanical regime of measuring instruments   883
W. Trylinski

24. Transducer practice: displacement   1037

P. H. Sydenham

25. Transducer practice: flow   1069

T. J. S. Brain and K. A. Blake

26. Transducer practice: thermal   1137
P. H. Sydenham

27. Transducer practice: chemical analysis   1173
R. S. Watts

30. Management of existing measurement systems   **1283**
J. W. Hobson

Editor's Preface

This second volume of the *Handbook of Measurement Science* continues the theme, expounded in the Preface to Volume 1, that there exists 'an underlying collection of fundamentals that is applicable, in part or whole, to all measurement situations regardless of how diverse the applications appear to be.' The background to the development of this work is described in the Preface to Volume 1.

Volume 1 includes theoretical fundamentals and covers the theory of measurement, where measurements fit in the general scheme of systems, the several kinds of standards that assist succinct definition of terms and units. Over several chapters it then deals with the processing of signals of analog and digital form, and how to transmit signals. The final chapter of Volume 1 gives an overview of the use of feedback in systems. This volume follows on from Volume 1 with the presentation of what can be loosely called *fundamentals of practice*. It covers the realization of measurement systems, their design, manufacture, and management.

Chapter 15, the first of Volume 2, deals with the building blocks that are used in measurement systems, thereby providing an approach to understanding their internal structure. Chapters 16 and 17 then provide an appreciation of how the static and dynamic behaviours of systems are described and the various modes of likely behaviour that will be encountered.

With this general background established, Chapter 18 then provides the tools for mathematical modelling of instrument system stages, models that enable performance to be assessed and optimized before a system is actually manufactured. The display interface between the measurement system and an operator is the topic of Chapter 19.

The majority of modern measurement systems involve electrical signals.

Chapter 20 covers measurement of the various electrical quantities. Measurement systems in the main, use combinations of elements from the mechanical, electronic, and optical hardware regimes. The design principles involved are the subjects of Chapters 21 to 23.

Chapters 24 to 27 are primarily included to provide exemplification of how the preceding principles are put into practice. They cover the dominant methods used to transduce displacement, flow, thermal, and chemical measurands. Knowing which principle to use to transduce a given measurement, how to optimize the performance, and how to design the necessary additional data processing system, is not all that might be required in the overall design of a total measurement system. Chapters 28 and 29 cover aspects of manufacture.

Many people who are involved with measurement systems do not need to design them; their requirement is to maintain already manufactured systems that are installed in monitoring and control applications. Chapter 30 is an account of the experience of a person in managing an extensive system of instrumentation in a collection of factories. If measuring instruments are to correctly and traceably map a variable into an equivalent signal they must be calibrated, and perhaps be evaluated and their performance accredited. Chapter 31 discusses these aspects. Finally, in Chapter 32, readers are provided with an introduction to the literature of measurement science and technology.

As I recorded in the Preface to Volume 1, I again wish to recognize the willing cooperation of the internationally located contributing authors to this volume. Their contributions show that measurement science is, indeed, a universal discipline that is practised in much the same manner in countries across the world.

My appreciation is extended to the library staff at the Levels Library of the South Australian Institute of Technology who helped in such matters as checking citations of contributors. Elaine Milsom and Elsie Clarke assisted with retyping text where needed. On the almost certain assumption that this volume will be taken into production in the same excellent manner as was Volume 1, I again thank the John Wiley and Sons staff at their Chichester, UK, location.

Adelaide   P. H. SYDENHAM
December, 1981

Handbook of Measurement Science, Volume 2
Edited by P. H. Sydenham
© 1983 John Wiley & Sons Ltd

Chapter

15 P. H. SYDENHAM

Physiology of Measurement Systems

Editorial introduction

Strange, around 1870, considered the general architecture of instruments under the heading 'the physiology of instruments'. This opening chapter discusses the building blocks that comprise a measurement system thereby beginning the main theme of this volume: the fundamentals for the realization and use of practical systems of measurement.

15.1 INTRODUCTION: STATICS AND DYNAMICS

Measurement systems are formed of several basic functional kinds of subsystem blocks. They can be described in a theoretical sense or their characteristics of performance can be expressed in a verbal manner. Each approach has its merits, as will be seen below, but neither is adequate for all circumstances. A balance is needed between the two extremes to suit the need.

In order to have an adequate understanding of a system stage the designer or user must have a conceptual idea of what the block is supposed to do. Knowledge of a subject generally begins with a linguistic description, the theoretical mathematical stage developing as greater depth of fundamental understanding is needed to improve or explain the performance. As a rule, persons working with the hardware manifestation of measurement systems tend to be pragmatic in approach, theory being applied and developed when it is clear that the effort required will pay off in a relatively short time or that costs can support that line of wide horizon pursuit.

In this work this, and the following chapters, lead up to actual measurement system design. They begin with a linguistic approach, largely reflecting the practice found in the measuring instrument application industry and the initial step at which a general understanding is reached before theoretical approaches can be applied.

The building blocks of instrument systems possess behaviour characteristics that can be divided broadly into those concerning their static (or steady state in the case of a.ç. based systems) performance and those pertaining to time or spatial dynamics of the stage in question. For example, an electronic amplifier possesses a zero frequency gain value—this being its gain to pseudo-static signals. The gain, however, is frequency dependent and may differ at non-zero frequencies.

Measurement systems designers and users can convey considerable information about the various functional blocks without needing to become deeply involved in their dynamic behaviour. There has evolved a common practice of specifying static performance parameters when describing an instrument unit. This practice reflects the fact that static or steady-state performance figures are much easier to procure, in practice and in theory, than those of dynamic behaviour.

This chapter deals with the general architecture, or as Strange, around 1870, called it, 'the physiology of instruments'. The following two chapters introduce static regime parameters and commonly met dynamic responses. Together the three chapters explain the majority of the common concepts that arise in measuring instrument systems. Together they introduce the terms used along with the methods of describing performance and they lead to an appreciation of what goes into a measuring system to obtain a desired measurand mapping.

Regardless of the energy regime involved, measurement systems can be broken down into several basic kinds of functional blocks. This feature enables a generalized approach, in practice and in theory, to be made to descriptions and understanding. This approach is well developed for handling the dynamics of systems through the method of systems theory—also sometimes referred to as the method of the *analogies*. It is also possible to describe static characteristics in a generalized manner although not with the same degree of mathematical rigor. The material presented here is not usually restricted (there are exceptions) to any particular energy regimes or their combination.

The most commonly met energy regimes used in instrumentation are those involving energy of mechanical, electrical, thermal, fluid and radiation forms. Thermal energy can be considered as a subgroup of radiation energy; radiation as a class might also include other forms—subclassification is not unique. However, in instrumentation, thermal and optical subclasses have emerged, with somewhat specific meanings, as dominant classes. Energy forms are often used in combination through the use of a suitable conversion stage. This especially so for instrument systems.

Stages using optical, and near-optical electromagnetic energy, tend not to be discussed in place with the other four, above-mentioned, energy forms as they do not have obvious generalized places in the whole of the framework of dynamics systems theory. A few analogous situations have been quantified but not enough to form a unified theory. Two of the dominant analogies between optical systems and others are the use of electrical transmission line theory to assist multiple-layer optical interference filter design and the use of Fourier techniques in analytical methods for obtaining the spatial optical transfer function (OTF) of an optical imaging system. Another reason for generally ignoring optical aspects of dynamics systems in systems theory is that the optical energy part of a stage has such fast response in the dynamic state that it can often be regarded as having instantaneous behaviour. Therefore, it can be ignored where another, slower to respond, energy regime is present. When working with signals that are conveying information at near-optical frequencies then clearly this simplistic approach cannot be used.

Measurement systems of the kind that this work is largely about, those used and found in the physical science and engineering fields, are generally concerned with designs that are *deterministic*: it can be assumed that they respond in a manner that can be explained by physical laws, albeit perhaps laws that may yet have to be established. The assumption usually exists that links set up through a measurement system are *causal* in nature; what happens at an input will cause outputs to occur. A measuring instrument is an assembly of hardware embodying a physical law, or laws, as perfectly as is needed, or can be achieved.

Although the philosphical impact of the above statement is of little consequence to practising designer and user it is well to remember that other kinds of system operation might be implemented in the not too distant future. There has certainly been interest in the possibility of self-organizing forms of system stage and data processing capability has reached a stage where lengthy and complex logical tasks can be undertaken with low design penalties.

It is not necessary to discuss the systems approach any further here; static performance can be reasonably well explained and described without needing a high level of systems expertise. Chapters 2, 4, 8, 14, 17, and 18 each provide greater detail about systems and measurements. General accounts of the nature of instrument systems are also to be found in Bosman (1978), Doebelin (1975), Finkelstein (1963, 1977), Finkelstein and Watts (1978), JSI (1973), Mesch (1976), and Neubert (1976). These each provide, in different ways, general descriptions of the static and dynamic regimes of instruments and instrument systems. Extensive papers by M'Pherson (1980, 1981) provide insight into systems design in general.

15.2 THE VARIOUS ROLES OF INSTRUMENT STAGES

A measuring system has the task, in the terms expressed in Chapter 1, of mapping a defined physical variable into a representational equivalent quantity. The system of blocks brought together to do this acts as a connecting link by which measurement data are captured from a subject. These data are inherently more than just a set of values; the quantities conveyed also have associated with them, by external definition, a coding that identifies the information with a measurement unit and a scale that tells the user what is being mapped. Thus measurement systems, as well as being efficient and accurate information carriers, also have specific meaning associated with the quantities so produced.

It is observable from experience that, for information to be extracted from a subject, there must exist an energy or mass carrier than can be modulated in some way from its steady state to take up message quantities for transfer from the system under study to the place of use. The magnitude of the energy acting as such a carrier, or the amount of mass being transported, is apparently of little consequence for the concepts of energy and mass are, apparently, not of the same form as meaningful information. Such is the state of ignorance about information and knowledge as philosophical entities. The relationship between energy and measurement systems is cursorily expanded in Stein (1970) and Sydenham (1977).

Thus, the prime purpose of a stage in an instrument system designed to be a link in a measurement process is that it must provide efficient and accurate message transfer. It must be kept associated with the required coding of

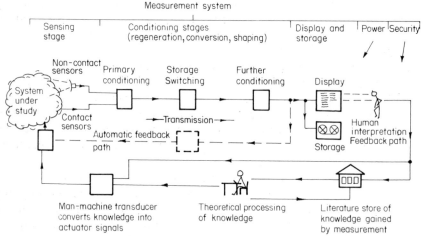

Figure 15.1 Schematic representation of the various stages that might be employed to form a measurement link in a control situation

meaning placed upon the data. It should, therefore, be able to transmit data within Shannon rates (from information theory aspects) in order to retain its messages intact but the data must also be interpreted correctly by the following stage. The design and use of a measurement system requires a broader knowledge and experience than communications engineering alone provides.

Measurement systems generally can be broken down conceptually into paths in which serially-connected blocks operate on the measurement information as it passes from the system under study to the place of use. This is inherently so because measurement requires unique mappings on the basis of one for each measurand of interest. In practice individual channels may be multiplexed in time, frequency or space onto a common carrier so that more than one measurand is mapped with a common piece of apparatus. It is, therefore, quite feasible to portray a complex multiparameter system in terms of singular paths for each measurand.

Several authors have published diagrams that schematically portray the kind of stages used in a generalized measurement link (see e.g. Doebelin, 1975; Beckwith and Buck, 1969). Figure 15.1 shows the one I use in teaching; it is not greatly different from others presented elsewhere.

This diagram shows how information about a subject is extracted from a subject through an interface. The information, coded in some way by the user to provide it with meaning as well as message properties, then passes through several basic kinds of stage to arrive eventually at the place of use. In the diagram it might be fed back, without being displayed, to close a control loop. Alternatively, it might be displayed to an observer who takes no closed-loop action on the original subject at that time. The knowledge so obtained (measurement data, their meaning and interpretation form knowledge) might be stored in a library for future use. It appears that the ultimate purpose of all measurements is, at some time, to gain control of a subject through knowledge about it.

Measurement systems proper are formed of those functions shown in the top line of the diagram, the use of the measurement then becoming part of control. If, however, measurement systems are seen only in context with the top line then there exists a danger that the proper and efficient generation and use of the data will not occur. Adequate closed-loop characteristics can only be obtained if the measurement data are appropriate in amplitude, phase, and speed of response.

This discussion is concerned with measurement systems of all kinds. The signals that convey measurement information can, therefore, be of many forms: electronic, mechanical, pneumatic, thermal, etc. These in turn can be either of analog or digital form. Philosophical and technological aspects of measuring instrument development are discussed in Sydenham (1979).

In a generalized loop, such as that given in Figure 15.1, distinct functions of sensing, conditioning, transmission, storage, and display of signals can be identified. Further to these, they generally each need a source of energy—the power supply—to generate them, plus protective facilities. In practice these functions often overlap and may not be physically isolated in a lumped sense. Before considering these functions in turn, classification of the term *transducer* is needed.

A transducer is fundamentally an element, using a suitable physical energy principle, that converts input energy of one regime into that of another. For example, a microphone element converts pressure variations (fluid energy form) into electrical signals (electrical energy form). An electric motor is also a transducer for it converts electrical energy into mechanical energy. Thus the term transducer is strictly, and it should be used this way, a description for any form of energy converter, including both measurement and power forms.

Where a transducer is primarily designed to produce measurement information it should be referred to as a *sensor*. Where its prime purpose is to convert energy, for energy purposes, in a measurement and control application it should be called an *actuator*. The practical design of sensors and actuators, although both based upon the same mathematical model, is quite different. The actuator is designed to convert energy with as little loss as is practicable and it is not usually designed as a link through which quality information passes. In a closed-loop control application the actuator can be quite inferior in terms of linearity, noise resistance, and non-linear effects to the sensor counterpart. Applications do exist however, where a transducer can act as either a sensor or an actuator; sometimes sensors inadvertently act as actuators!

15.3 SPECIFIC ROLES OF INSTRUMENT STAGES

Specific roles are now considered as a general introduction to measurement subsystems. The design and operation of each of these is more fully discussed in other relevant chapters.

15.3.1 Sensing

A sensing stage has the prime purpose of extracting, by an information filtering and selection process, the required latent information about a selected attribute. It must map the chosen measurand into faithful signals which bear a known relationship to the measurand in terms of scale form and magnitude. It should not pick up, or generate, additional stray data (called *noise*) that might be interpreted as signal. It usually has to operate at low energy levels; which makes the design more difficult. Signal magnitudes may be close to, or even much below, the level of noise sources. This problem cannot always be avoided because many transducer effects employed only receive low energy modulations from the process of interest. It is, however, often possible to reduce the presence of noise, or to process the signal to reduce noise or to enhance the signal relative to the noise level as a ratio. Methods are covered in Chapter 11. A sensing stage often transduces the energy form but not always: it can also convert variables of the same energy content.

15.3.2 Conditioning (Regeneration, Conversion, Shaping)

Information accepted from the subject via the sensor will often not be obtainable in the signal form that is ideally required to suit the final task requirement. Through the use of various kinds of conditioning the signal can be processed to make it compatible with the following stage.

It might be *regenerated* in the same form but with increased (amplified) or reduced (attenuated) amplitude. Regeneration becomes necessary in such cases where the information content of data degrades when sent over a transmission link; or when primary signals may not have sufficient amplitude to avoid noise pick-up being significant. It may need regeneration at a slightly different level for reasons of accuracy of calibration, which is the action of adjustment so that the numerical magnitude has an accurate correspondence with a declared standard for the unit of measurement being determined with the instrument.

Conditioning also includes operations that *convert* the signal into another energy form, as is needed, for example, when a primary sensing stage does not produce signals that are compatible with the next stage.

Conditioning also includes *shaping* stages that operate on the signal purposefully to alter its character, such as when changing the frequency spectral distribution to remove noise signals or increase gain at certain frequencies in order to flatten the frequency response. It might also be convenient to modulate the signal to allow beneficial methods of conditioning to be applied, or to prepare it for transmission over such a medium as radio waves.

There does not exist a common terminology for these concepts and different persons may use slightly different classifications and nomenclature. The designer or user must, through such descriptions, be able to define clearly what kinds of conditioning are needed for a given task. Tables 3.1 and 3.3 list standards documents that may be useful. Chapter 3 discussed the problem of terminology. These different roles established, the design or selection of stages in a hardware form, through the use of systematic methodology, becomes a more certain and reliable process.

Regeneration, conversion, and shaping may be needed many times at various places in a serial measurement path. The functions may also be combined. The important criterion that must be observed is that the measurement information is faithfully preserved as it passes through each stage.

15.3.3 Transmission

Connections are needed through which the information-bearing signal can pass from one stage to the next. The connection distance between stages can vary from micrometres to interplanetary ranges; numerous methods are in existence with which to transmit signals. Transmission links must be able to transfer enough data at fast enough rates without deteriorating the signal information by its inherent conditioning processes. It is often necessary to incorporate *repeaters* that regenerate the signal to make up for degradation.

Transmission of measurement data in particular, tends to be called *telemetry* but other terms that might be used include *communication link, on-line, channel, highway, bearer, bus, route,* and *carrier.* The transmission link may also be described by the name of the method used, such as coaxial cable or radio. Stein (1970) expands on methods for impressing information into a carrier.

A subgroup of the transmission subsystem stage is *switching* of data into appropriate channels. *Data loggers,* for example, switch a number of channels of measurand data into a common line or to a common storage medium. Where data are sent out to more than one location from a common position the unit is often called a *distributor,* the unit sending data in different directions is sometimes referred to as a *director.*

As well as accurately transmitting data a transmission link must also be properly interfaced with the sending and the receiving stages. Failure to do this results in shortcomings occurring such as loss of energy transfer, possible generation of reflections of energy which, being modulated, appear as extraneous measurement data, loading of the signal, and picking up of unwanted noise.

Measurement data can be replicated continuously, there being no concept of total quantity of a *bit* of data. Thus a distributor can send the same data to as many locations as is needed, provided the hardware has the capacity to do it. Chapter 13 covers the transmission of data.

15.3.4 Storage

Measurement signals convey information that may need to be *stored* for immediate use, or for use in the longer term. Numerous methods exist for storing data, the most prevalent, but not sole, methods being those based upon electronic principles. It is, on occasion, feasible to use other methods.

Data storage takes several forms. It can be permanently stored: loss of the power supply or malfunction of the recording medium will not destroy the data of the so-called *non-volatile* store. At the other extreme are· those storage methods that only retain data for as long as is needed. In some the process of reading out the data may erase that stored. Storage techniques are also used to produce time delays. Other names used for storage stages are *memory, data bank, stack, location,* and *file*.

15.3.5 Display

At any stage of the measurement system there may arise the need to observe the measurement data making use of human senses. *Displays,* apparatus that interface hardware systems to human sight faculties, come in many forms ranging from the simpler, single graph, of the cathode ray oscilloscope to the more complex graphic unit. Display can be in the form of numbers or as patterns; they can be composed of two-level digital, or multilevel analog, pixel elements in monochrome or colour. It is at this stage that the measurement system user is reminded that the original measurand is being mapped into a representation of the original attribute and that numbers are not the only form of useful mapping that can be made in order to improve knowledge of a subject through the use of measurement.

As with other kinds of stages, a display functions from the input data supplied and can, by its inherent signal conditioning capability, produce a display that is not a true mapping. For example, a pen recorder will plot a sine wave with attenuated amplitude if it does not possess adequate frequency response.

Depending on their form, displays are referred to as a *visual display unit* (*VDU*), *digital readout* (*DRO*), *monitor, recorder, graphics terminal, indicator,* they may also be referred to by the name of the device employed.

Displays should, however, properly be regarded as a subgroup of the *man–machine–system* (MMS) interface through which technological apparatus connects to human faculties through the use of the sensing mechanisms of man: hearing, seeing, touch, taste, and smell. For each of these, interfaces have been devised through which man becomes part of a man–machine–system. Chapter 19 addresses the MMS interface.

The information conveyed on the energy links established between the two machine forms can bring about change in the actions of the, more advanced, human machine. This concept has been expanded in Sydenham (1978).

15.3.6 Power Supply

As processes used in a measurement system dissipate energy by virtue of loss mechanisms the system must be operated by one or more *power supplies* that feed the system with energy to create the information links.

It is feasible to provide stages with power obtained from the carrier that conveys the information. This is called *signal-powering*. An example is the simple crystal set, radio receiver that obtains energy directly from the receiving antenna.

It is, however, more usually the case that stages will be operated from specially designed power supplies. These provide the appropriate form of energy, such as electricity, air pressure, and mechanical motion, to the stages.

Advances in the technology of electronic systems, those used most prolifically in measurement systems, have gradually altered the traditional view of a power supply from being a single central, large-capacity unit to one in which distributed smaller units are mounted throughout the circuitry. It will become progressively more difficult to isolate power supply functions from the rest of the apparatus, but the need for power will remain, albeit at increasingly lower capacities.

Common practice tends not to depict the presence of a power supply in descriptions of a system, the need for it being assumed. The design of the power supply, therefore, can often be overlooked, a practice that can be dangerous. The quality of the power supplied can influence the measurement data fidelity through such mechanisms as modulation of the information carrier, injection of spikes, and other noise. A falling supply potential with load can lead to inaccurate operation of stages. The cost of power supplies compared with the above-mentioned stages, can be very significant.

A good measurement system cannot be assembled without careful study of the characteristics of its power supply working under actual operating conditions.

A measurement system can always be broken down into, or built up from, assemblies of the stages having the functions described above. They form the building blocks of measurement systems in the same way as do the basic functions used in, say, a digital electronic system. The designer combines these (Figure 15.2) in varying ways to arrive at the required sensing performance. It is, therefore, necessary to ascertain that each stage has the ability to transmit measurement information without loss of fidelity, speed or quantity.

Assembly of a serial chain can be possible via many different combinations. Often there exists, by the time the first prototype complete system is finalized, considerable redundancy. There is often merit in appraising the whole design at the system level in order to eliminate any unnecessary stages that have somehow been included. As an example, it may be a requirement that the link filters out certain frequency components. Initially an electronic filter may have been added to achieve this. Later study of the whole chain may reveal that the system does not need this stage since a mechanical component, included for conversion reasons, may produce the desired filtering effect.

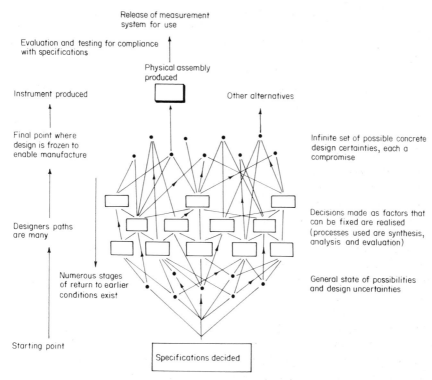

Figure 15.2 Measurement systems realized are often not a unique solution— schematic of the design process

Optimization of a measurement system would follow the usual design methods for controlling the design of an extensive, multistage system. This process is conducted by judicial application of both theoretical and practical procedures.

It is often possible to construct a stage that has the ability to play several different roles simultaneously. A suitably designed, feedback, operational electronic amplifier can, for instance, provide conditioning and storage. Whereas this may be efficient from a circuit *pin-count* cost viewpoint it may make development, servicing, and trouble-shooting difficult. It is often preferable to obtain required functions as separately identifiable units for then the signal can be more easily monitored.

15.3.7 Protection and Security of Correct Operation

Getting a measurement system to operate and carry out a given measurement task is the basic design and assembly problem. However, design for continued reliable operation must also recognize that components are not perfect and that external influences may, at times, give rise to system failure. Thus measurement systems may require, to an extent depending upon their importance, means to protect them from factors that can impair their performance. Systems with this role of protection are also known as *security systems*.

The designer of a security system must be able to predict what kinds of influence might arise and introduce technology to prevent such effects influencing the system; or design it to be able to accept their influence making alterations to the mode of operation such that the prime purpose is still upheld. It is impossible to design an absolutely reliable and secure system, one dominant reason being that man must first be able to visualize, or have observed, what circumstances might arise before strategies can be designed to combat them.

In a measurement system it is not sufficient to insist that the system continues to operate and that signals keep flowing from the output port. There should also be incorporated techniques whereby the quality of the data is checked to ensure that the data always represent a true mapping of the measurand. Numerous ways are available to do this. As an example, designs of thermal imaging instruments often include a facility whereby the sensor periodically views a standard, black-body, source. In this way calibration is maintained. Many instruments, however, provide data that are not continuously verified. Common design practice in digital electronic systems checks received data by the use of parity. Greater application of the self-checking philosophy can be expected in future designs.

Some forms of protection act to force the system to shut down in a safe mode when catastrophic departure from proper operation occurs. The simple fuse wire in an electrical circuit or the shear-pin or slip-clutch in the mechanical system are examples of this. There are, however, circumstances where disconnection of power or signal can be disastrous and secondary means are needed to secure *fail-safe* operation. In some cases it may be necessary to shut down the whole system in a certain order. Some measurement systems also must be shut down, and started up, under conditional control.

15.3.8 Servicing, Calibration, and Maintenance

Each of the above factors must be considered if a measuring instrument or system is to work properly and have a long, trouble-free, service life. Instrumentation tends to require a certain amount of servicing and recalibration at given time intervals. It should also be capable of simple and rapid repair when it becomes damaged or fails internally.

There is little doubt that measuring equipment could be designed to be virtually trouble free for what ever lifetime is needed—space missions have shown that. The factor that prevents this, of course, is the cost in terms of time and manpower expended. In some disciplines, instrumentation specifications include expressions of the *mean-time-between-failure* (MTBF) rate expected and the *mean time to repair* (MTTR) the unit but this is not always a stated factor of design. Instrumentation is, by no means, always reliable and there appears to be trend toward higher failure rates as world economies act to attempt to reduce the cost of equipment. Unfortunately there is not necessarily a correlation between quality and life of operation against the price paid. The consumer must be very wary, conducting tests or having *evaluations* (see Chapter 31) made to establish the quality of products purchased or built.

Many managers will opt for the cheapest tender for instruments. This does not necessarily save money for in the long term the cost of the measurement can run high when a vital measurand is lost due to instrument failure. Long-life, trouble-free, operation comes from good design (which in turn also hinges on adequate specification in the first instance) and from good manufacture. Many instruments are based upon a poor principle that is made to work by the incorporation of additional stages that eliminate the errors introduced by use of the inappropriate principle. Add to this the numerous possibilities of poor manufacture of the system and the manufactured products used to built it and the chances of obtaining quality performance at a cheap price are low.

The new generation of *smart* instruments will considerably improve security and protection aspects of operation. (Electronic integrated circuits can now withstand considerable abuse!) Mass-production methods, such as integration in electronics, mean that the quality is usually higher. Singly-made parts, once regarded as the finest that could be employed, are no longer necessarily the best that can be used. A significant factor to proper operation is likely to be the care and methods used to inspect the product and its components. The quality of an instrument can usually be gauged from the manufacturer's quality control efforts. Chapters 28 and 29 are relevant to these design aspects.

Each of the above topics form the substance of much of this Handbook; in-depth explanations are provided in the appropriate chapters.

15.4 AN INSTRUMENT DISSECTED

The following example shows how the foregoing remarks apply to an actual measuring instrument system, in this case one of relatively small extent.

The instrument selected to illustrate these and further points is built to transduce the relative moisture content (relative humidity (RH)) of ambient air providing an electrical control, or recording signal in the range 2–10 mV for 20–100% RH respectively. It also gives visual, direct-reading, display of the instantaneous value of relative humidity. The instrument is designed to sell into the low-cost market and provides ±2% RH accuracy. It quantifies a variable in a relatively simple manner, is not particularly complex or sophisticated but does provide a useful exemplification of the principles given in previous sections due to the many mechanical and electrical stages included in its design.

The actual instrument is shown in Figure 15.3. The scale and protective cover are removed to reveal the internal organization of the mechanical and electronic components.

Several stages progressively provide and condition the measurand, relative humidity, into both an electrical signal output and a mechanical visual display output.

The simplest block diagram portrayal would be that shown in Figure 15.4a. In many measurement and control systems a single block provides an adequate representation. To appreciate the internal operation, however, it is necessary to divide the single block into cascaded, isolatable, subsystem stages. This is done, again as a block diagram, in Figure 15.4b.

The sensing stage is a tensioned hygroscopic membrane that exhibits length change with varying moisture content. It interfaces the environment under study (for its % RH) providing a transduced mechanical signal in the form of a small positional change occurring at the unconstrained end. The membrane is chosen from many materials as one that is not affected significantly by any other influencing parameter; therefore, it responds sufficiently uniquely to moisture content changes. This not to say that it is not influenced in any other way: it most certainly is subject to such effects as temperature but they will be of little consequence in this use. It is suitable for this application because it adequately provides a sensing action to the order of discrimination and repeatability needed. The instrument is not expensive and the performance is not in a high-accuracy class for the % RH variable. Many sensing principles could have been employed instead but the designer has, no doubt, eliminated all others through study of their various physical factors and because of other design parameters such as supply and known experience with the material. Design of mechanical stages is described in a series of papers by Sydenham (1980) and in Chapter 21.

Although established procedures, such as *utility analysis*, have been developed by which an optimum choice can be determined for several alternatives, such methods are not always used. It would often be the case that a designer chooses a method because of existing experience with it or because of such other factors that do not get proper objective attention.

The membrane does not provide sufficient mechanical length change for the movement to be easily seen by the naked eye. Its *sensitivity* is too low for that use. Thus it is followed by a conditioning stage in which mechanical regeneration at a larger amplitude is achieved with a mechanical lever

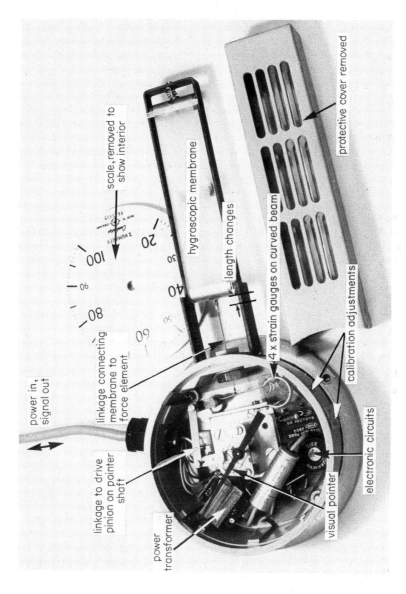

Figure 15.3 Internal view of an humidity sensor giving electrical and mechanical output measurement signals. (Reproduced by permission of Foster Cambridge Ltd, England)

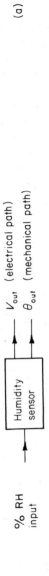

(a)

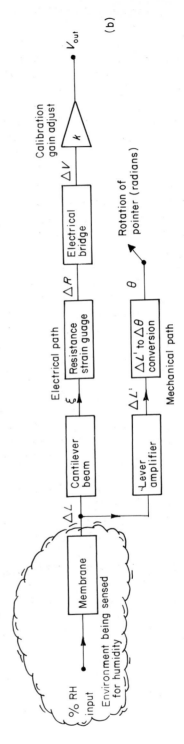

(b)

Figure 15.4 Block diagram representations of the operation of the humidity sensor shown in figure 15.3: (a) as used as part of a larger system; (b) broken down into identifiable stages

providing mechanical gain. Further conditioning, this time in the form of a transformation conversion in the same energy regime, changes the linear motion into a rotary equivalent. In the electrical path it can be seen that the membrane length change is used to vary the bending existing in a curved, spring beam. This beam also tensions the membrane via its compliance.

The beam acts to convert displacement into strain so that the electrical resistance strain gauge method can be operated to convert mechanical action into an electrical form. Four gauges are used, in bridge configuration, to reduce their temperature error and to reduce common-mode noise pick-up. Further conditioning, through another stage of conversion, converts resistance changes of the gauges into voltage variations. Each stage must convey equivalence to the next, preserving the measurement mapping requirement from the environment the instrument is sensing, right through to the output voltage and pointer position.

The instrument still does not provide a completely adequate signal transformation because the electrical bridge and the membrane will not, due to manufacturing and materials parameters, always provide the correct magnitude voltage corresponding to the calibration needed to relate % RH to millivolt output. Thus the last stage in the electrical path is one that regenerates the electrical signal with means to set the correct relationship needed for calibration. The mechanical path includes an adjustable length lever in the linear to rotary transformation stage.

No obvious signal shaping is evident in the instrument but the designer would have been conscious of the frequency response needs of each element in the two paths. The design is, however, based upon a low-pass, low-frequency response sensing membrane so following stages need not be particularly responsive. It is doubtful if the designer took any in-depth interest in the dynamic response for the measurand itself, in the applications for which the instrument was designed, cannot vary rapidly.

Transmission of signals can be identified here as the mechanical linkages and the electrical wiring of the components and the printed circuit board. The low-pass response of the instrument enables simple linkages and wiring to be used.

The instrument displays the measured variable in mechanical terms on the scale (removed in the illustration) but does not display the magnitude of the electrical signal. It is a simple matter to display the measurand at another location for the output is so set that % RH corresponds with a full-scale indication common to the range of many electrical analog and digital indicators.

Storage of the measurand values is not provided deliberately in the design. It only indicates the instantaneous value within its response time. There are, however, many storage mechanisms present in the system. They occur in the mass of the metal parts, giving storage of thermal energy, and in the membrane's slow response to take in and give up moisture, the two rates being different. These storage mechanisms give the instrument its slow response; they also assist integration of the higher-frequency noise and signal components. Mechanical elements undergoing strain and the membrane, each possess the ability to store information about the past history of the device's circumstance. Elastic *after-effect* and *hysteresis* phenomena give rise to creeping values and to back-lash when the measurand alters sign of direction.

The block diagram does not (as was mentioned earlier to be normal practice) show the presence of a power supply. In this case the mechanical path is signal-powered and the instrument will indicate, through the pointer, even when the electrical power is not connected. The electrical path, however, needs a low d.c. voltage supply to operate. This is provided, for a.c. mains use, by a small transformer, rectifier bridge, and smoothing capacitors followed by a voltage regulator stage.

Finally to protection and security: no internal fuse is supplied for the electrical supply; nor does the supply have over-demand protection in its regulator stage. There is no need for protection of the mechanical path because the instrument is totally enclosed and because movement of the needle by an external source (fingers?) cannot deflect the system past safe limits. The whole assembly is encased to protect it from fingers, from reasonable quantities of splashed fluids and from dust. In turn the user is protected from lethal electrical levels by the use of an insulating plastic case and by placing the high-voltage connections out of reasonable reach. Calibration, of span and zero, can be made externally through holes in the case.

Although the instrument is simple and not complex to understand it does, however, involve many design parameters that needed careful selection and integration. The complete design of such an apparently simple assembly, especially when both the static and dynamic performances are involved to exacting limits, can be a very challenging task of work but one in which systematic and rigorous methodology can be applied in order to realize a workable efficient measurement system.

REFERENCES

Beckwith, T. G. and Buck, N. L. (1969). *Mechanical Measurements*, Addison-Wesley, Reading, Mass.

Bosman, D. (1978). 'Systematic design of instrumentation systems', *J. Phys. E: Sci. Instrum.*, **11**, 97–105.

Doebelin, E. O. (1975). *Measurement Systems: Application and Design*, McGraw-Hill, New York.

Finkelstein, L. (1963). 'The principles of measurement', *Trans. Soc. Instrum. Tech.*, **15**, 181–9.

Finkelstein, L. (1977). 'Instrument science—introductory article', *J. Phys. E: Sci. Instrum.*, **10**, 566–72.

Finkelstein, L. and Watts, R. D. (1978). 'Mathematical models of instruments—fundamental principles', *J. Phys. E: Sci. Instrum.*, **11,** 841–55.

JSI (1973). 'Instrument science. What is the use of the subject?' (several short papers), *J. Phys. E: Sci. Instrum.*, **6,** 2–7.

M'Pherson, P. K. (1980). 'Systems engineering: an approach to whole-system design', *Radio and Electronic Engineer*, **50,** 11/12, 545–58.

M'Pherson P. K. (1981). 'A frame-work for systems engineering design', *Radio and Electronic Engineer*, **51,** 2, 59–93.

Mesch, F. (1976). 'The contribution of systems theory and control engineering to measurement science', Survey Lecture SL5, *IMEKO VII Congress, London, May.*

Neubert, H. K. P. (1976). *Instrument Transducers*, 2nd edn, Clarendon Press, Oxford.

Stein, P. K. (1970). 'Sensors as information processors', *Research/Development*, June, 33–40.

Sydenham, P. H. (1977). 'The nature and scope of measurement science', *Proc. Conf. on the Nature and Scope of Measurement Science'*, *Armidale, Australia, 1976*, IMEKO, Budapest.

Sydenham, P. H. (1978). 'Measurement-science in engineering education', *Proc. Conf. on Engineering Education*, Institution of Engineers, Australia, Publ. 78/6, July, Sydney.

Sydenham, P. H. (1979). *Measuring Instruments—Tools of Knowledge and Control*, Peter Peregrinus, London.

Sydenham, P. H. (1980). 'Mechanical design of instruments', *Measurement and Control*, series of papers commencing from **13,** 365.

Handbook of Measurement Science, Volume 2
Edited by P. H. Sydenham
© 1983 John Wiley & Sons Ltd

Chapter

16 P. H. SYDENHAM

Static and Steady-state Considerations

Editorial introduction

This chapter continues the explanation of the component parts of a measurement system.

For reasons of economy much of the description of the subsystems of measurement systems is provided in terms of static or steady-state descriptors.

After a brief general introduction to both the static and dynamic regimes of hardware, attention is given to the former.

16.1 TRANSFER CHARACTERISTICS

Stages used to form a measurement system can be represented as blocks of hardware that effectively transfer an input to their output either in the same energy form, like the voltage ratio of an amplifier, or as a converted quantity, such as that met in the previous chapter where a sensor converted relative humidity into a voltage and angular motion equivalents.

The ratio of output to input quantities is termed the *transfer characteristic.* The magnitude and phase between the two quantities are expressed as numerical quantities, in terms of the units of the two variables involved. For example the *sensitivity* of the previously mentioned humidity sensor was designed to be 0.1 mV per 1%RH.

When the system is investigated further it is seen that this transfer expression (the usual one quoted by manufacturers) is a great simplification of the full characteristic existing for it provides no information about the sensor's response to dynamic changes of relative humidity. What is missing is some form of statement as to how the system will respond to changing input signals. This can be done by adding a frequency-dependent component to the expression already given to represent the static behaviour.

As simplification, to cover the static region of use only, suffices for many tasks it is common practice to quote only the static transfer characteristic. The *steady-state* equivalent is given, in the case of a carrier system or a purposefully designed system made to operate at a given frequency. The steady state is also sometimes called the *quasi-static* condition.

In practice the *state-variable* form of relationship between the input and the output for a stage often cannot be easily arranged in the form of independent input and output signals. For this and other reasons it is common practice to convert that relationship into its Laplace transform equivalent which can then be arranged to produce a transformed transfer characteristic called the *transfer function.*

As an example, the relationship between an input variable x and an output variable y existing through a measurement instrument stage might be:

$$\frac{d^2 y}{dt^2} + \frac{3\,dy}{dt} + 2y = x + \frac{dx}{dt}$$

from which we require the ratio y/x to be formed to obtain the transfer characteristic. It will be noticed that measurement system behaviour is generally described in terms of linear differential equations, this being in line with general physical systems description.

As the system is linear in the mathematical sense, the Laplace transform can be taken obtaining:

$$s^2 Y(s) + 3sY(s) + 2Y(s) = X(s) + sY(s)$$

This can be rearranged to give:

$$\frac{Y(s)}{X(s)} = \frac{s+1}{s^2 + 3s + 2} = G(s) \quad \text{the transfer function}$$

If the process involves a time delay then terms of $e^{-s\tau}$ form will be included, for example:

$$\frac{Y(s)}{X(s)} = \frac{e^{-s\tau}}{s+1}$$

It is in the Laplace form that transfer characteristics are most usually expressed as they are immediately in a form that can be operated upon or studied directly.

It is generally permissible to use the operated D in place of s and some authors use p instead of s. When the system is operating with sinusoidal signals (which can usually be the case because complex signals can be broken down into sinusoidal components) it is permissible to replace s by $j\omega$ enabling the response at any chosen angular frequency ω to be calculated.

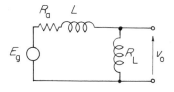

Figure 16.1 Electrical schematic circuit of tachogenerator

As an example the tachogenerator, a device that generates output voltage proportional to angular shaft velocity and, therefore, maps shaft speed into an equivalent voltage representation, has the schematic electrical circuit shown in Figure 16.1. Being a driven rather than driving mechanical device its mechanical inertia and friction are not of importance in obtaining the transfer function between angular velocity input θ and the generator voltage output v_0. It is designed such that voltage generated is proportional to the shaft speed according to $v_0 = K_g \, d\theta/dt$; K_g being the *generator*, or *transducer constant*.

The mechanical parts of the generator are forced (an assumption) to follow input demands and cannot cause lag or lead action due to the generator's compliance or angular mass. The electrical output circuit, however, contains storage in the form of inductance of the windings and this can influence rapidly changing signals. Energy is dissipated via the two resistances, one of which is the load resistance of the next stage that it drives: let $R = R_L + R_a$.

The output voltage v_0, that generated across R_L, is, in Laplace form:

$$v_0 = \frac{R_L K_g s\theta}{R + sL_a}$$

which can be rearranged to yield the transfer function:

$$\frac{v_0}{\theta} = \frac{R_L K_g}{L_a} \left(\frac{s}{s + R/L_a} \right)$$

Having developed the expression in this form, for this example, we can move on to explore the three regions of transfer characteristics that could be involved when describing the transfer characteristics of a stage in a measuring instrument, or of the whole instrument system.

It can be seen, from the above example, that when s (that is, $j\omega$ for sinusoidal excitation) is zero the system is operating with zero frequency in the electrical circuits. The generator is turning to produce an output related to the generator constant K_g modified by the, *loading*, divider effect of R_a and R_L. In this situation, all frequency-dependent components have no effect on the output. Transients that can arise in the electrical circuit when an input mechanical or output load transient occur have, in fact, settled down: the generator is producing its *steady-state* output. Thus the generator constant K_g is, in fact, the response existing after internal transients have settled to be insignificant. It represents the *steady-state transfer characteristic* and generally is the *sensitivity* figure quoted.

The generator will, however, not produce the simple steady-state transfer relationship during transients. It may be necessary, in application, to make allowances for the *transient characteristics*, which are more complex. This certainly is the case if rapid-response, closed-loop, operation with the tachogenerator is sought. Phase-shift and gain changes during transients can cause instability.

If the component is to be used in applications where transient response is important then it will sometimes be found that the manufacturer will quote the necessary additional component values (resistance and inductance in this case) so that the user can construct the full transfer function in order to obtain the transient behaviour characteristics needed to condition a tight, stable system.

It is, however, often necessary for the user to conduct tests to obtain the additional values, the only parameter quoted being the steady-state characteristic. For example, the previously mentioned humidity sensor data sheets contain no information about transient behaviour, it being unnecessary for the market intended.

In addition to these two regions of response there exists another that can cause confusion for it too is often quoted as though the component only possessed that characteristic mode of behaviour: this is the *d.c. response*. As the name suggests this is the response characteristic of the stage when the input, measurement-bearing, signal is at zero frequency—direct current or direct-coupled would appear to have formed the symbolism 'd.c.'. Not all systems possess a finite value d.c. characteristic, nor is it always relevant.

As an example consider the d.c. electronic amplifier. This has a certain gain value when operated at zero frequency. What is often not realized is that it has a transient response during which gain varies and it will exhibit differing gain with signal frequency. When used as an a.c. amplifier at a given frequency it then has a certain steady-state gain. Thus use of the apparently simple amplifier block must make allowances for the various modes of transfer characteristic. Furthermore the input signal level can often also alter the transfer characteristic.

It has been shown, by example, that stages used in a measurement system can possess three distinctly different transfer characteristic functions depending upon the time duration and form of the signal. Which are important must be decided for each application. The easiest to measure and quote are d.c. gain and/or the steady-state value. The transient characteristic is much more complex to assess and express and it also varies with signal conditions. The humidity sensor example, given in the previous chapter, easily breaks down, by analysis, to yield its d.c. characteristics but realization of complete dynamic transfer functions that would enable the transient response to be studied would require expenditure of considerably more effort. The development of transfer functions for very complex systems can often take many years of effort. Systems identification procedures attempt to obtain them from practical testing.

Further explanation about transfer functions, in general, is presented in DiStefano *et al.* (1976) and Doebelin (1975). Atkinson (1972) explains the difference between steady-state and transient characteristics in terms of the solution of linear differential equations having constant coefficients. These concepts are also expanded elsewhere in this Handbook.

Having shown how a stage can be modelled theoretically, and therefore be uniquely described on paper, in terms of its transfer function it is important next to realize that linear systems of interconnected transfer functions, no matter how complicated, can (in theory) be reduced to a single-stage transfer function. This means that a complex measurement instrument system, built from many stages as a serial path including feedback loops with other paths, can be reduced mathematically to just one transfer function that describes its complete overall static and dynamic behaviour. This reduction allows the transient, steady-state, and d.c. responses of any system to be calculated or simulated without need to assemble all stages. That this is so seems intuitively reasonable for any black box exhibits these three modes of behaviour.

There is no need here to elaborate on the techniques used in *block-diagram algebra;* they are well covered in DiStefano *et al.* (1976). It stems basically from the prime fact that transfer functions are commutative provided it can be assumed that they do not load each other when connected. Interaction of stages destroys the rule making reduction more complicated. Interaction significantly changes the response—refer to the following chapter for often-met situations.

When compounding stages together to build up a measurement system it is, therefore, necessary to consider if the stages interact to a significant degree. In electronic systems this problem has generally been eliminated by designing integrated circuits to connect on the basis of low output impedance being connected to high input impedance in order to preserve voltage levels. In cases where power must be transferred efficiently it is necessary to use stage impedances that are equal. For current transfer, another information transfer mode, the opposite to the first given voltage case, applies.

Electronic systems connections generally present few connection problems (provided well established rules are adhered to) but the same is not true of mechanical, thermal, fluid, and optical stages for in those disciplines component design does not enable ideal matching arrangements to be had so easily.

Transfer characteristics of a stage can be stated in terms of the response in either the time or the frequency domain (refer to Chapter 4). Conversion from one domain to the other is achieved via the Fourier transformation. When the signal is of a steady-state, time-domain condition, because it is periodic, it can easily be transformed to yield the singular set of frequency components. Signals of transient nature present greater difficulties in transformation for the spectrum changes during the transient.

It is, therefore, quite reasonable for the response of a stage to be stated in terms of the frequency domain. The same information could be expressed in the time domain, in terms of the stage's response to a suitable time variant input. In some applications it is more convenient to obtain or describe a transient in time response form (such as to a step or impulse input) than it is in the frequency domain.

Zero-frequency response is, in the frequency domain, the value of amplitude, or phase, at zero frequency. As frequency response plots are generally expressed using logarithmic scales, zero frequency cannot actually be represented. Amplitudes are generally expressed in decibel form but for some instrumentation unconverted ratios may be preferable.

The above remarks apply to stages having at least one each of a clearly identifiable two-terminal input and output port. In use a two-terminal device, such as a capacitor or a spring, becomes a two-port device, either as a series (through-variable) or as a parallel (across-variable) arrangement, by suitable connection with common lines. The above remarks then apply provided it is clearly stated what are the connection arrangements and the terminating conditions existing. Alternatively the two-terminal device can be described in terms of its terminal impedance response, called the *driving point impedance* (or admittance) which can also be fully expressed mathematically in Laplace form.

Simple common components, such as the electrical inductor or mechanical mass, are so well known that they usually do not need to be specified (for response reasons) in greater detail than the numerical constant of the parameter they provide, for example inductance in henry units or mass in kilogram units. This information is sufficient for users to construct the response characteristics required. When the component is uncommon a mathematical driving-point impedance expression is needed to provide adequate information about its static and dynamic behaviour.

In many cases it is not convenient to express the true response behaviour of a terminal port and approximations are used to map a many-variable parameter situation into a single one. A moving-coil loudspeaker (as an actuator or a sensor) mechanism is generally characterized by stating the impedance it presents at a given audio frequency this being expressed as a pure resistance of x ohms. In reality the terminal impedance is a complex quantity having both real and imaginary parts and varies greatly with frequency. This applies to the input and output impedances of many stage blocks used. This is, however, not always the case, a notable exception being the transmission line which provides constant purely resistive impedance at any length provided it works within certain assumptions of being lossless and properly terminated. Somewhat confusingly *impedance* is often used synonomously with *resistance*: correctly the former is the resistance to flow when both real and imaginary components exist as a *complex variable*.

The tacit assumption often exists that components and subsystem blocks are what they are defined to be for all regions of operation. In practice a component can behave quite differently for signal frequencies, levels, noise contents, and other parameters to those for which it was purposefully made to handle. For example, a capacitor becomes an inductor with rising frequency; a mechanical mass provides properties of spring compliance under certain conditions. Ideally all components and stages used in measurement systems should be defined to provide the user with information about their whole spectrum of response conditions but this is not practicable. The instrument designer, in particular, must be aware that simplistic definitions, such as those giving no more than steady-state or d.c. information, may not suffice. Lack of observance and understanding of this defect in subsystem description is a prime reason for failure of many designs when first assembled and for incorrect operation when fed with input and influence quantity signals that do not conform to those it is designed for.

It is, therefore, in principle possible to model any complete or part measuring system as a transfer function expression that enables its behaviour to be understood without needing to resort to actual hardware. Whether this is worthwhile, however, depends upon the application and the extent of the system. A very practical point is that many of the constants required to provide a boundary to the mathematical model must come from measurement of existing phenomena.

Reduction, in this way, of an arrangement of many subsystems' mathematical models into one, retains all information about the overall behaviour but does not allow interstage signals to be studied. This realization is not to be confused with *identification* processes and *simulations* that also can provide overall input–output relationships that apply for some modes of operation and not others and that may be realized by quite different internal processes. If the internal subsystem transfer functions are adequate models of the stage that they represent then so also will be the overall reduction obtained by block-diagram algebra (provided conditions of connection and range of operation are met).

Transfer function representation is only as good as the assumptions on which it is developed. Many are, and need to be, simplifications of the real situation; often they ignore the higher-frequency terms that can be considered to be of no practical consequence to response.

For example, the transfer function of a printed-armature d.c., instrument motor needs only to allow for response lag caused by the mechanical storage aspects of the armature disk, electrical storage time constants of the armature circuit being much smaller than those of mechanical sources (50 μs compared with 30 ms). A conventional style, field-controlled, d.c. motor, however, requires a more complex transfer function, one having a second storage term to allow for the delay caused by significant storage effects in the electrical field-coil circuit. The two transfer functions are:

Printed motor   Field controlled motor

$$G(s) = \frac{\omega_{out}}{v_{in}}(s) = C_1 \frac{1}{\tau_m s + 1} \quad G(s) = C_2 \frac{1}{(\tau_m s + 1)(\tau_f s + 1)}$$

where

ω_{out} = shaft velocity
v_{in} = drive voltage
τ_m = mechanical time constant
τ_f = electrical time constant of field winding
C_1, C_2 = constants related to generator and torque constants; and to mechanical friction and field-coil resistance.

This background enables this account to concentrate next on the static, steady-state, regime of operation for instruments. Dynamic performance is the subject of the following chapter.

Consideration of the many descriptors for static and quasi-static performance of an instrument stage shows that they generally group into terms concerned with measuring instruments in general and with a second group, process control instruments (which is overlapped somewhat by the first), that uses terms of its own particular form of application.

16.2 STATIC CONSIDERATIONS OF GENERAL INSTRUMENTATION

Through the use of the humidity sensor, described in the previous chapter, this section introduces terms used to describe the performance of an instrument from the point of view of its static and/or steady-state behaviour.

The terms so introduced are primarily based upon those defined in BS 5233 'Glossary of terms used in metrology;' in BS 2643:1955 'Glossary of terms relating to the performance of measuring instruments' in PD 6461 'Vocabulary of legal metrology' and in AS 1514 'Glossary of terms used in metrology—Part 1', documents introduced in Chapter 3.

Terms describing measurement performance, such as discrimination, sensitivity, repeatability, and reproducibility, have already been explained in Chapter 3. Several commonly used terms, to be described here, do not appear in those publications and have been drawn from glossaries given in texts in general instrumentation (Bell and Howell, 1974; Foxboro-Yoxall, 1972; Herceg, 1972; ISA, 1977; N. S. Corp., 1977; O'Higgins, 1966; Stata, 1969).

To represent the humidity sensor's static performance parameters it is first necessary to construct the block diagram showing the complete instrument broken down into serially connected paths. This was done in Figure 15.4. Two paths were identified through the instrument from the sensing membrane to the two outputs. These produce an electrical and a mechanical output signal.

An overall transfer sensitivity (S_1 and S_2) can be developed through each path by multiplication of the static transfer characteristic for each of the stages connected to form a link. This yields

$$S_1 = \frac{V_{out}}{\% \text{RH}} = \left(a \frac{\Delta L}{\% \text{RH}}\right)\left(b \frac{\xi}{\Delta L}\right)\left(c \frac{\Delta R}{\xi}\right)\left(d \frac{\Delta V}{\Delta R}\right)\left(e \frac{V_{out}}{\Delta V}\right)$$

$$S_2 = \frac{\theta}{\% \text{RH}} = \left(a \frac{\Delta L}{\% \text{RH}}\right)\left(f \frac{\Delta L'}{\Delta L}\right)\left(g \frac{\theta}{\Delta L'}\right)$$

where $a, b, \ldots, g$ are the static, or quasi-static, *transducer constants* of the various stages that relate input to output variables when the stage is in a settled state of response. In some cases the transducer constant is given a special name. For example it is called (in a normalized form) the *gauge factor* for a resistance strain gauge.

Thus the static sensitivity is found as the product of static transducer constants for all stages in a path. If different gain is needed it could be arranged anywhere in the path by altering any suitable one of the constants.

Range, the factor that decides the extent of measurement ability from the smallest to the largest signal amplitude that can be accommodated without the system being *overloaded* or signals becoming limited (*saturated*), is decided by the capability of all stages. If the signal level of any one moves into a region of improper operation the whole system will fail to operate correctly.

The *effective, measuring* or *working range* is the 'range of values of the measured quantity for which any single measurement, obtained under specific conditions of use of a measuring instrument, should not be in error by more than a specified amount'.

The *upper* and *lower range limits* form the limiting boundaries inside which the instrument has been constructed, adjusted or set to measure the specified quantity. The effective range may not necessarily be all of the instrument's interval between these two limits. *Rangeability* is the term sometimes used to relate, by some form of expression, the relationship between upper and lower limits of useful range.

In the humidity sensor example the manufacturer states that the instrument has a range between 20% to 100% RH, but on the instrument itself is issued a warning that use outside of the interval 20% to 85% RH may lead to temporary loss of calibration. Furthermore, careful reading shows that it is only within given accuracy limits over 30% to 85% RH, thus the range of possible operation is within 20% to 85% RH, rangeability being perhaps given (there is no standard methodology for expressing this), as providing a ratio of maximum and minimum limits of 4.25:1. Rangeability is more useful for use with instruments having unbounded measurand possibilities, such as occurs in flow metering where the two limits of practical flow meters are often rather restricted compared with the interval that might be called for in practice.

A *multirange* instrument is one in which the range of operation can be selected by *manual*, or more recently, by automatic means, the latter being referred to as an *auto-ranging* instrument.

When describing the effect on a calibration of influence quantities it is common practice to state the range over which such quantities can be tolerated. This is defined as the *reference range*. The humidity meter performance is stated for a reference range of 0 to 40 °C temperature variation.

As a guide, ambient environmental conditions may usually be taken as having a mean temperature and reference range of 25 ± 10 °C; relative humidity of 90% or less for general use (but 40% to 60% for storage of paper materials); and barometric pressure lying within 90 to 110 kPa (900–1100 mbar).

Influence quantities that might be important, would also include vibrations, power supply (electrical, mechanical, hydraulic, pneumatic) variations, radio frequency, ambient lighting and especially the modulation from fluorescent tubes, magnetic and electric fields, angle of tilt of instrument, ionizing radiation, and improper use by unauthorized persons. More detail is given in Paine (1974).

Many measurements can be performed more easily, or with improved accuracy, if the instrument is built to provide a limited effective range. This can be done by designing the lower range limit to be well above zero (the converse hold for measurands having values less than zero). Such systems are called *suppressed-zero* systems. A common example is the voltmeter used to indicate mains voltages in the range 200 to 250 volts.

For such systems sensing signal levels will not have a zero level coinciding with the output zero, that is, in the above example a 200 V input may deliver 0 V to the indicator. Any other system of levels might be chosen. The degree of suppression is expressed as the ratio of scale interval omitted, above or below zero, to that presented. A similar and wider concept is when the range is expanded at any point. This is often termed *band* or *scale spreading*.

Range of operation for static and quasi-static use is often different to that in the instrument's dynamic transient mode. The term *dynamic range* is used to indicate which regime the range quoted relates to. It is generally stated in the same ratio form as is given above for rangeability, being presented in decibel form. At times its common use would, however, appear to really be defining the static and quasi-static performance, not the time-dependent dynamic behaviour.

Related to the range of an instrument is the *scale factor* that might need to be applied to the actual numerical value obtained in order to arrive at the correct magnitude. Scaling is often practised to allow a normalized instrument to be applied to many ranges of task. Except where the instrument is multiranging the system should be marked accordingly. Multiranging instruments usually indicate automatically the appropriate scale factor as the range changes.

Many instrument systems require an operation to set the position of the zero. For example, a gas analyser and its sampling line may need to be *purged* with a suitable clearing gas before use. Electronic, d.c., systems usually need periodic *zero setting* to compensate for drift that has occurred.

Drift tends to be regarded as a d.c. property of a system but this depends upon the response needed. In many electronic systems frequencies of operation less than 10 Hz are regarded as being at d.c. In some disciplines, however, examples being Earth tides or temperature variations in a cool room, responses may be concerned with frequencies of the order of cycles per day or per year. Drift, if studied closely enough, will be found to be a dynamic phenomenon and may need description to allow for its actual behaviour. The simplest form of expression is statement of the linear component in terms of drift occurring per variable of interest—such as time or supply level. In electronic amplifiers drift with time is often quoted as though it were such a linear-time function but in reality it is often formulated on a square-root basis, for drift in this case does not accumulate linearly. Drift is a complex, poorly understood, parameter, one that defies generalization. The problems of specifying drift in a more detailed manner are left to the later section on descriptions of linearity.

Related to the problem of drift specification is how to adequately describe *aging* effects. These also defy generalization. Aging is a factor, like drift, that brings about lack of reproducibility in an instrument for over a time-interval the values of components, and hence transducer constants, may change. Aging can also arise because of use of the components. There are no general rules to follow to reduce aging but the problem usually reduces to knowledge of material properties. In the humidity sensor example aging is to be expected in the membrane, the curved spring strip, in the strain gauges, and in the electronic components. Each of these changes will gradually give rise to a shift in calibration and in the position of the zero.

Internal adjustments are provided so that these can be reset. Being an inexpensive design, intended for low-accuracy use, no figures of aging or drift are quoted. Higher priced instruments usually quote such figures. The task of deciding when an instrument should be recalibrated, to make up for errors of this kind, is very subjective. The history of an instrument's performance is an important factor in being able to predict when it will have drifted, or aged, out of the calibration tolerance band.

Measuring instruments do not respond in a completely continuous manner, nor do they always give the same value measurand for both directions of closure to a point. The term *threshold* is used to signify the smallest level of signal to which the system will respond. All measurements, no matter how well designed, will ultimately be limited by some form of discrete phenomenon. Another example is seen as the threshold of the zero-suppressed voltmeter mentioned above, which occurs at 200 V. Threshold is either caused by a feature of the physical principle involved (for example a turbine flowmeter will not respond correctly until there is sufficient fluid in the pipe) or it may be deliberately introduced (such as to suppress the zero). It provides a *dead-band* of operation.

Many instruments exhibit dead-band effects that are caused by *backlash* and *play* in the drives. These will provide different values for each direction of approach to the point of interest. Where it cannot be removed it is common practice to approach the final value in the same direction each time. Dead-band can become particularly significant as a cause of instability in closed-loop systems: some, however, make use of it to allow the system to settle in the dead-band.

Dead-band produces a kind of *hysteresis* but that latter term is more generally used to describe the magnitude of the hysteresis phenomenon arising internally in a material. Figure 16.2 shows a typical hysteresis loop with its measured quantities. Magnetic material, for instance, does not show the same magnetic field strength for rising and falling induction levels. The magnitude of the strain condition for a cyclically stressed mechanical member will vary for the direction of the stress. In each, hysteresis magnitude increases with increasing excursion. Hysteresis magnitude can be quoted as the ratio of maximum difference of the upscale and downscale variations to the full-scale excursion. Other expressions are used. In some applications the total area is more important than this difference.

In the humidity sensor, mechanical hysteresis arises in the strained member carrying the strain gauges and certainly in the membrane. Backlash arises in the pin-joints of the mechanical system (but is kept in control by a biasing tension), in the sector-gear driving the pointer pinion, and because of slack in the bearing system. The art of good design is to ensure that these are not significant for the intended purpose and that they do not increase significantly as wear occurs. A highly-damped, resonant system (see follow-

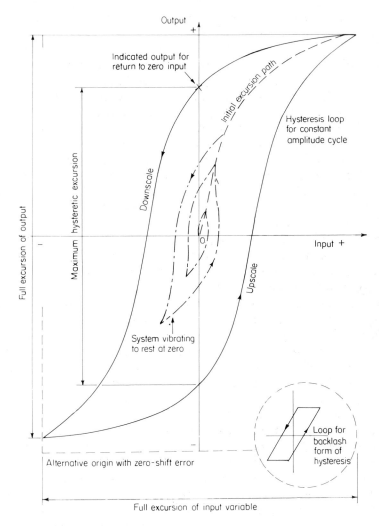

Figure 16.2   Schematic of a generalized hysteresis loop

ing chapter) and slowly moving signal systems will exhibit greater hysteretic effects than more rapidly moving arrangements.

Larger excursions also increase hysteresis error. Systems that can resonate to rest will usually show less error from this source because the hysteresis loop traversed in each cycle will become progressively smaller, the system coming to rest in the centre. Electronic systems generally do not pose these problems but some circuits purposefully introduce backlash. Storage effects in semiconductor elements can lead to backlash in fast-acting circuits. Backlash, hysteresis, drift, creep, and aging are all subclasses of a general *noise* problem that instruments of measurement must cope with. Noise is the general name given to unwanted sources of apparent signal that could lead to error in the measurement mapping process. Every energy regime has its own sources of noise; details are to be found in the chapters on the various regimes that follow later in this Handbook. The user of an instrument will often need to know the tolerance of the instrument to common noise sources.

By suitable design a stage can often be made such that noise quantities (of certain kinds) influence two features of the system in such a manner that their combined effect on the true signal is differenced. Thus deliberate introduction, through a second influence point, of correlated noise into the system can be used to dramatically reduce the original, impossible to eliminate, noise effect. As no process is perfect these *common-mode rejection* techniques need some means of stating their effectiveness. This is done through the use of the *common-mode rejection ratio* (CMRR), which is defined as the ratio of the common-mode noise signal amplitude to the common-mode error signal remaining after rejection. The term is used extensively in electronic systems description where it is usually quoted in decibels. It is, however, a quite general concept and can be applied to any common-mode signal-to-noise rejecting system. As with many parameter statements, the need to keep the statement simple can lead to the assumption that CMRR is a simple ratio. This is not always so; it will often be a function of many system variables. Average figures are frequently quoted. It can, at least in electronic systems, vary with temperature and usually decreases with decreasing frequency of the noise signal. Often the CMRR quoted is for d.c. input signals.

In some instances the need is for definition of how well system signals exceed noise levels: this can be assessed from the *signal-to-noise ratio* (S/N or SNR). This is the ratio of the signal amplitude to that of the noise, again usually expressed in decibels. The amplitude used could be *average, root mean square* (r.m.s.), or *peak-to-peak*. Steady-state a.c. signals are involved and they may be of sinusoidal or complex waveform. Rarely are the waveshape or amplitude criteria quoted, leaving considerable room for imprecision, especially if the signal and the noise do not have the identical waveforms that a ratio properly demands.

The introduction of a stage into a measurement system usually degrades the SNR to some degree. The ratio, in decibels, of SNR at the output to SNR at the input is termed the *noise figure*.

Measuring instruments are designed to fulfil a certain task at a specified accuracy level. As increasing the accuracy of a measurement implies an increasingly more costly design the general rule is to choose an instrument that just copes with the task needed. It is quite pointless to specify an instrument that has accuracy exceeding the task; its capability will not be utilized and the maintenance costs will be greater than necessary. To assist the specification and selection of instruments of appropriate accuracy many disciplines issued standards documents that state *accuracy classes*. There is no general rule to the development of such classes, a dominant reason being that each kind of measurement presents different ranges of possibility and need. For example, it is quite feasible to call up time measurements to errors of parts in 10^{12} but photometry can only be best achieved to parts in 10^4. Where the need has arisen, agreed accuracy classes are defined, within which 'the instrument will have certain declared metrological properties within specified limits of error'. A class is 'usually identified by one of a set of symbols, adopted by convention, and usually referred to as the "class index".' Definition of accuracy often ties in with that of linearity which is covered in Section 16.4. The term *conformance* is sometimes used to relate an instrument to a specified accuracy class. *Interchangeability* is the term used to describe how well an instrument can be exchanged with another having the same specifications; tolerances of manufacture and differences in the various manufacturers' designs causing units to be slightly different to each other yet still be suitable for a given task.

Although the response of a stage is dependent upon its dynamic behaviour the *frequency response* of a stage is a steady-state characteristic that is taken after the stage has settled to the final value after excitation has been applied at each frequency.

In the simplest form frequency response can be stated in a verbal manner describing it in such terms as 'flat to x Hz' (or 'between', or 'above'). In such cases the extremes are denoted, by convention, to be where the response has fallen off (or risen to) the point where the ratio of input to output signal magnitudes has changed by 0.707. In some cases the 3 dB points are used, both ratios being the half-power points. Some times the response obtained is a function of signal amplitude. In such cases it is necessary to state the amplitude of the test.

A more adequate description is to provide a magnitude and a phase *Bode diagram*. Examples are given in Figure 16.3. The true Bode diagram uses only straight line segments to approximate the actual response but this distinction is not always made. The upper curve is, by convention, the amplitude response. Step, and other time-domain plots, are transient statements and are, therefore, covered in the next chapter.

Systems involving continuous cyclic signals, require description of amplitude, phase angle, phase shift, power factor, and true and apparent power. Each of these terms is used in instrumentation in the same way as it is in electrical power systems. Probably the greatest error that arises in their use in instrumentation is the lack of care in ascertaining that the system is

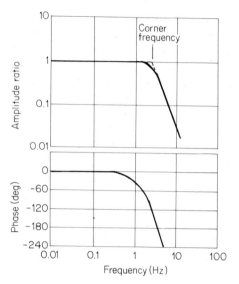

Figure 16.3 Amplitude and phase responses expressed in Bode diagram form

operating with sinusoidal waveforms (which electrical power systems generally are assumed to be) for the parameters are often quoted in terms of sinusoidally based quantities such as r.m.s. and average amplitudes and power levels.

Constants quoted are often far from being such. Many are functions of several operational parameters. Noise figure for an electronic amplifier, for example, is a function of signal amplitude, frequency of operation, and the actual device used. Constants may vary with time. There do exist, however, certain constants, called *physical constants*, that are believed to be unchanging. It is, therefore, often necessary to know more about the formulation of a given constant if more intelligent use is to be made of the number quoted. The designer or user is usually left to explore such ramifications through reading the primary literature. In essence, consensus obtained from practice provides the published information generally required. The difficulty is who shall decide what is general demand and how does the user know if the factors given are entirely adequate for the task in hand. Overspecification can be useful but will increase the cost.

16.3 PROCESS CONTROL INSTRUMENTATION

A very considerable part of the production-run content of manufactured instruments is for use in the process-control industry. In this area of application the presence of needs for similar basic control operations has given rise to the development of terms that have become reasonably standardized. There are, however, many slightly different standards documents and to these the manufacturing companies have added their own standardized glossaries. Despite the differences it is possible in this area of measurement systems to describe many systems features in a way that finds reasonably general acceptance and usage. The intrinsic requirement for well expressed tenders and contracts has considerably assisted the development of standardized nomenclature. The reader must decide the correct source document to be used at any given time.

To the terms introduced in this section must be added those described in Section 16.2 for it is not possible to entirely isolate those of general instrumentation from those of the process industries. Further information on process instrumentation is to be found elsewhere in this publication, notably in Chapters 28, 29, 30, and 31 and in Considine (1974).

Publications of the Instrument Society of America (S5.1 on symbols, S5.4 on instrument loop diagrams, S37.1 on electrical transducer terminology, and S51.1 on process instrumentation terminology) also are relevant. ISA (1977) is a prime reference.

The discussion of the previous chapter showed that instrument systems break down into several basic stages, these apply quite clearly to process-control instrumentation. In that application it is possible to identify (Figure 16.4) the widespread usage of the so-called *loop* (one for each controlled

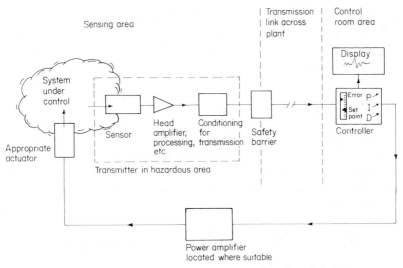

Figure 16.4 Identifiable stages of instrumentation found in the process control industry

measurand) in which a *sensor* (often also called a *pick-up, primary element, transducer, gauge, transmitter* or *cell,* each of which should be deprecated unless specified otherwise) sends a measurement signal to a *controller* unit via a *transmitter* or *relay* unit. It may be necessary to operate upon the signal before transmission, the transmitter in such cases including circuitry and correction mechanism for this purpose. The transmitter also prepares the original sensor signal for use on a suitable transmission bearer to the controller unit.

The controller, usually a defined stage of hardware, is a general purpose hardware system arrangement that enables, among other functions, the received sensor signal to be used as the feedback signal operating into the summing junction of the loop. Controllers generally accept either millivolt signals, direct thermocouple connections or the resistance connections as are needed in a platinum resistance bridge. Built into the controller is usually to be found the facility to set up the control loop in terms of the required degree needed for proportional, integral, and derivative control action, commonly referred to as P, I, D. A range of controller output forms is provided to suit the type of actuator used, examples being on–off switches, or silicon controlled rectifier units.

Loops like this originally developed from a totally analog signal system in which linear control theory can be applied for conditioning the loop performance. With time they have also evolved into digital and hybrid forms. In some cases the controller unit, as such, has become part of a digital computer system and cannot be isolated.

In order to *set up* a loop and to *monitor* its action it is necessary to include a continuous or discrete display of the error signal and other pertinent data such as the measurand. In some systems this is achieved using the controller mechanism to directly display the *error signal* and *setpoint,* the latter being set into the unit manually. A dedicated display is not always used; the plant operator can make use of *trend recorders* that are plugged in, where and when needed.

Most systems would also include *monitors* of the *limits* of operation. These raise *alarms* when limits are exceeded, thereby advising the operator that intervention is needed. Addition of a trend recorder or an *indicator* is then used to diagnose the difficulty. The variety of control loops is very wide and inroads made by digital methods have greatly changed much of the hardware that is used. Many digital systems, however, retain the analog control-loop philosophy actually presenting the operator with information in the traditional form through the creation of displays on VDU's.

A preset, fixed set-point is not always satisfactory. Many processes require a changing reference signal that will cause the control loop to follow a prearranged programme. For this *programming units,* formed by motor-driven cams, electronic function generators, and computer generated functions, are employed to progressively change the reference input signal to the controller summing point. This causes the loop to follow, altering the controlled system *state variable* (another term synonomous with *measurand* in this case) as required. Examples of these are found in annealing furnace controls and in the ovens of gas chromatographs in which temperature is raised according to a preset programme in order to speed up transmission of the slower moving gases through the separation column.

The most general method for sending industrial sensing data to the controllers has been by way of individually dedicated hardwired circuits bearing analog signals. However, in the 1970's other methods have been added, based on digital methods—see Chapter 13. Controller units are generally placed in a central control room for reasons of operator convenience. This practice requires extensive cabling in a plant where hundreds of loops are needed. Increasing plant sizes and greater need for redundant circuits to improve reliability of safe and efficient operation have resulted in wiring costs rising to be in the vicinity of some 15% of plant initial capital cost. Experience has also shown that improved means were sometimes needed to obtain an adequate level of plant security of operation. The use of digital bus systems as *data highways* has been introduced to reduce the cabling costs.

A sensor signal must be suitably conditioned to suit the transmission system of the plant. Long-established transmission systems include the electrical systems of $0-100 \, \text{mV}$, $4-20 \, \text{mA}$, and $1-5 \, \text{V}$; $3-15 \, \text{psi}$ pneumatic signals and others. Each has its own advantage over the others: they each attempt to standardize sensor output requirements and solve the problem of noise pick-up that can occur over long plant cables. Process instrument manufacturers supply the sensor along with the appropriate conditioning, the whole being mounted in a robust enclosure that has become commonly known as the transmitter.

Sensors seldom provide enough signal energy to operate later stages and *amplifiers* will be needed at many places in the signal chain. By suitable single or multiple feedback connections, these can be set up to provide gain, shaping, and regeneration. These are known as *instrumentation amplifiers*. Often the use of an *isolating amplifier* or *isolating transformer* is dictated to electrically separate one circuit from another, or to disconnect a d.c. path.

Although theoretically a sensor needs only two terminals at a port practical techniques often incorporate additional leads. These extend the operation of a stage to make allowance for such defects from the basic principle applied as connecting lead resistance or temperature effects. Thus many sensors use *multiple-wire* connecting systems.

Once conditioned as a unique singular measurement data path a number of sensor signals may be multiplexed, in frequency or time, onto a common transmission bearer.

Energy sources are needed in a sensing stage to make a measurement and sensors often must be placed in intimate proximity with the process they help to control. Thus a sensor might be the cause of an explosion if it fails or generates sparks in operation. For this reason safety codes must be adhered to and the equipment designed such that explosion will not be caused. A concise introduction to the practice adopted in process industries is presented in Jones (1974) where reference is made to several standard specification documents.

Several approaches may be adopted. A first is to enclose the sensor, transmitter, and the other apparatus that must be used in the *hazardous area*, in a *flameproof enclosure*. Another philosophy aims to design the apparatus with greater than normal concentration on ensuring that it operates within safe limits and will not overheat or fail when experiencing energy levels beyond normal. In yet another procedure the equipment is continuously maintained in a condition, by the use of purging or pressurizing gases for example, such that flammable gas cannot build up within the instrument enclosure (see ISA (1969) for further explanation).

A more fundamental, and overall easier to apply, approach is that termed *intrinsic safety*. In this concept the apparatus is only allowed a given level of energy under the worst case that can be envisaged. The energy quantity is limited to well below that which can initiate ignition. For example, the minimum energy needed to ignite propane gas is around 0.3 mJ. Intrinsic safety, however, is not yet a universally accepted principle and, as is usually the case with any form of standardization, the appropriate code must be consulted.

For intrinsic safety to work efficiently the measurement signals must pass from the hazardous zone to one where the limited energy capacity requirement can be relaxed. This interface point is called the *safety barrier*. As barriers often are formed using zener diodes to restrict energy levels that can be drawn from the source, industry has adopted the term *zener barrier*. This barrier acts to limit, or disconnect through a fuse, the energy that can be drawn by the instrumentation that lies in the hazardous areas. Further details on use of instruments in hazardous areas are to be found in Jones (1974), Magison (1978), ISA (1960), and Chapter 13.

Finally on the particular subject of the instrumentation of the process control industry it is appropriate to discuss briefly the terminology and practice of the controller unit: proper understanding of measurement requires a basic appreciation of the total practice adopted.

The controller is identified as a distinct unit in a process-control loop but may itself be comprised of *modular units* that are assembled to provide the operation desired. Each measurand of a process needs its own controller so it is quite common for a plant to have as many as a thousand or more of them. They are generally mounted together in the plant operating room. Because of these practical requirements manufacturers have concentrated on standardization of design so that they are suitable for high-density panel mounting with simplified maintenance and adjustment. Digital control methods have begun a break from this traditional practice and the very large size panels of the past are gradually being displaced by small VDU's that provide the operator with the ability to monitor, upon demand, any loop and to study its past and predicted performance.

Although the controller, as a distinctly separate unit, is declining in use for large installations it is still appropriate for small plants: the nomenclature, therefore, continues. Now follows explanation of several commonly used terms.

The term *offset* has several similar meanings. The first is when used in a general context to describe a signal that is not exactly where it is intended to be, an example being the *offset voltage* that might occur when a sensor does not provide *zero output* for zero state of the input measurand. A second, more specific use, is to describe the steady-state difference that exists between the desired operating point of a control loop and the actual value. This is also called *droop*. It may arise as a fundamental property of the kind of feedback loop used, or be the result of equipment being unable to provide enough *regulation*. It is generally a load-related variable. In electronic device description it has other related meanings each of which is generally specified by an additional word, for instance the *offset calibration* of a semiconductor active device is the error band defined by the maximum error in calibrating the *offset voltage*. A *zero-shift* in the value of a variable provides a constant parallel displacement of its entire calibration curve. In a controller stage zero-shift gives rise to offset from the desired operating point. The instantaneous difference between the desired *set point* and the actual *control point* is termed the *deviation*.

In selecting a controller it is necessary to choose it so that its set point can be set over the range required. Tuning a complete plant, or setting it to produce an alternative product, may require different control point settings. The range over which the controller can be set is decided by the sensor connected and the controller itself. This range is termed the *span* or *scale* and is the algebraic difference between the lower and upper limits of satisfactory operation. In order that the span can be set to lie within given tolerance limits controllers and allied sensors generally have a *span adjust* facility. Span is closely related to calibration but possession of a correct span does not ensure that the span relates to the variable in an accurate manner. As an example, a temperature controller may be adjusted such that span is 50 °C within say ±1% but the span may not lie accurately onto a desired 100–150 °C range. Thus both span and offset influence calibration.

Span, the control variable limits, and the error are often all shown in some way in the display part of the controller module (if it exists). Additionally to these there may also be monitor and alarm settings that can be adjusted in the display to show the values to which they are set. In this way the operator can observe the instantaneous value of the error, the actual set-point required, the deviation, and if the signal is between limits. However, incorporation of D, I and (also called *reset*) actions can cause the deviation signal displayed to be somewhat misleading. This is one reason why trend recording is often needed. A further reason is that the loop may operate with an inherent time delay of the form of a *transport lag* wherein the signal is delayed, arriving without time distortion at a time later than it was generated—compare the delay resulting from a simple water container with small outlet pipe and that of water beginning to travel along a long pipe at uniform velocity. Terms *dead-time, dead-band*, and *dead-zone* relate to such phenomena. These effects are described in the following chapter.

Control loops can sometimes be used without the need for compensation but often improved transient and steady-state performance are obtained by the use of derivative and integral action. Although these compensation methods (see Chapter 14 or Atkinson, 1972; Coughanowr and Koppel, 1965), are based on well-organized theory, the practising plant operator is able to follow quite simple rules for tuning a system without knowledge of this theory. These rules are based on pioneering work by J. G. Ziegler and N. B. Nichols in 1942 (see Coughanowr and Koppel, 1965 for an exposition). Thus controllers contain adjustments for the degree of proportional, derivative, and integral action. These are progressively adjusted to a set schedule whilst observing the loop response. Several of the larger process instrument manufacturers have each published books on the tuning of controllers; some maintain training apparatus for customers' use. The practice of use of standardized controllers and sensors and simple tuning routines rests entirely on the assumption that the loop is linear in operation for small error signals, an assumption that is not always valid in practice.

The above concepts have been developed for both electrical and pneumatic control loops, the terminology and schemes of operation at the conceptual level being mostly identical. There is a clear swing away from pneumatic methods for new installations but it will be many years before pneumatic methods have no place in the process-control industry; in some areas they will continue to be preferred. The majority of explanatory texts on industrial control relate to electrical methods, examples that are oriented to practising engineers, rather than to academicians, being Atkinson (1972), Bryan (1967), and Coughanowr and Koppel (1965). Pneumatic control instrumentation is explained in the trade publication, Foxboro-Yoxall (1972).

16.4 ACCURACY, LINEARITY, DRIFT, AND THEIR DESCRIPTION

For reasons of simplicity of specification and description the system's performance features of *accuracy* and related *linearity* are often quoted in a simplistic manner by the use of a single numerical value. This practice might hide, from those persons who are not aware of the facts, that these are often gross simplifications of a much more complex reality. To illustrate this point the following paragraph is taken from the terminology section of Foxboro-Yoxall (1972):

> Accuracy includes the combined conformity, hysteresis, and repeatability errors. The units being used are to be stated explicitly. For example, expressed in output units, a typical expression might be ±1 °C; expressed in percent of output span, the expression could be ±0.5 percent of output span; expressed in percent of the upper range value, it might read ±0.5 percent of upper range value. For indicators, it might read ±0.5 percent of scale length; or in some cases it is expressed as a percent of actual output reading, such as ±1.0 percent of actual reading.

Confronted with this variety of simplifications leads one to refer to approved standards documents, such as BS 5233:1975 or AS 1514, to find that accuracy is defined only in a general sense and that no suggestion for a uniform methodology is provided. The ISA document, formulated by the SP37 subcommittee, on transducer specification, provides greater depth in the manner given above. (It is issued as S37.1 or ANSI MC6.1:1975.)

In reality the concept of accuracy is more properly discussed in terms of errors and the uncertainty of a measurement made with the measuring instrument. The term accuracy, one source suggests, should only be used as a general descriptor of system performance.

Implicit in any statement of uncertainty of an instrument is the apparent assumption that the instrument will provide the same absolute or proportionate performance at any part of its usable range. This is rarely a totally valid assumption.

The manner in which error varies over the effective range of an instrument is expressed as the *linearity* of the instrument (but which correctly actually relates to the *non-linearity* or lack of linearity between desired and actual performance).

A process is said to be linear when the relationship between two quantities is such that a change in one quantity is exactly proportional to the associated change in the other quantity.

Expression of the non-linearity of an instrument is achieved in many ways most of which, again, result in a simple statement that conveys information about how the signal deviates as it ranges over the available span. Fundamentals of measurement indicate that a single numerical value can only characterize a single attribute of a situation. This is, therefore, the case in the statements of linearity: each method of description conveys different information about the linearity. Seldom is an actual curve presented, the exception being when a calibration chart is provided.

Linearity is expressed as deviation from a specified straight line. Four predominant reference lines might be used.

The first uses a line placed so that it lies in the position satisfying the *least squares* criterion calculated for the data set available. It is often called the *best straight line* (BSL). It might also be referred to as a least squares line. Figure 16.5a might be an example of this.

In many instances the system provides zero output for zero input, in which case a more appropriate line to use is a BSL that is computed so that it is constrained such that the line passes through the zero point. In suppressed or elevated-zero systems the point of interest will be the pseudo-zero used. Figure 16.5b illustrates this line. This method describes the zero-based linearity of the system. One variation of this is to use not a zero-based line, but some other suitable reference point. This is termed BSL with forced reference linearity.

In terminal-based linearity a reference line is adopted that links the upper and lower limits of the span, a line drawn between its terminal points. It is also called terminal linearity of end-point linearity. The line so made is sometimes referred to as the end-point line. Figure 16.5c shows this line.

Not defined in BS 5233:1975 but which is in AS 1514:1980, another line that can be used is that which (Figure 16.5d) lies in the theoretical position of the principle invoked. This is called the theoretical slope linearity.

Having first decided which line is the most appropriate to use, the next step is to select a method of description that will express the magnitude of deviations from the line. In general this will usually be the greatest departure from the line of the output variable. Linearity of instruments is generally determined as the result of a calibration test in which only a limited number of values would be determined. Some calibration processes provide a considerable number of values and if they are of random nature the deviation might then be expressed in terms of the standard deviation or some other statistical descriptor.

A little thought reveals that the above methods of expressing linearity and accuracy of an instrument are subject to many causes of error in their determination and interpretation. A process with significant hysteresis can exhibit quite different linearity with each direction of signal change and with signal amplitude. The magnitude of influence variables at the time of the determination also can alter the error magnitudes. The method of calculation of the best fit line and the number of data samples available will further influence the value that will be stated as the singular value of linearity of the instrument.

Expressions of the linearity of a system are invariably determined for a system in the steady or static state. Rarely will the same relationship apply when the stage is in the dynamic state. Storage parameters will alter the instantaneous transfer characteristic according to the signal amplitude and

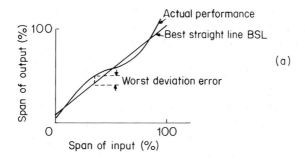

(a)

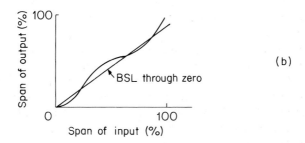

(b)

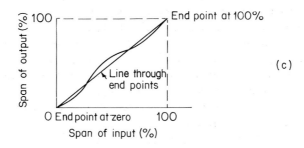

(c)

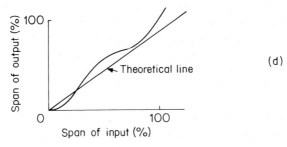

(d)

Figure 16.5   Four methods of specifying linearity of
a response

history of previous signal excursions. This point is often overlooked when accuracy and linearity figures are taken from data sheets for application to sensors that are used for rapidly changing measurands. Pressure gauges, accelerometers, position sensors, thermometers, and the like will give very different instantaneous performance figures to those measured in the steady state.

Another factor to be considered may be that the stage transfer characteristic may possess adequate linearity of response but that the BSL may slope at the wrong rate because of an inaccurate d.c. sensitivity coefficient for the stage.

Where possible it may be preferable to provide a mathematical expression that describes a curve fitted to the data, an example being the relationship for the output of a thermocouple.

It can be seen, therefore, that expressions of accuracy and linearity are invariably simplifications. Specifications of these parameters must be assembled, and then interpreted, with care if ambiguity is to be avoided.

Extending these ideas further, to cover other non-linear responses, again introduces likely confusion as the direct result of seeking simplified statements of performance. Specification of *drift* is a related area where such problems arise.

Drift is the feature of a system that characterizes how a system variable that is intended to be constant, changes with time. The drift of a gravity meter is one example. The term is also used to describe effects of influence quantities on the output, an example being the temperature drift of an electronic amplifier. It is also a term applied to explain how a measurand varies with time or other variables. An example of this use is to characterize settlement of a structure with time. It is sometimes used synonomously with the term *stability*. Some kind of qualifying statement is needed to uniquely define which context is being used.

Drift conveys very-low-frequency response information about the measuring instrument or the measurand. It can, on occasion, be very difficult to separate the two.

Figure 16.6 shows several possible drift curves. For each the need is to formulate a simple expression that will convey information about the trend of the output away from a chosen reference value.

When the drift is linear it is a simple matter to express the slope in terms of the two parameters involved. The gravity meter example of Figure 16.6a might be expressed as drifting at x milligals per hour. Being entirely linear it would not matter where the slope was determined along the curve.

The situation becomes a little more difficult when needing to express the drift for the case shown on Figure 16.6b. If the characteristic possesses a cyclic component of constant amplitude then a mean line is easy to determine. If, however, the curve does not have obvious linear features, like that

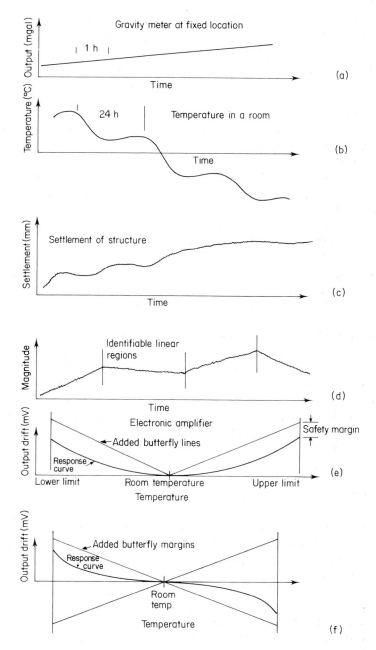

Figure 16.6 Typical drift curves for example instruments and
measurands

of Figure 16.6c, then a best fit line must be fitted from which the drift can then be assessed as for the previous cases.

When the curve clearly possesses two or more linear regions (Figure 16.6d), each of different slope, a more extensive statement will be required. No standards of practice appear to exist for such cases so description should carry a statement of the additional features rather than simply quote a fixed numerical drift coefficient.

If the drift curve returns to the original value (Figure 16.6e), quoting the mean drift that has occurred between the two end points would imply that the system is perfectly stable. Further, non-linear curves, such as in Figure 16.6f, would be grossly misrepresented. One approach, that outlined in Stata (1969) for example, is to use a *butterfly characteristic* curve. Typical boundary curves have been superimposed onto the two amplifier drift curves. Drift is calculated as that occurring from the reference point (which is often room temperature at which the amplifiers were adjusted for the test) to the extremities of workable range, plus a margin of safety. This coefficient then provides a conservative figure within which drift can be expected to lie for a given temperature excursion. Such simplification, however, is at the expense of possible overdesign on the part of the user.

These examples clearly show that specification of drift can be a complicated matter and that simplistic definitions, as with those for non-linearity, can be misleading. It has been known for manufacturers to use these shortcomings of description to their advantage when specifying equipment.

16.5 SPECIFICATION OF MEASUREMENT SYSTEMS

It can be seen, from Chapters 3, 15 and in this one, that description of a measuring instrument stage can involve many terms from a very large possible nomenclature. The terms themselves often possess synonomous meanings and adequate description (or interpretation of already prepared specifications) can easily get out of hand in an overenthusiastic attempt to ensure that an adequate description has been committed to paper.

There exists a saying '$pecify $anely or $uffer'. It well illustrates the importance of spending enough time and effort on specification writing (or reading) for these are the written statements that form the cornerstones of the designer tasks or the user's selection procedures. Too often instrumentation just happens through an ill-controlled design or selection process. An essential step at the start of an instrument design, or selection for a given application, is a good set of 'specs'.

In the designer's area they must be raised to a well-developed stage before manufacturing effort is expended. In the user's area they must adequately communicate the performance required of the eventually delivered product. They should not be finalized after manufacture or installation, when problems arise, for by then errors of specification, or choice, cannot be so easily rectified.

Specifications, where possible, should always be prepared with both the designer (or supplier) and the user being party to common discussions. A third party, the component vendor, may also need to be involved. Figure 16.7 shows how Wheeldon (1974) depicts the situation existing when a new instrument system is to be made. His two papers expand the points now presented.

Various kinds of specification exist, each having a different purpose of definition.

System user and system designer relationships
   Operational requirements specifications
   Functional and technical requirements specifications
   Design specifications (to implement and check design)
   Manufacturing and quality control specifications
   Factory test specifications
   Acceptance test specifications (factory)
   Documentation specifications
   Packaging and delivery specifications
   Installation specifications
   Commissioning specifications
   Acceptance test specifications (field)

System designer and component vendor relationships
   Standard item specifications
   Modified 'standard' item specifications
   Special subcontract specifications
   Incoming inspection specifications (system designer)
   Incorporation of vendor specifications in system design specifications

The designer is in the position where it is necessary to take the somewhat imprecise design ideas of others and transform them into quite hard and specific facts. A realizable, producible design at a satisfactory cost must be achieved. Value judgments are needed about what the customer says is wanted (which is often what are then thought to be wanted but are often not at the outcome); what the vendor can offer (or what is said to be available but may not be when the actual need for parts arises some time later); the designers own ability; facilities for design and the allowed time and money available to complete the task.

These factors have been expressed above in terms of a designer creating a new instrument. Similar lines of reasoning apply to the person given the task of selecting proprietary equipment, for which appropriate kinds of specifications are needed, to form a required system function.

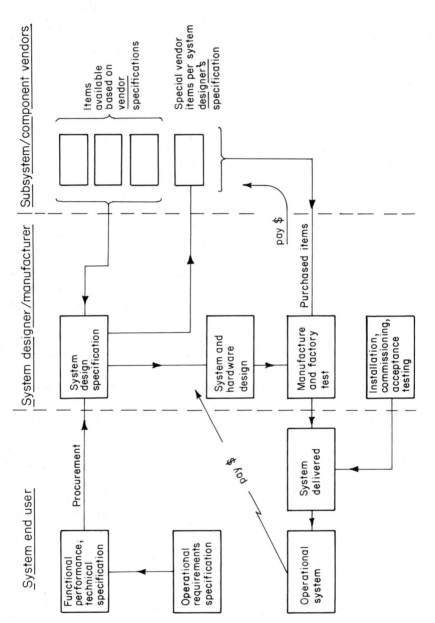

Figure 16.7   Relationship between the parties involved in a system design process (Wheeldon, 1974).

To completely specify a product requires powers of prediction plus enormous effort in attempting this virtually impossible task. A total specification may well satisfy one party in the situation but someone has to pay the price of obtaining the information. The result, in practice, is a balance between exactitude as the perfect ideal and no prethought at the other extreme. It is this process that gives rise to the generation and use of simplified statements of the kind that have been introduced in this and other chapters.

There are no short cuts to establishing the happy medium at which specifications should be aimed. Where possible it is advantageous to make use of standards codes and documents, as were introduced in Chapter 3. Written reference to these can reduce the need for much verbal detail needing to be expressed in a specification.

The Instrument Society of America provides preprinted check lists that are designed to assist uniformity of specification for over twenty common instruments (S20 series 'Specification forms for process measurement and control instruments, primary elements, and control values', 1975).

As has been well illustrated previously, care must be exercised in the actual interpretation of terms. Vague expressions such as 'surface to be a smooth finish' are virtually worthless. Never leave out (and be wary if they are not given in an existing statement) major factors that might seem patently obvious—they may not be so obvious to others.

Specifications should be clarified before, not after, the signing of a contract or manufacture or purchase. Watch out for specifications that appear to break a fundamental law of physics.

Specifications have tendency to *creep* in some organizations as each level in the contracts process adds a little more safety margin on the figures of desired performance. This can raise the final price considerably for cost of instrumentation generally rises more rapidly than gains in performance so obtained.

Specifications can sometimes come to the designer as the result of the horse–camel syndrome. This describes the creation of specifications by the process of selecting feature points from different suppliers catalogues of the same-purpose instrument, combining them all into the new specification that none has already offered. Such features often are developed by a manufacturer at the expense of another parameter. The combined instrument may well be a physical impossibility to make, or at the least, be a very much more expensive version. Many suppliers are, unfortunately, unable to discern if this practice has been used and they may well take on the task only to find out eventually, to the customer's dismay, that they cannot supply the product.

The use of graphical descriptions of linearity, responses, drift, etc. can often be a better method of description than mere words alone. Specific reference to the glossary source used would be a considerable improvement in many cases of instrument specification.

When very large sums of money are involved it will be usual to have available the services of expert specification writers and system designers. Unfortunately, instrumentation is so broad in nature and so diffuse in application that, more often than not, it is treated as a host of minor specification tasks within a larger system. For this reason it is to be expected that instrument specifications will generally be less satisfactory than desired and that actual performance will often fall short of that stated in the specification statements.

16.6 TESTING STATIC REGIME PARAMETERS

It is appropriate to include a brief introduction to testing in this chapter as the procedures of testing influence the specification and the terminology used in describing and verifying instrument stages. Testing products for conformity requires the expenditure of resources and, as with the raising and interpretation of specifications, it too tends to seek the most simplified procedures that will fulfil the task as it is formulated at the time. Thus testing and a proper understanding of terminology are complementary topics.

Although the basic concepts involved in testing do not differ from those of measurements in general, testing is seen more as a subclass, within the total measurement class, that is concerned with arbitration procedures. One definition, a rather restricted use, declares it is 'to ascertain the performance characteristics of a meter or instrument while the device is functioning under controlled conditions'.

A dictionary definition states that a test is a 'trial determining a things existence or quality or genuineness or fitness for a purpose, standard or other means serving this end'.

It can be taken that a test in contrast to a measurement aims to establish conformity to an already declared standard by an accepted procedure of measurement and is possibly not the best term to use when describing the use of measurement for the purpose of seeking knowledge in a more open-ended manner. As an example establishing the compressive breaking load of a standard-sized concrete casting in a standardized manner to an established code of practice constitutes a test. Observing what happens to a concrete beam under load is more appropriately called a measurement when there are no standardized procedures laid down about the manner in which the observation should be made.

Those actively involved in the testing industry appear to be quite clear about the distinction but those who use measurements in a more general way, such as a scientist does, often use the term synonomously with the word measurement.

Tests of properties can be obtained through the use of destructive and non-destructive procedures. In the latter area large industrial demand and specific institutions devoted to it have helped establish it as a specific measurement discipline. It has grown to include those methods for establishing the properties of materials, and their use in assemblies, in applications where the more obvious testing method would impair, or destroy, the performance. Non-destructive testing obtains test data without damaging the product. As an example a steel strut can be tested for its safety strength by conducting a tensile test on a sample specimen, stressing it to failure. This is obviously of no use if the same strut must be used; other methods have been found that provide the strength from tests that obtain data that will yield similar information—in this case ultrasound, strain gauging, and ionizing radiation procedures.

Many direct test procedures are non-destructive by nature; measuring the gain of an electronic amplifier does not damage its performance. These are not, however, generally described as non-destructive methods, that term having the quite specific meaning described above.

The need for simplification brings with it the need for care in test result interpretation. Often assumptions are used about the system under test, assumptions that considerably relax the effort needed.

Steady-state models are often formed of complex stages in order to avoid needing to conduct dynamic-regime tests. As an example attenuation (a steady-state parameter) of an RC, low-pass, electric filter stage at a given frequency can easily be obtained, in theory and in practice, by replacing the capacitor with a resistor having a value of the impedance the capacitor has at the frequency of interest. Thus it only applies at that specific frequency.

In practical measurement the use of an a.c. voltmeter will provide static regime readings because the instrument converts the dynamic, steady-state, a.c. signal into a d.c. equivalent. This holds true only if the a.c. to d.c. converter used has flat frequency response over the range of frequencies of signals that might be measured. Also the waveform must be adequately sinusoidal (there are exceptions to this but this simplification illustrates the point) for a correct value to be indicated.

Loading effects, which may occur in any of the energy regimes with which measurements are conducted, and mismatch at terminations, can give rise to test values that vary from those of the situation occurring when the testing device is not connected.

Recorders are useful as means of verifying that the transient state of the signal has indeed died away to leave an isolated steady state. They also assist verification that a steady state actually exists. Spot measurements in time with simple measuring instruments, such as the multimeter, can provide inadequate information about the true state of affairs.

Care is needed to ensure that transient phenomena are not operating unrevealed to the observer. They can significantly alter the steady-state value so obtained. High levels of a.c. noise can saturate an amplifier reducing the gain to low-frequency signals; noise spikes can charge peak-detector systems.

Test equipment must be maintained in a good state of repair and be calibrated to a traceable procedure. Calibrations, d.c. zero-set, and other adjustments needed by the operator before use, plus correct environmental control must be observed before a test is made. If any parameter of the specified test procedure is not properly followed then the test can only be declared as invalid.

Testing follows such well established procedures that it might appear that untrained personnel can be used to conduct the tests. This assumption is often wrong for the procedures laid down cannot take into account all of the possible factors that need to be allowed for in order to make a good measurement. Trained staff can sense when things are not in order and be able to act accordingly. Untrained persons will often make assumptions that are incorrect providing a calibration, or acceptance agreement, which should not have occurred.

In large organizations it is often common practice for all instrumentation to be tested to the declared specifications on delivery. In many, and certainly in the smaller ones where expertise often does not exist, no delivery check is made and faults go undetected. This has not helped maintain standards of supply because manufacturers are not caused to always test their apparatus as well as it should be done.

There appear to be few general works in print that describe testing at a general philosophy and practical level, Gilmore (1969) being an example. Many texts, however, have been prepared that describe testing for specific areas, such as microwave, mechanical, non-destructive, high voltage, and others. The bibliography (IMEKO, 1980) provides a list of several hundred texts many of which relate to testing. Other chapters of this work also provide appropriate references.

As an immediate source of titles, electronic testing is covered in Edwards (1971), Herrick (1972), Malvino (1967), Oliver (1971), Turner (1963), and Waller (1972). Elements of mechanical testing can be obtained in Beckwith and Buck (1979), Birchon (1975), and McGonnagle (1973), the latter two being specifically on non-destructive testing. The Sira Institute has published (Sira Institute, 1978) a study on optical testing. These are but a few of the many texts available.

Additionally these exists a wealth of information about testing in the standards specification literature. Such sources are quite specific about procedures (compared with texts that attempt to generalize) and should be followed when relevant. As with specification document preparation, these should be considered for adoption when preparing statements about test procedures as their use will help standardize the practice on behalf of both parties to a test; to the test operator and to the user of the results.

REFERENCES

Atkinson, P. (1972). *Feedback Control Theory for Engineers*, Heineman, London.

Beckwith, T. G. and Buck, N. L. (1969). *Mechanical Measurements*, Addison-Wesley, Reading, Mass.

Bell and Howell (1974). *The Bell and Howell Pressure Transducer Handbook*, CEL/Instruments Div., Bell and Howell, Pasadena, USA.

Birchon, D. (1975). *Non-destructive Testing*, Clarendon Press, Oxford.

Bryan, G. I. (1967). *Control Systems for Technicians*, University of London Press, London.

Considine, D. M. (1974). *Process Instruments and Controls Handbook*, McGraw-Hill, New York.

Coughanowr, D. R. and Koppel, L. B. (1965). *Process Systems Analysis and Control*, McGraw-Hill, New York.

DiStefano III J. J., Stubberud, A. R., and Williams, I. J. (1976). *Feedback and Control Systems: Schaum's Outline Series*, McGraw-Hill, New York.

Doebelin, E. O. (1975). *Measurement Systems: Application and Design*, McGraw-Hill, New York.

Edwards, D. F. A. (1971). *Electronic Measurement Techniques*, Butterworths, Sevenoaks.

Foxboro-Yoxall (1972). 'Process control instrumentation—with explanatory notes', Publ. 105E, Foxboro-Yoxall, Reading, UK.

Gilmore, H. L. (1969). *Integrated Product Testing and Evaluation: a System Approach to Improved Reliability and Quality*, Wiley, New York.

Herceg, E. E. (1972). *Handbook of Measurement and Control*, HB-72, Schaevitz Engineering, Pennsauken, USA.

Herrick, C. N. (1972). *Instruments and Measurements in Electronics*, McGraw-Hill, New York.

IMEKO (1980). *Bibliography of Books Published on Measurement Science and Technology—with Emphasis on the Physical Sciences*, P. H. Sydenham (Ed.), TC-1 Committee IMEKO, Dept. Applied Physics, Delft Tech. Univ., Delft.

ISA (1960). 'Electrical instruments in hazardous atmospheres', *Publ. No.* RP12.1, Instrument Society of America, Pittsburgh.

ISA (1969). 'Instrument purging for reduction of hazardous area classification', *Publ. No.* S12.4, Instrument Society of America, Pittsburgh.

ISA (1977). 'Standards and practices for instrumentation', *Publ. No.* SP5, Instrument Society of America, Pittsburgh.

Jones, E. B. (1974). *Instrument Technology*, Vol. 1, Newnes-Butterworths, Sevenoaks.

Magison, E. C. (1978). *Electrical Instruments in Hazardous Locations*, Instrument Society of America, Pittsburgh, USA.

Malvino, A. P. (1967). *Electronic Instrumentation Fundamentals*, McGraw-Hill, New York.

McGonnagle, W. J. (1973). *Non-destructive Testing*, Gordon and Breach, London.

N.S. Corp. (1977). *The Pressure Transducer Handbook*, National Semiconductor Corp., Santa Clara, USA.

Oliver, B. M. (1971). *Electronic Instrumentation Fundamentals*, McGraw-Hill, New York.

O'Higgins, P. J. (1966). *Basic Instrumentation—Industrial Measurement*, McGraw-Hill, New York.

Paine, A. (1974). 'The uncompromising environment', in *Stathmology*, P. H. Sydenham (Ed.), Department of Continuing Education, University of New England, NSW, Australia.

Sira Institute (1978). *Optical Sources, Detectors, Components and their Fabrication and Testing*, Sira Institute, Chislehurst, UK.

Stata, R. (1969). *A Selection Handbook and Catalog Guide to Operational Amplifiers*, Analog Devices, Inc., Cambridge, USA.

Turner, R. P. (1963). *Basic Electronic Test Instruments*, Holt, Rinehart and Winston, New York.

Waller, W. F. (1972). *Electronics Testing and Measurement*, Macmillan, Basingstoke, UK.

Wheeldon, R. (1974). 'Specifications—the identifying facts' and 'Generalised approach to selection and design (of instruments)', in *Stathmology*, P. H. Sydenham (Ed.), Department of Continuing Education, University of New England, NSW, Australia.

Handbook of Measurement Science, Volume 2
Edited by P. H. Sydenham
© 1983 John Wiley & Sons Ltd

Chapter

17 P. H. SYDENHAM

Measurement Systems Dynamics

Editorial introduction

Although a considerable amount of instrument subsystem blocks can be adequately characterized and appreciated in terms of their static or steady-state behaviour it often is essential to appreciate how they respond under dynamically changing conditions.

This chapter extends the material of the previous two chapters into the dynamic regime of instrument systems. Linearly responding systems are emphasized, it being considered that this treatment will suit the bulk of response situations met in practice.

17.1 INTRODUCTORY REMARKS

As was explained in previous chapters the dynamic regime of an instrument stage is generally a more complex aspect to characterize and test than the stage's static performance. To fully understand the behaviour during the dynamic state it is necessary to allow additionally for the transient solution of the transfer characteristic when forced by an input function, a factor that was not important to understanding of the steady-state and static regime behaviours.

Dynamic behaviour of systems has been well explained at the theoretical level if the performance remains linear during the dynamic state. It has been shown that the energy relationships of electrical, mechanical, thermal, and fluid regimes can each be characterized by the same mathematical description and that a system containing a cascaded chain of a mixture of these, as is typical of instrument systems, can be studied in a coherent manner by the use of such mathematical techniques. This body of knowledge and technique lies in the field of *systems dynamics.* It has progressively grown from isolated, unconnected explanations of the analogous situations existing mainly in electrical, mechanical, and acoustical areas, into one cohesive systematic approach to the study of the dynamics of linear physical systems.

The approach has been widened considerably (see Klir, 1972), to take in philosophical concepts, such as are found in social systems, this broader assembly of knowledge usually being referred to as *general systems theory*. General systems theory certainly has application and potential in some instrument systems but most designers and users of instruments will find that the more confined and mathematically rigorous assembly of technique and knowledge (systems dynamics) will be that found to be usefully applicable in practical realization and understanding of measuring systems. The use of the word *systems* is extremely diffuse and it carries numerous connotations ranging from a totally general concept to the quite specific use that is based in mathematical explanation.

As indicators of the part that is useful to measuring instrument dynamic studies the reader is referred to Karnopp and Rosenberg (1975), MacFarlane (1964), Paynter (1961), Shearer *et al.* (1967), and Zadeh and Desoer (1963) for general expositions, and to Coughanowr and Koppel (1965), Fitzgerald and Kingsley (1961), Koenig and Blackwell (1961), and Olson (1943) for more specific uses in areas of chemical plant, electrical machines, electromechanical devices, and the so-called *analogies*, respectively.

Several authors have extracted, from the total systems knowledge contained in such works, the much smaller part that is needed by measuring systems interests, presenting this as chapters in their works. Such accounts are to be found in Beckwith and Buck (1969), Doebelin (1975), and Neubert (1976).

Notable papers have also been published that further condense the information (see Bosman, 1978; Finkelstein and Watts, 1978). Various chapters of this Handbook, and especially so the following chapter (Chapter 18) on transducer fundamentals, make use of systems theory in their explanations. Finally, many control-theory-based works are of relevance.

There exists, therefore, a considerable quantity of well organized knowledge about the physical behaviour that may be encountered in measurement systems studies. When response is linear it is possible to make use of mathematical models to obtain a very workable understanding of the dynamic behaviour. If it is non-linear, however, then the situation is not so well catered for because no such widespread and generally applicable mathematical foundation has been forthcoming. In practice, however, designers and users of instruments can find considerable value in assuming linearity, if only for a limited range of operation. For this reason it is important to know how to recognize if a system functions in a linear manner and which techniques apply in such cases.

The purpose of this chapter is not to expound the mathematical modelling and design of transducers in a concise and rigorous manner (that is provided in Chapter 18) but to present the commonly met characteristics of linear dynamic systems that will be encountered so that their class can be recognized and operated upon using a fundamental approach that is based on knowledge of their characteristics. It also provides the basic descriptive terminology needed in specification of response characteristics.

17.2 FORCING FUNCTIONS

A stage in a system is forced into its transient state by one or more external inputs. These are termed *forcing input* or *excitation functions*. They may be applied to the intentional input of the stage or cause transient behaviour by entering, as influence quantities, through numerous other unintentional *ports*. In reality the true two-port, four-terminal, instrument stage can rarely be realized for virtually all designs are influenced to some extent by unwanted noise perturbations entering through many mechanisms.

When considering the dynamic response that might arise for a stage it is, therefore, necessary to first decide the ports through which a forcing function signal might enter. This decided, the next step is to assess which kind of forcing function is relevant and apply this to the real physical device, or to its correct mathematical model expressed in the state-variable or transfer function form. Alternatively it might be simulated on an analog or digital computer. When the transfer function model is used the product of the Laplace transform of the forcing function and the transfer function can be solved to provide the transient dynamic behaviour in a reasonably simple manner. This method is well developed in numerous electrical engineering and systems texts. Chapter 14, Section 14.2.2, contains examples of response evaluation of a simple positioning system. The section presented here describes typical forcing functions that might be used to test or study an instrument stage.

Certain types of forcing function lend themselves to analytical solution being easy to apply to the Laplace method of response evaluation. They are also simple to procure as practical test signals that, although not perfect, come close enough to the mathematical ideal. For these reasons testing and evaluation of systems tends to attempt first to make use of one or more of several basic forcing functions.

These functions (see Figure 17.1) are the discontinuous *unit step* and the *unit impulse*, the *ramp* plus the continuous *sine wave*. They are described as mathematical entities in Chapters 4 and 14 and, therefore, do not need further theoretical expansion here.

These functions are easy to produce and apply in practice but it must be

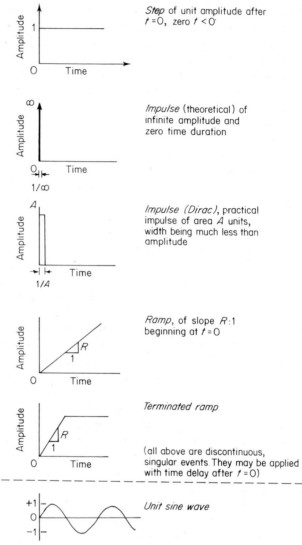

Figure 17.1   Typical forcing functions used in testing and
study of the dynamic response of systems

recognized that they may not provide an adequate simulation of the real
forcing signals existing. This point is not always made clear in the general
treatment of the dynamics of common linear systems stages.

When such simplifications are not adequate other more suitable ones must
be applied. If they can be transformed with the Laplace transform expression, and if the product with the transfer function can be arranged so that

solution into time-variant terms is obtained, then a theoretical study can be made. In many cases, however, this is not possible within bounds of realistic adequacy and alternative methods of study must be employed if no way can be seen to simplify satisfactorily the forcing transfer function. In such instances, simulation on a computer often provides the means to a solution.

A step or impulse will excite a system into its dynamic transient response as might a sudden change of demand in a control loop, or a sudden change in a measurand as exists when an indicating voltmeter is connected to a live circuit. Many physical systems, however, cannot provide such rapidly changing signals. Mechanical and thermal systems often cannot supply a rate of rise that is great enough to be regarded as a step this being due to the presence of significant storage of energy within the components. For example attempting to square-wave modulate the dimensions of a piezoelectric crystal at relatively high frequency will produce quasi-sinusoidal output response, not a square wave—the crystal acts as a filter.

Faster acting systems, such as found in the electronic and optical disciplines, can supply such rapid changes. When the response to a slow-to-rise input signal is needed the use of the step or impulse function may provide misleading information. A ramp function is more applicable in such cases. Many of the published systematized transient behaviour descriptions have not included this particular forcing function, the solutions usually presented being for impulse, step, and sine wave inputs. Sinusoidal excitation can, however, sometimes be used to approximate ramp responses.

The unit impulse function represents the input provided by a sudden shock in a mechanical system—it might be due to play in a drive link taking up. In electronic systems it represents such events as a high-voltage pulse generated by lightning or the switching surges from a non-zero crossing controlled silicon controlled rectifier. In practice a pseudo-impulsive function, called the *Dirac delta function*, is used instead of the true impulse. As will be seen the transient solutions of typical linear systems to impulse and step functions are somewhat similar in transient shape.

The third most commonly used input function is the sine wave (or cosine wave, which is the same function, phase-shifted in time). This acts to excite a system in a continuous manner forcing it to be excited in both the transient and steady state when the sine wave is initially applied, the former dying away to leave the latter as the solution most usually discussed. In practice systems are often likely to be disturbed by a continuous complex input waveform. As complex waveforms can be broken down by Fourier techniques, into a set of sine waves of different frequencies, amplitudes and phases, the use of sine waves of the correct amplitude and frequency enables the system to be studied one component at a time. Sine wave response is also called *frequency response.*

The concept that a complex continuous waveform can be so resolved into separate components rests on the assumption that the system is linear and that superposition applies. Non-linear systems can behave quite differently to complex signals, creating, for instance, harmonics of lower frequency than exist in the original signal.

It is often considerably easier, and more reliable, to obtain the transient response of a system by practical testing than it would be to develop a mathematical model. This is one of the reasons why data sheets of instrument products often include graphical statements of transient response. Further detail about forcing functions can be found in chapters of Atkinson (1972), Coughanowr and Koppel (1965), and Shearer *et al.* (1967).

17.3   RESPONSES OF COMMON INSTRUMENT STAGES

17.3.1   General Remarks

The following discussion applies only if the dynamic behaviour of the system of interest can be adequately characterized by a suitable linear ordinary differential equation, this often being found to be the case but by no means always so.

These equations will have the general form (called the non-homogeneous linear differential equation) for input variable x and output variable y and unit amplitude input:

$$a_n \frac{d^n y}{dt^n} + a_{n-1} \frac{d^{n-1} y}{dt^{n-1}} + \cdots + a_0 y = x(t)$$

The coefficients $a_n, \ldots, a_0$ are constants. They could be functions of t but not of y. Powers of derivatives do not occur in this class of differential equation. Not all orders of derivatives less than the highest are necessarily present.

The expression on the left-hand side is the *characteristic equation* of the differential equation. It contains the information that characterizes the physics of the system of interest and how it will provide output when it is excited. That on the right-hand side characterizes the forcing function, the input signal, applied to that system. Thus both sides contain independent information about, on the one hand, the transfer characteristic of the stage and, on the other hand, the forcing function applied to the stage. As both functions are linear expressions, by the above definition of requirement, they can both be transformed into transfer functions by Laplace methods.

System performance can thereby be obtained theoretically if the non-homogeneous linear differential equation for the specific system can be solved to yield the output versus time relationship.

In providing explanation of dynamic response, authors can make use of the differential equation format given here but often using other symbols: they may make use of the operator D to replace derivatives; the explanation might be presented directly in transfer function form; differential equations might be presented in complex number form, plus other variations. These tend to confuse the issue somewhat but each produces the same result.

The above given general differential equation can be written in the specific sense as an ascending order of equations when additional derivative orders are progressively added. For unit amplitude input these become:

zero order $\quad a_0 y = x(t)$

first order $\quad a_1 \dfrac{dy}{dt} + a_0 y = x(t)$

second order $\quad a_2 \dfrac{d^2 y}{dt} + a_1 \dfrac{dy}{dt} + a_0 y = x(t)$

nth order $\quad a_n \dfrac{d^n y}{dt^n} + a_{n-1} \dfrac{d^{n-1} y}{dt^{n-1}} + \cdots + a_0 y = x(t)$

Zero-, first- and second-order systems are all that need be considered in this discussion of the transient response of instrument stages because higher-order differential equations can be reduced, if need be, to these, through suitable manipulation. Also, establishing analytical solutions, which requires the roots to be found, to higher orders can be far more difficult, if at all possible, than for these three cases. This may seem rather short of real requirements but in general a very large part of instrument testing and characterization can be done with understanding of these three types of system alone.

Transformed into typical unitized transfer functions the differential equations become:

zero order $\quad \dfrac{Y(s)}{X(s)} = 1$

first order $\quad \dfrac{Y(s)}{X(s)} = \dfrac{1}{\tau s + 1}$

second order $\quad \dfrac{Y(s)}{X(s)} = \dfrac{1}{(\tau_1 s + 1)(\tau_2 s + 1)}$

nth order $\quad \dfrac{Y(s)}{X(s)} = \dfrac{1}{(\tau_1 s + 1)(\tau_2 s + 1) \cdots (\tau_n s + 1)}$

where $\tau_1, \ldots, \tau_n$ are termed time constants. This ground work laid now enables the three types of system to be studied in order.

17.3.2   Zero-order Systems

The zero-order system is so trivial mathematically that it is rarely mentioned in texts on dynamic response. In practice, however, it is often (but not always) a most desirable response, one that the designer strives to achieve but rarely obtains in a perfect sense.

The differential equation form, given above, shows that there are no derivatives, meaning that the system does not alter time features of a time-dependent function introduced to the system by the forcing function. Thus the zero-order stage cannot introduce phase shifts between the input and output, alter frequencies, or provide changing amplitude with frequency. It can, however, provide constant attenuation or gain and it can transform energy variables. The numerical coefficient a_0 is, of course, the static sensitivity of the stage, the constant of transduction. In principle the zero-order stage provides perfect dynamic response for situations where the input variable must not be processed with respect to its time features as it passes through the stage.

Considering the transfer function form it can be seen, but this will become clearer when the higher-order systems are covered below, that the stage contains no storage mechanism because there is no time constant associated with it.

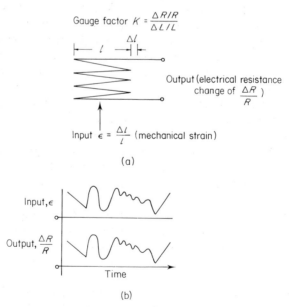

Figure 17.2   Example zero-order system—a resistive strain gauge: (a) gauge and its mathematical model; (b) responses

In many design situations a zero-order response is highly desirable. For example consider a resistance strain gauge (Figure 17.2a) in which extension of the electrical resistance wire, attached to a sample, alters the resistance proportional to the strain so induced. In normal use the gauge, being apparently purely resistive electrically, will respond completely in phase with strains of the sample. No matter how rapid the input signal (Figure 17.2b), the output follows without delay, without signal amplitude change, and without frequency change.

As the gauge is presumably being used to monitor the strain behaviour of the sample, zero-order response will yield the observer true information about what is actually happening to the sample with time and in amplitude. In control-loop design zero-order sensing components often help the designer obtain a very tight, responsive system.

In practice the designer or user must decide if zero-order characteristics are what is needed (see below). Assuming this to be so it is then necessary to ascertain that the device does indeed produce them to the dynamic performance level desired, for in reality a perfect zero-order system is a mathematical abstraction. All systems include some measure of storage effects; these will degrade the zero-order response into a higher order when they become significant.

In the strain gauge example storage might arise through thermal capacity of the gauge and sample; through after-elastic effects in the gauge material, through parasitic electrical inductance and capacitance when the frequency of response or excitation increases sufficiently; through mechanical compliance (and its capacity to enable the mass of the gauge to move out of phase with the sample) in the bonding of the gauge to the sample via the thin layer of adhesive; plus other reasons. These, in practice, will each introduce time-dependent features placed over the frequency spectrum of response from virtually d.c. to radio frequencies.

The art of good instrument design and application is to select a stage that performs with the response that just suffices. Returning to the relative humidity sensor example, of the previous chapters, it can now be seen that the hygroscopic membrane, the strained metal beam, the strain gauges, and the electrical resistance, Wheatstone bridge used can each be regarded as zero-order elements for the purpose of studying and designing the sensor for the application it was built to fulfil. If response faster than several minutes is needed then the various elements must then be viewed as higher-order components and a more complex model must be used to study and improve the dynamic behaviour.

Design of instrument systems does not, however, always require zero-order performance. In many instances there is need for storage in a system. Storage helps provide such desirable effects as integration and differentiation which, in turn, enable filtering to be devised. Applicability of the relative humidity sensor, mentioned above, is enhanced in many applications by the long storage time of the membrane. This helps average (integrate) fluctuations that might occur in a room in which a fan circulates pockets of air of differing %RH that would otherwise cause a closed-loop controller to attempt to smooth to the detriment of the real need.

Of the texts already referenced concerning dynamic response only Doebelin (1975) makes specific mention of zero-order systems. His description includes another example, that of a resistance potentiometer used as a position-sensitive sensor.

It will probably now be clear that it is not possible to provide a list of zero-order instrument devices and stages for that attribute is decided by the use to which the stage is applied. As a guide if the energy storage effects are minimal compared with energy levels existing and if only relatively-low-frequency forcing functions are to be applied to the stage then zero-order response probably occurs. The important factors are the *relative* values of these parameters, not their absolute magnitudes. A device operating at picowatt energy levels can provide delays of hours and conversely a megawatt energy system can respond in milliseconds.

It can be very dangerous in practice to assume the response order for a given system. Simple testing will easily establish the order for the application in question. This is to be recommended where possible. Section 17.5 discusses testing of dynamic response. Many of the problems that arise in measuring system design and application occur as a result of assuming that a sensor or stage provides zero-order response when, in the specific application, it actually operates with higher-order response.

17.3.3   First-order Systems

The differential equation for a first-order system is, in the symbols as expressed above and with a unit amplitude forcing function:

$$a_1 \frac{dy}{dt} + a_0 y = x(t)$$

Rearranging the coefficients to yield a unit coefficient for y gives:

$$\frac{a_1}{a_0} \frac{dy}{dt} + y = \frac{1}{a_0} x(t)$$

Taking Laplace transforms produces:

$$\frac{a_1}{a_0} sY(s) + Y(s) = \frac{1}{a_0} X(s)$$

Rearrangement again, gives the transfer function as:

$$G(s) = \frac{Y(s)}{X(s)} = \frac{1}{a_0} \cdot \frac{1}{(a_1/a_0)s + 1}$$

At zero frequency, and with sinusoidal excitation, the operator 's' validly becomes $j\omega$ and is equal to zero. Thus:

$$G(s) = \frac{1}{a_0}$$

The coefficient $1/a_0$ represents the static regime *transduction constant*, often denoted as K. It may involve either attenuation (or gain) of a common input–output variable or relate one form of variable to another.

The ratio a_1/a_0 is termed the time constant for reasons that will become clear below. It has dimensions of time and it is generally denoted as τ. First-order systems all have the same characteristic kind of time response to any given forcing function and τ specifies the actual numerical magnitude of that response this being regardless of the energy regime of the system.

The mercury-in-glass thermometer is that chosen by many authors to illustrate a first-order system. In use the thermometer is immersed in a fluid bath and is intended to transduce fluid temperature into an equivalent capillary-tube mercury movement without introducing delay or being affected in amplitude by the speed of the input temperature change. In practice the thermometer can only be regarded as a zero-order system when very slow changes are to be monitored. For changes taking place in around seconds the thermal storage effect of the mercury and the heat flow restriction of the thermal film transfer effect, that thermally couples the fluid to the mercury, provide first-order action.

Figure 17.3 is a schematic of the system including all variables that experience shows to be relevant in order that the system can be adequately treated by a first-order model.

The physical operation of the system must first be studied to discover assumptions that can be made in order to obtain an adequately descriptive model that is also reasonably workable, mathematically.

Assumptions in this example relate to such factors as the majority of resistance to heat transfer between the fluid and the mercury being provided by the heat transfer film, the glass being of insignificant thermal resistance. The mercury is presumed to have perfect heat conducting properties and to respond to heat changes producing instantaneous corresponding volume changes. Other factors assumed are that the glass envelope changes volume at the same rate as the mercury, an assumption that is not exactly true when high discrimination is needed since the glass takes a considerable time to return to a smaller volume after being expanded. Other factors that must be

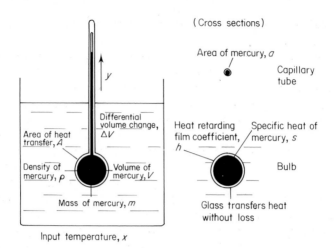

Figure 17.3   Basic physical system of a mercury-in-glass
thermometer modelled at a first-order system level

presumed to hold are that heat transfer is achieved over a constant surface
area and that it is not a function of the actual temperature. At least two
more important requirements, that do not strictly hold, are that heat is only
stored in the mercury, not in the glass, and that no heat is lost to the
environment via the heat conduction path of the thermometer stem.

Despite the obvious presence of these many imperfections the first-order
model, that can be developed for the thermometer, does provide a quite
reasonable estimate of the dynamic performance of the thermometer under
changing bath temperature conditions.

Derivation of the differential equations, on the basis of energy balance
and knowledge of the physical process by which heat changes cause a rise in
the mercury meniscus, leads to the actual development of numerical quantities for the time constant and the transduction constant. This example is
worked through in both Coughanowr and Koppel (1965) and Doebelin
(1975) each providing a slightly different form of expression of the resulting
numerical factors.

In terms of the parameters defined in Figure 17.3 it can be shown that:

$$\left.\frac{\Delta y}{\Delta x}\right|_{\omega=0} = K = \frac{\Delta V \cdot V}{a}$$

and

$$\tau = \frac{\rho s V}{hA} = \frac{\text{heat stored}}{\text{resistance to heat flow}}$$

Note that the time constant parameter concerns one storage and one energy dissipative parameter.

It is, therefore, possible to determine theoretically not only the transduction constant as a numerical value (as also could a practical test) but it also provides knowledge of the factors of the system upon which it depends. Similarly so for the time constant. Knowing these parameters enables the system to be tuned to provide whatever static and dynamic response is needed or to learn that existing materials and other practical constants might not allow a specified performance to be realized. They also enable the system to be easily specified in terms of three statements: the system order, K, and τ. The constants, however, vary greatly depending on the system conditions into which the thermometer is coupled. They cannot be quoted without reference to the model conditions that are applicable.

The importance of τ can now be considered in its capacity as a unique descriptor of the transient behaviour of a first-order system when excited by certain given forcing functions.

Of the forcing functions introduced above, the step (the impulse, being similar, need not be covered in the same detail here), the ramp, and the sine wave are the most useful inputs to consider. Other input functions will require a special mathematical study of the transfer function along the lines now outlined.

17.3.3.1 Response of first-order system to a step function

The Laplace expression for a step function of amplitude A is:

$$X(s) = A/s$$

Response of a first-order system, which will have $G(s) = K/(\tau s + 1)$ (where K is the transduction constant and τ the time constant), to this is given by:

$$G(s)X(s) = \frac{K}{\tau s + 1} \frac{A}{s}$$

Expansion by partial fractions yields:

$$Y(s) = \frac{AK}{s} - \frac{AK}{s + 1/\tau}$$

Using tables of Laplace transformations of common functions to invert this expression gives, for t from zero onward:

$$y(t) = AK(1 - e^{t/\tau})$$

This expression describes the dynamic behaviour of the first-order system to which a step function of a given amplitude is applied at $t = 0$. For the above to hold when the step is applied, the system must already be in the steady state, meaning that the storage element is completely discharged through the dissipative path; the initial conditions are thus satisfied. If energy is already being discharged, that is, the system is still in the transient state of a previous forcing function application, then the response will be the result of two forcing functions not the step applied here.

Being generally applicable, this time function can usefully be drawn as a graph in which the input is normalized to a unity maximum and the time scale is expressed in terms of a time variable ratio. This is shown in Figure 17.4.

If the step is applied as a decrement, or fall, rather than the rise shown here, the curve is simply reversed, the shape is identical. It will then follow the form:

$$y(t) = AKe^{-t/\tau}$$

It can be seen that τ is an important descriptive constant of the system, deciding the curve's actual magnitude. It also directly describes the *initial rate of rise*; a tangent to the original line will intercept the final value level at the one time-constant point. This implies that if the system were to maintain (which it does not) the initial slope the output would reach the final value in the time of 1τ. In general this slope, for any order of system, is termed the *slewing rate*.

It can be seen from Figure 17.4 that the continued response reaches what might be considered to be close enough to the final value, at levels 86.5, 95.0, and 98.2%, levels for 2τ, 3τ, and 4τ, respectively. By 5τ it is certainly close enough to a settled value for most applications. Although the time

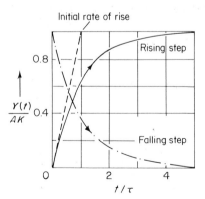

Figure 17.4 Normalized response of a first-order to a step input function

constant is the key parameter of definition it must be realized that final values are only reached at a longer time given by the appropriate factors presented above. In 1τ the response has only reached 63.2% of the final value. Furthermore it rises 63.2% of the remainder in the next 1τ period and so forth.

Respective values for a falling unit step are that it falls to 36.8% in 1τ, to 13.5% in 2τ, to 5% in 3τ, and 1.8% by the time 4τ has passed.

In order to provide a general descriptive term that applies to any instrument response, linear or non-linear, instrument users have adopted the term *settling time*. This is usually quoted as the time needed for the system to settle to within given percentage limits of the final value. These limits may be expressed as an error band. For example, it will take a first-order system, excited by a unit step function, a time of 3τ to come within $\pm5\%$ limits. Note that a first-order system cannot overshoot the final value but approaches it from one side only.

Detailed analysis of impulse input function response is to be found in Coughanowr and Koppel (1965).

17.3.3.2 Response of first-order system to a ramp function

Step inputs, as has already been mentioned, may give an unrealistically severe rate of rise. Using a step or impulse in such cases will provide an output response that is much greater in transient deviation than actually occurs. Slower rise rates, as represented by the ramp input function, are often more realistic forcing functions to use. As will be seen the output response to a ramp follows the input more faithfully but a quite unexpected delay and possibly a progressively increasing deviation error results.

Using the same mathematical technique as was applied in the previous step response explanation it is first necessary to obtain the Laplace transform for a ramp function of rate of rise, R. Forming the product of this with the transfer function for a first-order system having a transduction constant of K yields:

$$Y(s) = G(s)X(s) = \frac{K}{\tau s + 1}\frac{R}{s^2}$$

Rewritten this becomes:

$$Y(s) = \frac{RK}{\tau}\frac{1}{s^2(s + 1/\tau)} = \frac{a}{s^2(s + b)}$$

where

$$a = \frac{RK}{\tau} \quad b = \frac{1}{\tau}$$

Expanding this into partial fractions gives:

$$Y(s) = \frac{a/b}{s^2} - \frac{a/b^2}{s} + \frac{a/b^2}{s+b}$$

reverting a, b to physical constants:

$$Y(s) = \frac{RK}{s^2} - \frac{RK\tau}{s} + \frac{RK\tau}{s+1/\tau}$$

These are now in the form from which inverse Laplace transforms can be recognized from standard tables. Three time functions result:

$$Y(t) = RKt - RK\tau u(t) + RK\tau e^{-t/\tau}$$

i.e. output = ramp + step + exponential transient.

This is interpreted to be a ramp function response of slope rate RK. It will have a different rate to the input when the stage has either attenuation or gain for the common input and output variables (that is, volts in and volts out as would occur in a RC stage) and the rate will differ when the rate is not linked to the input because the transduction constant is one of conversion (that is, temperature change yielding a length change output). If the transduction constant is unity then input and output slopes will be identical.

The second term, a step of $RK\tau$ amplitude, and negative in sign, tells us that the output response begins to occur at $t = 0$ after a time lag. If the rates of the input and output are identical there will exist a constant lag with time but if they differ, the lag magnitude will be proportionately related to the time that has elapsed after initiation of the input signal. This lag effect has been interpreted by some authors as meaning that this combination of input in a first-order system will produce a system in which the output measured value at a given time is that of a fixed time before. This observation only holds true if the rates of the input and output ramps are identical. If not, the steady-state error changes with time.

Finally, the third component of the total response is a transient signal of exponential form that occurs at the commencement of the ramping action to die away to virtual zero within a short time.

The total response is shown diagrammatically in Figure 17.5. The form of the transient part of the response, when isolated, follows that given as the falling curve in Figure 17.4.

Clearly if the output is to be a faithful replica of the ramp input then the time constant of the first-order stage must be sufficiently small in magnitude. Furthermore, the transduction constant K needs to be unity. These conditions met, a ramp will be followed with negligible *droop*, with minimum time delay, with minimal transient component, and with a small fixed, rather than changing, magnitude error. (Under such conditions it virtually becomes a zero-order system.)

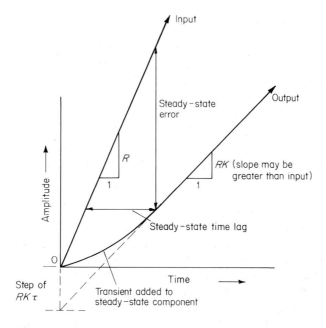

Figure 17.5   Response of a first-order system to a ramp
input function

17.3.3.3   Response of first-order system to sine waves (frequency response)

The requirement for testing, or understanding, may relate to the behaviour under conditions of continuous steady-state excitation of the system stage. This is fundamentally approached using the sine wave signal train as the forcing function since all other continuous complex signals can be reduced to the sum of such responses. The forcing function in this case in the untransformed time domain is:

$$X(t) = A \sin \omega t$$

In Laplace form:

$$X(s) = \frac{A\omega}{s^2 + \omega^2}$$

Thus the response of a first-order network to this will be:

$$Y(s) = G(s)X(s) = \frac{K}{\tau s + 1} \frac{A\omega}{s^2 + \omega^2} = \frac{AK}{\tau} \frac{\omega}{(s^2 + \omega^2)(s + 1/\tau)}$$

This can be shown to be (Coughanowr and Koppel, 1965), after rearrangement, taking partial fractions and then inverse Laplace transformation:

$$Y(t) = \frac{AK\tau\omega e^{-t/\tau}}{\tau^2\omega^2 + 1} + \frac{AK}{\sqrt{(\tau^2\omega^2 + 1)}} \sin(\omega t + \phi)$$

where $\phi = \tan^{-1}(-\omega\tau)$; i.e.

$$\text{output response} = \text{transient} + \text{steady-state components}$$

The transient term is short lived, decaying to leave only the steady-state component that is the contribution described as the frequency response of a system. The steady-state term is a sine wave of the same frequency as the forcing function but it is phase shifted as a lag, by angle ϕ. It will have the so-called d.c. gain decided by the amplitude of sine wave at that frequency and the transduction constant. The gain factor, however, will progressively decrease (be attenuated) as frequency rises, the magnitude of this attenuation being dependent upon the time constant and the frequency.

It is clear, from the presence of the transient term, that a first-order system will not immediately respond in the steady-state form but will undergo initial exponential behaviour. Thus the system will not *ring* (oscillate) but does need a certain amount of time to establish its intended frequency response action.

The steady-state normalized frequency responses of a first-order system with transduction constant K, to input sine wave frequencies of amplitude A are plotted in Figure 17.6, being given with respect to the amplitude and phase of signals.

A first-order system will, therefore, not alter the applied frequency but may attenuate the amplitude and introduce phase shift rising, in a lagging sense, to a maximum of 90°, the degree of these effects depending upon the time constant and frequency.

By way of an example that shows how these dynamic features can provide incorrect measurement action by a first-order sensor, consider a temperature sensor. The example is fully worked through in Coughanowr and Koppel (1965) so only the results need be given here. A thermometer with a time constant of six seconds (that for its use in a defined application; it is not a feature of the thermometer alone) is placed in a bath to detect bath temperature variations of ±1 °C amplitude that are sinusoidal at around three cycles per minute. These would be indicated to be half their real magnitude, the sine wave output being of the same frequency but delayed by some four seconds.

A second example is found when considering the mechanical pen recorder. When adjusted for critical damping the response of a plotter can be regarded as being essentially of first-order character. If the recorder response is inadequate it will plot signals at reduced amplitude. If the signal of

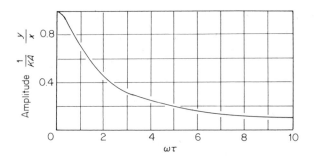

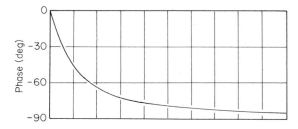

Figure 17.6 Normalized response of a first-order system to sinusoidal signals of varying frequency

one channel is of lower frequency than the other but of same amplitude, the recorder will not record a true indication of the amplitudes and phase existing between the two signals being recorded.

For similar reasons filters, added to smooth out noisy signals (in any energy regime), may also lead to similar inaccuracies when more than one frequency component exists in the forcing function, that is, any complex continuous signal. The response can be found by making use of superposition, which is applicable when the system is linear. This allows each component to be evaluated separately for magnitude and phase values, the set then being added to obtain the total response. A worked example for a forcing function with two components is given in Doebelin (1975).

Complex signals will not, therefore, give exactly the same normalized responses as are given in Figure 17.6 but they will always produce attenuation and phase shift as τ and ω increase. It can be seen that a square wave, being a set of sine wave components, will produce a response in which each cycle follows a step response profile as the square wave rises and falls, the start point of each new transition depending upon the level to which it rose, or fell, in the previous cycle.

As has already been mentioned the time constant used must be the effective value present when the stage is interfaced. It will not be the same value as that of the stage in isolation unless the input source and output load added to the stage do not interact through loading mechanisms. For example a simple electrical low-pass RC stage will have, in isolation, $\tau = RC$ but when the input source impedance and the output load impedances are connected they each modify the effective value of resistance because the load resistance is added in parallel with C and the source resistance is added in series with R.

This situation has already been met when the thermometer was considered; in that example the properties of the fluid can greatly vary the effective time constant experienced.

It might be thought that the type of a stage can be recognized uniquely from the shape of the step response. However, as will be explained in the following two sections, this is not necessarily a unique indicator because systems of any order can, if suitably conditioned, exhibit somewhat similar transient responses. This occurs when the time constants of the respective order contributions are such that the components producing them effectively work in phase. As the order rises, however, the shape of the curve changes more noticeably enabling the non-conformity to a first-order response to be more easily recognized.

17.3.3.4   Examples of first-order systems

The suitably thermally lagged thermometer has already been discussed. From that example it is possible to recognize that a first-order system results when there exists a significant energy storage mechanism and inability for that store to discharge through a dissipative element in a relatively short time. The direct combination of these two system properties decides the all-important time constant.

As the dissipating property (often called the *loss*) decides how well a response is damped (assuming, in such cases that a fixed level of energy storage capacity exists) this is often called the *damping* property. The simplest form, mathematically, is that in which the rate of damping is fixed independently of the rate of the signal. In mechanical systems this is called Coulomb friction. Almost always, however, it will be found that the damping property in a system is dependent upon influence variables, such as temperature. Friction that is proportional to the velocity of the signal is termed *Newtonian*. Add to these two the fact that the friction coefficient for static systems is generally much higher than for those after *breakaway* has occurred (non-linear friction) and it will be seen that practical use of linear systems theory is a much simplified version of reality. Although expressed in now deprecated units the practical information on damping contained in Drysdale and Jolley (1924) is very relevant to fine mechanical systems.

In electrical systems the obvious first-order example is the RC combination, such as is met in a low-pass filter stage. As will be discussed later, in

Section 17.3.6, cascading first-order stages raises the order of the system producing a different result from that obtained by the same components arranged to produce the same time constant as a single stage.

In fluid systems first-order stages occur in tanks of fluid being drained, the resistance to outward flow and the capacity of the tank forming the time constant pertinent to level of the fluid. Chemical systems provide this order action when solutes are mixed into solution. In acoustics an air enclosure having a *baffled* (resistance to airflow) output port, as is used in some speaker enclosure designs, is another fluid regime example. The same applies to a similar situation met in pneumatic instrumentation; much of pneumatic equipment, however, is not linear in operation.

Further details of practical examples are to be found in Atkinson (1972), Coughanowr and Koppel (1965), Olson (1943), and Shearer *et al.* (1967). First-order systems are also referred to as providing, or being, an *exponential lag* stage.

17.3.4 Second-order Systems

Second-order systems can arise when two energy storage mechanisms exist in a system and act with some degree of interaction. They need not be of the same energy type, for energy conversion mechanisms make them mathematically and practically compatible. Dissipative qualities are, again, important as this is the factor that decides the damping of responses.

A most noticeable feature of second-order systems is that they can overshoot final values, even providing oscillation (when zero damping exists).

Following the procedures used in the previous section, the characteristics of a second-order stage will first be derived followed by study of its response to step, ramp, and continuous sine wave forcing functions.

The second-order system is described by the non-homogeneous ordinary linear differential equation:

$$a_2 \frac{d^2 y}{dt^2} + a_1 \frac{dy}{dt} + a_0 y = x(t)$$

Taking Laplace transforms for $f(0) = 0$ (where the system is initially in the steady state) gives:

$$a_2 s^2 Y(s) + a_1 s Y(s) + a_0 Y(s) = X(s)$$

Rearrangement into the transfer function form gives:

$$G(s) = \frac{Y(s)}{X(s)} = \frac{1}{a_0} \frac{1}{(a_2/a_0)s^2 + (a_1/a_0)s + 1}$$

For this expression three parameters of practical implication can be defined. They are:

$$K = \frac{1}{a_0}$$
transduction constant, the d.c. gain or conversion constant

$$\omega_n = \left(\frac{a_0}{a_2}\right)^{1/2}$$
the angular natural frequency (with zero damping present); it has time dimensions

$$\zeta = \frac{a_1}{2\sqrt{(a_0 a_2)}}$$
damping ratio, a dimensionless number relating, in practice, the magnitude of actual damping to that at $\zeta = 1$, called *critical damping*

These parameters occur as the result of interpretation of the solution of the above equations that show that the system is oscillatory in nature and that damping, due to losses, controls the degree of oscillation. Being properties of the stage and not of the stage coupled to any particular forcing function, they can be used as descriptors of a second-order block.

The *period* of the natural angular frequency of the system can be expressed instead of the angular frequency as its 'reciprocal' and be given the symbol τ. This is done by some authors and is adopted here. It must be made clear that τ is not used in the same context as it was for the first-order system, nor is it exactly the period of the oscillating waveform (it is only for a zero damping situation).

Using these alternative symbols enables the transfer function to be expressed either as:

$$G(s) = \frac{K}{(1/\omega_n^2)s^2 + (2\zeta s/\omega_n) + 1}$$

or as:

$$G(s) = \frac{K}{\tau^2 s^2 + 2\zeta\tau s + 1}$$

both forms of which are commonly used.

As the denominator of the transfer function is a quadratic function it can be factored into a form that shows that the system could be formed by using two suitable first-order systems placed in cascade, but more of this aspect in Section 17.3.6.

Note that the natural frequency and damping ratio are interdependent, each also being functions of the transducer constant. Thus gain changes and damping ratio changes may alter the natural frequency.

Changes of K matched by appropriate changes to the damping ratio can retain a given natural frequency.

The implication of these constants becomes clear when the step response of such a system is studied.

17.3.4.1   Response of second-order system to step function

Forming the output expression in Laplace form from the Laplace transforms of the transfer function given above and the unit step function yields:

$$Y(s) = \frac{1}{a_0} \frac{1}{s} \frac{1}{(a_2/a_0)s^2 + (a_1/a_0)s + 1}$$

This reveals that the response will be one involving roots of a quadratic. As the actual working is not needed to understand the output response and because it is fully presented elsewhere, such as in Coughanowr and Koppel (1965), the method of solution only needs to be considered.

Forming the partial fractions and inverting the Laplace expressions gives the time-domain solution. The solutions can only be obtained mathematically as three separate regions, these depending upon the nature of the roots of the quadratic. The damping ratio ζ is the dominant factor here. The roots are complex when $\zeta < 1$, real and equal when $\zeta = 1$ and real when $\zeta > 1$. These three regions are termed *underdamped* or *oscillatory*; *critically damped*; and *overdamped* or *non-oscillatory* for the three regions, respectively.

It must be clear that the response range itself passes through a gradual gradation from the under- to overdamped response. It is a need of the mathematical method of solution that dictates three different regions, not a property of the real physical system.

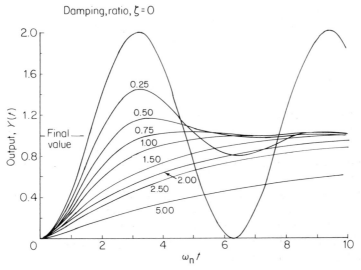

Figure 17.7   Normalized response of a second-order system to a unit step function

A family of response curves is presented, in normalized manner, as Figure 17.7, these show the response as the damping ratio is altered from zero to around 5.0. Note that at large damping ratios the response curve takes on a similar form to that of a first-order system the two being slightly different in shape at the origin—see Section 17.3.6.

It can be seen that second-order system response can overshoot the final value and if inadequately damped it can continue to oscillate after a step is applied. It can be seen that the second-order system, with zero damping, acts as a sinusoidal signal generator if the losses are truly zero or are actually supplied to the system.

A step function is the most severe form of input level change that can be applied. In practical systems this rate of rise may be unrealistic and in such cases a step input would produce a response that is considerably more oscillatory than reality might yield. The transient is sometimes called *ringing*. A slower rate of rise, as would be simulated by a ramp that reached a terminated level at given amplitude, would often be more reasonable to apply. This situation is worked through in Doebelin (1975) to show that with a *terminated ramp* forcing function, the underdamped $(\zeta < 1)$ response can be made to be virtually error-free if the natural frequency of the system is large compared with the reciprocal rise-time duration of the ramp applied, this also applying for virtually zero damping. Doebelin makes the point that this means that stages having little damping may be satisfactory if they have high natural frequencies and are not excited in use by rapid rate of rise steps. The example cited as such a case is the piezoelectric accelerometer when used to measure responses of relatively low-pass mechanical functions.

The general relevancy of the step response for many practical systems has given rise to several descriptive terms that relate to the underdamped responses given in Figure 17.7.

Overshoot is defined as the, less than unity, ratio of the magnitude to which the first overshoot rises over the final value line to the final value. Numerically this is given by:

$$\text{overshoot} = \exp\left(\frac{-\pi\zeta}{\sqrt{(1-\zeta^2)}}\right)$$

and is, thus, decided only by the damping ratio. Note that the half magnitude of the overshoot cannot exceed the final value.

Decay ratio is defined as the, less than unity, ratio of successive amplitudes of decay cycles. It happens to be, mathematically, the square of the overshoot function given above.

These two expressions enable the damping for a given system to be calculated from a given record or the overshoot to be estimated from the tail-end of a record such as occurs when phase delay loses the first part of a record in an oscilloscope.

Their behaviour with varying damping ratio is plotted in Coughanowr and Koppel (1965).

Rise-time does not follow the same simple relationship with the defined τ as did the first-order case. It is quoted here as the time for the signal to rise from zero to the first crossing of the final value.

Settling time, defined in the previous section, applies with the same meaning to these waveforms. It is also called *response* time.

The period of oscillation, T, the time duration from one peak to the next of the cyclic transient, is not given directly by τ when damping is present but is found from:

$$T = \frac{2\pi\tau}{\sqrt{(1 - \zeta^2)}}$$

the term τ being the period of the undamped second-order system. As earlier hinted at, the natural frequency of the system is a function of the damping ratio. As damping increases, natural frequency falls according to:

$$\frac{\omega}{\omega_n} = \sqrt{(1 - \zeta^2)}$$

As with first-order systems, although these factors appear to define the system as a stand-alone unit they are subject to modification by the terminations used. For example the damping ratio that would be used when studying the response of an electrical mirror galvanometer (as used in the ultraviolet chart recorder) depends very much upon the signal source resistance applied to it.

The impulse response is not covered here, the reader being left to refer to Coughanowr and Koppel (1965) and Doebelin (1975) for worked derivations and normalized response curves that match those given in Figure 17.7. A somewhat similar oscillatory response, described in identical terms results.

Although the treatment has covered the response in a general manner many texts, such as those often referred to in this chapter, include worked examples of specific practical systems of mechanical, electrical, electromechanical, acoustical, and pneumatic nature. In the discipline of physics it is covered most usually under the titles of *wavemotion* and *simple harmonic motion*. Closed-loop control systems often result in second-order open-loop transfer functions and can be regarded sometimes more simply as a stage with this form of response, instead of needing to be seen as a closed-loop stage.

17.3.4.2 Response of second-order systems to ramp functions

The response of the second-order system to a ramp forcing function is derived by the same procedure as was used for that of a first-order system. It is worked through in Doebelin (1975) and Atkinson (1972). In Chapter 14, Section 14.2.2, Atkinson has discussed the response of a mechanical positioning system that has a first-order equation. (Care is needed to differentiate between the orders used here to describe differential equations and the type 0, or 1, 2, 3, and so forth, used to describe the classes of behaviour of linear closed-loop control systems.)

The response is similar to that of a first-order system, (Figure 17.4 shows the form of response as that for a type-1 control system) fed with a ramp, the exception being that now the system might oscillate about the final ramp line during the initial transient portion of the response curve. Again there exists a step function component that delays the output ramp by an amount decided by the damping ratio ζ and natural angular frequency ω_n. The resulting steady-state time lag and error are both reduced by increasing the natural frequency and reducing the damping ratio. Doebelin (1975) provides a normalized chart of how deviation error varies with varying damping ratio as the transient solution dies away.

17.3.4.3 Response of second-order systems to sine waves (frequency response)

The forcing function here is that used to establish the first-order frequency response in Section 17.3.3.3. Solution is needed for the Laplace form of function:

$$Y(s) = \frac{A\omega}{s^2 + \omega^2} \frac{K}{\tau^2 s^2 + 2\zeta\tau s + 1}$$

This involves solution of the roots of two quadratics. That solution is worked through in several texts (see Beckwith and Buck, 1969; Coughanowr and Koppel, 1965; Shearer et al., 1967). The resulting time-domain parameters of phase and amplitude are obtained from the output expression:

$$Y(t) = \frac{AK}{\sqrt{\{[1-(\omega/\omega_n)^2]^2 + (2\zeta\omega/\omega_n)^2\}}} \sin(\omega t + \phi)$$

where

$$\phi = -\tan^{-1}\left(\frac{2\zeta}{\omega/\omega_n - \omega_n/\omega}\right)$$

or in the form where $\tau = 1/\omega_n$:

$$Y(t) = \frac{AK}{\sqrt{\{[1-(\omega\tau)^2]^2 + (2\zeta\omega\tau)^2\}}} \sin(\omega t + \phi)$$

where

$$\phi = -\tan^{-1}\left(\frac{2\zeta\omega\tau}{1-(\omega\tau)^2}\right)$$

These expressions show that the output is a sine wave of the same frequency as the input signal but that the magnitude can now be greater, or less, in magnitude as the frequency rises. When $\omega = \omega_n$ the amplitude can rise to a theoretical infinity magnitude, the magnitude being decided by the damping ratio.

The phase component of the equations shows that it lags from zero to a maximum of approaching, but never quite reaching, $-180°$ if the damping ratio approaches zero. It is $-90°$ at the natural frequency, ω_n. The expressions are plotted in normalized form as the two plots of Figure 17.8. In this regard second-order systems are quite different to first-order stages. These curves are plotted in a particularly accurate manner in Shearer *et al.* (1967).

Because the system can produce output signals larger than provided to it it has the ability to provide signal gain or amplification; the output amplitude variable is, therefore, termed the *magnification ratio.*

Whether magnification is to be deliberately adopted or not, depends upon the application. In some detection systems a second-order sensor is purposefully designed to ring (by virtue of absence of damping) at its natural frequency so that magnification is gained. In other systems it may be desirable to keep the frequency response flat for as wide a bandwidth as possible.

The degree of resonance can be described in terms of the *Q-factor*, a quality factor or figure of merit. It can be shown (Shearer *et al.*, 1967) to be:

$$Q = \frac{1}{2\zeta\sqrt{(1-\xi^2)}}$$

or, for low levels of damping it approximates to:

$$Q \approx 1/2\zeta$$

Alternatively, it is defined in more direct terms of the peakiness of the resonance curve as the bandwidth at the *half-power points* (where the power level has dropped to half that at the peak) divided into the resonant frequency. Both methods give the same result; direct measurements on the amplitude–frequency response curve can be used to determine the damping ratio. A word of caution is needed here. In many second-order systems resonances need time to develop when fed with a swept frequency forcing function. Response curves swept too rapidly may, therefore, yield resonance peaks below their true value.

The factor Q is also a measure of the *selectivity* of the system. This usage applies when the system is used to deliberately detect signals near to the resonant frequency, this being achieved by attenuating all others with respect to that small bandwidth.

Quotation of flat-response bandwidth is usually achieved by specifying the range of frequencies over which the response remains within the maximum and the half-power (3 dB) points. From Figure 17.8 it can be seen that the

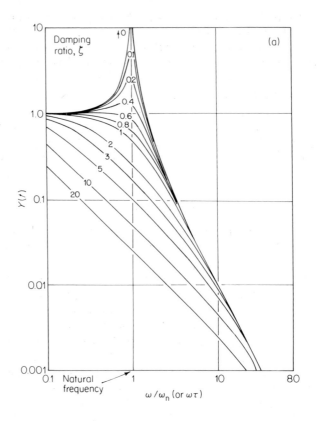

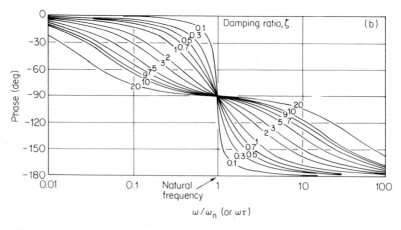

Figure 17.8   Normalized response of second-order system to sinusoidal
excitation: (a) amplitude; (b) phase

widest flat response is obtained with a damping factor of around 0.5. By comparison with the frequency response curve of the first-order system (Figure 17.6) it can be seen that a second-order system will provide a flat response of similar bandwidth but that the first-order system will not *fall off* as rapidly. Additionally the phase shift of each might be important. At the 3 dB point of each a first-order system introduces approximately $-60°$ shift, the second-order stage some $-110°$ of lag.

Complex signals fed to a second-order stage can be treated as the sum of sinusoids. If the original character of the signal is to be preserved then each frequency must be transmitted through the stage without attenuation or phase shift.

17.3.5 Higher-order Systems

In the previous sections methods have been outlined to show how, at least in principle, any system that can be described by any order linear differential equations system can be investigated theoretically.

It can be seen that as the order rises general explanation becomes progressively more complicated but, nevertheless, solutions for orders higher than two can be obtained. Doebelin (1975) outlines the procedures for handling the responses of a general instrument stage. In a very large part of measurement systems the highest order for a single stage that needs to be considered, because approximation can be used, is the resonant, second order. With the information provided here the reader should be in a very reasonable position to describe, test, and specify the dynamic performance of measurement system stages.

Although general methods of handling nth-order systems have been developed (see Shearer *et al.*, 1967) there is little point in continuing to study each. The intention is not to review the state of the art of system mathematics but to provide an introduction to the dynamics of systems as will be met in practical measurement engineering. It would appear that no author has developed orders higher than second providing standard solutions and curves for general application.

Higher-order systems result when stages are coupled together. It is therefore, practical and useful, to study several often met situations of this type in order to see how responses might be modified when a block is added to others.

17.3.6 Cascaded Stages

In Chapter 15 it was shown how individual stages of a system can be cascaded to form a chain. It was previously stated that the overall response can be found by multiplying together the transfer characteristic functions to form a new function that represents the whole.

The dynamic and static behaviour of cascaded stages can be found in this manner by considering the product of the transfer functions of each as a single transfer function. This, however, only holds true if the energy storage mechanisms of each stage do not interact with each other. For example, when connecting two RC, low-pass, filter stages together their overall transfer function is only the product of the two first-order functions if the second stage does not load that connected to it. The extent of loading decided the degree of modification of the dynamic response for all order stages except that of zero order. As zero-order systems contain no energy storage elements they cannot interact in any other way, other than to modify the signal level or energy form according to their numerical product. Thus adding a zero-order stage to the frequency-dependent stage does not alter the dynamic part of the performance (other than to alter the amplitude). Thus, a zero-order buffer amplifier, that might be used to isolate the two above mentioned RC stages, would prevent interaction and would not alter the dynamic transient behaviour of the stages. If this were the case the two first-order stages could then be regarded as one through their joint product transfer function. In many applications it is often preferable to purposefully add zero-order buffer stages so that interaction does not occur, thereby keeping the behaviour within workable mathematical and practical bounds and, as is explained below, to obtain the most responsive system. It is, however, a practical observation that the interactive system often can provide the best engineering solution in terms of resource used—natural systems abound with them.

If two first-order stages are cascaded and they do not interact then:

$$G(s) = G_1(s)G_2(s)$$

$$= \frac{1}{(\tau_1 s + 1)(\tau_2 s + 1)} = \frac{1}{\tau_1 \tau_2 s^2 + (\tau_1 + \tau_2)s + 1}$$

If, however, they do interact it will be found (Coughanowr and Koppel, 1965 present a worked example), that the coefficient of the s term will be increased by an additional factor that is a constant of the system to yield the general form:

$$G(s) = \frac{1}{\tau_1 \tau_2 s^2 + (\tau_1 + \tau_2 + C)s + 1}$$

Thus the roots of the expression are no longer as simple and the behaviour is thus different. The constant, here denoted C, is a function of the degree of loading that occurs between the two stages. It is zero for zero interaction. As a general observation Coughanowr and Koppel (1965) state that interaction acts to reduce the responsiveness of the system compared with one having the same two stages operating without interaction. They have presented a worked example to illustrate this conclusion, showing how the effective τ of the combined system is increased when interaction occurs. Interaction effectively changes the time constants τ_1, τ_2 tending to make one more dominant than the other. Considerable interaction, therefore, can cause the system to change from what is a second-order system to one that can be adequately considered as a first-order one having the dominant time constant.

The reduced responsiveness (also termed *sluggishness*) caused by cascading stages is referred to as *transfer lag*. This is not to be confused with *transport* lag, the effect produced when a signal arrives intact but after delay time. Transport lag is easily dealt with mathematically (Coughanowr and Koppel, 1965) to yield a transfer function:

$$G(s) = \frac{Y(s)}{X(s)} = e^{-\tau s}$$

Although simple to describe mathematically its presence in a system can greatly affect performance, especially when the system is part of a closed-loop controller. It occurs in such places as the *dead-time* in mechanical system caused by slop in linkages or the delay caused when heated fluid has to pass through a pipe to the place where it joins the next stage.

The most general manner by which to obtain the response of cascaded stages is to derive the differential equations solving them to obtain the time response. Each combination will yield a differing response. Total generalization leads to complexity that is often not warranted by the study that has arisen. General solution of higher-order equations is given in Shearer *et al.* (1967).

Other options are to reason out the expected responses from the information given here. To this end certain cases are now studied. In many instances it is more expedient to build and test the system rather than study it as a mathematical model. The virtue of the mathematical model is that it enables the system to be tuned in an efficient manner, an aspect that must be derived otherwise through ordered careful experimental procedures.

Cascading zero-order systems does not give rise to higher-order systems but with the first and higher orders cascading the order is always raised. If the time constant of one first-order stage is much less than those of the others then it probably will not significantly alter the performance: it can be considered to be of zero order, in which case it will only change the static characteristics.

Adding first-order systems together forms higher-order systems. As a special, and useful case, the effect of cascading several identical first-order stages in a non-interacting manner can be considered. This is done by Coughanowr and Koppel (1965) to yield the step responses shown in Figure 17.9. The graphs are actually those of progressively higher-order, critically

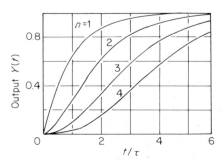

Figure 17.9 Step response of non-interacting, cascaded, identical first-order systems. n is the number of stages in cascade

damped, systems: compare that of second order with that shown in Figure 17.7 for a damping ratio of unity. With that degree of damping the second-order system has identical, real roots which is the case considered in Figure 17.9.

As the degree of energy interaction increases to a point where energy can be shunted back and forth between the energy sources of the complete system the dynamic behaviour changes from the first-order like response to one of second order where oscillation takes place.

If the energy is shared by the storage mechanisms in a stable manner, such as it is by two capacitors in an electrical circuit that are connected in parallel, then only the non-oscillatory response can occur.

If, however, the two stores hold energy significantly out of phase, as do a coupled capacitor and an inductor or a spring and a mechanical mass, then oscillation may arise as the energy shuttles back and forth. Damping gradually dissipates the initial energy provided by the transient forcing function and the oscillation dies away. If it is continuously provided (*pumped*) as is the case for continuous excitation then the response is held in oscillation at a magnification ratio decided by the degree of losses present.

This physical approach to understanding the dynamic behaviour can be followed through into systems of higher order than two. Physically the energy can be shunted from store to store forming coupled oscillations between existing dominant energy couples. Thus a series spring–mass system coupled to a second such system will exhibit behaviour wherein the first system is resonating with the other one initially static, the state gradually reversing with time in a cyclic manner until all energy is dissipated. If one second-order system is more dominant than the other the resonances to step inputs will occur with greater amplitude in one of them. The presence of such coupled systems can, therefore, be identified by studying the response to a step input, looking for time-changing resonant amplitudes at various points of the system.

Satisfactory mathematical modelling of such systems may lead to the need to use *partial differential equations* in order to allow for the non-ideal nature of real objects such as masses, springs, inductors that often cannot be regarded as lumped elements. The solution of the equations for higher-order systems involves *eigenvalues* which are the roots of the characteristic equation of the problem. The level of mathematical skill needed is beyond that of many people involved in use and specification of measuring systems and, therefore, its applicability to aid system description is limited.

It can be seen that higher-order systems, formed by cascading given stages, may raise the overall order to one having undesirable response. By the addition of carefully chosen cascade (or parallel) stages it is often possible to largely negate the unwanted effects, thereby regaining a satisfactory or even improved response. This process is called *compensation*. It has been most extensively developed in the area of closed-loop systems; Chapter 14 contains a section in that respect.

The general principle of compensation is quite widely applicable. It can be used to extend the frequency response of amplifers by altering the response at high frequencies, to tighten up the dynamic response of a second-order system by making it appear as though it were of first order. Electrical circuit theory methods have been developed that enable a designer to directly synthesize the desired compensation. This is more direct than the cut-and-try method of analysis that is often used to improve a system. Methods of electrical circuit synthesis can be applied to other energy systems but it will not always be useful because not all regimes can realize the form of components that the methods finally dictate are needed to form a real, practical compensation stage.

17.4 NON-LINEAR SYSTEMS

Much of the work published on the performance of systems stages is relevant only to linear systems. This is largely because no equivalent generalized theories have yet been developed to handle non-linear situations. This does not mean that non-linear systems are not encountered for they are. It must be remembered that many of the descriptors used to specify instrument performance are based on the implicit assumption that the system acts in a linear manner. This is easily tested by observing if doubling of a step input yields a doubled output. If the system parameters run into regions of non-linearity, such as when saturated against a voltage supply rail or a mechanical stop, then linear explanation ceases to be relevant beyond that point.

Non-linear systems can produce most unexpected dynamic results. The phenomenon of *ferroresonance*, for example, can produce frequencies lower than the excitation. Truncated, linear, continuous functions can *alias* to produce lower frequency components.

As linear techniques are so workable one method of handling non-linear situations is to treat the system in small enough steps by bounding its solution to small regions in which response is adequately linear. This method is known as *piece-wise linearization*. The solution of one region can then be used to excite that adjacent.

Handling non-linear systems is covered in Shearer *et al.* (1967). In a few cases the non-linear system can be converted into a linear one by substitution of a transformed variable. The closed-loop system with backlash can be handled in this way. Such instances are, however, more the exception than the rule.

17.5 TESTING DYNAMIC RESPONSES

Testing was introduced in the previous chapter in Section 16.6. There the prime interest was concerned with giving an introduction to static, steady-state parameter testing which forms the bulk of performed tests.

Throughout this chapter certain system dynamic parameters have become obvious as those that can be measured in order to well specify a system. These include time constant, natural frequency, damping, system order, settling times, slew rate, rise- and fall-times, and more.

The concepts embodied apply to most systems that will be encountered. Although the systems of interest may not be in the electrical domain it will often be necessary to add sensors that convert the original variables into electronic means because they can provide an ever-widening range of methods for testing systems.

Whichever kind of instrument is used, be it mechanical, electronic, optical, thermal, etc., it is most important to first ascertain if the testing device itself has an adequate response. This usually means verifying that it performs as a zero-order system in comparison to that being tested. If not, then correction to results obtained will be needed to allow for the effects of amplitude and phase distortion as were discussed previously. Error commonly arises in the use of pen recorders with insufficient bandwidth or when the operator does not allow a system enough time to build up the amplitude of resonances, two examples mentioned earlier. It is usually prudent to first test the testing apparatus with known input forcing functions.

A choice often exists between use of a time-domain or frequency-domain based test. A step or impulse test is often easy to apply but capturing the transient may not be so easy, especially if the time of occurrence is not known. Special transient recorders are available to overcome this problem.

Furthermore, a more realistic transient input may be a modified rate of rise step which may not be so easy to programme as an input.

Conversely, measuring the frequency response is sometimes more easily done as it involves steady-state, rather than dynamic, measurements. However, it takes time to set up unless sophisticated scanning, sweep-frequency generators are available.

In principle it does not matter which domain the test is performed in for Fourier transformation can be used to transform the signal obtained in one domain into the form of the other domain. Apparatus to handle Fourier transformation is becoming cheaper and more available with time.

Whichever type of test is made it is important to create a test that relates to real needs of the equipment. Some tests used are adopted because they are easy to implement; they often have little practical bearing on the use of the instrument being tested.

It can be seen that whichever domain is used it is possible to identify if the system is linear and to decide its effective order. If it is not of a simple kind then *systems identification* methods can be adopted.

The texts listed in Section 16.6 are relevant to dynamic testing to a varying degree. In general, however, dynamic testing is poorly covered in the literature, this probably reflecting the general level of understanding and apparent need of the majority of persons concerned with measurement systems.

REFERENCES

Atkinson, P. (1972). *Feedback Control Theory for Engineers*, Heineman, London.

Beckwith, T. G. and Buck, N. L. (1969). *Mechanical Measurements*, Addison-Wesley, Reading, Mass.

Bosman, D. (1978). 'Systematic design of instrumentation systems', *J. Phys. E: Sci. Instrum.*, **11**, 97–105.

Coughanowr, D. R. and Koppel, L. B. (1965). *Process Systems Analysis and Control*, McGraw-Hill, New York.

Doebelin, E. O. (1975). *Measurement Systems: Application and Design*, McGraw-Hill, New York.

Drysdale, C. V. and Jolley, A. C. (1924). *Electircal Measuring Instruments, Part 1*, Ernest Benn, London.

Finkelstein, L. and Watts, R. D. (1978). 'Mathematical models of instruments—fundamental principles', *J. Phys. E: Sci. Instrum.*, **11**, 841–55.

Fitzgerald, A. E. and Kingsley, C. (1961). *Electrical Machinery*, McGraw-Hill, New York.

Karnopp, D. and Rosenberg, R. C. (1975). *System Dynamics: A Unified Approach to Physical Systems Dynamics*, MIT Press, Cambridge, Mass.

Klir, G. J. (1972). *Trends in General Systems Theory*, Wiley, New York.

Koenig, H. E. and Blackwell, W. A. (1961). *Electromechanical Systems Theory*, McGraw-Hill, New York.

MacFarlane, A. G. J. (1964). *Engineering Systems Analysis*, Harrap, London.
Neubert, H. K. P. (1976) *Instrument Transducers*, 2nd edn, Clarendon Press, Oxford.
Olson, H. F. (1943). *Dynamical Analogies*, Van Nostrand, New York.
Paynter, H. M. (1961). *Analysis and Design of Engineering Systems*, MIT Press, Cambridge, Mass.
Shearer, J. L., Murphy, A. T., and Richardson, H. H. (1967). *Introduction to Systems Dynamics*, Addison-Wesley, Reading, Mass.
Zahed, L. A. and Desoer, C. A. (1963). *Linear System Theory*, McGraw-Hill, New York.

Handbook of Measurement Science, Volume 2
Edited by P. H. Sydenham
© 1983 John Wiley & Sons Ltd

Chapter

18 L. FINKELSTEIN and R. D. WATTS

Fundamentals of Transducers: Description by Mathematical Models

Editorial introduction

A byproduct of research conducted for early 1930's broadcast radio was the realization that mathematical models of such equipment as microphones loudspeakers and pick-up cartridges enabled designers to systematically improve the overall system quality.

From this arose, by the 1940's, the method of the *analogies* by which electrical, mechanical, and acoustic energy systems could be modelled by the same forms of mathematical description.

Although this fundamental approach began to help instrument design needs its general use in measurement systems design has only just begun to be exploited.

In this chapter the authors provide an authoritive keystone account of the fundamental principles and operation of measuring instruments and their elements, in particular of transducers and sensors, in terms of their description by mathematical models.

Systematic breakdown and rigorous mathematical description of the building blocks involved provides for an efficient approach to the understanding, evaluation and design of instruments.

This account is based on reports and papers published earlier by the authors.

18.1 INTRODUCTION

18.1.1 General

The object of this chapter is to present some of the fundamental principles of operation and construction of measuring instruments and their elements, in particular of transducers and sensors, in terms of their description by mathematical models.

The approach is based on the consideration that measuring instruments,

The material in this chapter is based on a paper entitled Mathematical Models of Instruments—Fundamental Principles—*J. Phys. E:* Vol. 11, No. 9, September 1978, subsequently reproduced in *Instrument Science and Transducer Technology*, Vol. 1, edited by B. E. Jones and published by Adam Hilger in 1982. It is reproduced here by courtesy of the Institute of Physics.

as well as instruments for computation, communication, and control, constitute a special class of devices or machines concerned with *handling of
information.* This is in contradistinction to other machine classes which are
primarily concerned with the handling of matter or energy.

18.1.2 Information Machines

Brief explanations of the concepts of *machine, information, signal,* and
information machine are useful here.

A *machine* may be described in general terms as a contrivance which
transforms a physical input into a physical output for a definite purpose. The
physical inputs and outputs of machines are *flows* of *matter* or *energy* or
both. One example of a machine is a lever, which transforms a mechanical
power input at one of its ends into a mechanical power output at the other
end for the purpose of achieving either a mechanical advantage, or a
magnification of movement. Another example of a machine is an electrical
generator, which transforms a rotational mechanical power input into an
electrical power output, electrical energy being more convenient for the
purpose of utilization. A further example of a machine is a pump, which
uses a mechanical power input to transform the fluid flow at one pressure
into a fluid flow at a higher pressure.

Information and its relation to measurement is discussed in Chapter 1. To
recapitulate: if an object or event (termed a *symbol*) bears a defined relation
to some referent object or event, then the symbol together with the defined
relation carries information about the referent. If the relation is one-to-one
then the information is unambiguous. For example, in a lever used for
movement magnification, the output displacement, together with the arm
ratio of the lever, carry information about the input displacement.

The information-carrying symbols are features of material objects or of
energy flow. The physical quantity, the magnitude or time variation of which
carries information, is termed a *signal.* In the above example of a lever, the
input and output displacements are signals.

Information machines such as a measuring instrument or a measuring
instrument element, are devices which transform a physical input into a
physical output for the purpose of either providing the output information in
a more convenient form than the input information, or for processing the
input information in some way. They function by performing a prescribed
functional transformation of the information-carrying feature of an input
signal into the information-carrying feature of the output signal.

A lever used for displacement magnification is an information machine, its
function being to transform the input displacement into a larger output
displacement which bears a precise ratio to the input, the larger displace-

ment being the more convenient form of information carrier. The principle of operation of the lever in this case is the maintenance of the *precise relation* between input and output. An electrical generator as described previously may act as a tachogenerator and is then an information machine performing a transformation of the velocity of rotation into a precisely proportional output voltage, the electrical signal being the more useful form for further processing. Similarly the pump may be used as a metering pump maintaining a precise relation between mechanical rotation and fluid volume output.

It is this basic principle of their *functioning* which distinguishes information machines from other machines. In the above examples the lever, the generator or the pump when used to supply power have as their function to supply the required output power with a particular capacity and with an optimal efficiency. When used as information machines their principal function is a precise maintenance of the required transformation relation.

This basic principle of functioning of information machines determines the principles of their design and construction. Information machines are commonly built up as assemblies of simpler components.

Also there is a limited number of important structures of interconnection of components employed in all instruments, feedback being perhaps the most striking example. A wide variety of instruments can be constructed from a much smaller variety of component types and an even smaller variety of basic interconnection structures. Chapter 2 provides a general introduction to systems aspects of measurement systems.

The present chapter is concerned in particular with the class of instrument elements termed *instrument transducers*. They are devices which *transform* an input signal associated with one form of *energy* into an output signal associated with *another* form of energy maintaining a functional relation between the two. Commonly transducers form the front element of an instrument system and are then known as *sensors*. They may, however, also form the final *actuation*, or *effectuation* element, or an intermediate element. In some sensors both the input and the output are of the same energy form, say in a current transformer used in an ammeter system for high currents. They are not transducers but in this chapter the term transducer will be used to embrace such sensors. Chapter 15 introduces the building blocks of measurement systems.

18.1.3 Mathematical Models as Descriptions of Instruments

The mathematical model of an instrument is a set of equations which relate the inputs of the instrument to the outputs. The inputs are the independent physical effects acting on the instrument, the outputs are the physical actions exerted by the instrument on the environment. Modelling is discussed in Chapter 2.

Mathematical models may be:

(a) *Functional models:* these describe relationships between outputs and inputs without explicit correspondence between physical features of the device and particular features of the model.

(b) *Physical models:* these describe the behaviour of the instrument relating the behaviour to its construction; particular features of the model correspond to particular features of the form, dimensions, and material properties of the device.

One may distinguish between:

(a) *Energy-flow* models, which relate all the significant energy flows into and out of the device and which thus represent a complete description of the physical behaviour of the device.

(b) *Signal-flow* models, which relate only the information-carrying signals of the instrument. They are basically models of a particular aspect only of the full energy-flow model.

(c) *Information-flow* models, which basically represent only the symbols (the information-carrying features of the signals) and the transformations performed on them by the instrument.

Mathematical models of instruments always constitute an abstraction from reality. They represent essential chosen features of an instrument with omission of irrelevant details. The level of abstraction differs according to the purpose for which a model is required. Two particular levels of model abstraction may be mentioned. For the purpose of describing the principles of operation of a particular type of instrument, one may use an archetype physical model which retains only those features essentially characterizing the principle of operation. At the other end of a design process (Watts, 1966) a detailed physical model for use in design retains all those features which enter into the performance requirements, as well as those features which affect the performance and which the designer must specify for the purposes of construction (Watts, 1968; Abdullah *et al.*, 1977).

The mathematical model of an instrument is first of all a concise and rigorous description of instrument performance. The calibration curve of an instrument when expressed as an equation is, after all, a simple mathematical model. A full energy-flow model of an instrument enables its performance to be predicted under a wide range of conditions and incorporates the effects of dynamic response, the effect of the disturbance of the system under observation introduced by the instrument, the influence of any energizing power supplies and also effects of environmental disturbances.

A mathematical model of the physical type makes clear the relations between the function of an instrument and particular features of its construction. Thus, for example, those features which determine dynamic response and gain show up distinctly in the equations. Similarly an analysis of the model can reveal those components which critically determine performance and those to which the instrument is less sensitive.

The *first* and foremost use of mathematical models is to *provide insight*. In organizing information about instruments, mathematical models, in particular those of the archetype physical form, provide a powerful means of understanding the principle of the instrument and the factors governing its capabilities and limitations. Essential similarities between devices of apparently different kinds and basic properties of structures, such as feedback, are made clear and explicit. Thus the immense quantity and diversity of information in the art of instrumentation can be arranged into a manageable structure.

In design, idealized mathematical models can be used to evaluate, simply, alternative candidate design concepts, while detailed models can be used to choose the specific form, dimensions and materials of components (Watts, 1968; Abdullah *et al.*, 1977).

18.1.4 Systemic Description of Instruments

Since, as previously mentioned, instruments are built up of a relatively small variety of basic types of components using a few basic types of interconnection structures, they are usefully analysed and synthesized as systems. This gives insight into essential similarities between superficially very different devices, highlights the possible use of a number of alternative components to perform a particular function, simplifies the design of a complex instrument by decomposing the total function into a number of standard simple partial functions for which convenient realizations are known and so on. In mathematical modelling, in particular, the systems approach eases the task by decomposition of the total system to be modelled into components with simple models and then combining these component models into a total model by a structure of interconnection equations. Some basic concepts and terminology of the systems approach will be used here and will be defined and explained.

In the context of this chapter, and in light of the fact that no standard terminology exists, a *system* will be defined as an assembly of components which is organized to perform a particular function and which can be considered as a unit.

Components of a system which can be considered to be a unit from the point of view of their function or construction, and which themselves are systems of simpler components are termed *subsystems*.

A component of a system which cannot be further decomposed is termed an *element* of the system.

A system may itself form part of a larger system; the latter is termed its *supersystem*. An instrument system generally forms part of a larger information acquisition and processing system; in the case of measuring instrumentation the supersystem comprises the object under measurement and possibly off-line data processing.

Those parts of the universe which affect the system but do not form part of the system or its supersystem are termed the *environment*.

The set of interconnections between two systems or two systems components is termed their *interface*.

18.2   A SYSTEMATIC SCHEME OF INSTRUMENT DESCRIPTION

18.2.1   General

The insight into the behaviour of instruments (and indeed of any complex physical systems), as well as the formulation of their mathematical models, are greatly assisted by the adoption of a systematic methodology of physical system description.

The methodology presented here is based on an approach first rigorously and completely formulated in Paynter (1961) and subsequently developed and presented in MacFarlane (1964, 1970), Shearer *et al.* (1967), Karnopp and Rosenberg (1968, 1975) and van Dixhoorn and Evans (1974). While based on the above, this account represents the approach of the present authors (Finkelstein and Watts, 1969, 1978a, b).

The basis of the approach is the description of the interaction between physical objects, in this case machines and their components, in terms of energy flows. The physical quantities which characterize the energy flow form the physical variables of the description. The quantities which describe the storage, interconversion and distribution of energy within the object and which are determined by its geometry (form and dimensions) and material properties are used as parameters of the description.

18.2.2   Variables of the Scheme of Description

Energy flow between elements or systems is described in terms of a pair of variables. In generalized form the variables are denoted by $\dot{x}$ and $\dot{y}$ so that the instantaneous energy flow rate, (power), is given by

$$\dot{E} = \dot{x}(t)\dot{y}(t)$$

where t denotes time. The variables $\dot{x}$, $\dot{y}$ are thus a *power conjugate* pair: they are termed *energy rate variables* or *intensities*.

An $\dot{x}$-variable involves two points (or regions) in space for its measurement and can be said to act *across* them. It is termed a *two-point, across,* or *trans-* variable.

A $\dot{y}$ variable involves one point (or region) in space for its measurement and can be said to act *through* an element. It is termed a *one-point, through* or *per-* variable.

Time and its inverses are termed *zero-point* variables.

Related to each of the energy *rate* variables is an energy *state* variable. These variables are denoted by x and y and are also termed *extensities*. They may be used to describe the energy stored in elements or systems. We have

Table 18.1   Variables of physical system description (Reproduced by permission of The Institute of Physics)

Class of system ╲ Variables	Through- or per-		Across- or trans-	
	Energy state (extensity)	Energy rate (intensity)	Energy rate (intensity)	Energy state (extensity)
General	y	$\dot{y}$	$\dot{x}$	x
Mechanical translation	h momentum	f force	$\dot{x}$ velocity	x displacement
Mechanical rotation	H angular momentum	T torque	$\dot{\phi}$ angular velocity	ϕ angular displacement
Electrical	q charge	i current	v voltage	λ flux linkage
Fluid flow	g volume	$\dot{g}$ volumetric flow rate	p pressure	P pressure momentum
Thermal	S entropy	$\dot{S}$ entropy flow rate	θ temperature	—
Thermal (pseudo)*	Q heat	$\dot{Q}$ heat flow rate	θ temperature	—
Chemical	n mole no.	$\dot{n}$ mole flow rate	μ chemical potential	—
Time: t				

* Modelling of thermal system interactions is usually in terms of the pseudo-variables $\dot{Q}$, θ, although their product is not power.

the relations:

$$\text{rate variable} = \frac{d}{dt}(\text{state variable})$$

$$\text{change in state variable} = \int_{t_1}^{t_2}(\text{rate variable})\,dt$$

Table 18.1 shows (horizontally) a classification of variables according to this scheme. Table 18.1 also shows (vertically) variables appropriate to various forms of energy. The scheme of description considers mechanical energy as either translational in one dimension or as rotational, coupling between them being considered as an element.

It is possible to classify the variables of description of physical systems into *effort* and *flow* variables. For all systems other than the mechanical ones, the effort variable is the same as the *across* variable and the *through* variable is the same as the flow variable. However, for mechanical systems the through variables become effort variables and the across variables become flow variables. The general result is that *dual analogs* replace *direct analogs* in mathematical descriptions: for example, in the force–voltage, velocity–current analogies which arise, mechanical node connections are analogous to electrical mesh connections.

18.2.3 Ports and Terminals

Instruments or instrument elements are connected to other devices or the system environment at points (regions) in space named *terminals*.

Energy flows or other physical actions are described with reference to a *pair* of terminals. A pair of terminals associated with a physical action is termed a *port*; it constitutes an interface. Elements or systems with a single port are termed *uniports*. Those with more than one port are termed *multiports*.

When energy is not exchanged the port is said to be *non-energic*. The pair of power conjugate variables acts at such a port. One of these is independent, impressed on the port from the environment. The other is dependent, determined by the object.

When energy is not exchanged and the port is said to be *non-energic*. The variable which is active at a non-energic port is termed an *active variable*, or *active signal*. Active variables may, in general, be any one of the x, $\dot{x}$, y or $\dot{y}$ variables.

A variable acting at a non-energic port is independent. If it is impressed by the environment on the object it is termed a *control* or *command* variable and is independent of the object. If it is impressed by the object on the environment it is termed an *observable* variable and is independent of the environment.

18.2.4 Laws of Interconnection at Terminals

The across and through variables, in acting at terminals, obey generalized Kirchhoff's laws.

(a) Kirchhoff's Generalized Vertex Law (expresses continuity of the through extensity):

> The algebraic sum of through rate variables acting
> at a point in a physical system is zero.

(b) Kirchhoff's Generalized Circuit Law (expresses continuity of the across extensity):

> The algebraic sum of across rate variables round any
> closed circuit of a physical system is zero.

Kirchhoff's laws taken conjointly express the principle of conservation of energy.

18.2.5 Terminal Equations

The behaviour of an instrument is described in terms of *terminal equations* which relate the through and across (or flow and effort) variables at its ports.

In general a terminal equation is a non-linear integro-differential equation. In the case of multiports the terminal relation for one port (say $\dot{x}_1$, $\dot{y}_1$) may involve variables at other ports ($\dot{x}_k$, $\dot{y}_k$, . . .) as indicated in Figure 18.1.

For the case of linear terminal relations we define the concept of a generalized impedance which is given by the *transfer function:*

$$Z(s) = \frac{\dot{x}(s)}{\dot{y}(s)} \qquad \begin{aligned} \dot{x}(s) &= \mathcal{L}\dot{x}(t) \\ \dot{y}(s) &= \mathcal{L}\dot{y}(t) \end{aligned}$$

where $\mathcal{L}$ is the Laplace transform. The generalized admittance is defined as:

$$Y(s) = Z(s)^{-1}$$

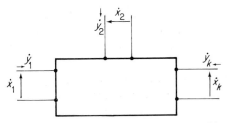

Figure 18.1   Block diagram of a multiport

It should be noted that in the case of mechanical energy, impedance is conventionally defined as the ratio of force or torque to velocity, that is, the ratio of the through to the across variable. It may be convenient to say that for all systems impedance is the *ratio* of the effort to the flow variable.

As stated, most practical systems have non-linear terminal relations. However, the terminal variables $\dot{x}, \dot{y}$ at a port, often vary only by small increments $\delta\dot{x}, \delta\dot{y}$ from the steady operating point $\bar{x}, \bar{y}$. If the terminal characteristic is say:

$$\dot{x}_i = f(\dot{y}_1, \dot{y}_2, \ldots, \dot{y}_n)$$

one can write approximately:

$$\delta\dot{x}_1 = \frac{\partial f}{\partial \dot{y}_i}\, \delta y_i + \cdots + \frac{\partial f}{\partial \dot{y}_n}\, \delta\dot{y}_n$$

The above relation is linear. It can be termed the *incremental* (or small-signal) terminal relation. When using an incremental or small-signal model we shall normally write $\dot{x}, \dot{y}$ for $\delta\dot{x}, \delta\dot{y}$.

For a general linear multiport the terminal relations can be written in some general form such as:

$$\begin{bmatrix} \dot{x}_1(s) \\ \dot{x}_2(s) \\ \cdot \\ \cdot \\ \cdot \\ \dot{x}_k(s) \end{bmatrix} = \begin{bmatrix} Z_{11}(s) & Z_{12}(s) & \cdots & Z_{1k}(s) \\ Z_{21}(s) & Z_{22}(s) & \cdots & Z_{2k}(s) \\ \cdot & & & \cdot \\ \cdot & & & \cdot \\ \cdot & & & \cdot \\ Z_{k1}(s) & Z_{k2}(s) & \cdots & Z_{kk}(s) \end{bmatrix} \begin{bmatrix} \dot{y}_1(s) \\ \dot{y}_2(s) \\ \cdot \\ \cdot \\ \cdot \\ \dot{y}_k(s) \end{bmatrix}$$

$$Z_{ij}(s) = \frac{\dot{x}_i}{\dot{y}_j} \quad (\dot{y}_p = 0 \text{ for } p \neq j)$$

For $i = j$

$$Z_{ii}(s) = \frac{\dot{x}_i}{\dot{y}_i} \quad \text{terminal impedance at port } i$$

For $i \neq j$

$$Z_{ij} = \frac{\dot{x}_i}{\dot{y}_j} \quad \text{transfer impedance } j \text{ to } i$$

The above terminal relations are written in terms of all the independent variables being through variables and the dependent variables being across variables. More generally at some of the ports the independent are through and others across. Denoting the independent variables by U and the dependent by W we have for the terminal relations of a linear multiport:

$$
\begin{bmatrix} W_1(s) \\ W_2(s) \\ \cdot \\ \cdot \\ \cdot \\ W_k(s) \end{bmatrix} = \begin{bmatrix} G_{11}(s) & G_{12}(s) & \cdots & G_{1k}(s) \\ G_{21}(s) & G_{22}(s) & \cdots & G_{2k}(s) \\ \cdot & \cdot & & \cdot \\ \cdot & \cdot & \cdot & \cdot \\ \cdot & \cdot & & \cdot \\ G_{k1}(s) & G_{k2}(s) & \cdots & G_{kk}(s) \end{bmatrix} \begin{bmatrix} U_1(s) \\ U_2(s) \\ \cdot \\ \cdot \\ \cdot \\ U_k(s) \end{bmatrix}
$$

where G_{ij} is the transfer function from U_j to W_i

$$
G_{ij}(s) = \frac{W_i(s)}{U_j(s)} \quad (U_p = 0 \text{ for } p \neq j)
$$

For $i = j$

$$
G_{ii} = \frac{U_i(s)}{W_i(s)}
$$

G_{ii} is the impedance at port i if W is an effort and U a flow and the admittance at port i if W is a flow and U an effort.

For $i \neq j$

$$
G_{ij} = \frac{U_i(s)}{W_j(s)}
$$

is a transfer impedance if W is an effort and U a flow, a transfer admittance if W is a flow and U an effort or a coefficient if U and W are both either flows or efforts.

The terminal relations may be a physical model if the parameters of the relations are expressed in terms of the form, dimensions and material properties of the machine components. Commonly, however, these relations are used as a functional model, with impedances and the like expressed in an aggregated form unrelated to the physical feature of the machine. Usually impedances and the like are derived from measurements on the machine or component and cannot be related to constructional features.

While linear models are frequently adequate for describing the behaviour of instruments, it may be necessary for some purposes to consider the complete non-linear model. The *state variable* representation is then most convenient. This takes the general form of the differential equations:

$$
\frac{d\mathbf{z}}{dt} = \mathbf{f}(\mathbf{z}, \mathbf{m}) \quad \mathbf{w} = \mathbf{g}(\mathbf{z}, \mathbf{m})
$$

where $\mathbf{m}$ is the vector of input variables, $\mathbf{w}$ the vector of output variables, and $\mathbf{z}$ the vector of state variables. The input and output variables are chosen to be rate variables appearing at the terminals, while energy state variables can conveniently be chosen as state variables.

Detailed analysis of the functioning of physical elements as instrument elements is provided in Section 18.4.

18.2.6 Physical Elements

Instrument elements are most conveniently modelled in terms of idealized physical elements performing a single non-decomposable operation on the energy flow.

Such elements can always be described in terms of a relationship between energy rate and state variables which depends only on the *geometry* and *material properties*. This relation is termed the *constitutive relation* of the element.

The energy relations at a port of an element are summarized in Figure 18.2. Energy in an element can be *stored*, *converted*, or *transmitted*. Elements can thus be classified into:

(a)   Energy stores: that is elements which store energy. Distinguishable are: (1) through variable stores—storing a through variable; (2) across variables stores—storing an across variable; (3) coupled stores or storage fields—storing energy flows from several ports.

(b)   Energy converters: that is elements which convert or transform energy. Distinguishable are: (1) bilateral converters; (2) unilateral converters.

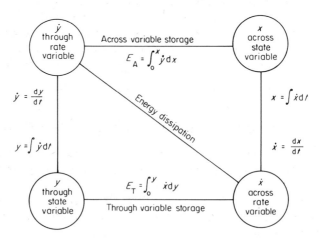

Figure 18.2   Energy relations at a port of an element

(c)   Power sources: that is elements supplying energy to a system. Distinguishable are: (1) through variable sources; (2) across variable sources.
(d)   Interconnection elements: that is elements which conduct energy without store or conversion. Distinguishable are: (1) transmitters; (2) junctions.

It is emphasized that this classification refers to *idealized* elementary functions only; practical elements will in general perform several of these functions, (say, both store and convert energy) and may be viewed as systems of ideal elements.

The characteristics of the elements listed above are summarized in the following subsections. Table 18.2 provides constitutive relationships in graphical form and examples of realizations.

18.3   INSTRUMENTS AND INSTRUMENT ELEMENTS: FUNCTION AND MODEL

An instrument, or instrument element such as a transducer, was described in Section 8.1.2 as a machine which maintains a prescribed functional relation between an input signal and an output signal. It is convenient to discuss this functioning in terms of the scheme of physical descriptions described in Section 18.2 and, in particular, the terminal relations discussed in Section 18.2.5.

An instrument maintains a relation between an input terminal variable and an output terminal variable. In terms of the terminal relations of linear multiports, an instrument maintains a relation between some input U_j at port j and some output W_i at port i. An off-diagonal transfer function G_{ij} describes the interaction. In the case of transducers G_{ij} is termed the *transduction coefficient.* A diagonal transfer function G_{ii} between the terminal variables at the same port, maintains a functional relation between variables at that port: it may also be used as an instrument functional relation.

18.4   INSTRUMENT ELEMENTS: MODEL AND FUNCTION

The archetype mathematical models of many important instrument components (in particular, transducers) can be considered as physical system elements.

The principles of the functioning of instruments and in particular of transducers can best be considered first in terms of *mathematical models* of *physical system elements.*

These will now be reviewed.

18.4.1 Stores

18.4.1.1 Through variable store

These are elements which *store* the *through* variable. The general constitutive relation for them is:

$$y = \Phi_C(\dot{x})$$

Energy stored in the element:

$$E_T = \int_0^y \dot{x}\,\mathrm{d}y$$

Co-energy stored in the element:

$$E_T' = \int_0^x y\,\mathrm{d}\dot{x}$$

$$\dot{x} = \frac{\partial E_T}{\partial y}, \quad \dot{y} = \frac{\mathrm{d}}{\mathrm{d}t}\left(\frac{\partial E_T'}{\partial \dot{x}}\right)$$

For a linear element we have

$$\text{\textit{constitutive relation }} y = Cx$$

$$\text{\textit{terminal relation }} \dot{y} = C\frac{\mathrm{d}\dot{x}}{\mathrm{d}t}$$

In Laplace transform form

$$\dot{y}(s) = sC\dot{x}(s) \quad Z(s) = \frac{1}{sC}$$

A typical through variable store element is an electrical capacitor (Figure 18.3). The constitutive relation for a linear capacitor is

$$q = Cv$$

The terminal relation is

$$i = C\frac{\mathrm{d}v}{\mathrm{d}t}$$

with v as the independent and i as the dependent variable.
 Alternatively we can write

$$v = \frac{1}{C}\int i\,\mathrm{d}t$$

with i as the independent and v the dependent variable.

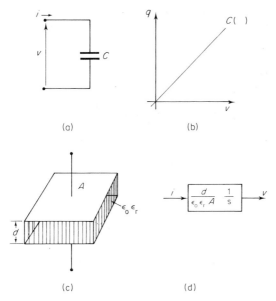

Figure 18.3 Electrical capacitor: (a) schematic; (b) constitutive relation; (c) schematic of archetype model; (d) signal flow diagram of current integrator

The impedance of the capacitor is given by

$$Z(s) = 1/Cs$$

A typical realization of a capacitor consists of two parallel plates with a dielectric medium between them. If A is the area of the plates, d the separation and $\varepsilon_0 \varepsilon_r$ the permittivity of the medium, we have an archetype model of the store:

$$C = \varepsilon_r \varepsilon_0 A/d$$

A typical instrument use of the element is to integrate current flow. The element then maintains a functional relation between current flow and voltage.

18.4.1.2 Across variable store

These are elements which *store* the *across* variable. The general constitutive relation for them is:

$$x = \Phi_L(\dot{y})$$

Energy stored in the element:

$$E_A = \int_0^x \dot{y} \, dx$$

Co-energy stored in the element:

$$E'_A = \int_0^{\dot{y}} x \, d\dot{y}$$

$$\dot{y} = \frac{\partial E_A}{\partial x}, \quad \dot{x} = \frac{d}{dt}\left(\frac{\partial E'_A}{\partial \dot{y}}\right)$$

For a linear element we have:

$$x = L\dot{y}$$

The terminal relation is

$$\dot{x} = L\frac{d\dot{y}}{dt}$$

In Laplace transform form we have

$$\dot{x}(s) = sL\dot{y}(s) \quad Z(s) = sL$$

A typical across variable store element is a spring (Figure 18.4). The constitutive relation for a linear spring is

$$f = kx$$

The terminal relation is

$$f = k\int \dot{x} \, dt$$

with $\dot{x}$ as the independent variable and f as the dependent variable.
   Alternatively we can write

$$\dot{x} = \frac{1}{k}\frac{df}{dt}$$

with f as the independent and $\dot{x}$ as the dependent variable.
   The impedance of the spring is

$$Z(s) = k/s$$

   A typical realization (one of many forms) of the spring is a cantilever of rectangular cross section. If the length of the cantilever is l, the cross section has width w, thickness t, and the Young' modulus of the material is E, then

$$k = \frac{Ewt^3}{4l^3}$$

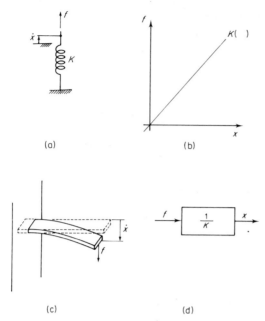

Figure 18.4 Mechanical translational spring: (a)schematic; (b) constitutive relation; (c) schematic of archetype model; (d) signal flow diagram relating output displacement to input force

A typical instrument use of the spring is to transform an input force signal into an output displacement signal proportional to the input force.

18.4.1.3 Coupled stores or energy storing fields

An element which has more than one port and stores, without loss, the energy supplied at these ports is termed a *coupled store* or *energy storage field*. The functioning of these elements exhibits energy conversion and also control action: discussion is therefore deferred until conversion and control have been considered.

18.4.2 Energy Conversion

18.4.2.1 Bilateral converters

Bilateral converters are elements with two or more ports which transfer, without storage or loss, energy supplied at one or more of the ports into an energy output at the other ports.

Elements in which the ports between which the energy transfer takes place have the same energy form are known as *transformers*.

Elements in which the energy transfer between ports involves a change of energy form are known as *converters*.

A distinction is made between *normal* transformers or converters (also known as transforming converters) and *gyrating* transformers or converters (also known as *gyrators* and *gyrating converters*).

Consider a normal transformer or converter having two ports, indexed 1 and 2. In such a transformer or converter the state *across variables* or the state *through variables* at the two ports are related by a constitutive relation:

$$x_1 = \Phi_T(x_2) \quad \text{or} \quad y_1 = \Phi_T(y_2)$$

The power flow balance equation is:

$$\dot{x}_1 \dot{y}_1 + \dot{x}_2 \dot{y}_2 = 0$$

For a linear transformer we have from the above,

$$\dot{x}_1 = N\dot{x}_2 \text{ and } -N\dot{y}_1 = \dot{y}_2$$

A typical example of a *normal converter* is an electrodynamic element, consisting of a coil rotating in a magnetic field. This is shown in Figure 18.5 which gives the notation. If the element is ideal, that is, there is no momentum storage in the rotor or magnetic storage in the coil due to the flow of current, we have the constitutive relation which describes the variation of flux linkages in space.

$$\lambda = \lambda(\phi)$$

The power flow balance equation is

$$T\dot{\phi} + vi = 0$$

If the flux distribution is designed to be linear, as is usual,

$$\lambda = BAN\phi$$

where A is the area of the rotor coil, N is the number of current conductors on the coil, and B is the density of magnetic flux threading the coil. Hence the terminal equations of the transducer are

$$v = BAN\dot{\phi} \quad T = -BANi$$

In Laplace transform form we have:

$$\begin{bmatrix} T(s) \\ v(s) \end{bmatrix} = \begin{bmatrix} 0 & -BAN \\ BAN & 0 \end{bmatrix} \begin{bmatrix} \dot{\phi}(s) \\ i(s) \end{bmatrix}$$

The element can then be used as a transducer realizing either a current to

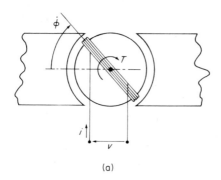

(a)

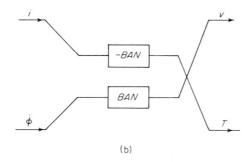

(b)

Figure 18.5 Normal bilateral converter—
coil moving in magnetic field: (a) schematic;
(b) signal flow diagram for ideal element

torque transformation or a velocity to voltage transformation with transduction coefficient $-BAN$ and BAN respectively.

Consider now a *gyrating transformer* or *converter* having two ports indexed 1 and 2.

In such a gyrator the state *across variable* at one port is related to the state *through variable* at the other port by a constitutive relation:

$$x_1 = \Phi_G(y_2) \quad \text{or} \quad y_1 = \Phi_G(x_2)$$

The power flow balance equation is given by:

$$\dot{x}_1 \dot{y}_1 + \dot{x}_2 \dot{y}_2 = 0$$

For a linear gyrating transformer we have:

$$\dot{x}_1 = N\dot{y}_2 \quad \text{and} \quad -N\dot{y}_1 = \dot{x}_2$$

A typical example of a gyrating tranformer is a bellows. This is shown diagrammatically in Figure 18.6 which gives the notation. If the element is ideal, that is, if there is negligible elastic energy stored in either the stiffness

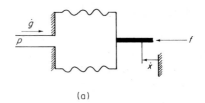

(a)

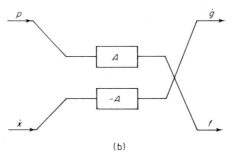

(b)

Figure   18.6   Gyrating   bilateral
converter—bellows: (a)   schematic;   (b)
signal flow diagram for ideal element

of the bellows or the compression of the fluid, we have the constitutive
relation which connects the change of fluid volume in the bellows with the
movement of the bellows:

$$g = g(x)$$

The power balance equation is

$$f\dot{x} + p\dot{g} = 0$$

If the effective area A of the bellows is constant, as is usual, we have

$$g = -Ax$$

Hence the terminal equations of the transducer are

$$f = pA \quad \dot{g} = -A\dot{x}$$

In Laplace transform form we have:

$$\begin{bmatrix} f(s) \\ \dot{g}(s) \end{bmatrix} = \begin{bmatrix} 0 & A \\ -A & 0 \end{bmatrix} \begin{bmatrix} \dot{x}(s) \\ p(s) \end{bmatrix}$$

The element can be used as a transducer maintaining a functional relation
between a pressure input and a force output. Alternatively it may maintain a
functional relation between an input velocity and an output flow. The
transduction coefficients are both A and $-A$.

18.4.2.2 Unilateral converters (dissipators)

Unilateral converters transform the *input energy* into *thermal energy*.

In general, the thermal energy is removed from the instrument: the unilateral converter then acts as a dissipator.

The general constitutive relation for a unilateral converter is:

$$\dot{x} = \Phi_R(\dot{y})$$

The power balance equation is given by:

$$\dot{Q} + \dot{x}\dot{y} = 0$$

The content is:

$$G = \int_0^{\dot{y}} \dot{x}\,\mathrm{d}\dot{y}$$

The co-content is:

$$G' = \int_0^{\dot{x}} \dot{y}\,\mathrm{d}\dot{x}$$

$$\dot{x} = \frac{\partial G}{\partial \dot{y}}, \quad \dot{y} = \frac{\partial G'}{\partial \dot{x}}$$

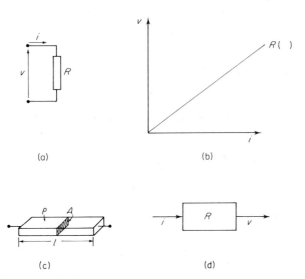

(a) (b)

(c) (d)

Figure 18.7 Unilateral converter (dissipator)— electrical resistor: (a) schematic; (b) constitutive relation; (c) schematic of archetype model; (d) signal flow diagram relating output voltage to input current

For a linear element we have:

$$\dot{x} = R\dot{y}$$

In some cases the thermal energy generated in the unilateral converter enters into the instrument operation and needs to be taken into account. An example is the self-heating effect in a strain gauge.

A typical example of a dissipator is the electrical resistor shown in Figure 18.7. The terminal relation for a linear resistor is

$$v = Ri$$

An archetype realization of a resistor is a metallic conductor. For a conductor of length l, cross-sectional area A and resistivity ρ, we have the archetype model

$$R = \rho l / A$$

18.4.3 Power Sources

Power sources are elements which deliver power at their output port. They must, of course, be either stores or transformers. They are in fact idealized elements, that is, stores of infinite capacity or transformers of infinite power input capacity and in which the power input is not considered.

There are two types of power sources: *through variable sources* and *across variable sources.*

A through variable source has a through variable output which is independent of the across variable at the port.

The constitutive relation is:

$$\dot{y} = \dot{y}(t)$$

A typical example is a mass exerting gravitational force and acting as a source of constant force.

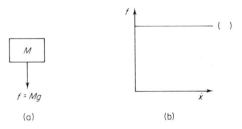

Figure 18.8 Through variable source—weight: (a) schematic; (b) constitutive relation

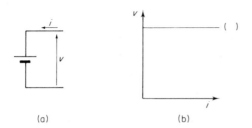

Figure 18.9 Across variable source—electric cell: (a) schematic: (b) constitutive relation

An across variable source has an across variable output which is independent of the through variable at the port.

The constitutive relation is:

$$\dot{x} = \dot{x}(t)$$

A typical example is an electric cell of very low internal resistance which functions to provide a constant voltage irrespective of current drain.

Through and across variable sources are illustrated in Figures 18.8 and 18.9, respectively.

The elements having a *single signal* only are not of course, transducers but have a function in transduction which will be described later in connection with controlled elements.

18.4.4 Interconnection Elements

Interconnection elements, as their name implies, join other elements of a physical system to establish a system structure. Their models are trivially simple, being merely applications of the general Kirchhoff's laws. Constructively, however, cables, shafts, pipes, switches, and the like, are substantial components and need to be considered in design. These elements can be classified into *transmitters* and *junctions*.

A transmitter shown in Table 18.2 is a two-port device in which the through and across variables applied at one port appear unchanged at the output port.

The constitutive relation for a transmitter with ports indexed 1 and 2 is

$$\dot{x}_2 = \dot{x}_1$$

The power balance equation is:

$$\dot{x}_1 \dot{y}_1 + \dot{x}_2 \dot{y}_2 = 0$$

Table 18.2 System elements (Reproduced by permission of The Institute of Physics)

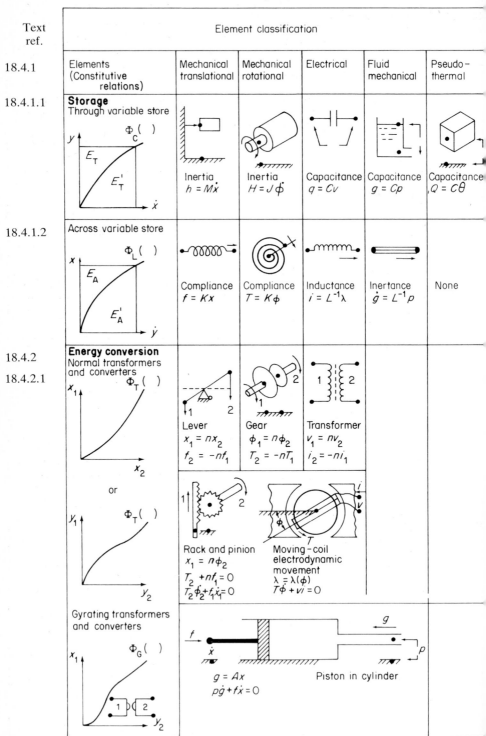

Text ref.	Element classification					
18.4.1	Elements (Constitutive relations)	Mechanical translational	Mechanical rotational	Electrical	Fluid mechanical	Pseudo–thermal
18.4.1.1	**Storage** Through variable store $\Phi_C(\ )$	Inertia $h = M\dot{x}$	Inertia $H = J\dot{\phi}$	Capacitance $q = Cv$	Capacitance $g = Cp$	Capacitance $Q = C\theta$
18.4.1.2	Across variable store $\Phi_L(\ )$	Compliance $f = Kx$	Compliance $T = K\phi$	Inductance $i = L^{-1}\lambda$	Inertance $\dot{g} = L^{-1}p$	None
18.4.2 18.4.2.1	**Energy conversion** Normal transformers and converters $\Phi_T(\ )$ or $\Phi_T(\ )$	Lever $x_1 = nx_2$ $f_2 = -nf_1$ — Rack and pinion $x_1 = n\phi_2$ $T_2 + nf_1 = 0$ $T_2\dot{\phi}_2 + f_1\dot{x}_1 = 0$	Gear $\phi_1 = n\phi_2$ $T_2 = -nT_1$ — Moving–coil electrodynamic movement $\lambda = \lambda(\phi)$ $T\dot{\phi} + vi = 0$	Transformer $v_1 = nv_2$ $i_2 = -ni_1$		
	Gyrating transformers and converters $\Phi_G(\ )$	Piston in cylinder $g = Ax$ $p\dot{g} + f\dot{x} = 0$				

Text ref.						
8.4.2.2	**Unilateral converters (dissipators)** $\dot{x}\dot{y} + \dot{Q} = 0$	Damper friction viscous drag $f = B\dot{x}$	Damper friction viscous drag $T = B\dot{\phi}$	Resistance $v = Ri$	Friction Capillary friction $p = R\dot{g}$	Thermal conductance $\theta = R\dot{Q}$
8.4.3	**Power sources** Through variable sources	Force generator (driver)	Torque generator	Current generator	Constant-flow pump	Heat flow generator
	Across variable sources	Velocity generator	Angular velocity generator	Voltage generator	Constant-pressure pump	Constant-temperature source
8.4.4	**Interconnection elements** Transmitters $\dot{x}_1 = \dot{x}_2$ $\dot{y}_2 = -\dot{y}_1$	Rigid rod $\dot{x}_1 = \dot{x}_2$ $f_2 = -f_1$	Rigid shaft $\dot{\phi}_1 = \dot{\phi}_2$ $T_2 = -T_1$	Lossless wire pair $v_1 = v_2$ $i_2 = -i_1$	Lossless pipe $p_1 = p_2$ $\dot{g}_1 = -\dot{g}_2$	Thermal contact $\theta_1 = \theta_2$ $\dot{Q}_1 = -\dot{Q}_2$

Text
ref.

Table 18.2   Systems elements (cont.)

18.4.4

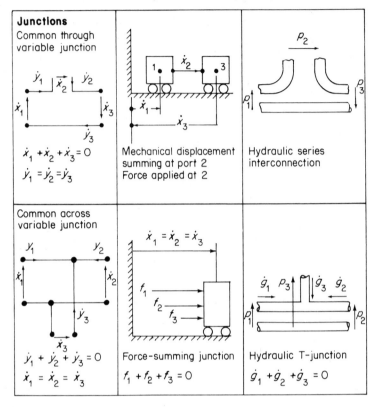

Junctions		
Common through variable junction	Mechanical displacement summing at port 2	Hydraulic series interconnection
$\dot{x}_1 + \dot{x}_2 + \dot{x}_3 = 0$ $\dot{y}_1 = \dot{y}_2 = \dot{y}_3$	Force applied at 2	
Common across variable junction	Force-summing junction	Hydraulic T-junction
$\dot{y}_1 + \dot{y}_2 + \dot{y}_3 = 0$ $\dot{x}_1 = \dot{x}_2 = \dot{x}_3$	$f_1 + f_2 + f_3 = 0$	$\dot{g}_1 + \dot{g}_2 + \dot{g}_3 = 0$

and hence

$$\dot{y}_1 = -\dot{y}_2$$

Typical examples of transmitters are a cable or a shaft.

A common through variable junction is shown in Table 18.2 and is a three-port element with the same through variable at each of the three ports. Topologically it is a series connection.

Indexing the three ports by 1, 2, and 3 respectively we have the constitutive relations of the element:

$$\dot{y}_1 = \dot{y}_2 = \dot{y}_3$$

The power balance equation is:

$$\dot{x}_1 \dot{y}_1 + \dot{x}_2 \dot{y}_2 + \dot{x}_3 \dot{y}_3 = 0$$

and hence

$$\dot{x}_1 + \dot{x}_2 + \dot{x}_3 = 0$$

This element can be best understood by examples of its realizations in Table 18.2.

Common across variable junctions are also shown in Table 18.2. They are three-port elements with the same across variable at each of the three ports. Topologically it is a parallel connection.

Indexing the three ports by 1, 2, and 3 respectively we have the constitutive relations of the element:

$$\dot{x}_1 = \dot{x}_2 = \dot{x}_3$$

The power balance equation is

$$\dot{x}_1\dot{y}_1 + \dot{x}_2\dot{y}_2 + \dot{x}_3\dot{y}_3 = 0$$

and hence

$$\dot{y}_1 + \dot{y}_2 + \dot{y}_3 = 0$$

They are best understood in terms of interconnections of the kind where there is a meeting of a through variable, for example, force-summing points in mechanical systems, current nodes in electrical networks, and fluid flow pipe junctions as illustrated in Table 18.2.

18.4.5 Controlled Elements

As has already been stated, the constitutive and, hence, the terminal relations of the physical elements described above depend on their geometry and material properties.

A physical input at one port of an element which alters either or both of these features of an element may, without energy conversion taking place, alter the terminal relations of the element at another port (Finkelstein and Watts, 1978b).

A multiport in which a through or across rate variable acting at one port affects or controls the terminal relations at one or more of the other ports, without energy transformation between them, may be termed a controlled element. It is also frequently termed a *modulated element* in the literature.

The variable which affects the terminal relations at the other port or ports will be termed the controlling variable and the port at which it acts the controlling port. The controlling variable is frequently, but not always, a state variable. The ports at which the controlling variables affect the terminal relations are termed the *controlled ports*.

A controlled element in which no storage, dissipation or transformation of energy takes place at the controlling port is termed a *pure controlled element*. A pure controlled element in which the terminal relations at the controlled ports are linear, and in which the constant of proportionality in these relations varies linearly with the controlling variable, will be termed an ideal controlled element.

There are controlled versions of each of the classes of elements discussed above, specifically stores, converters, interconnection elements and hence, of course, sources. Examples of controlled converters and interconnection elements will be given in this section to illustrate the functioning of controlled elements. Controlled stores will be considered in the next section.

A good example of a controlled transformer used in instrumentation is an electrical transformer consisting of two coaxial coils coupled by a movable iron core: (Figure 18.10 gives the notation). The movement of the core controls the transformation ratio. For the purpose of the present analysis the electromechanical energy will be neglected, the core velocity and hence the mechanical power input being considered to be small.

The constitutive relation of the transformer is

$$\lambda_2 = n(x)\lambda_1$$

showing that the fraction of the primary coil flux which is linked with the secondary coil 2, is a function of x.

The power balance equation is

$$v_1 i_1 + v_2 i_2 = 0$$

Assuming, for simplicity of explanation, negligible power output the terminal equation is given by

$$v_2 = n(x)v_1$$

For a fixed energizing voltage v_1 the element maintains an output signal v_2 which is a function of the core position x. It is thus a *position* to *voltage* transducer.

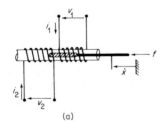

(a)

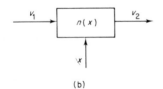

(b)

Figure 18.10   Controlled transformer: (a) schematic; (b) signal flow diagram

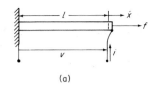

(a)

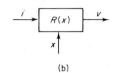

(b)

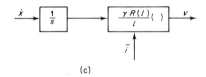

(c)

Figure 18.11 Controlled dissipator—electrical resistance strain gauge: (a) schematic; (b) signal flow diagram; (c) small-signal flow diagram

As another example of a controlled element consider an unbonded resistance strain gauge (see Figure 18.11 giving the notation).

We have the constitutive relation

$$v = R(x)i$$

The power balance equation is

$$f\dot{x} + vi + \dot{Q} = 0$$

If again, for simplicity of explanation, we neglect the mechanical power input to the gauge and the heating effect, we have as the terminal relation

$$v = R(x)i$$

The linearized small signal model

$$v(s) = \frac{\gamma R(l)}{l}\, \bar{i}\, \frac{1}{s}\, \dot{x}(s)$$

where γ is the *gauge factor*:

$$\gamma = \frac{\delta R/R}{\delta l/l}$$

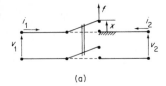

(a)

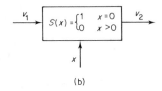

x

(b)

Figure 18.12   Electrical switch: (a) schematic; (b) signal flow diagram

For a fixed energizing current $\bar{i}$ the *voltage output* of the gauge is a function of the *mechanical displacement x*.

As a final example consider an electrical switch which is basically a controlled transmitter. With reference to the illustration of Figure 18.12 we have

$$v_2 = S(x)v_1$$

where the switch function $S(x)$ is given by

$$S(x) = \begin{cases} 1 & \text{for} \quad x = 0 \\ 0 & \text{for} \quad x > 0 \end{cases}$$

Thus with an energizing voltage v_1 the output signal v_2 is a function of x. The device is thus a *position* to *electrical signal* transducer.

In conclusion we thus see that in a controlled element the input signal is applied at the controlling port and maintains a functional relation to an output signal at the controlled port. The element must be energized from an outside source. The energy flow is from the energizing port to the signal output port. (They may be the same, as in the strain gauge). There is no energy flow from the signal input to the signal output port; the element may thus function as a *power amplifier*.

18.4.6   Coupled Stores

Coupled stores or storage fields, that is, elements that store energy from several ports in a common store will now be analysed in terms of energy storage, conversion and control.

We may distinguish between *through*, *across*, and *mixed coupled* stores.

18.4.6.1 Through coupled store

A through storage field stores the through variables at each port. The general constitutive relation for a through storage field is:

$$y_i = \Phi_{Ci}(\dot{x}_1, \dot{x}_2, \ldots, \dot{x}_n)$$

Energy stored in the element is given by:

$$E = \sum_{i=1}^{n} \int_0^y \dot{x}_i \, dy_i$$

and

$$\dot{x}_i = \partial E/\partial y_i$$

Consider as an example of a through storage field a cylinder of perfect gas (Figure 18.13). It has two ports: the *fluid port* (at which flow is stored) for

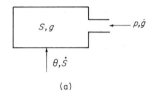

(a)

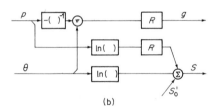

(b)

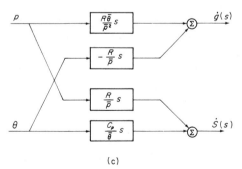

(c)

Figure 18.13 Thermal fluid transducer: (a) schematic; (b) signal flow diagram; (c) small-signal flow diagram

which we have the constitutive relation

$$g = -R\theta/p$$

and the *thermal port* (at which entropy is stored) for which the constitutive relation is

$$S = C_p \ln\theta + R \ln p + S'_0$$

In these equations θ is the absolute temperature, R the gas constant, C_p the specific heat at constant pressure, and S'_0 is a constant; the negative sign is due to the fact that by convention in modelling, inward flow, that is contraction, is taken as positive.

It it can be seen that the element can act as a thermal to fluid transducer, in that for a constant pressure it maintains an expansion g linearly related to the temperature θ: (gas thermometer). The pressure to entropy relation can also be maintained but does not have common instrument application.

The small signal model in Laplace transform form is given by

$$\begin{bmatrix} \dot{g}(s) \\ \dot{S}(s) \end{bmatrix} = \begin{bmatrix} \dfrac{R\bar{\theta}}{\bar{p}^2}s & -\dfrac{R}{\bar{p}}s \\ \dfrac{R}{\bar{p}}s & \dfrac{C_p}{\bar{\theta}}s \end{bmatrix} \begin{bmatrix} p(s) \\ \theta(s) \end{bmatrix}$$

We can see from the off-diagonal elements of the matrix (which give the transduction coefficients) that the device can be considered as a normal converter transforming temperature variations into flow and pressure variations into entropy flow. The diagonal elements represent, respectively, *through storage* at the fluid port controlled by temperature and *through storage* at the thermal port controlled by pressure. These controlled stores have no instrument applications but illustrate an important phenomenon.

18.4.6.2 Across coupled store

An across storage field stores the across variables at each port. The general constitutive relation for an across storage field is:

$$x_i = \Phi_{Li}(\dot{y}_1, \dot{y}_2, \dots, \dot{y}_n)$$

and the energy stored in the element is given by:

$$E = \sum_{i=1}^{n} \int_0^x \dot{y}_i \, dx_i$$

and

$$\dot{y}_i = \partial E/\partial x_i$$

Consider, as an example of an across storage field, an iron core (the

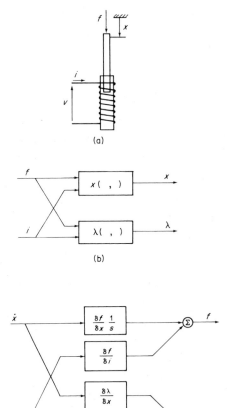

Figure 18.14 Transducer consisting of
coil with movable core: (a) schematic;
(b) signal flow diagram: f, i as inputs. (c)
small-signal flow diagram: x, i as inputs

momentum of which can be neglected) moving in an electric coil (Figure
18.14).

There are two ports: an electrical one at the coil and a mechanical one at
the core. One can consider that a current is impressed on the coil and a
force on the core. There are then two constitutive relations:

$$x = x(f, i) \quad \lambda = \lambda(f, i)$$

The first describes the fact that the position of the core is a function of the
impressed force on the core and of the current which generates another

force on the core. The second describes the fact that the flux linked with the coil is a function of current and position which in turn is a function of current and force. The constitutive relations show that the device is a coupled across store.

It can be seen that the device can act as a transducer maintaining, for example, for a constant applied force a functional relation between current and position. An example of an instrument application is a moving-iron meter.

In another common instrument application of this form of element, displacement rather than force is the independent variable. This does not alter the nature of the element.

We have then

$$f = f(x, i) \quad \lambda = \lambda(x, i)$$

The small-signal model of the element is then

$$\begin{bmatrix} f(s) \\ v(s) \end{bmatrix} = \begin{bmatrix} \dfrac{\partial f}{\partial x} \dfrac{1}{s} & \dfrac{\partial f}{\partial i} \\ \dfrac{\partial \lambda}{\partial x} & \dfrac{\partial \lambda}{\partial i} s \end{bmatrix} \begin{bmatrix} \dot{x}(s) \\ i(s) \end{bmatrix}$$

The off-diagonal elements represent a *transformation* between electrical and mechanical signals as in the electrodynamic transducer. The diagonal elements represent respectively at the mechanical port an elastic, that is an across, store and at the electrical port an inductive, that is again an across, store. Since $\partial f / \partial x$ and $\partial \lambda / \partial i$ are functions of $\bar{x}, \bar{i}$, the stores at each port are controlled by the input signal at the other.

The device is most commonly used with zero velocity and high frequency current energization as a displacement controlled inductor.

18.4.6.3 Mixed coupled stores

A mixed storage field is an element which stores the through variables from some ports and the across variables from others, In the n-port store, considered above, let ports 1 to j be through storage ports and ports $j+1$ to n, across storage ports. Then the form taken by the general constitutive relations is:

$$y_i = \Phi_i(\dot{x}_1, \ldots, \dot{x}_i, \ldots, \dot{x}_j, \dot{y}_{j+1}, \ldots, \dot{y}_n) \quad 1 \leqslant i \leqslant j$$

and

$$x_k = \Phi_k(\dot{x}_1, \ldots, \dot{x}_j, \dot{y}_{j+1}, \ldots, \dot{y}_k, \ldots, \dot{y}_n) \quad j+1 \leqslant k \leqslant n$$

The energy stored is given by:

$$E = \sum_{i=1}^{j} \int_0^y \dot{x}_i \, dy_i + \sum_{k=j+1}^{n} \int_0^x \dot{y}_k \, dx_k$$

and

$$\dot{x}_i = \frac{\partial E}{\partial y_i} \qquad \dot{y}_k = \frac{\partial E}{\partial x_k}$$

As an example of a mixed coupled store consider the parallel-plate capacitor given as an example of a through electrical store in Section 18.4.1.1. Consider further that in this case the plates move normally with respect to each other (Figure 18.15). Let a voltage v be impressed on the capacitor and a force f on the plates.

We can then write for the electrical port

$$q = \frac{\varepsilon_r \varepsilon_0 A}{x} v$$

which stores current in the form of charge as a through store.

At the mechanical port we can write

$$f = \tfrac{1}{2} \varepsilon_r \varepsilon_0 A (v/x)^2$$

reflecting the balance between the impressed force and electrostatic attraction. At this port then the device stores energy due to displacement acting as an across store.

The element may be used as a transducer generating, for a fixed plate separation, a force proportional to voltage.

We can rewrite the above equations in a conventional form to show that it is a mixed coupled store:

$$q = \sqrt{(2\varepsilon_r \varepsilon_0 A)} \sqrt{f}$$

$$x = \sqrt{(\tfrac{1}{2} \varepsilon_r \varepsilon_0 A)} \frac{v}{\sqrt{f}}$$

The small-signal model at the operating point $\bar{x}, \bar{v}$ in Laplace transform form is

$$\begin{bmatrix} i \\ f \end{bmatrix} = \begin{bmatrix} \dfrac{\varepsilon_r \varepsilon_0 A}{\bar{x}} s & -\dfrac{\varepsilon_r \varepsilon_0 A \bar{v}}{\bar{x}^2} \\ \dfrac{\varepsilon_r \varepsilon_0 A \bar{v}}{\bar{x}^2} & -\dfrac{\varepsilon_r \varepsilon_0 A \bar{v}^2}{\bar{x}^3} \dfrac{1}{s} \end{bmatrix} \begin{bmatrix} v \\ \dot{x} \end{bmatrix}$$

The off-diagonal elements of the model represent energy transformation in the device; the diagonal elements represent a displacement-controlled

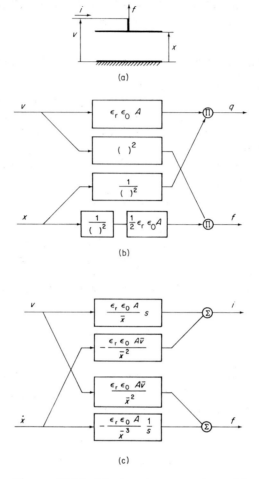

Figure 18.15 Moving-plate capacitor: (a) schematic; (b) signal flow diagram; (c) small-signal flow diagram

capacitive or through storage element at the electrical port and a voltage controlled elastic element, that is across storage, at the mechanical port.

The device may be used as a transducer in the following modes. At constant voltage and high velocity, that is, high frequency vibration, it can be used as a velocity to current transducer. At zero velocity it can be used as capacitance controlled by the displacement x and hence as a mechanical to electrical signal transducer.

18.5 RELATIONS BETWEEN THE TRANSDUCTION COEFFICIENTS

18.5.1 Maxwell's Thermodynamic Relations

From the principle of conservation of energy we may derive important relations between the transduction coefficients connecting the terminal variables of converters and coupled stores. They are generalizations of the well known *Maxwell's thermodynamic* relations. We review these briefly here on the basis of Watts (1981).

Consider the expression of the first law of thermodynamics which describes conservation of energy:

$$dU = \theta \, dS - p \, dV$$

where U is total internal energy.

Here V is used as the classical symbol for volume instead of g. We also take contraction to be negative as in thermodynamic convention rather than positive as in the general analysis of dynamic systems.

Since U is a function of the extensities S, V, dU is a perfect differential and we have:

$$\left(\frac{\partial \theta}{\partial V}\right)_S = -\left(\frac{\partial p}{\partial S}\right)_V$$

We may then define the other thermodynamic functions by Legendre transformations of U and derive from them the remaining Maxwell's relations. We have thus the Helmholtz free energy F:

$$F = U - \theta S$$

$$dF = -S \, d\theta - p \, dV$$

and hence

$$\left(\frac{\partial S}{\partial V}\right)_\theta = \left(\frac{\partial p}{\partial \theta}\right)_V$$

The Gibbs free energy is

$$G = F + pV$$

$$dG = -S \, d\theta + V \, dp$$

and hence

$$\left(\frac{\partial S}{\partial p}\right)_\theta = -\left(\frac{\partial V}{\partial \theta}\right)_p$$

Enthalpy is

$$H = U + pV$$

$$dH = \theta \, dS + V \, dp$$

$$\left(\frac{\partial \theta}{\partial p}\right)_S = \left(\frac{\partial V}{\partial S}\right)_p$$

We note that the above Maxwell's thermodynamic relations in fact describe relations between the transduction coefficients in the coupled thermal–fluid store discussed in Section 18.4.6.

18.5.2 Generalization of Maxwell's Relations

Consider an energy conserving multiport. The energy supplied at a port is either *through energy*

$$dE_{T_1} = \dot{x}_1 \, dy_1$$

or *across energy*

$$dE_{A_1} = \dot{y}_1 \, dx_1$$

For the complete multiport we have

$$dE_T = \sum_{i=1}^{n} \dot{x}_i \, dy_i$$

$$dE_A = \sum_{i=1}^{n} \dot{y}_i \, dx_i$$

The Hamiltonian function of general dynamics is then defined as the total energy stored in the system, analogous to U:

$$H = E_T + E_A$$

State functions for various forms of energy conserving element may then be derived from the Hamiltonian function using Legendre transformations.

We shall consider here a two-port element only for simplicity of explanation and because two port transduction represents the majority of significant elements.

The elements to be considered then are:

(a) Elements in which through energy is input at both ports—that is, normal converters coupled by the through extensity and through coupled stores. Maxwell's relations apply to a special case of such an element.

(b) Elements in which across energy is input at both ports—that is, normal converters coupled by the across extensity and across coupled stores.

Table 18.3 Maxwell's relations, energy functions, and analogs

Classical thermodynamics	Dynamics		
θ temperature	$\dot{x}_1$ $\Big\{$ Through coupled store	$\dot{y}_1$ $\Big\{$ Across coupled store	$\dot{x}_1$ $\Big\{$ Mixed coupled store
S entropy	y_1 $\Big\{$ Normal through	x_1 $\Big\{$ Normal across	y_1 $\Big\{$ Mixed coupled,
p pressure	$\dot{x}_2$ $\Big\{$ coupled	$\dot{y}_2$ $\Big\{$ coupled	$\dot{y}_2$ $\Big\{$ gyrator
V volume	y_2 $\Big\{$ converter	x_2 $\Big\{$ converter	x_2 $\Big\{$ converter
Maxwell I $\left(\dfrac{\partial \theta}{\partial V}\right) = -\left(\dfrac{\partial p}{\partial S}\right)$ $dU = \theta\, dS - p\, dV$ $U = \delta Q + \delta W$ $\quad = U(S, V)$ $\quad =$ internal energy	$\left(\dfrac{\partial \dot{x}_1}{\partial y_2}\right)_{y_1} = \left(\dfrac{\partial \dot{x}_2}{\partial y_1}\right)_{y_2}$ $dE_T = \dot{x}_1\, dy_1 + \dot{x}_2\, dy_2$ $E_T = E_{T_1} + E_{T_2}$ $\quad = E_T(y_1, y_2)$ $\quad =$ through stored energy	$\left(\dfrac{\partial \dot{y}_1}{\partial x_2}\right)_{x_1} = \left(\dfrac{\partial \dot{y}_2}{\partial x_1}\right)_{x_2}$ $dE_A = \dot{y}_1\, dx_1 + \dot{y}_2\, dx_2$ $E_A = E_{A_1} + E_{A_2}$ $\quad = E_A(x_1, x_2)$ $\quad =$ across stored energy	$\left(\dfrac{\partial \dot{x}_1}{\partial x_2}\right)_{y_1} = \left(\dfrac{\partial \dot{y}_2}{\partial y_1}\right)_{x_2}$ $dH = \dot{x}_1\, dy_1 + \dot{y}_2\, dx_2$ $H = E_{T_1} + E_{A_2}$ $\quad = H(y_1, x_2)$ $\quad =$ Hamiltonian
Maxwell II $\left(\dfrac{\partial \theta}{\partial p}\right)_s = \left(\dfrac{\partial V}{\partial S}\right)_p$ $dH = \theta\, dS + V\, dp$ $H = U + pV$ $\quad = H(S, p)$ $\quad =$ enthalpy	$\left(\dfrac{\partial \dot{x}_1}{\partial \dot{x}_2}\right)_{y_1} = -\left(\dfrac{\partial y_2}{\partial y_1}\right)_{\dot{x}_2}$ $dE = \dot{x}_1\, dy_1 - y_2\, d\dot{x}_2$ $E = E_{T_1} - E'_{T_2}$ $\quad = E(y_1, \dot{x}_2)$ $\quad =$ through stored (energy − coenergy)	$\left(\dfrac{\partial \dot{y}_1}{\partial \dot{y}_2}\right)_{x_1} = -\left(\dfrac{\partial x_2}{\partial x_1}\right)_{\dot{y}_2}$ $dE = \dot{y}_1\, dx_1 - x_2\, d\dot{y}_2$ $E = E_{A_1} - E'_{A_2}$ $\quad = E(x_1, \dot{y}_2)$ $\quad =$ across stored (energy − coenergy)	$\left(\dfrac{\partial \dot{x}_1}{\partial \dot{y}_2}\right)_{y_1} = -\left(\dfrac{\partial x_2}{\partial y_1}\right)_{\dot{y}_2}$ $dL' = -\dot{x}\, dy_1 + x_2\, d\dot{y}_2$ $L' = -H + x_2 \dot{y}_2$ $\quad = E'_{A_2} - E_{T_1}$ $\quad = L'(y_1, \dot{y}_2)$ $\quad =$ co-Lagrangian
Maxwell III $\left(\dfrac{\partial S}{\partial p}\right)_\theta = -\left(\dfrac{\partial V}{\partial \theta}\right)_p$ $dG = -S\, d\theta + V\, dp$ $G = H - \theta S$ $\quad = G(\theta, p)$ $\quad =$ Gibbs' function	$\left(\dfrac{\partial y_1}{\partial \dot{x}_2}\right)_{\dot{x}_1} = \left(\dfrac{\partial y_2}{\partial \dot{x}_1}\right)_{\dot{x}_2}$ $dE'_T = y_1\, d\dot{x}_1 + y_2\, d\dot{x}_2$ $E'_T = E'_{T_1} + E'_{T_2}$ $\quad = E'_T(\dot{x}_1, \dot{x}_2)$ $\quad =$ through stored coenergy	$\left(\dfrac{\partial x_1}{\partial \dot{y}_2}\right)_{\dot{y}_1} = \left(\dfrac{\partial x_2}{\partial \dot{y}_1}\right)_{\dot{y}_2}$ $dE'_A = x_1\, d\dot{y}_1 + x_2\, d\dot{y}_2$ $E'_A = E'_{A_1} + E'_{A_2}$ $\quad = E'_A(\dot{y}_1, \dot{y}_2)$ $\quad =$ across stored coenergy	$\left(\dfrac{\partial \dot{y}_1}{\partial \dot{y}_2}\right)_{\dot{y}_1} = \left(\dfrac{\partial x_2}{\partial \dot{x}_1}\right)_{\dot{y}_2}$ $dH' = y_1\, d\dot{x}_1 + x_2\, d\dot{y}_2$ $H' = E'_{T_1} + E'_{A_2}$ $\quad = H'(\dot{x}_1, \dot{y}_2)$ $\quad =$ co-Hamiltonian
Maxwell IV $\left(\dfrac{\partial S}{\partial V}\right)_\theta = \left(\dfrac{\partial p}{\partial \theta}\right)_V$ $dF = -S\, d\theta - p\, dV$ $\quad = H + \delta W$ $\quad = F(\theta, V)$ $\quad =$ Helmholtz free energy	$\left(\dfrac{\partial y_1}{\partial y_2}\right)_{\dot{x}_1} = -\left(\dfrac{\partial \dot{x}_2}{\partial \dot{x}_1}\right)_{y_2}$ $dE = -y_1\, d\dot{x}_1 + \dot{x}_2\, dy_2$ $E = E_{T_2} - E'_{T_1}$ $\quad = E(y_2, \dot{x}_1)$ $\quad =$ through stored (energy − coenergy)	$\left(\dfrac{\partial x_1}{\partial x_2}\right)_{\dot{y}_1} = -\left(\dfrac{\partial \dot{y}_2}{\partial \dot{y}_1}\right)_{\dot{x}_2}$ $dE = -x_1\, d\dot{y}_1 + \dot{y}_2\, dx_2$ $E = E_{A_2} - E'_{A_1}$ $\quad = E(x_2, \dot{y}_1)$ $\quad =$ across stored (energy − coenergy)	$\left(\dfrac{\partial y_1}{\partial x_2}\right)_{\dot{x}_1} = \left(\dfrac{\partial \dot{y}_2}{\partial \dot{x}_1}\right)_{x_2}$ $dL = y_1\, d\dot{x}_1 - \dot{y}_2\, dx_2$ $L = -H + \dot{x}_1 y_2$ $\quad = E'_T - E_{A_2}$ $\quad = L(\dot{x}_1, x_2)$ $\quad =$ Lagrangian

(c) Elements in which through energy is input at one port and across at the other; that is gyrating converters and mixed coupled stores.

Table 18.3 gives the relations between the transuction coefficients, together with their derivations, by analogy to the thermodynamic Maxwell relations.

It also shows the relation between the thermodynamic state functions and the state functions of classical dynamics: the Hamiltonian $H = E_A + E_T$, the co-Hamiltonian $H' = E'_A + E'_T$, the Lagrangian $L = E'_T - E_A$ and co-Lagrangian $L' = E'_A - E_T$.

18.5.3 Examples and Application

Application of the above generalization can be explained by examples.

As already explained, Maxwell's relations give the connections between the transduction coefficients of the thermal fluid transducer (see Section 18.4.6.2) based on a through coupled store.

For converters, both normal and gyrating, the generalized Maxwell's relations express the connection between the transduction coefficients which emerge from the definition given in Section 18.4.2.1.

For coupled stores, however, the relations are most significant.

For the across coupled store (an iron core in a coil) considered in Section 18.4.6.2 we can (from row 2 column 3 of Table 18.3) conclude that, for the transuction coefficients:

$$\left(\frac{\partial f}{\partial i}\right)_x = -\left(\frac{\partial \lambda}{\partial x}\right)_i$$

For the mixed coupled store (a moving plate capacitor) considered in Section 18.4.6.3 we see, from row 4, column 4 of Table 18.3, that

$$\left(\frac{\partial q}{\partial x}\right)_v = -\left(\frac{\partial f}{\partial v}\right)_x$$

The transduction coefficient between force and voltage can then be deduced from the relations between charge and plate separation, that is, the equation for the capacitance C in terms of x.

The relations enable us then to deduce one transduction coefficient from the other. Only one needs to be deduced theoretically or measured experimentally.

18.6 INSTRUMENT ELEMENT FUNCTIONING IN A SYSTEM

We shall consider, by means of two examples, the functioning of an instrument element in a system. The examples taken will be an electrodynamic element illustrating the functioning of a converter and a potential divider illustrating the functioning of a controlled element.

18.6.1 Electrodynamic Element

A model of the element in ideal form has been discussed in Section 18.4.2.1.

Functionally the transducer can be used either to transform a current input into a torque output or to transform a velocity input into a voltage output maintaining a prescribed relation between input and output. For the purpose of the example we shall consider the *velocity to voltage transformation*, the transducer being used essentially as a tachometer element. The following refers to Figure 18.16.

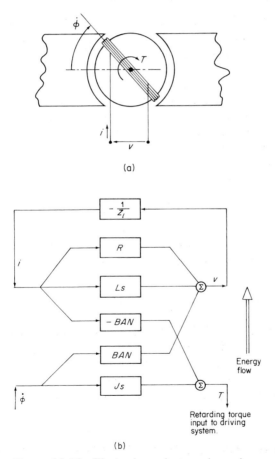

Figure 18.16 Electrodynamic transducer in a system: (a) schematic; (b) small-signal flow diagram

We shall now write the terminal equations of the instrument in the s domain as:

$$\begin{bmatrix} T(s) \\ v(s) \end{bmatrix} = \begin{bmatrix} Z_{11}(s) & Z_{12}(s) \\ Z_{21}(s) & Z_{22}(s) \end{bmatrix} \begin{bmatrix} \dot{\phi}(s) \\ i(s) \end{bmatrix}$$

For a non-ideal element having an appreciable rotor inertia J, as well as coil self-inductance L, resistance R, we have

$$\begin{bmatrix} T(s) \\ v(s) \end{bmatrix} = \begin{bmatrix} Js & -BAN \\ BAN & Ls + R \end{bmatrix} \begin{bmatrix} \dot{\phi}(s) \\ i(s) \end{bmatrix}$$

The functional meaning of the matrix elements is

$$\text{transduction coefficients} = -Z_{12} = Z_{21} = BAN$$

$$\text{mechanical impendance on open circuit} = Z_{11} = Js$$

$$\text{electrical impedance with clamped rotor} = Z_{22} = Ls + R$$

We may note that in this application the *flow of energy* is from the mechanical to the electrical port, so that there is a mechanical power input and an electrical power output. In terms of *information-carrying* signals, however, the input is indeed the velocity impressed at the mechanical port, but torque, (the other variable at the mechanical port) is an output variable. At the electrical port voltage is the output signal; current is, however, independently determined by the electrical impedance of the *load* on the transducer.

Let us now consider that the transducer forms part of a supersystem. Let the electrical impedance of the load on the transducer by given by Z_l.

Then we have

$$v = -Z_l i$$

The object of the instrument is to realize a prescribed transformation of $\dot{\phi}$ into v. When no power is abstracted from the instrument, this is given by

$$v = Z_{21} \dot{\phi}$$

or

$$v = BAN \dot{\phi}$$

However, when the supersystem is connected and abstracts power, we have:

$$v = \frac{Z_{21}}{1 + Z_{22}/Z_l} \dot{\phi}$$

If Z_l is a resistive load, say R_l, we have:

$$v = \frac{BAN}{1 + (Ls + R)/R_l} \dot{\phi}$$

The effect of the load imposed by the supersystem is to reduce the sensitivity and to introduce a dynamic lag due to L. The less Z_{22} and the greater Z_l, the less the effect.

Now, as stated above, the transducer will provide a torque output T, which will act in such a way as to oppose the rotation of the mechanical system driving the transducer.

With no energy output to the supersystem:

$$T = Z_{11}\dot{\phi}$$

or

$$T = Js\dot{\phi}$$

With power abstracted at the signal output port we have

$$T = \left(Z_{11} - \frac{Z_{12}Z_{21}}{Z_{22} + Z_l}\right)\dot{\phi}$$

or

$$T = \left(Js + \frac{B^2 A^2 N^2}{Ls + R + R_l}\right)\dot{\phi}$$

There is thus a retarding torque produced by the inertial effects and an additional retarding torque generated due to the power output to the supersystem. The additional torque is determined by

$$\frac{B^2 A^2 N^2}{Ls + R + R_l}$$

The degree to which the retarding torque affects the driving system depends on the driving capacity of the latter.

We can examine briefly the way in which the model demonstrates the factors governing the dynamic response of the instrument. We see that, in principle, no lag is introduced between velocity and voltage by the transduction. However, there is a lag on load due to the effect of the coil and impedance. We can also see that the inertia of the rotor implies that the retarding torque increases proportionally with frequency, so that the rapidity of the changes of $\dot{\phi}$ is limited by available mechanical driving power.

Thus we conclude in general that in a converter element the power output at the signal port must be supplied from the signal input port and affects the system from which power is taken, that is, in the case of a transducer the system under measurement on which it generally acts as a disturbance. The variables conjugate to the information-carrying signal at a port act as a disturbance and their effects must be kept small. In the case of a converter, storage and dissipation at the signal port enhance the disturbance.

18.6.2 Potential Divider Element

To illustrate the functioning of a controlled element we shall consider one of the most common forms of such an element used in instrumentation—the potential divider shown in Figure 18.17. It can be considered as consisting of two-ganged controlled resistors.

The terminal equations of the element can be written as

$$v_1 = Ri_1 + R\frac{x}{l}i_2$$

$$v_2 = R\frac{x}{l}i_1 + R\frac{x}{l}i_2$$

$$f = M\frac{d\dot{x}}{dt}$$

where R is the resistance of the track, M the mass of the slider, and l the length of track.

The friction at the slider is neglected for simplicity of explanation and only its mass considered.

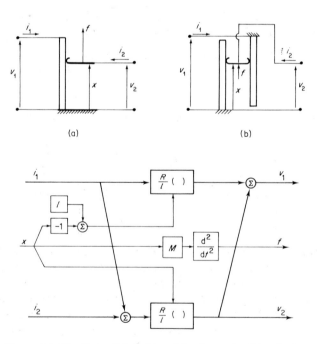

(a)   (b)

Figure 18.17   Potential divider: (a) schematic; (b) principle of potentiometric transducer as two controlled resistors; (c) signal flow diagram (If $i_2 = 0$, $v_2 = i_1(R/l)x$)

The function of the element as a transducer is to transform an input displacement of the slider into a proportional voltage. To perform this, the energizing current i, must be maintained constant and the current i_2 must be small compared with i_1 giving the signal transformation relation:

$$v_2 = \left(\frac{R}{l}\bar{i}_1\right)x$$

The transduction coefficient is a function of the energizing current $\bar{i}_1$.

If the energy taken out at the output port is appreciable we can analyse its effect in terms of a load, say a resistance R_e connected at the output port.

Then we have by Kirchhoff's laws

$$v_2 = -R_e i_2$$

and

$$v_2 = \frac{(R/l)i_1 x}{1 + (R/R_e)(x/l)}$$

The effect of substantial energy abstracted at the ouput port, that is of R_e being small compared with R, is to make the relation between v_2 and x non-linear.

One can conveniently discuss the functioning of the element in terms of its small-signal model which is, in Laplace transform form,

$$\begin{bmatrix} v_1(s) \\ v_2(s) \\ f(s) \end{bmatrix} = \begin{bmatrix} Z_{11}(s) & Z_{12}(s) & Z_{13}(s) \\ Z_{21}(s) & Z_{22}(s) & Z_{23}(s) \\ Z_{31}(s) & Z_{32}(s) & Z_{33}(s) \end{bmatrix} \begin{bmatrix} i_1(s) \\ i_2(s) \\ \dot{x}(s) \end{bmatrix}$$

In terms of the construction parameters of the element we have

$$\begin{bmatrix} v_1(s) \\ v_2(s) \\ f(s) \end{bmatrix} = \begin{bmatrix} R & \dfrac{R}{l}\bar{x} & \dfrac{R}{l}\bar{i}_2\dfrac{1}{s} \\ \dfrac{R}{l}\bar{x} & \dfrac{R}{l}\bar{x} & \dfrac{R}{l}(\bar{i}_1 + \bar{i}_2)\dfrac{1}{s} \\ 0 & 0 & Ms \end{bmatrix} \begin{bmatrix} i_1(s) \\ i_2(s) \\ \dot{x}(s) \end{bmatrix}$$

In a controlled element such as this the energy abstracted at the signal output does not affect the signal input port as can be seen from the fact that $Z_{31} = Z_{32} = 0$. The power is supplied from the energizing port and is determined by impedances Z_{12} and Z_{13}. The smaller the impedances (that is, in effect the smaller R/l) the smaller the effect on the energizing supply.

The transduction coefficient which relates $\dot{x}$ and v_2 is given by Z_{23}.

Z_{21} determines the effect on the output signal v_2 of i_1, the change of energizing current from the operating point $\bar{i}_1$. It must be kept small and hence R/l should be small.

Z_{22} determines the effect on the output signal v_2 of the changes i_2 of output current. Again it should be kept small and hence R/l should be small.

Z_{23} is the transduction coefficient: $\bar{i}_2$ must be effectively kept small and $(R/l)\bar{i}_1$ large.

18.7   ERROR MODELS

18.7.1   General

Errors in transducers arise from two sources:

(a)   Influence variables—that is, physical variables other than the input signal, which affect the output signal and thus act as a disturbance.

(b)   Inherent changes—that is, variations of the parameters of the transducer due to effect of ageing, wear, damage, and so on.

The effects can be examined by a study of models of instrument elements.

18.7.2   Influence Effects from the Supersystem

We must first consider as influences the variables conjugate to the signals at the input and output port. We can generalize the consideration of Section 18.6.

At the input port, the signal conjugate to the signal is a dependent variable determined by the instrument characteristics and, for some devices, power output. The variable acts on the system to which the instrument is connected and may thus alter the input signal. It is desirable, other things being equal, that this variable be minimized. Another way of looking at it is to require that power demand at the input signal port be minimized. In a converter, that power supplies the output signal power and the stored and dissipated energy at the port. In a controlled element the input signal power supplies only energy stored or dissipated at the signal port. In a converter the input power required may then be minimized by minimization of output power demand. For all elements the stored and dissipated power at the port is described by the input impedance which characterizes these effects. It is desirable that these effects be minimized. If the input signal is an effort, then to minimize the conjugate variable the input impedance should be maximized. If the input signal is a flow, then the input impedance should be minimized.

At the output port the variable conjugate to the output signal is an independent variable determined by the system connected at that port. It affects the output signal and thus is an influence variable. It is desirable that, other things being equal, it should be minimized. In other words, to minimize the influence of the output signal, output power should be minimized. Output power effects in a converter were considered above. In a controlled element, output power is supplied from the energizing source which may be affected by power demanded from it with resultant errors. Output power affects the output signal through storage and dissipation effects at the output signal port which should be minimized. They are characterized by the output impedance. If the output signal is an effort, the effect on it of the conjugate variable is minimized by minimizing the output impedance. If the output signal is a flow, then the output impedance should be maximized.

Further consider energization in controlled elements as considered in Section 18.6.2. In such elements the functional relation between input and output signal depends on an energizing variable. Variations of the energizing variable may be considered as in influence variable. The energization may vary either as a result of changes in the energizing source, or as a result of effects on it of variations of power demand caused by output signal power. The effects are minimized by making the energizing source as nearly ideal as possible and the output signal power small compared with energizing power.

18.7.3 Influence Effects from the Environment

Instrument elements are affected by actions from the environment.

We may distinguish two kinds: those acting at a port, *additive* to signals acting at the port, and those *modifying* geometry or material properties of elements and thus in effect acting as additional controlling variables.

To illustrate these we may refer to the example of a spring discussed in Section 18.4.1.2 and shown in Figure 18.4. We now analyse the functional spring model and consider that the spring, the upper end of which is considered fixed under reference conditions, moves under the effect of a vibration. Let x' represent the movement and let it be taken in the same direction as x. Then

$$f = k(x - x')$$

If the element is used to transform a force into a displacement, then we have a nominal force displacement relation under reference conditions

$$x = \frac{1}{k} f$$

With the effect of the influence we have

$$x = x' + \frac{1}{k} f$$

x' is associated with the same port as f, and acts as an additive influence error.

If we now consider the effect of temperature on the spring, we may in the first instance analyse the functional model, assuming that the constant k is a function of temperature $k(\theta)$. Under reference conditions $\bar{\theta}$, the force displacement relation is

$$x = \frac{1}{k(\bar{\theta})} f$$

If the temperature changes by a small amount θ from the reference conditions we have

$$x = \frac{1}{k(\bar{\theta})} f - \frac{1}{k^2(\bar{\theta})} \frac{\partial k}{\partial \theta} \theta f$$

The influence error is then

$$-\frac{1}{k^2(\bar{\theta})} \frac{\partial k}{\partial \theta} \theta f$$

Note that the effects of θ and f are *multiplicative*.

18.7.4 Inherent Effects

If we now consider inherent error effects in the resistor discussed in Section 18.4.2.2 and shown in Figure 18.7 let the resistor be a function of geometrical and material parameters $p_1, p_2, \ldots, p_n$. Consider further that these parameters change from nominal conditions $\bar{p}_1, \bar{p}_2, \ldots, \bar{p}_n$ by $p_1, p_2, \ldots, p_n$. Then at nominal conditions

$$v = R(p_1, p_2, \ldots, p_n) i$$

After change of parameters we have

$$v = R(p_1, p_2, \ldots, p_n) i + \left(\frac{\partial R}{\partial p_1} p_1 + \frac{\partial R}{\partial p_2} p_2 + \cdots \frac{\partial R}{\partial p_n} p_n \right) i$$

where the second term is the inherent error.

For the archetype model of the realization of the resistor illustrated we have

$$v = \frac{\bar{p}\bar{l}}{\bar{A}} i + \frac{\bar{p}\bar{i}}{\bar{A}} \left(\frac{p}{\bar{p}} + \frac{l}{\bar{l}} - \frac{A}{\bar{A}} \right) i$$

Note that the inherent error is a function of the input signal. We may compare it with the effect of the temperature effect on the spring which is essentially similar in form.

REFERENCES

Abdullah, F., Finkelstein, L., and Rahman, M. M. (1977). "The application of mathematical models in the evaluation and design of electromechanical instrument transducers', *J. Appl. Sci. Engng*, **A2**, 3–26.

van Dixhoorn, J. J. and Evans, F. J. (1974). *Physical Structure in Systems Theory*, Academic, London.

Finkelstein, L. and Watts, R. D. (1969). 'Measurement as a systematic study', *Measurement Education:IEE Conf. Publ. No. 56*, pp. 101–5.

Finkelstein, L. and Watts, R. D. (1978a). 'Mathematical models of instruments—fundamental principles', *J. Phys. E: Sci. Instrum.*, *11*, 841–55 (Reprinted in 1982, B. E. Jones (Ed.), *Instrument Science and Technology*, Vol. 1, Adam Hilger, Bristol, pp. 9–27.)

Finkelstein, L. and Watts, R. D. (1978b). 'Control elements as basic elements of physical systems', *Systems Structures in Engineering* O. Bjørke and O. I. Franksen (Eds), Trondheim, Tapir, Norway.

Karnopp, D. and Rosenberg, R. C. (1968). *Analysis and Simulation of Multiport Systems: the Bond Graph Approach to Physical System Dynamics* MIT, Cambridge, Mass.

Karnopp, D. and Rosenberg, R. C. (1975). *System Dynamics: A Unified Approach*, Wiley-Interscience, New York.

MacFarlane, A. G. J. (1964). *Engineering Systems Analysis*, Harrap, London.

MacFarlane, A. J. J. (1970). *Dynamic Systems Models*, Harrap, London.

Shearer, J. L., Murphy, A. J. and Richardson, H. H. (1967). *Introduction to System Dynamics*, Addision-Wesley, Reading, Mass.

Paynter, H. M. (1961). *Analysis and Design of Engineering Systems*, MIT, Cambridge, Mass.

Watts, R. D. (1966). 'The elements of design', *The Design Method*, S. A. Gregory (Ed.), Butterworths, London, pp 85–95.

Watts, R. D. (1968). 'Systematic design procedures', *Electronic Design: IEE Conf. Publ. No 45*, pp. 240–57.

Watts, R. D. (1981). 'Maxwell's relations: instrument transduction coefficients: analogies', *Department of Systems Science, The City University, Res. Mem. No. DSS/RDW/222*.

Handbook of Measurement Science, Volume 2
Edited by P. H. Sydenham
© 1983 John Wiley & Sons Ltd

Chapter

19 D. BOSMAN

Human Factors in Display Design

Editorial introduction

It must not be overlooked that in situations where the measurement output data are conveyed to an human observer, that person forms a cascaded link in the measurement chain. The observer greatly decides, as his *inner representation*, the meaning to ascribe to the data presented and, therefore, is a key part of the mapping process that underlies all measurements.

This chapter deals with this measurement systems building block, the emphasis being upon the visual display method of the man–machine interface.

19.1 INTRODUCTION

Many measurands in the scientific and technical fields require instrumental intervention in order to be properly perceived by man. The process of measurement is carried out with the aid of a specific structure of functions, designed to comply with particular needs. The functions include signal acquisition, conditioning, conversion, transmission, processing, and last, but not least, display of data; each function performing a transformation such that the distortion of the information carried by the signal remains within accepted limits. The display function is to bring about the final transformation, its output conforming to among others the psycho-physical requirements of man's senses.

However, the displayed data only convey information insofar as they contribute to the state of knowing of an apprehending mind. Therefore, the total transformation should also meet requirements of cognitive nature. The equipment must facilitate a mapping of the object or process to be measured or controlled, into an inner representation of same in the mind of the observer or operator. Constraints imposed by the equipment on monitoring and control strategies, which may be dependent on prevailing conditions in, for instance, the process, in the environment or in the operator, may also distort perceived information to the point that proper identification or even recognition (of changes of state in the process to be monitored) is hampered.

Thus, the machine–man interface (MMI) includes both perceptive and task aspects which need to be carefully designed, just as inter-instrument interfaces are.

More than eighty per cent of data display is in visual form, because the eye is a highly developed sensor, better supported by the brain than all other senses (Cornsweet, 1970; Fogel, 1967). It affords sufficient resolution and linearity to accommodate such refined concepts as the ratio scale, see Chapter 1 (Finkelstein, 1974; Suppes and Zinnes, 1963).

Other forms of display, for example, auditory and tactile, have severe shortcomings in this respect. Even with today's proliferation of displayed variables, audition of symbolic signals other than the spoken word has very limited application: usually only as a non-directional alarm signal or, at most, in a back-up mode for a low resolution nominal or ordinal scale. Similarly, kinesthetic and touch stimuli are used for warning (stick shaker in aircraft) and for confirmation or identification of a selected control knob or handle: tactile coding (Moore, 1976; Shackel, 1974; Woodson and Conover, 1973; Chapanis and Kinkade, 1972).

Table 19.1 shows a number of important design factors for visual displays. This chapter discusses some of the listed human (or *ergonomic*) factors, under the assumption that the instrumental factors are given sufficient attention elsewhere. Unlike many other texts concerned with visual displays, the emphasis is here on (visual) perception of the behaviour of a process, rather than being limited to ophthalmic factors involved.

There exists considerable authoritative literature on the psycho-physical and anthropometric characteristics and constraints of man. Handbooks (McCormick, 1970; Van Cott and Kinkade, 1972; Woodson and Conover, 1973; Kraiss and Moraal, 1976) and textbooks, such as Poole (1966) and Luxenberg and Kuehn (1968), provide most of the required design data.

The perception of the value of a displayed variable depends on many factors: remembered values and measurand behaviour, known relations to other variables, expected process operation, and the like. Thus, apart from the psycho-physical factors, also task-related factors have influence: in Table 19.1 a number of important task aspects are included. In this chapter one section is devoted to explain some important notions of task design and task experience. In this area much is as yet insufficiently understood to give explicit rules or advice, the reader is referred to the literature on that subject.

The person operating or monitoring a process has an inner representation of the operation of that process and of his own task therein. Its influence on

Table 19.1   Machine–man communication by visual means

Instrumental factors	Human factors

Scale aspects
metrical information
  scale type: nominal, ordinal, inter-
  val, ratio, cardinal
  format: symbolic, bar graph, pointer
  and dial
structural information
  format: pictorial, graphic, cell or-
  ganized, symbolic, dynamics
	Cognitive aspects
	process architecture ⎫
	measurand dynamics ⎬ mental model
	image structure (pattern), data struc-
	  ture
	functional groups
	trending

Presentation
display configuration
  size, viewing angle
  sequential vs parallel
  duration of presentation
  special effects (blinking
  picture or scale redundancy)
properties
  foreground, background luminance
  contrast, resolution
  colour: hue, saturation
  raster, dot matrix, random scan
  refresh rate, persistence
	Psycho-physical aspects
	legibility
	  visual acuity
	  brightness
	  contrast discrimination
	  fotopic or mesopic vision
	temporal characteristics
	  adaptation
	  critical flicker frequency, strobo-
	    scopic effects
	  peripheral response
	miscellaneous
	  focus range
	  colour perception
	  Gestalt laws
	  expectancies
	  stress
	  pathologies
	  vigilance

Quality aspects signal-to-noise ratio MTF (blur, smearing) cross talk angle of visibility reflection symbol or graph continuity	*Anthropometric aspects* sizes, distances field of view posture
	Task aspects complexity, number and frequency of decisions and prorities tempo repetition rate training

the accuracy of his perception is humorously interpreted by Bo Bojesen (Figure 19.1a). Some other interpretations are to be seen in Figure 19.1b and in McKay (1963), Smallwood (1967), Bainbridge (1969, 1974), Rasmussen (1974), Kalsbeek (1978).

One important reason that suboptimality in this respect of MMI design often does not show in the performance of the system, is the great flexibility

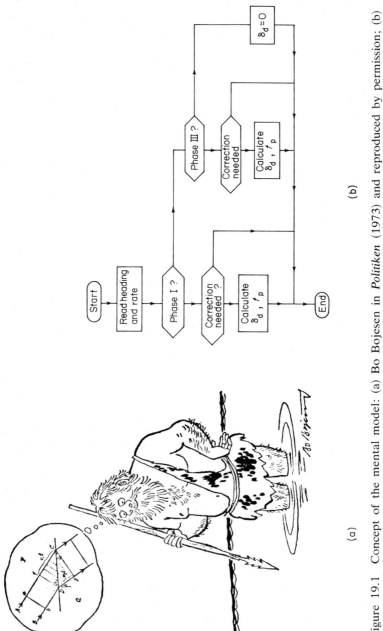

(a)

(b)

Figure 19.1   Concept of the mental model: (a) Bo Bojesen in *Politiken* (1973) and reproduced by permission; (b) flowchart of the decision making element (Stassen and Veldhuyzen (1976), reproduced by permission of Plenum Publishing Corp.)

and adaptability of the human observer or controller, which when combined with his internal feedback can selectively improve the act of perception. In the long run, too much reliance on this human faculty implies a risk for the operation of the *man–machine system* (MMS) and for the mental health of the operator.

To maintain a specified integrity, the magnitude and the frequency of the errors must remain within tolerance limits, notwithstanding foreseeable changes in equipment properties and environmental conditions and/or small, unintended, deviations in procedures, algorithms, and programs. With respect to the latter, integrity sometimes calls for fail-safe or even safe-life operation (Section 19.2).

Error sources control is well developed in the technical engineering of systems; it is also quite practicable for the psycho-physical aspect of a measurement task relating to such things as observation errors (Section 19.3), and for the anthropometric aspect contributing to operations errors (Section 19.4), see also Meister (1971) and Swain (1970) on human reliability.

19.2 TASK, ARCHITECTURE, INNER REPRESENTATION

For very simple instruments and systems, explicit design of the measurement activity or task may not be worth bothering about, except when the task requires very fast response or is of a highly repetitive nature. In that case one should consider whether automation is the more humane solution. In more complex measurement and control situations, such as exist in many sophisticated scientific experiments, in the process industry, in aircraft operations, and the like, proper design of tasks can be essential to system integrity and to how the operator experiences his task.

In a process wherein the operations are the result of the combined action of equipment (parts) and man, the designer must decide which activities (not necessarily functions in the engineering sense) shall be allocated to the machine and which to the operator (Rijnsdorp and Rouse, 1977; Moraal, 1976; Singleton, 1974; Woodson and Conover, 1973; Jordan, 1963). Many aspects are involved in that choice, for example, man's dexterity, vulnerability, capability, wellbeing. Fitts (1951) provides a list of statements which compare man with machine. Updated versions can be found in many ergonomic handbooks (Moraal, 1976; Woodson and Conover, 1973; Van Cott and Kinkade, 1972; McCormick, 1970). Many MMS simply are improvements over previous designs so that function allocation analysis is quite practicable and operator experience can or should be included.

Activities are not isolated events, they pertain to the processes they operate on, and thus they are interrelated to form an *activity structure*. This is the network of all activities, structured in combinatorial and sequential fashion in both the equipment (as embodied in machine operations) and in the personnel (monitoring and control actions, data processing, equipment handling).

In the sequential sense, an ordering can be distinguished between programs, algorithms, and procedures. The first two are not difficult to design: the program being a mere sequence of actions and events, the algorithm moreover including feedback loops and decision points where branching into two or more alternative actions is allowed, albeit only those governed by recently acquired facts. However neither programs nor algorithms are 'human-friendly'. Jordan (1963) states: 'Men are flexible, but cannot be depended upon to perform in a consistent manner'. By contrast, the procedure provides ample opportunity for utilization of man's flexibility and ability to handle unforeseen events. The procedure is a mode of conducting business or a specific action (Fowler and Fowler, 1963). At one or more branching points of the activity network, some form of intelligence (usually the human mind) provides unforeseen information in addition to, or even *in lieu* of, recently acquired facts which are logically derived from the process itself. Thus the procedure can account for the unpredictable, for unforeseeable (by the designer!) clues. In MMI design, charting such tasks may be done following Folley (1960); examples of sequential analysis are given in Kidd and Van Cott (1972), Döring (1976), and in Figure 19.2 which depicts a programme, in this case the foreseeable part of a procedure, of an air traffic controller (Noll *et al.*, 1972).

Being involved in measurement and control, the operator perceives controls and displays in particular settings and states. Different shapes, sizes, colours, and other presentation aspects (see Section 19.3) assist in the identification of partial functions of controls and displays, thus aiding the execution of the task. The appearance, the apparent operation of machines and equipment, the interrelations between them, the controls and displays and their location, all add up to a *sensory architecture* of the system (Bosman, 1978). Usually not user-programmable (except for some forms of electronically programmable displays), this architecture is designed-in and therefore is mostly invariant to the operator. That is also true for those activities allocated to him, which are designed to complement and to control machine operations; although he must, to a certain extent, be allowed to carry them out his own way. The inclusion of so-called *safe-life* operations, that is ruggedized equipment and overtrained operators who never err or fail, is not operator-friendly and, to say the least, very fictitious (Jordan, 1963). It is better to design *fail-safe* procedures, with equipment allowing sufficient time to correct for human errors and to replace faulty equipment well before the situation may become critical.

During training and in early stages of actual operation the operator needs frequent feedback about the states of the MMS, with display of

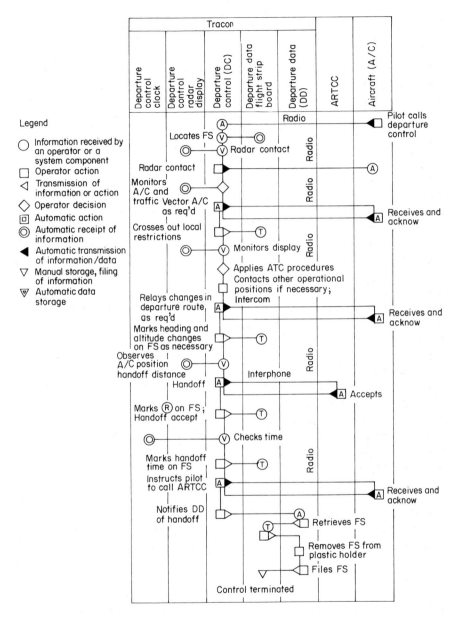

Figure 19.2 Operational sequence for Instrument Flight Rules IFR departure (Reproduced by permission of Advisory Group for Aerospace Research and Development)

back-up information to promote confidence. Gradually the function of composite activities in the operator's task and their importance relative to the process are apprehended to the point where the operator takes short-cuts. He starts to concatenate simple activities to form longer procedural strings, ignoring detailed back-up data. Also the dynamics of the MMS become sufficiently familiar to use trend data in connection to past observations combined with expected plant behaviour. These facts again influence the accuracy of perception.

Such functional aspects perceivable in the networks of activities allocated to the operator, including his own monitoring control strategies, form the *functional architecture* of the MMS. Transparency of the functional architecture allows for quick familiarization and, more important, promotes insight as to the correct decisions in the procedures under abnormal circumstances or untrained, critical conditions. In Figure 19.3, these notions are depicted in a block diagram starting from the activity analysis.

Albeit also designed-in, the functional architecture can and should be less rigid to the operator than is the sensory architecture; for instance, the trainee requires more frequent feedback (displays should allow for that), while the experienced operator relies more on trending. Similarly, it should

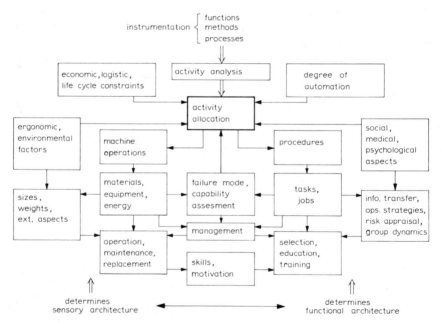

Figure 19.3   General synthesis diagram with activity allocation in the central position

accommodate differences between operators depending, among other considerations, upon education, previous experience, dexterity, training efficacy, and developmental stage of the inner representation related to the monitoring, control task in the particular setting.

It is postulated that, in the early stage of formation of the inner representation, the operator learns and uses the sensory architecture to assess, step-by-step, the state of the equipment in order to guide his own action. The functional architecture is the more difficult and takes longer to master, because of its pronounced sequential character combined with the dynamic characteristics of the equipment. The designer should take that into consideration: for example, strings of activities such as occur in a program, should be not too long without being interruptable in order to provide opportunities for quality monitoring and internal feedback. At such interrupt points the equipment should be relatively stable, that is, it should not run towards a critical state but provide fail-safe operation.

Furthermore, activities should not be in conflict with stereotypes (Moore, 1976; Shackel, 1974; Woodson et al., 1972). Finally, programs and algorithms, embedded in procedures, should never lead to a conflict, however improbable, with possible states in the environment or within the equipment itself unless an emergency procedure exists which provides for, say, safe shutdown.

Summarizing, the display system is a compound communication channel which maps process properties (that is, functional architecture, behaviour, state) into the perceptive system of the operator (Bosman and Umbach, 1978). As such it is a window on the process to be monitored and/or controlled and should provide the operator with sufficient insight about the most likely activity to occur next, or to be performed by him. To convey such notions the display system should not only accommodate such properties as range, accuracy, and slew rate but also the properties of the process as a whole: for instance among others the functional architecture.

19.3 DISPLAY AND PERCEPTION

19.3.1 Perception of Displays

Perceptual capabilities are rate-limited. The channel capacity of visual auditory senses for different unidimensional stimuli (identification of one out of many discriminable categories, that is, decision of combinatorial character) where single motoric responses are required, varies between 2.5 and 3.5 bits/second, occasionally up to 7 (Van Cott and Warrick, 1972). When the motoric response is not the limiting factor, and for multidimensional stimuli, it is reported that the upper limit may reach even 40 to 50 bits/second, this being under the premise that the stimuli map into a fully developed and rich inner representation descriptive of the sources emanating the stimuli.

Because human data rate capacity varies so much with psycho-physical function, with training and experience, with architectural transparency of the process to be monitored or controlled, and so forth, the designer of instrumentation is advised either to keep to the low end of the scale, or to request expert advice in this matter. The latter becomes more important because computer applications in combination with visual display units (VDU's) can easily, and often unnecessarily, burden the operator with long duration tasks involving internal processing at high data rates, even if the required response rate remains rather low.

19.3.2   Response to Displays

The response time (RT) in the reading of visual displays with a simple response (the actuation of a button), was experimentally shown to be between 0.2 and 1 s, with an average of 0.6 s, under given experimental conditions with regard to such variables as display size, viewing time, brightness and contrast, resolution, and image motion.

Other experiments with reading random numbers have also shown that for analog dials the mean reading time increases logarithmically with precision required, whereas reading time is independent of precision for digital presentation (Nason and Bennett, 1973). With a scale divided in 100 discernible intervals (6.7 bits), and process dynamics limiting the change of indication between readings to, say, 10% of that range, the actual reading uncertainty is about 3.3 bits, providing that this 10% interval of the scale does not extend much beyond the foveal angle of the detection lobe (see Figure 19.10 presented later). Since the foveal angle (field of view (FOV) of high resolution and colour perception) is only about 2° of arc, at a 50 cm viewing distance the area of sharp viewing has a diameter of about 2 cm. Reading an instrument with a face of, say, 8 cm requires a number of eye fixations to locate the exact position of the pointer relative to the dial (see Figure 19.4, from White and Ford, 1960).

Each fixation takes time, the *dwell* time, which varies between 0.25 s and 1 s depending on the circumstances (dial design, illumination, contrast, change of indication) (see Fitts *et al.*, 1950; McRuer *et al.*, 1967). Remembrance of the previous position of a pointer can reduce the number of fixations required for precise reading. Allowing in the example for three fixations of average 0.35 s during viewing time, the effective data rate, including response time would be 3.3/1.7 = 2 bits/second.

In the multi-instrument display situation, remembrance of previous pointer positions may also increase instrument reading frequency. In dynamic situations, such as approach and landing of transport-type aircraft

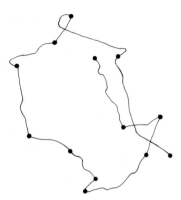

Figure 19.4 Slow muscular
scanning in visual search

(Fitts *et al.*, 1950), an average instrument reading frequency of 0.25 per second was experienced. Presumably, these figures are also representative for processes other than flying, since human capability already is a significant limiting factor.

19.3.3 Optimal Conditions for Displays

Optimal conditions for the human senses are very variable (Cornsweet, 1970; Poole, 1966). Visual acuity varies with background luminance and contrast. It is expressed in milliradians (mrad) or minutes of arc. Resolving power is, by normal vision (100%), about 0.3 mrad at high background luminance levels and with a contrast ratio (dark-on-light) better than 20% (see Figure 19.5a). Focus range varies with luminance, colour, and age (Kraiss, 1976; see Figure 19.5b).

Through the resolving power and the distance to the display, the minimum dimensions of alphanumeric characters and scale markings are determined. Minimum reading distance must be compatible with writing distance (if writing or keying is required), that is from 30 to 40 cm. That means that average people older than 40 shall need glasses if minimum dimensions of important markings indeed conform to 0.3 mrad. Increasing the size, and distance if possible, can mean more comfortable reading for elderly people. Another reason not to miniaturize down to the limit is the spatial frequency response (*modulation transfer function* (MTF)) of the eye (see Figure 19.6; Cornsweet, 1970).

It shows that the contrast sensitivity is attenuated considerably at line widths smaller than, say, 1 minute of arc (beyond 30 cycles per degree). Furthermore, the brightness sensation varies logarithmically with luminance

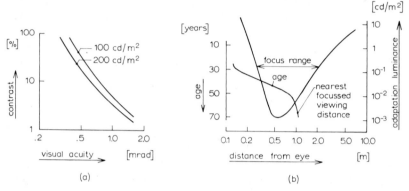

Figure 19.5   Discriminable size and focus range (Reproduced by permission of Heyden Book Co. Inc. and Verlag TÜV Rheinland)

(Cornsweet, 1970) so that moderate light levels of, say, 50 cd/m² can be accommodated with occasional variations between 5 cd/m² and 500 cd/m².

Murrell (1969) has shown that the human observer can interpolate quite accurately into five intervals, which gives the possibility to remove scale clutter caused by too many markings. All this leads to recommendations for symbolic displays such as found in national standards and in literature (Kraiss, 1976; Bottomley, 1976; Shackel, 1974; Grether and Baker, 1972; Woodson and Conover, 1973; McCormick, 1970). For instance, identifiable intervals on a scale should occupy at least 3 mrad; major markings, black-on-white should be 0.3 mrad wide and 10 mrad long, with intermediate markings also 0.3 mrad wide and 6 mrad long, and minor markings 0.2 mrad wide and 4 mrad long. For 50 cm viewing distance, scale dimensions conforming to these recommendations are as shown in Figure 19.7.

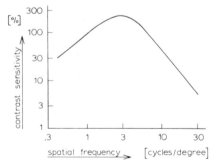

Figure 19.6   MTF of the eye for lower luminance levels

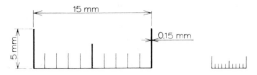

Figure 19.7 Recommended scale dimensions for high illumination level

19.3.4 Symbolic Indicators

Symbolic indicators come in types (Figure 19.8), each of which has a certain advantage in some applications, being less suitable for others. A good evaluation is given in Grether and Baker (1972). This is summarized in Table 19.2 and in Pearce and Shackel (1979).

The fact that suitably structured complex stimuli score higher data rates than arrays of instruments each displaying a single number is an important indication that proper preprocessing and formatting of the data, to fit the

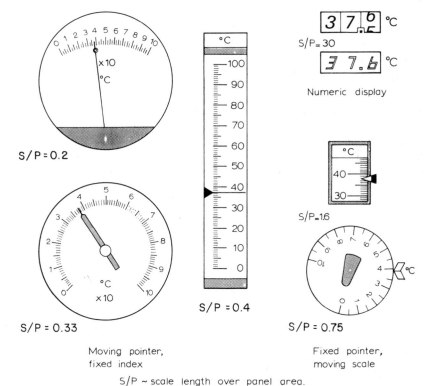

Figure 19.8 Basic types of symbolic indicators

Table 19.2   Relative evaluation of basic symbolic indicator types (Reproduced from Grether and Baker (1972) by permission of Department of Commerce, USA)

For	Counter is	Moving pointer is	Moving scale is
Quantitative reading	Good (requires minimum reading time with minimum reading error)	Fair	Fair
Qualitative and check reading	Poor (position changes not easily detected)	Good (location of pointer and change in position is easily detected)	Poor (difficult to judge direction and magnitude of pointer deviation)
Setting	Good (most accurate method of monitoring numerical settings, but relation between pointer motion and motion of setting knob is less direct)	Good (has simple and direct relation between pointer motion and motion of setting knob, and pointer-position change aids monitoring)	Fair (has somewhat ambiguous relation between pointer motion and motion of setting knob)
Tracking	Poor (not readily monitored, and has ambiguous relationship to manual-control motion)	Good (pointer position is readily monitored and controlled, provides simple relationship to manual-control motion, and provides some information about rate)	Fair (not readily monitored and has somewhat ambiguous relationship to manual-control motion)
Orientation	Poor	Good (generally moving pointer should represent vehicle, or moving component of system)	Good (generally moving scale should represent outside world, or other stable frame of reference)
General	Fair (most economical in use of space and illuminated area, scale length limited only by number of counter drums, but is difficult to illuminate properly)	Good (but requires greatest exposed and illuminated area on panel, and scale length is limited)	Fair (offers saving in panel space because only small section of scale need be exposed and illuminated, and long scale is possible).

operator's inner representation, must be considered seriously (Loewe, 1968). Figure 19.9 shows typical display formats. With increasing complexity the size of display panels should not just be increased to accommodate more, one-dimensional, metering instruments. This is especially so when the displayed measurands are related. One important reason is that static visual

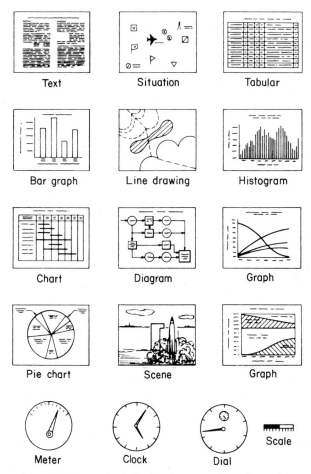

Text Situation Tabular

Bar graph Line drawing Histogram

Chart Diagram Graph

Pie chart Scene Graph

Meter Clock Dial Scale

Figure 19.9 Typical display formats, from Luxenberg and Kuehn (1968). (Reproduced by permission of McGraw-Hill Book Co. Inc.)

acuity, in polar coordinates represented by the visual detection lobe, decreases sharply outside the foveal area (to about 10% and less) (Figure 19.10).

Although in most situations a first fixation in the direction of a desired instrument is rather accurate (within 2° or 3° of arc), the operator may choose to process visual data with a fixation point 10° away (still within parafoveal vision) when time is of importance and momentary correlation is required between the two displays (Robinson, 1978). Peripheral vision cannot be used for perception of quantities, only for events. There is, therefore, a limit to the size of the multi-instrument display above which

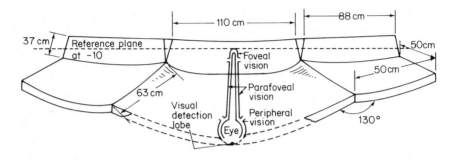

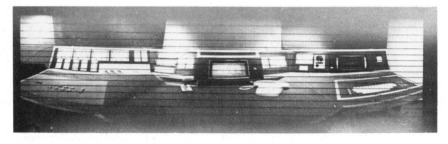

Figure 19.10   Image distortion due to optical mapping of the spatial geometry, as depicted by a 180° rotating camera. It shows the visibility of a wrap-around console which seems spread out flat by this technique. Also shown is the visual detection lobe (separability in min^{-1}) as function of retinal angle

integrated displays (more than one usually correlated variable combined in one display) become more effective.

In contemporary technology, the electronic integrated display is often built around the cathode ray tube (CRT), although for some applications liquid crystal displays (LCD), light emitting diode displays (LED) or plasma displays are preferred. For applications requiring high resolution, that is, images with a very high picture element (*pixel*) count, the CRT is as yet the only practicable solution.

An integrated display can be confusing (multipointer instrument, table of numbers on a CRT) when the operator has to perform mental operations to obtain an indicated value (Grether, 1949). Mental additions such as are required in three pointer altimeters, mentally combining numbers such as occur in projected tables, give rise to high error frequencies. The data should be arranged to form substructures (Kraiss, 1976; Grether and Baker, 1972) which require simple search strategies and appeal to the inner representation of the operator (Bainbridge, 1969; Rasmussen, 1974). Also the coding symbology should be considered (Kraiss, 1976; Dallimonti, 1976; Grether and Baker, 1972). An already effective method is to design groups, with compatible coding, such that data within a group have one specific aspect in common (Rasmussen, 1968; Kraiss, 1976; Danchak, 1976).

The link between the arrangement of the display and the operator's inner representation is not yet understood. It is recommended that design decisions are based on feedback data from a number of operators with, in addition, later training of the operators with the objective to mould the pattern of their inner representation to better fit the MMS characteristics.

The human mind is much better trained to grasp the meaning of a whole image or figure (pictorial displays) than it is to combine individual presentations of relevant parts, such as is the case with the display of many process variables on separate instruments. Certain suggestions to take advantage of this fact, as derived from Gestalt theory, have been made (Kraiss, 1976; Rasmussen, 1974; Woodson et al., 1972; Coekin, 1969; Loewe, 1968). The use of bar graphs (Reynold, 1978; Milam, 1973; Rasmussen, 1968) for the display of such things as pressure profiles and temperature families (Figure 19.11) is obvious but also favourable over say, mimic displays for small numbers of other correlated measurands. However, mimic displays and other pictorial devices also have their merits and much research is as yet necessary to obtain clearer insight in the problem of display type selection.

In dynamic situations, especially for higher order dynamic systems, it has been shown advantageous (Smith and Kennedy, 1975) to present additional information on the system state changes in the form of so-called *quickened* displays. Slow response of the system or operator is partly compensated for by derivative signals, in small proportions added to the actual variable to be controlled.

When a perfectly accurate model of the process is available for manual control of the process, the display of the actual magnitude of the variable is not necessary preferable display being of deviations from setting and trend information. This is also a type of derivative display, but easily integrated as Bowen (1967) has shown in his Figure 2 where 147 variables, shown in non-figurative form on one single screen, are monitorable without clutter.

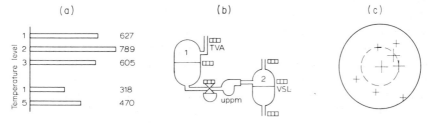

Figure 19.11   Three display types: (a) bar graph; (b) mimic display; (c) cross-sectional temperature distribution. (Reproduced by permission of Risø National Laboratory)

19.3.5 The Cathode Ray Tube Display

Light emitting devices such as the CRT are driven at a certain refresh rate for frame or image. It is difficult to obtain the desired light intensity levels without the sensation of flicker. Choosing long persistence phosphor tends to go at the expense of picture sharpness, so, for conventional CRT's video bandwidth must be increased (at the time of writing other electronic display types are under development with inherent storage characteristic). At an actual refresh rate of 50 Hz (not interlaced 25 Hz), the screen luminance should remain less than $100 \, cd/m^2$, at 60 Hz less than $500 \, cd/m^2$ (Meister and Sullivan, 1969; Turnage, 1966). Although $50-100 \, cd/m^2$ is acceptable, it seems desirable that for display purposes a revised video standard, better suited to adaptation to changes in environmental illuminance, should be internationally introduced. Apart from flicker, there exists a host of other questions about the widespread use of the CRT. Vibrating text, non-uniform sharpness, the background luminance, abuse of colour, glare, legibility of characters and other problems have led to much debate and research. A number of national preliminary recommendations have emerged. Umbers (1976) made a compilation of the whole field, showing that in CRT display design one should come well prepared to account for the human factor.

19.4 DISPLAY PANEL AND CONSOLE DESIGN

19.4.1 Interfacing to the Operator's Eyes

In Figure 19.10 the visual outlines as extracted from an actual 180° photograph taken with a visibility camera are depicted, when the operator's eye is in a central position (0° horizontal, $-10°$ vertical). It shows the geometrical distortions experienced by the optical (visual) channel without corrections introduced by the brain. Also the visual detection lobe (polar representation of static visual acuity as function of retinal angle) is shown superimposed on the figure. In the foveal area of nominally 2° of arc (circle with diameter of 20 mm at normal viewing distance) images are sharp, gradually becoming less recognizable for larger displacements from the central position. Combined with the optical geometrical distortion shown it will be clear that important, high attention value displays should not be mounted outside the area ±30° horizontal. It has been shown that operators can still actively control variables which are displayed in parafoveal (±12° outside foveal) view, but at a 10 to 20 dB lower control loop gain (Levison *et al.*, 1971). It is indeed possible to monitor signals located in an unfavourable position, though at degraded performance (Dreyfuss, 1966) provided that for larger displacements the signal is properly coded for attention value (moving or rotating object, blinking light). Recent work (McLeod, 1977) supports these findings. He has shown that in the parafoveal view up to 12° of arc, symbols can be identified with a high probability, reaction time only marginally increasing with angle. In the horizontal plane the performance is somewhat better than in the vertical plane, but as a generalization the performance drops off for angles greater than 12° of arc.

Eyes move ±15° of arc in the horizontal direction, +25° of arc (upward) and −35° of arc (downward) in the vertical direction. The normal declination of the eye is −5° of arc (downward); for the standing operator a comfortable

Figure 19.12   Field of view with fixed head, due to eye
rotation

Figure 19.13 Panel layout laid out to desired data—see text. MS, main system; AS, auxiliary system; SR, switch register (for start-up etc.); C, controls

average declination angle is $-10°$ of arc, for the seated person $-15°$ of arc (Dreyfuss, 1966; Woodson *et al.*, 1972). With head fixed and at 50 cm distance, the observer can thus accurately monitor displays in an area of width 28 cm and height 58 cm, with parafoveal vision extension, at reduced accuracy, of about 13 cm in all directions (see Figure 19.12). To the engineer this may suggest that CRT's could be placed with the longer dimension in the vertical direction if one is dealing with mutually correlated data; similarly panel areas in multi-instrument displays should be laid out with vertical emphasis.

Further utilization of panel area is made by head rotation. Easy rotation in the horizontal plane is within $\pm45°$ of arc, maximally $\pm60°$ of arc. In the vertical plane, easy head rotation is $\pm30°$ of arc, maximally $\pm50°$ of arc. In Figure 19.13 a panel lay-out is suggested which conforms to these data. The MS2 area in that figure can pertain to a separate main system and also to a back-up mode display that is used when the MS1 display or control system unexpectedly becomes unserviceable. For instance, MS1 being a fully electronic integrated display, MS2 composed of multi-instrument conventional displays.

19.4.2   Design of Consoles

Given these visual angle and head movement data, display console design depends on size, visual acuity, and comfort data. Sizes can vary considerably; for instance, the height of the eyes of a seated small woman (2.5 percentile) is 102 cm, of a seated large man (97.5 percentile) is 133 cm (Bialoskorski, 1978). A difference of 31 cm in height means, at a viewing distance of 50 cm that, looking at a display designed for the average person, the smallest person must continuously look upward at $2°$ of arc, the largest $32°$ of arc downward, in order to observe the display area of greatest importance. A downward declination of more than $17°$ of arc for prolonged periods may cause headache and possibly complaints of pain in the lower dorsal regions. Because problems of bodily sizes must not become a constraining criterion for persons doing this type of work, it is recommended to design some form of adjustment into the console and into the chair. With

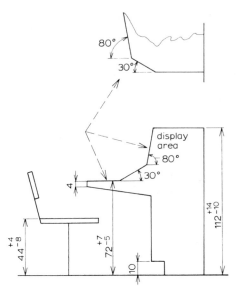

Figure 19.14 Dimensions (in cm) and geometry of display console for seated person

this in mind, dimensions in the vertical sense for a console for seated operation are as for Figure 19.14. It is assumed that the operator must have unobstructed view in the forward direction (low top of, or a window in, the console). Other anthropometric requirements, such as room for elbow and leg movements and correct seating position, must also be met. For further information see national anthropometric data and the extensive literature on this subject.

With electronic displays and controls, the dimensions of the console can be much smaller compared to conventional displays. Integrated and programmable displays, programmable electronic controls, shrink size with the effect that the operator does not have to leave his chair. This is not physiologically desirable. Solutions to such problems lie, if limited movement may be sought, in proper adaptation of task structure. Also with this lightweight equipment it is no longer necessary to build a strong support function into the console. The display area and control area may be designed as separate entities and each adjustable, giving much more freedom to adapt to the individual operator.

Finally, a small horizontal tablepiece is required where objects can be put such as an arm rest, or to occasionally place a separate device such as typewriter, calculator, writing pad or drinking cup.

19.5 CONCLUDING REMARKS

In this chapter a number of ergonomic factors concerning the man–machine interface have been presented, with special reference to display design. Beside that, man commands many more faculties. Kalsbeek (1978) points out that:

(a) A display is only part of a complex system of transformations which constitutes the dialogue between a machine or a process and the operator.

(b) A human individual is itself a complex biological system, which also needs internal control. One of its main goals, pursued irrespective of his task, is continuing and shaping its existence.

Other needs and desires, therefore, also seek to be fulfilled. Furthermore, the discussion was also limited to the visual channel. Often the other sensory functions can, and will, be used by the operator to aid or corroborate a visual stimulus, a fact which can be taken advantage of in the design of MMI, and which is also being extensively researched.

REFERENCES

Bainbridge, L. (1969). 'The nature of the mental model in process control'. *Proc. Int. Symp. on Man–machine Systems, Cambridge, UK*.

Bainbridge, L. (1974). 'Problems in the assessment of mental load', *Le Travail Humain*, **37**, 279–302.

Bialoskorski, J. St. (1978). *Revue Bibliographique*, Conservatoire National des Arts et Metiers, Paris.

Bosman, D. (1978). 'Systematic design of instrumentation systems', *J. Phys. E: Sci. Instrum.*, **11**, 97–105. (Reprinted in 1982, B. E. Jones (Ed.), *Instrument Science and Technology*, Vol. 1, Adam Hilger, Bristol, pp. 53–64.

Bosman, D. and Umbach, F. W. (1978). 'A design methodology'. Part 1 of Bosman, D., Kalsbeek, J. W. H., and Umbach, F. W. 'Display research and design methodology', in *Proc. Symp. on The Operator–Instrument Interface, 19–20 April, Middlesbrough*, The Institute of Measurement and Control, London.

Bottomley, S. C. (1976). 'Scales and graticules', *Measurement and Control*, **9**, 245–50.

Bowen, H. M. (1967). 'The imp in the system', *Ergonomics*, **10**, 112–19.

Coekin, J. A. (1969). 'A versatile presentation of parameters for rapid recognition of total state', in *Int. Symp. on Man–Machine Systems, September, Cambridge: IEEE Conf. Rec.*, Vol. 69, pp. 8–12.

Chapanis, A. and Kinkade, R. G. (1972). 'Design of controls', in H. P. Van Cott and R. G. Kinkade (Eds), *Human Engineering Guide to Equipment Design*, US Government Printing Office, Washington DC, pp. 345–37.

Cornsweet, T. N. (1970). *Visual Perception*, Academic Press, New York, pp. 330–42.

Dallimonti, R. (1976). 'Human factors in control center design', *Instrum. Tech.*, **23**, No. 5, 39–44.

Danchak, M. M. (1976). 'CRT displays for power plants', *Instrum. Tech.*, **23**, 29–36.

Dreyfuss, H. (1966). *The Measure of Man, Human Factors in Design: Library of Design*, Whitney, New York.

Döring, B. (1976). 'Analytical methods in man–machine system development', in K.

F. Kraiss and J. Moraal (Eds), *Introduction to Human Engineering*, Verlag TÜV Rheinland, Cologne.

Finkelstein, L. (1974). 'Fundamental concepts of measurement: definition and scales', *Measurement and Control*, **8**, 105–11.

Fitts, P. M., Jones, R. E., and Milton, J. L. (1950). 'Eye movements of aircraft pilots during instrument landing approaches', *Aero. Engng Rev.*, **9**, 24–29.

Fitts, P. M. (Ed.) (1951). *Human Engineering for an Effective Air Navigation and Traffic Control System*, Natl Res. Council, Washington DC.

Fogel, L. J. (1967). *Human Information Processing*, Prentice Hall, Englewood Cliffs, NJ.

Folley, J. D. (1960). 'Human factors and methods for system design', *American Institute for Research, Rep.* AIR-B90-60-FR-225, Pittsburgh, USA.

Fowler, H. W. and Fowler, F. G. (Eds) (1963). *The Concise Oxford Dictionary of Current English*, Oxford University Press, London.

Grether, W. F. (1949). 'Psychological factors in instrument reading. I. The design of long-scale indicators for speed and accuracy of quantitative readings', *J. Appl. Psych.*, **33**, 363–72.

Grether, W. F. and Baker, O. A. (1972). 'Visual presentation of information', in H. P. Van Cott and R. G. Kinkade (Eds), *Human Engineering Guide to Equipment Design*, US Government Printing Office, Washington DC, Chap. 3.

Jordan, N. (1963). 'Allocation of functions between man and machines in automated systems', *J. Appl. Psych.*, **47**, 161–5.

Kalsbeek, J. W. H. (1978). 'The human side of the interface'. Part II of Bosman, D., Kalsbeek, J. W. H. and Umbach, F. W., 'Display research and design methodology', in *Proc. Symp. on the Operator–Instrument Interface, 19–20 April, Middlesbrough*, The Institute of Measurement and Control, London.

Kidd, J. S. and Van Cott, H. P. (1972). 'System and human engineering analysis', in H. P. Van Cott and R. G. Kinkade (Eds), *Human Engineering to Equipment Design*, US Government Printing Office, Washington DC, pp. 1–16.

Kraiss, K. F. (1976). 'Vision and visual displays', in K. F. Kraiss and J. Moraal (Eds), *Introduction to Human Engineering*, Verlag TÜV Rheinland, Cologne, pp. 85–147.

Kraiss, K. F. and Moraal, J. (Eds) (1976). *Introduction to Human Engineering*, Verlag TÜV Rheinland, Cologne.

Levison, W. H., Elkind, J. I., and Ward, J. L. (1971). 'Studies of multi-variable manual control systems: a model for task interference', *NASA Rep. NASA-CR-1746*, May.

Loewe, R. T. (1968). 'System design, coding, formats and programming', in H. R. Luxenberg and R. L. Kuehn (Eds), *Display Systems Engineering*, McGraw-Hill, New York, pp. 24–69.

Luxenberg, H. R. and Kuehn, R. L. (1968). *Display Systems Engineering*, McGraw-Hill, New York.

McKay, D. M. (1963). 'Internal representation of the external world', *AGARD Avionics Panel Symp. on Natural and Artificial Logic Processors, 15–19 July, Athens*.

McCormick, E. J. (1970). *Human Factors Engineering*, McGraw-Hill, New York.

McLeod, S. (1977). 'Identification of alphabetic symbols as a function of their location in the visual periphery', *Rep. AMRL-TR-77-37, Aerospace Medical Division, Air Force Systems Command*, WPAFB, Dayton, Ohio.

McRuer, D. T., Jex, H. R., Clement, W. F., and Graham, D. (1967). 'Development of a systems analysis theory of manual control displays', *Rep. TR-163-1, Systems Technology Inc.*

Meister, D. (1971). *Comparative Analysis of Human Reliability Models*, Bunker Ramo Corp., Canoga Park, California.

Meister, D. and Sullivan, D. J. (1969). *Guide to Human Engineering Design for Visual Displays*, Bunker Ramo Corp., Canoga Park, California.

Milam, W. G. (1973). 'Advanced concepts in control room planning', *Instrum. Tech.*, **20**, No. 7, 43–8.

Moore, T. G. (1976). 'Controls and tactile displays', in K. F. Kraiss and J. Moraal (Eds), *Introduction to Human Engineering*, Verlag TÜV Rheinland, Cologne, pp. 172–93.

Moraal, J. (1976). 'What is human engineering', in K. F. Kraiss and J. Moraal (Eds), *Introduction to Human Engineering*, Verlag TÜV Rheinland, Cologne.

Murrell, K. (1969). *Ergonomics*, Chapman and Hall, London.

Nason, W. E. and Bennett, C. A. (1973). 'Dials v counters: effects of precision on quantitative reading', *Ergonomics*, **16**, 749–58.

Noll, R. B., Zvara, J., and Simpson, R. W. (1972). 'Analysis of terminal A.T.C. systems operations', *Air Traffic Control Systems*, AGARD CP-105.

Pearce, B. G. and Shackel, B. (1979). 'The ergonomics of scientific instrument design', *J. Phys. E: Sci. Instrum.*, **12**, 447–54. (Reprinted in 1982, B. E. Jones (Ed.), *Instrument Science and Technology*, Vol. 1, Adam Hilger, Bristol, pp. 93–101.)

Poole, H. H. (1966). *Fundamentals of Display Systems*, Spartan Books, Washington DC; Macmillan, London.

Rasmussen, J. (1968). 'On the communication between operators and instrumentation in automatic process plants', *Danish Atomic Energy Commission Research Establishment, Rep. Risö-M-686.*

Rasmussen, J. (1974). 'The human data processor as a system component, bits and pieces of a model', *Danish Atomic Energy Commission Research Establishment, Rep. Risö-M-1722.*

Reynolds, B. (1978). 'Operator interfaces for centralized control rooms', in *Proc. Symp. on The Operator–Instrument Interface, 19–20 April, Middlesbrough*, The Institute of Measurement and Control, London.

Rijnsdorp, J. E. and Rouse, W. B. (1977). 'Design of man–machine interfaces in process control', *Symp. on Digital Computer Applications to Process Control, 14–17 June, Delft*, paper S7, pp. 705–20.

Robinson, G. G. (1978). 'Dynamics of the eye and head during movement between displays: a qualitative and quantitative guide for designers', *Tech Rep. TR-78-2*, University of Wisconsin, DDC-AD-A052753.

Shackel, B. (1974). *Applied Ergonomics Handbook*, IPC, Guildford, UK.

Singleton, W. T. (1974). *Man–machine systems*, Penguin, Harmondsworth, UK.

Smallwood, R. D. (1967). 'Internal models and the human instrument monitor', *IEEE Trans. Human Factors Electron.*, **HFE-8**, 181–7.

Smith, R. L. and Kennedy, R. S. (1975). 'Predictor displays: history research and applications', *Dunlap and Associates Rep.*, Contract N00123-74-C-0952, US Naval Missile Center, Pt. Mugu, California.

Stassen, H. G. and Veldhuyzen, W. (1976). 'The internal model, what does it mean in human control?' in T. B. Sheridan and G. Johannsen (Eds), *Monitoring Behavior and Supervisory Control*, Plenum, New York.

Suppes, P. and Zinnes, J. L. (1963), 'Basic measurement theory', in *Handbook of Mathematical Psychology*, Wiley, Chichester.

Swain, A. D. (1970). *Development of a Human Error Data Bank*, Sandia Laboratories, Albuquerque, New Mexico.

Turnage, R. E. (1966). 'The perception of flicker in cathode tube displays', *Information Display*, **3**, 38–52, May–June.

Umbers, I. G. (1976). 'CRT/TV displays in the control of process plant: a review of applications and human factors design criteria', *Warren Spring Lab. Rep.*, Stevenage, UK.

Van Cott, H. P. and Kinkade, R. G. (Eds) (1972). *Human Engineering Guide to Equipment Design*, US Government Printing Office, Washington DC.

Van Cott, H. P. and Warrick, M. J. (1972). 'Man as a system component', in H. P. Van Cott and R. G. Kinkade (Eds), *Human Engineering Guide to Equipment Design*, US Government Printing Office, Washington DC, pp. 17–39.

White, C. T. and Ford, A. (1960). 'Eye movements during simulated radar search', *J. Opt. Soc. Am.*, **50**, 909–12.

Woodson, W. E. and Conover, D. W. (1973). *Human Engineering Guide for Equipment Designers*, University of California Press, Berkeley.

Woodson, W. E., Ranc Jr., M. P., and Conover, D. W. (1972). 'Design of individual workplaces', in H. P. Van Cott and R. G. Kinkade (Eds), *Human Engineering Guide to Equipment Design*, US Government Printing Office, Washington DC, pp. 381–418.

Handbook of Measurement Science, Volume 2
Edited by P. H. Sydenham
© 1983 John Wiley & Sons Ltd

Chapter

20 L. SCHNELL

Measurement of Electrical Signals and Quantities

Editorial introduction

The impact made on signal processing by electronic hardware is such that most systems will be designed to make use of electrical forms of signal. Once converted from the original energy form into a suitable electronic signal, information usually continues through a measurement system in this energy domain. The mapping, at some stage, will require quantification of the signals.

For this reason it is appropriate to provide here a review of the measurement of electrical signal parameters, including treatment of error sources that occur in the methods. Magnetic measurements are also covered.

20.1 ELECTRICAL UNITS AND BASIC METHODS FOR MEASURING ELECTRICAL QUANTITIES

20.1.1 Electrical Units

If a given field of physics can be described by m independent equations and there are n unknown quantities in these equations, then one can choose units arbitrarily for

$$n_0 = n - m$$

quantities. This done, the units for the rest of quantities in the equations will be found.

We are free to choose $n_0 = 3$ units for the description of mechanics and $n_0 = 4$ units for a joint description of mechanics and electricity.

The worldwide accepted SI system of units (refer to Chapter 3, Section 3.4) gives a compulsory prescription for the units of free choice by defining the dimension and magnitude of each of these units. Accordingly, the three units required for the description of mechanics are:

(a) *length* with *metre* as its unit;
(b) *mass* with *kilogram* as its unit; and
(c) *time* with *second* as its unit.

Of these, mass is a *prototype* unit, while length and time can be connected to physical constants.

A fourth unit must be chosen for the description of electrical phenomena, and this is

(d) *current* with *ampere* as unit.

According to its definition in the SI system: 'The ampere is that constant current which, if maintained in two straight parallel conductors of infinite length, of negligible circular cross section, and placed 1 metre apart in vacuum, would produce between these conductors a force equal to 2×10^{-7} newton per metre of length'.

This definition connects the unit of current with the metre, the kilogram, and the second, and contains the quantity μ_0, the *permeability* of the vacuum, fixing its value as $4\pi \times 10^{-7}$ N/A^2. (It is possible to express this in other units such as H/m.) The unit of current is realized experimentally by means of the current balance (Golding, 1949).

From the point of view of measurement, current is not the best choice, since its value can be reproduced only by means of an experiment providing the realization, and thus it does not have the advantage of *portability*.

Based upon the unit of current the other electrical units can, in principle, be determined. From the viewpoint of measurements a problem arises here again, since one obtains the other units through power. Based on the relationships $VI = I^2R = P$, V and R can be determined in principle; however, the uncertainty in the measurement of the thermal energy is too great to make the accuracy of the units derived in this way acceptable.

Therefore, it became necessary to derive another electrical unit, independent of the ampere. This is the unit of *resistance*, which can be reproduced in a circuit containing mutual inductance. In this case the resistance is compared with the units of length and time, using the value of μ_0 stated above (Golding, 1949). Resistance is a portable unit; this is exceedingly advantageous from the viewpoint of measurement. Instead of mutual inductance the *Thompson–Lampard capacitor* (Thompson, 1959)—the value of which can be calculated with a high accuracy—became the starting point. The unit of resistance (Elnekavé, 1975) can be determined from the reactance of the capacitor, appearing at alternating voltage. Calculation of capacitance uses the units of length and time, as well as the value of *dielectric constant* ε_0 derived from μ_0, and the *velocity of light*, c_0.

The unit of voltage is derived from the above units and is practically maintained by means of the *Weston cell* (Golding, 1949). This is a Hg–Cd cell which can be made with good reproducibility and has excellent long term stability. The electromotive force (e.m.f.) of a cell has a value of close to 1.01865 V. The temperature coefficient is $-40 \, \mu V/°C$; therefore it must be kept in a thermostated enclosure. Its internal resistance is high meaning it can be used only in a potentiometric circuit, where practically zero load can be ensured.

It is worth examining the uncertainty of the units produced by the above methods. Their definition in the SI system, in fact, supposes ideal conditions; however, practical *realization* of the units requires experiments having uncertainties.

Figure 20.1 (Melchert, 1979) illustrates the relation between mechanical units and electrical units, along with indication of the uncertainty in realization of the units. As can be seen, the relative uncertainties realizing the units ampere, ohm, and volt are 6×10^{-6}, 10^{-7}, and 6×10^{-6}, respectively.

Reduction of the absolute uncertainty of electrical units is aimed at in the experiments dealing with the connection of the unit of voltage to a natural constant. Such a possibility is offered by the *Josephson effect* (Josephson, 1962), which makes use of the relation between frequency f, voltage V, and h/e (Planck's constant/electron charge):

$$V = \tfrac{1}{2}fh/e.$$

Here f is a (very high) frequency that is accurately known in the experiment and h/e is a natural constant, V thus having a constant value (Taylor, 1967). The uncertainty of the realization of the relation expressed in the equation is not sufficient, because the value of h/e determined independently of the above equation is not entirely satisfactory, and so the Josephson voltage standard is used in the first place for checking the long term stability of the Weston cells.

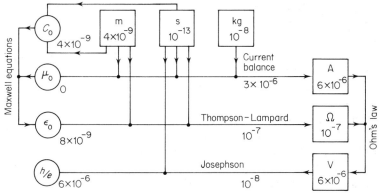

Figure 20.1   Relation between SI units and the natural constants, with indication of the uncertainty of realization

From the four physical units mentioned above all other electrical units can be derived. To produce a new unit not included among the basic units, the physical relationship must be found that connects the known and unknown quantities, then it must be realized by an experiment. In this way, all the electrical units can be produced step by step.

The important electrical and magnetic units are now summarized:

(a)   The *volt* (V) is the difference of electric potential between two points of a conducting wire carrying a constant current of 1 A, when the power dissipated between these points is equal to 1 W.

(b)   The *ohm* (Ω) is the electric resistance between two points of a conductor when a constant difference of potential of 1 V, applied between these two points, produces in this conductor a current of 1 A, this conductor not being the source of any electromotive force (e.m.f.).

(c)   The *coulomb* (C) is the quantity of electricity transported in 1 s by a current of 1 A.

(d)   The *farad* (F) is the capacitance of a capacitor between the plates of which there appears a difference of potential of 1 V when it is charged by a quantity of electricity equal to 1 C.

(e)   The *henry* (H) is the inductance of a closed circuit in which an e.m.f. of 1 V is produced when the electric current in the circuit varies uniformly at a rate of 1 A/s.

(f)   The *weber* (W) is the magnetic flux which, linking a circuit of one turn, produces in it an e.m.f. of 1 V as the flux is reduced to zero at a uniform rate in 1 second.

(g)   The *tesla* (T) is a flux density of 1 Wb/m^2.

The International Bureau of Weights and Measures (BIPM), situated in Sevre, is responsible for the uniformity of electrical units as it is for other units. This provides a central basis for a consistent system of physical measurements that is coordinated with those of other nations.

There is ample literature on the hierarchy and derivation of units in Coombs (1972).

20.1.2   Basic Methods of Comparing the Electrical Quantities

Comparison methods

Given the known value of an electrical quantity it can measured against the unknown one by comparison. Accordingly, the *comparison method* has an important role in measurements. The basic circuits used for comparison are collected in Figure 20.2. The relationships for evaluation can be derived easily and they indicate how the parameters of the circuit influence the *uncertainty* of the quantity to be measured.

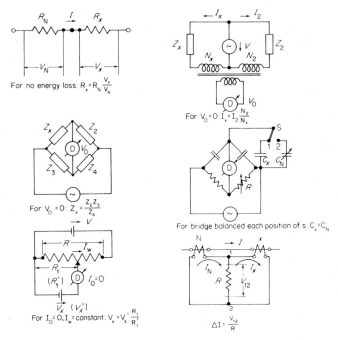

Figure 20.2   Basic comparison circuits

In measurements of resistance by comparing two voltages the uncertainty of R_x is dependent on that of R_N, V_x, and V_N.

The following three methods are *null methods*. This means that between appropriate points of the measuring circuit a zero value of current or voltage magnitude must be ensured. In these cases the sensitivity of the zero detector influences the uncertainties of the measurements. In addition, in a bridge circuit errors of three impedances, in a potentiometer network the uncertainty of a voltage (which may be for instance that of a Weston cell) and of the two resistors R_1 and R_1' while in a *current comparator* the uncertainty of the value of the current I_2 and the number of turns N_x and N_2 have an influence on the value of the quantity measured.

The uncertainty of comparison can be decreased by applying the *substitution method*, the principle of which becomes clear from study of the appropriate part of Figure 20.2. Provided that the elements of the bridge remain constant during the measurement, the uncertainty of C_x being measured depends only on C_N if $C_x = C_N$.

The *difference method* is advantageous for use in the case where the difference of two quantities very close to each other is to be measured. A simple example is portrayed in Figure 20.2. Two current transformers of identical ratio are used. The secondary currents are I_x and I_N, where I_N is known and I_x must be measured. The difference of these currents ΔI produces across the resistor R a voltage V_{12} and this is to be measured. If ΔI is small compared with I_x, the uncertainty of the measurement related to I_x will appear small as a second-order error.

The methods presented will be encountered in connection with a number of measurement problems.

Analog and digital indication

Measurement methods can also be classified depending on whether the measurement result appears in *analog* or *digital display form*.

The principle of analog indication is shown in Figure 20.3a. With an ideal scale and pointer the measurement result indicated by the pointer could be read with arbitrary accuracy, as for example $70.123456 \cdots$ V. However, due to various uncertainties affecting the measurement, the reading 70.1 V has only practical meaning, this being the valuable information. Storage and repeated display of analog signals is relatively difficult; further processing involves inevitable loss of measurement information.

The principle of digital indication can be seen in Figure 20.3b. Times t_1, t_2, t_3, and t_4 are the instants of measurement (sampling). In the case shown in the figure the quantum size is 1 V, which means that, if the voltage at time t_1 has a value of $105.5 < V < 106.5$ V, the device will indicate 106 V. If the size of the quantum is chosen in accordance with the uncertainty of measurement, the indicated value will furnish the really valuable information.

While analog indication is continuous in time, digital indication gives information generally at discrete moments.

It is characteristic of instruments with digital indication that they are fast

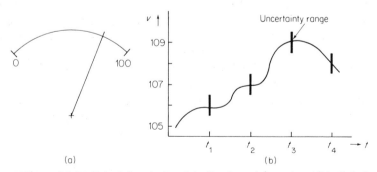

Figure 20.3   Principle of signal indication: (a) analog; (b) digital form

and accurate, their measurement results can easily be stored in digital form, and there is no loss of information in the course of further processing. Digital indication is *computer-oriented*. Digitally based measuring instruments are gradually replacing the traditional totally analog systems. It must, however, be realised that the initial stages of measurement within such instruments generally must be analog in nature.

20.2 COMPONENTS OF ELECTRICAL MEASURING DEVICES

20.2.1 Electromechanical Instruments

This section deals with the instruments and with the components of instruments that play a role in the measurement of the basic electrical quantities of voltage, current, impedance, power, and frequency.

In addition to the electromechanical instruments suitable for measuring voltage, current, resistance, and power, some important electronic elements of the instruments will be discussed and some examples of their application presented. We begin by considering the long established electromechanical instruments. In most of these devices (d'Arsonval moving-coil, electrodynamic, soft-iron, and induction meters) the torque necessary for the deflection of the pointer is produced by the interaction of the current to be measured and a magnetic field. In electrostatic instruments this torque results from electric field. This type of meter is predominantly used for direct measurement of high voltages (thousands of volts); they are not discussed here.

Characteristics of the construction and operation of electromechanical instruments are now considered that are important for the user.

Principle of operation

Depending on type and circuitry, the electromechanical instruments serve for measuring *direct* or *alternating current* or voltage. In the measurement of alternating signals the frequency limit is low, a few hundred or thousand cycles per second being the upper limit of their use.

The device consists of fixed and movable parts. The static deflection of pointer is determined by the equilibrium of two torques.

In the clockwise direction the *deflecting* torque M_d is produced by an *electromagnetic* force. If the current being measured is i and the deflection is α, then

$$M_d = f(i, \alpha)$$

The dependence on α arises because of the uneven distribution of the flux density. The measuring system stores an amount of magnetic energy W and the energy involved in deflection $d\alpha$ of the moving part is $M_d\,d\alpha$; thus the torque is equal to the specific change of the energy:

$$M_d = dW/d\alpha \quad \text{with } i = \text{constant}$$

Expressions of the deflecting torque for instruments of various types and the most important applications of the instruments are given in Figure 20.4. In the figure circles symbolize the coils of the meter; rectangles refer to iron cores. L_{12} is the mutual inductance between fixed and movable coils, while L_1 and L_2 are the inductances of the respective coils.

Type of instrument	Fixed	Moving	Deflecting torque	Remarks
	system			
Permanent magnet, moving coil, instrument D'Arsonval type	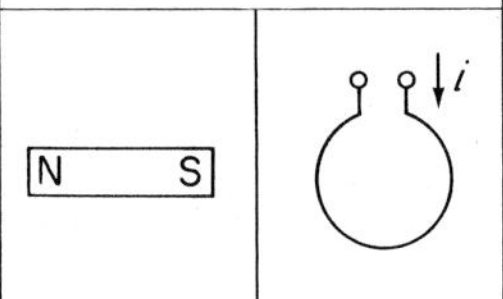 N S		$M_d = \dfrac{d\psi}{d\alpha}\,i$	For measuring d.c. current and voltage
Electrodynamic type moving coil instrument				For measuring d.c. and the r.m.s. value of a.c. quantities and power
iron cored			$M_d = \dfrac{dL_{12}}{d\alpha}\,i_1 i_2$	
air cored				Transfer standard
Soft iron instrument	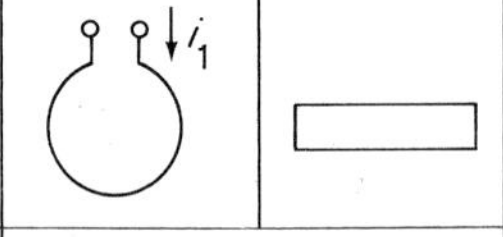		$M_d = \dfrac{1}{2}\dfrac{dL}{d\alpha}\,i^2$	For measuring d.c. and the r.m.s. value of a.c. quantities
Induction type energy meter	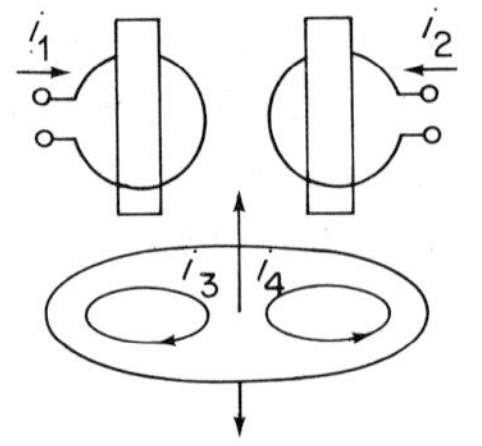		$M_d = \dfrac{dL_{14}}{d\alpha}\,i_1 i_4 +$ $+\dfrac{dL_{23}}{d\alpha}\,i_2 i_3$	Watt hour meter
Electrostatic instrument	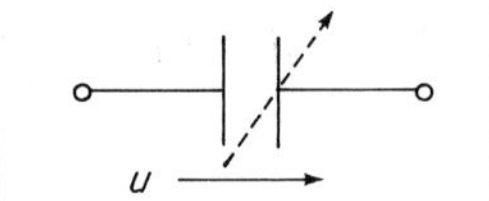		$M_d = \dfrac{1}{2}\dfrac{dC}{d\alpha}\,u^2$	For measuring very high d.c. and a.c. voltages

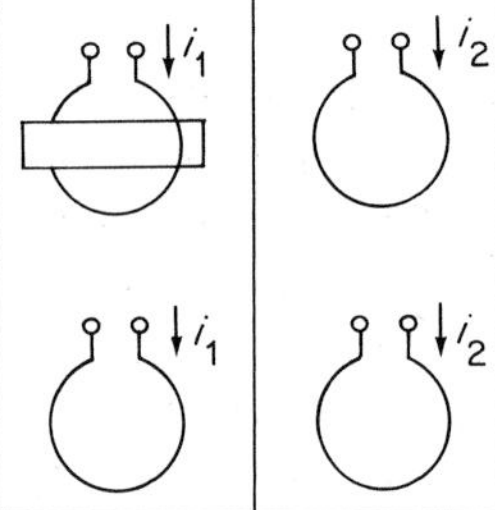

Figure 20.4   Working principles of electromechanical instruments

The deflecting torque is opposed by the *restoring* torque M_r produced by additional control springs or the flexure of the taut band suspension. Thus

$$M_r = -c_s \alpha$$

where c_s is the torque constant of the control spring.

In *quotient meters* the restoring torque is provided electrically in the same way as is M_d. In this case the moving part consists of two coils (having a common shaft and usually placed in differing planes), and the deflection of the moving part depends on the quotient of the currents flowing in the two coils. This is how the quotient meter of d'Arsonval type (cross-coiled instrument) measures resistance and the electrodynamical type measures say, phase angle ($\cos \varphi$) at industrial frequencies.

Characteristic properties

The aspects of the accuracy, sensitivity, power consumption, and overloadability of the instruments are now discussed briefly.

Accuracy

Accuracy of an instrument is determined by its error sources and characterized by the uncertainty of the indication. It is important to remember that an increase of accuracy is equivalent to a decrease of uncertainty. The errors of a meter may be caused by mechanical phenomena (friction, spring hysteresis, calibration error), by the variation of temperature (temperature dependence of the circuitry elements, that of the air-gap induction, and of the spring constant) or by electrical and magnetic phenomena (errors dependent on the frequency, effect of the signal shape, iron loss and hysteresis, external disturbing fields).

Accuracy of meters is usually expressed as a percentage of full scale value. Considering for example, for a 100 mA full scale (f.s.d.) current range meter, 1% accuracy means a ±1 mA uncertainty anywhere on the scale. For best accuracy it is recommended, therefore, that measurements be made using ranges where nearly full scale deflection is obtained. The accuracy level of a meter is given by the manufacturer, and is valid only within fixed reference conditions of temperature, frequency, position, and permissible external magnetic field.

Sensitivity

The scale characteristic of an electromechanical device is described by the equation:

$$\alpha = f(x)$$

where α is the deflection of the pointer and x the quantity to be measured.

The derivative of this equation:

$$S = d\alpha/dx = f'(x) \qquad (20.1)$$

is the *sensitivity* of the meter. Obviously, in the case of a linear scale, the sensitivity is the same at any point of the scale. Sometimes the reciprocal of sensitivity, the so-called *instrument constant*, is stated for characterization of an instrument as $C = 1/S$.

Power consumption

This is the power consumed by the instrument that is drawn from the circuit to be measured. The d'Arsonval type meter dissipates the smallest amount of power (10^{-6}–10^{-2} W) while the power dissipated by other electromechanical instruments may approach, or even exceed, 1 W. The small power consumption justifies the wide application of the d'Arsonval type meter. Power drawn from the measured circuit can be reduced to an arbitrary extent by the use of an electronic amplifier.

Overloadability

This is an important characteristic of indicating instruments. The overload may be of short duration (dynamic) or be long-lasting. The damage caused by overload may be a breakdown of insulation due to overvoltage, an overheating of the coils due to overcurrent, or mechanical damage due to dynamic inertial effects (IEC, 1973).

Main types of electromechanical instruments

The construction and operation of the d'Arsonval moving-coil, the iron-cored electrodynamic, and the soft-iron instruments are now outlined.

The schematic diagram of the permanent magnet, moving-coil instrument d'Arsonval type is shown in Figure 20.5.

The position $\alpha = 0$ indicates the starting point of the moving coil. The air-gap flux density B is intended to be uniformly distributed along the circumference. The current i feeds into the moving coil, having N turns, through the control springs. The equation which describes the scale characteristic of the instrument can be determined simply as

$$\alpha = \frac{BANi}{c_s} = \frac{\psi_0}{c_s} i = Si \qquad (20.2)$$

where $\psi_0/c_s = S$ is a constant. This is the sensitivity of the instrument.

It can be seen from this that the direction of the deflection of the instrument depends on the direction of the current. When alternating

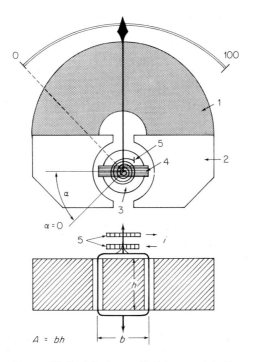

Figure 20.5 Moving coil d'Arsonval indicating instrument: 1, permanent magnet; 2, soft-iron poles; 3, soft-iron core; 4, moving coil; 5, control springs

current is applied to the instrument, it will measure its integral mean value, that is only the d.c. component.

The usual connections of the instrument are shown in Figure 20.6. In Figure 20.6a current is measured directly; this version can be used up to about 500 mA. In Figure 20.6b the instrument is *shunted* and so

$$I_0 = I \frac{R_s}{R_s + R_0}$$

where I_0 is the current passing the meter and I the total through-current. For this the deflection is given by

$$\alpha = \frac{\psi_0}{c_s} \frac{R_s}{R_s + R_0} I$$

In Figure 20.6c a series resistance is connected to the instrument, making it

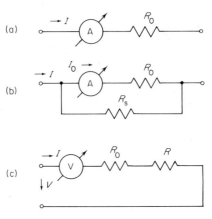

Figure 20.6 Circuitry of the d'Arsonval instrument. R_0, internal resistance of the instrument; R_s, shunt
resistance; R, series resistance. (a)
Equivalent circuit; (b) current extension; (c) voltage extension

suitable for measuring voltage. Here deflection is given by

$$\alpha = \frac{\psi_0}{c_s} I = \frac{\psi_0}{c_s} \frac{1}{R_0 + R} V$$

Where R is much greater than R_0 the instrument constant can be expressed
in ohms per volt.

The d'Arsonval instrument can also be used for measuring resistance. In
the arrangement shown in Figure 20.7 deflection of the meter depends on
R_x if V is constant, in accordance with the relationship

$$\alpha = \frac{\psi_0}{c_s} \frac{V}{R_0 + R_x} \tag{20.3}$$

The scale is hyperbolic and covers the resistance range from 0 to ∞. Note
that centre scale corresponds to the situation where $R_x = R_0$.

Two special purpose d'Arsonval instruments need mentioning here.

The *galvanometer* serves for measuring very small direct currents or
voltages. The expression $S = \psi_0 / c_s$ shows the possibilities of increasing the
sensitivity. They were important as null detectors before the appearance of
electronic amplifiers for such purposes. They were made with an optical
lever output. The current and voltage sensitivities obtainable from an optical
read-out with a length of 1 m are approximately

$$S_i = 10^7 - 10^{11} \text{ mm/A}$$

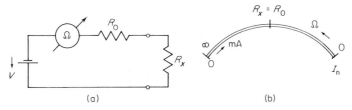

Figure 20.7 Use of moving coil meter for resistance measurement: (a) series ohmmeter arrangement; (b) scale of the instrument

and

$$S_v = 10^6–10^8 \, mm/V,$$

respectively.

The *fluxmeter* is suitable for measuring magnetic flux, or flux density. If the moving part of the d'Arsonval instrument is suspended on a support fibre free of torque ($c_s = 0$) and a test coil of N turns is attached to the instrument, in which the flux changes by $\Delta\phi$, then the voltage pulse arising on the change of the flux will change the deflection of the instrument by $\Delta\alpha$ according to the following relationship (Golding, 1949):

$$\Delta\alpha = \frac{1}{\psi_0} N\Delta\phi = \frac{1}{\psi_0} \Delta\psi \tag{20.4}$$

The electrodynamic instrument differs from the d'Arsonval instrument essentially in that, instead of a permanent magnet an electromagnet, excited by the signal being measured, produces the magnetic field required for the deflection.

In the construction shown in Figure 20.8 the magnetic field is produced by current i_1 in the iron core, hence this device is also termed a *ferrodynamic* instrument. Similarly to the d'Arsonval instrument, the air gap flux density also has a uniform distribution along the circumference.

In this case, there are two currents in the device: i_1 is the current of the fixed part and i_2 that of the moving part. The relationships stated in Figure 20.4 can be used for calculating the instantaneous value of the deflecting torque:

$$M_d = Ki_1 i_2, \tag{20.5}$$

where K is a constant, dependent on the actual parameters of the instrument.

Because of the integrating effect of the mass of the moving part, the mean of the instantaneous values has to be taken into account. For sinusoidal

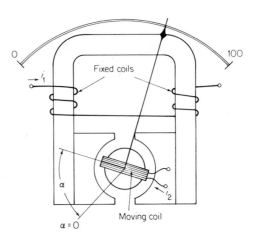

Figure 20.8   Schematic of iron-cored electrodynamic indicator

signals, starting from equation (20.5):

$$\bar{M}_d = KI_{1m}I_{2m} \frac{1}{T} \int_0^T \sin \omega t \sin (\omega t + \varphi)\, dt$$

$$= KI_1 I_2 \cos \varphi$$

where I_{1m} and I_{2m} are the peak values, I_1 and I_2 the r.m.s. values of the currents passing in the fixed and moving parts respectively, and φ is the phase angle between them. Based on $\bar{M}_d + M_r = 0$, the relationship

$$\alpha = \frac{K}{c_s} I_1 I_2 \cos \varphi \tag{20.6}$$

gives the deflection of the pointer. When $I_1 = I_2 = I$, and $\varphi = 0$, that is, the fixed and the moving coils are connected in series,

$$\alpha = \frac{K}{c_s} I^2 \tag{20.7}$$

Since any periodical signal can be composed from sinusoidal signals, equation (20.7) shows that the instrument truly measures the r.m.s. value of periodical signals of any shape. This holds as long as the air-gap flux density is proportional to I_1, that is, as long as the iron has linear characteristics.

On the basis of equation (20.6) *effective power* can be measured by the electrodynamic instrument.

With substitutions $I_1 = I$ and $I_2 = V/R$ (Figure 20.9) the relationship

$$\alpha = \frac{K}{c_s} \frac{1}{R} IV \cos \varphi = \text{constant} \times P$$

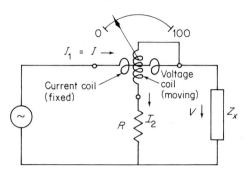

Figure 20.9 Measurement of power with
the electrodynamic indicator

will be obtained, where P is the effective power. In industrial application such devices are employed to measure the power.

It is to be noted that this type of instrument is made also in a construction without iron. This provides higher accuracy and facilitates the comparison of r.m.s. values of low frequency a.c. voltages and d.c. voltages, and is therefore suited for transferring the unit of voltage (defined in the form of direct voltage) into the domain of alternating signals; with a relative uncertainty of about 10^{-3}. This is why it is termed a *transfer standard*. The thermocouple instrument also exhibits similar properties (see Section 20.3.2).

There is no need to conduct current into the moving part of the soft-iron instrument. The moving part is usually an iron plate fixed eccentrically to the shaft. A similar plate is fastened to the inside of the fixed coil. Due to the current passing through the fixed coil the two plates become magnetized in the same direction, thus repelling each other with a force proportional to the square of the current. The equation describing its operation can be found in Figure 20.4. The device measures the r.m.s. value of the current passing through the fixed coil as long as the iron has linear characteristics. It is also suitable for measuring direct current.

The simple design makes it cheap and reliable in operation. It is primarily used for measuring currents and voltages at mains frequency. Literature dealing with electromechanical instruments is listed in Stout (1960).

20.2.2 Electronic Components of Instruments

In the measurement of electrical signals instruments containing electronic components play an important role. In this section a brief review is given of some of the important electronic components employed, in the first place, in amplifiers.

Characteristic features of amplifiers

Amplifiers serve for amplifying the signal (voltage, current) or power to be measured. Gain is defined as the ratio of output signal to input signal. From the point of view of measurement it is of paramount importance that the relationship describing the gain, the *transfer characteristic* of the amplifier (Figure 20.10a), be exactly known and well defined. In feedback amplifiers the magnitude of gain is dependent on feedback elements (generally passive) and thus can be defined with high accuracy.

With regard to the *frequency spectrum* of the input signal, gain for individual frequency components is not necessarily equal. This property is described by the frequency transfer characteristic of the amplifier.

The d.c. amplifier is of low-pass type (Figure 20.10b). Having direct coupling it is suitable also for amplifying direct voltage. The a.c. amplifier is coupled through a capacitor and can be used only for amplifying alternating signals. Typical examples are the *broadband* amplifier (Figure 20.10c), used for amplifying a broadband signal with the smallest possible distortion and the frequency selective amplifier (Figure 20.10d), serving to amplify a signal of only a given frequency f_0. Bandwidth of the amplifier is given as the frequency range between the points exhibiting a gain decrease of 3 dB.

A further important feature of the amplifier is the *impedance transformation*. This conveys information as to the degree of conversion to small output resistance from high input resistance.

In measuring devices d.c. amplifiers play an important role. For amplifying not too small ($>1\,\mu\text{V}$) voltages, d.c. amplifiers with direct coupling are

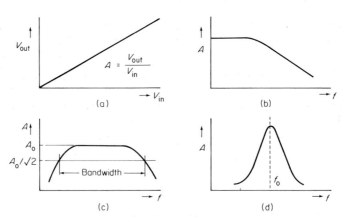

Figure 20.10   Characteristics of amplifiers: (a) amplifier with linear transfer characteristic; (b) frequency transfer characteristic typical of d.c. amplifier; (c) Broadband amplifier; (d) frequency selective amplifier

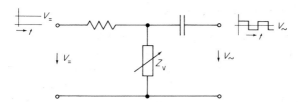

Figure 20.11   Principle of d.c.–a.c. conversion. Z_v is
a periodically varying impedance

used consisting of discrete semiconductors (FET, bipolar transistor) or integrated circuits.

The detection limit of the amplified voltage is limited by the *low-frequency noise* of semiconductor devices and the drift of the amplifier. The problem can be obviated by means of the so-called *chopper stabilized amplifier*, converting the small direct connected voltage to be amplified into an alternating signal by chopping it. This periodic signal is then amplified by an a.c. amplifier that behaves more favourably from the viewpoint of noise and drift; it is finally transformed back into a d.c. signal. Chopper amplifiers can be made only with smaller bandwidth than that of the d.c. amplifier with direct coupling.

The principle of d.c.–a.c. conversion is shown in Figure 20.11.

The periodically varying impedance Z_v may be a mechanical switch, a FET switch, a photoresistor or a vibrating capacitor.

Main properties of operational amplifiers

Operational amplifiers play an important role in the family of d.c. amplifiers. As can be seen in Figure 20.12 they are also suitable, in addition to amplification and impedance conversion, for performing various operations which make their application in measuring instruments particularly important. Figure 20.13(a) shows an inverting operational amplifier, which amplifies the input signal as $A = -V_{out}/V_{in}$. The negative sign means a 180° phase shift.

The differential amplifier, according to Figure 20.13b, has both an inverting and a non-inverting input. The relationship of voltages is given by the equation

$$V_{out} = A(V_{in}^+ - V_{in}^-) = AV_s,$$

where V_s is the symmetrical voltage to be amplified.

Important properties of the operational amplifier are that its open-loop gain is very large (60 to 120 dB), its input resistance is very high, and its output resistance very low. Maximum output voltage is limited by the supply

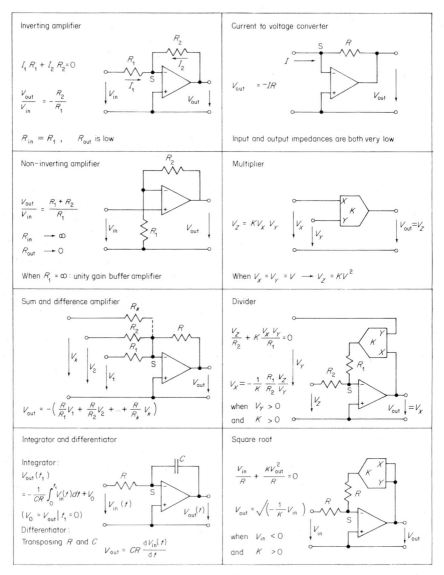

Figure 20.12   Basic circuits using the operational amplifier and analog multiplier

voltage. With zero input voltage the output voltage can be different from zero; this is termed the *offset voltage*. It can be compensated for but compensation is dependent on temperature and time. Slow variation of the output voltage for given constant input voltage is termed the drift.

For differential amplifiers the effectiveness of the *common mode rejection* (CMR) is of importance. Common mode signal is the voltage $\frac{1}{2}(V_{in}^{+} + V_{in}^{-})$

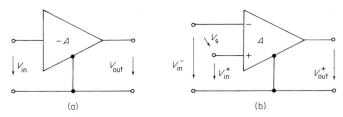

Figure 20.13   Operational amplifier connections: (a) inverting amplifier; (b) differential amplifier

(see Figure 20.13b), which gives the deviation of the floating V_s from the zero voltage point (earth, guard). In ideal cases the output signal V_{out} depends only on V_s, not on the common signal. Practically this is only approximately true. The higher is the common mode rejection performance, the more favourable are the properties of a differential amplifier. The degree of rejection is stated by the *common mode rejection ratio* (CMRR), which, in the case of an unloaded output, is the quotient of the gains pertaining to the symmetrical signal V_s and the common signal. The higher is this number, for example 120 dB, the better is the amplifier from the viewpoint of CMR.

The field of application of operational amplifiers to measurement is widened by the availability of analog multipliers, the scale factor, K, of the analog multiplier, is a constant by which the product is multiplied.

Application of operational amplifiers and analog multipliers

Some important applications are listed in Figure 20.12. At the summing junction point, S in the figure, Kirchhoff's nodal law has to be written with consideration to the fact that no current is flowing in the operational amplifier branch because of its approximately infinite input resistance. The relationship obtained in this way will yield the equation describing the operation. When examining the operation, it is necessary to remember that, due to the very great open-loop gain, the voltage between the input terminals of the amplifier can be considered as zero in the case of finite V_{out}.

For the subjects discussed in this section refer to Oliver and Cage (1971), Coombs (1972), and Herpy (1980), and to Chapters 11 and 22.

20.3 MEASUREMENT OF VOLTAGE AND CURRENT

Voltage and current are the two most frequently measured electrical quantities. On the basis of their time functions they can be classified as follows:

(a) *Deterministic signals,* the time function of which can exactly be described analytically.

(b) *Stochastic signals*, which can be described only by statistical methods (Bendat, 1971).

Stochastic signals may be carriers of important information, but they can also be undesirable *noise* making the measurement difficult. Noise can arise from within the measuring device, especially when very small signals are being measured, or come from external sources.

This chapter deals primarily with deterministic signals. However, frequently occurring situations may have noise superimposed on the signal to be measured. In the following sections all undesired deterministic or stochastic signals will be termed noise. In practice effort is made to eliminate the noise at source or to apply circuits that enhance the signal whilst reducing the noise.

Information about a signal is contained either by its time function or by its frequency spectrum (Chapter 4). However, it is often expedient to characterize the magnitude of the signal by a single number. In the case of a perfect d.c. signal such a statement contains the whole information. In other cases, however, it means a reduction of information. The time function of a signal can be visualized by means of a cathode ray tube (CRT) oscilloscope and plotted by a recorder, while its frequency spectrum can be examined by a spectrum analyser (Bendat, 1971).

In the measurement of voltage and current two points need to be mentioned:

(a) If it is desired to measure the voltage between two points of a network, then it is reasonable to claim that the connection of the voltmeter should not change the voltage to be measured. This claim is fulfilled if the internal resistance of the voltmeter is significantly higher than that of the measured circuit. Ideally this resistance is infinitely high; practically one thousand times higher usually suffices.

(b) If the current flowing in one branch of a network is to be measured, it is important that the insertion of the ammeter should not change the current to be measured. This is achieved by making the resistance of the ammeter small compared with other resistances of the network. The internal resistance of an ideal ammeter would be zero.

Devices suitable for measuring voltage can also be used to measure current, and vice versa.

20.3.1 Measurement of Direct Voltage and Current

The voltage constant in terms of time is relevant to *ideal direct voltage* (d.c.). From the viewpoint of measurement, however, signals varying slowly enough with time and the average of periodic or other signals are also called direct signals.

For the measurement of direct voltage and current by the *analog method* the moving-coil instrument is generally used, as illustrated in Figure 20.6. Its normal construction has full scale deflection (f.s.d.) values of $10\,\mu\mathrm{A}$ to $100\,\mathrm{mA}$ and $1\,\mathrm{mV}$ to $600\,\mathrm{V}$. Further increase of its sensitivity can be achieved but only at the expense of portability for the movement becomes mechanically delicate.

In voltage measurements it is common to characterize the device by the value ohms per volt (Ω/V). Considering Ohm's law in reciprocal form, $R_i/V_n = 1/I_n$, it can be seen that the Ω/V value is equal to the reciprocal of the current I_n existing for the full scale value of the voltage V_n to be measured. R_i is the *internal resistance* of the meter. For example, a voltmeter of $10^4\,\Omega/\mathrm{V}$ specification has a current of $10^{-4}\,\mathrm{A}$ at f.s.d.. Both sensitivity and input resistance can be effectively increased by using a d.c. amplifier between the signal source and the meter. In this respect refer to Section 20.2.2 and Figure 20.12.

The *digital voltmeter* (DVM) is a widely used instrument. It has the advantageous properties of digital indication (excluding an error of reading), wide measuring range, competitive cost, high measuring speed, possibility of automatic range selection as well as the possibility of automatic self-calibration. The digital output facilitates coupling to computers or measuring systems. Such instruments will, to an increasing extent, replace the majority of fully analog equivalents.

This section deals with the DVM types shown in Figure 20.14.

DVMs measuring the instantaneous value	DVMs measuring average value (integrating DVMs)
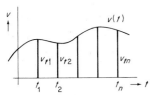	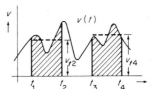
Measurement at instants $t_1, t_2, \ldots, t_n$	The indications v_{t2} and v_{t4} give the average voltages in the intervals of $t_2 - t_1$ and $t_4 - t_3$
Types	Types
successive approximation linear (voltage to time) ramp staircase ramp continuous balance	voltage to frequency conversion dual-slope

Figure 20.14   Types of digital voltmeter (DVM)

Successive approximation technique

Figure 20.15a shows the classical d.c. potentiometer suitable for comparing an unknown voltage V_x with the voltage of a Weston standard voltage cell. By means of automatic control an analog recorder of high accuracy (compensograph) can be obtained. In the scheme of Figure 20.15b the same measuring principle is realized by means of a digital voltage divider. A constant current source drives the current I_w through the resistor chain representing $2^0, 2^1, \ldots, 2^n$ values, any element of which can be shorted by a

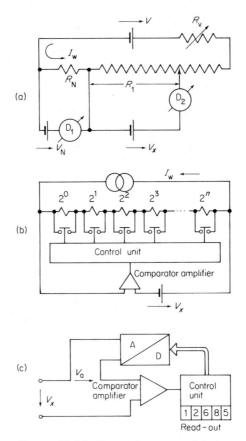

Figure 20.15   Successive approximation DVM method: (a) precision voltage divider (potentiometer): V_N, Weston Cell; I_w, working current; D_1, D_2, zero detectors; (b) digitally balanced potentiometer; (c) successive approximation method with D/A converter

switch controlled by the logic control unit. The comparator amplifier senses the difference of voltages appearing on the digital divider and V_x to be measured and a decision is issued to the logic control unit to control the digital divider until the difference voltage becomes zero. When a measurement process is completed it is followed by zeroing before starting the next measurement.

With adequate programming, a *follower-mode* can be accomplished in which the device follows variations of the voltage. In the arrangement shown in Figure 20.15c the control unit produces a set of digital codes which are converted into the analog voltage V_a by the digital-to-analog (D/A) converter. Depending on the output state of the comparator amplifier the control unit varies the value of V_a with a suitable strategy until the state $V_x + V_a = 0$ is reached. The value of V_a is displayed digitally by the instrument.

Linear ramp-type DVM

Frequency and time interval can be measured digitally in the simplest way. Essentially, a ramp signal of high linearity is generated (Figure 20.16) and compared with the signal V_x to be measured. At the zero voltage transition of the ramp signal, the ground comparator amplifier C_1 changes its binary state and as a consequence the gate closes and remains closed, until C_2

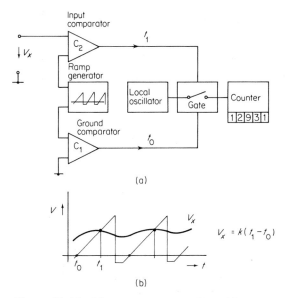

Figure 20.16   Linear-ramp type DVM: (a) block diagram of circuitry; (b) comparison of the ramp with input voltage V_x to be measured

changes its state at the instant when the ramp voltage equals V_x. In the closed state of the gate the counter counts the periods of the clock oscillator. Because of the direct relationship between slope of the ramp, number of counted periods, and value of the voltage to be measured, the latter can be directly displayed.

Staircase ramp-type DVM

The staircase ramp-type generator produces a voltage rising by steps (Figure 20.17), corresponding to the quantum size of the instrument. When the voltage produced in this way equals V_x, the comparator amplifier changes its state and stops the staircase ramp generator. The steps are counted by a counter and the value of voltage proportional to the counts is displayed. The control unit serves to control the procedure, it resets to zero and restarts after the measurement.

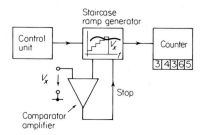

Figure 20.17   Block diagram of the staircase ramp-type DVM

Continuous balance-type DVM

In the arrangement shown in Figure 20.18 the instrument is not set to zero after a measurement cycle but, by means of a reversible counter, follows the variation of the voltage.

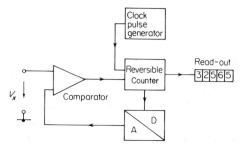

Figure 20.18   DVM of continuous balance type

The counting direction, up or down, of the reversible counter depends on the output state of the comparator. The content of the counter is converted by the D/A converter into a voltage proportional to the pulse number and this voltage is compared with V_x.

The DVM's mentioned so far measure the value of V_x at discrete time instants. If the signal is, for instance, noisy, the evaluation of the measurement performed in this way may be problematic. Such problems can largely be overcome by using integrating DVM's which measure the *average* value of signals during the sampling interval.

Digitally averaging DVM

The principle of the digitally averaging DVM is shown in Figure 20.19a. The voltage-to-frequency (V/F) converter converts the voltage to be measured into a proportional frequency. Passing through the gate, closed for interval T, pulses feed the counter the content of which is, in this way, proportional to the average of the signal during T.

Figure 20.19b shows the case when *a priori* information on the normal mode noise is available. If this comes from a source of, say, 60 Hz, then it can be filtered out with a choice of $T = k/60$ s where k is an integer number.

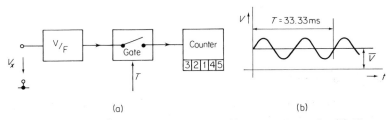

Figure 20.19   Digitally averaging DVM: (a) block schematic; (b) illustration of noise suppression

Dual-slope DVM integrating in analog mode

Switch S, in Figure 20.20a, is in position 1 for an interval T_1. During this time (Figure 20.20b), the output of the integrator is a rising voltage V_i, this being the integrated value of V_x. At the end of period T_1 switch S changes to position 2. The polarity of the reference voltage V_r is selected to be opposite to that of V_x so the voltage V_i' at the output of the integrator moves toward zero. Reduction continues until the zero comparator detects the 0 V state. This process takes time T. Change of the output state of zero comparator puts switch S back again into position 1.

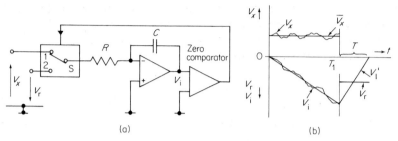

(a)   (b)

Figure 20.20   Dual-slope DVM: (a) block diagram; (b) waveforms

Based on the above:

$$V_i = -\frac{1}{RC} \int_0^{T_1} V_x \, dt = -\frac{1}{RC} T_1 \bar{V}$$

$$V_i' = -\frac{1}{RC} \int_0^T -V_r \, dt + V_i$$

In the state $V_i' = 0$

$$\bar{V} = T \frac{V_r}{T_1}$$

$\bar{V}$ being the averaged V_x voltage sought.

Since T_1 and V_r are known (the control unit is not shown in the figure) period T can be measured being proportional to the time average value of V_x.

Further information on DVM techniques is to be found in Oliver and Cage (1971), Coombs (1972), Dudley and Laing (1973), and in Chapter 12.

20.3.2   Measurement of Alternating Voltage and Current

All of the information carried by a.c. signals appears in the time function (time domain) or frequency spectrum (frequency domain) of the signal (refer to Chapter 4).

Examination of one period of a periodic signal can yield the waveform, frequency, and phase relative to some reference. Such an examination can be performed, for instance, by means of a cathode ray oscilloscope. However, it is not always necessary to examine the signal in this complete and detailed way.

In many cases characterization of a periodic signal may only require a number that informs about either the magnitude, the frequency or the phase of the signal. This means measurements can be performed according to three different methods. This account deals primarily with information condensed into a single number expressing the magnitude of the voltage. This number can represent one of the following features:

(a) The *direct current component* hidden in the alternating signal is given by the so-called *electrolytic mean*:

$$V_e = \frac{1}{T} \int_0^T v(t)\,\mathrm{d}t \tag{20.8}$$

where $v(t)$ is the instantaneous value of the periodical signal. Essentially, the measurement of V_e is a d.c. measurement, discussed earlier.

(b) The *absolute mean* value can be measured using a rectifying device:

$$V_{av} = \frac{1}{T} \int_0^T |v(t)|\,\mathrm{d}t \tag{20.9}$$

This expression gives the mean value of the alternating voltage rectified by the *full-wave* method.

In the case of *half-wave* rectification

$$V'_{av} = \frac{1}{T_0} \int_0^T v^+(t)\,\mathrm{d}t \tag{20.10}$$

where $v^+(t)$ is the positive half-period of $v(t)$.

(c) The *root mean square* (r.m.s.) value is obtained from the mean of the square of the instantaneous value of the alternating voltage:

$$V_{rms} = V = \sqrt{\frac{1}{T} \int_0^T v^2(t)\,\mathrm{d}t} \tag{20.11}$$

(d) The *peak* value V_p is the maximum value of the voltage. If the signal is not symmetrical about the zero axis, then it is meaningful to speak of V_p^+ and V_p^-. The distance between V_p^+ and V_p^- is the *peak-to-peak voltage* V_{pp}.

(e) d.c. voltages proportional to the *vector components* of a sinusoidal signal (one coinciding with an optional reference direction the other perpendicular to it) can be obtained and measured by means of a *phase-sensitive rectifier*.

There exist two possibilities for measuring different parameters of a.c. voltages:

(a) After a.c.–d.c. conversion a device suitable for measuring d.c. is employed (measurement of mean value, of peak value or of the r.m.s. value).

(b) An instrument is used, the deflecting torque of which is proportional to the square of the instantaneous value of the alternating signal and which is capable—because of its ballistic property—of producing the average value. There exist several electromechanical instruments suitable for measuring the r.m.s. value by this means.

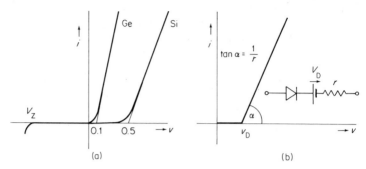

Figure 20.21   Diode and its equivalent circuit: (a) characteristics of
germanium and silicon diodes; (b) equivalent circuit

Two important types of a.c.–d.c. converters have to be mentioned: the *diode* and the *thermocouple converters.*

The important properties of a diode are shown in Figure 20.21a. Assuming the (in reality) non-linear resistance r to be linear, the equivalent circuit shown in Figure 20.21b will be obtained. As can be seen, the rectifying effect is present in the case of $v > v_D$, in which v_D is the *threshold voltage* of the rectifier. For a silicon diode $v_D > 0.5$ V. This causes problems only when low voltages have to be rectified.

The thermocouple converter (Figure 20.22) consists of a heating wire placed in an evacuated glass bulb along with a thermocouple sensing its temperature. The direct voltage V_t is proportional to the temperature of the filament, that is, to i^2. It follows that V_t is proportional to the square of the r.m.s. value of the current flowing in the heating wire.

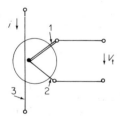

Figure 20.22   Thermocouple converter: 1, 2, thermocouple
elements; 3, heating wire

Measurement of the mean value by a diode rectifier

The circuit schematic and the time functions for sinusoidal signals are as shown in Figure 20.23. The scheme shown (Figure 20.23a) serves for measuring the mean value defined by equation (20.10).

The figure shows a half-wave rectifying circuit. The rectified voltage v_r can be measured by any d.c. meter having high internal resistance. In Figure

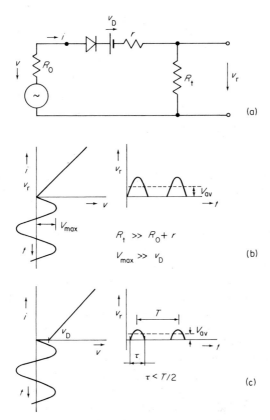

Figure 20.23 Average-responding detector
with diode: (a) circuit diagram; (b) large signal;
(c) small-signal characteristics, v_r is the instantaneous value of the rectified voltage, V_{av} its
mean value

20.10b the maximum of the voltage to be measured is $V_{max} \gg V_d$, therefore,
the relationship between v_r and v can be considered linear. In Figure 20.10c
V_{max} is not much larger than v_D, whereby an error arises.

Figure 20.24 shows full-wave circuits measuring the mean value defined
by equation (20.9).

For a sinusoidal signal, with a full-wave network ($v = V_{max} \sin \omega t$)

$$V_{av} = \frac{2}{\pi} V_{max} = \frac{1}{1.11} V_{rms}$$

where 1.11 is the *form factor* of the sinusoidal signal.

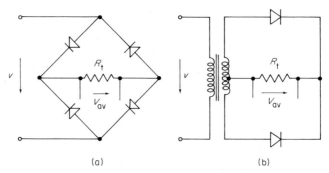

(a)   (b)

Figure 20.24   Full-wave rectifying networks: (a) bridge network; (b) centre-tapped transformer

Instruments responding to the mean value are generally scaled to indicate the r.m.s. value of the sinusoidal signal. They give erroneous readings in the case of non-sinusoidal waveshape signals. If the value indicated in such cases is divided by 1.11, the exact value of V_{av} will be obtained.

When measuring low voltages, the error effect of the threshold voltage v_D and of the non-linear resistance r can be considerably reduced by the use of an operational amplifier. In the conducting state of the diode, the gain of the non-inverting operational amplifier shown in Figure 20.25 is $(R_1 + R_2)/R_1$. The diode is open when v_r is positive and $v_r' - v_r > v_D$. When $v_r' - v_r \leqslant v_D$, the feedback stops and, due to the very high open-loop gain A_0 of the operational amplifier, the condition $v_r' - v_r > v_D$ is valid again in the case of a very low value of the input voltage v. Thus, the effect of the operational amplifier is such as if the threshold voltage of the diode had decreased to v_D/A_0. Oliver and Cage (1971) provides another account.

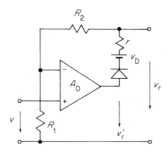

Figure 20.25   Compensation for threshold value v_D
by use of an operational amplifier

Measurement of the peak value by a diode rectifier

The circuit of a series diode peak value meter can be seen in Figure 20.26a.

The circuit waveforms (Figure 20.26b) demonstrate that, in the case of $v < v_c$, the capacitor charged to peak voltage is separated from the network

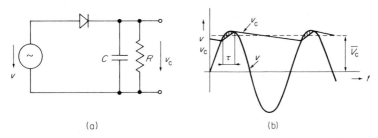

(a)   (b)

Figure 20.26 Series-diode peak value meter: (a) circuit; (b) waveform of the voltage

by the reverse-biased diode. Charging current flows only in the time interval marked by τ.

It is easy to see that the average $\bar{V}_c$ of voltage v_c will lie nearer to V_{max} the larger is the time constant RC of the system and the higher the frequency of the signal being measured is in the case of a fixed RC.

More frequently the *series capacitor* scheme is used. This is obtained in Figure 20.26a by exchanging the diode and the capacitor. One of the advantages of this circuit is that the series capacitor separates the d.c. component of the voltage to be measured.

As the diode is not ideal errors arise in measuring low voltages. The error can be considerably reduced by the use of an operational amplifier in a similar arrangement as is shown in Figure 20.25. However, it must be remembered that in this case the upper frequency limit of the voltage to be measured is determined by the frequency characteristic of the operational amplifier. This limit depends largely upon the closed-loop gain used. It typically begins to fall off at 10 kHz.

Measurement of the r.m.s. value

One of the most characteristic information quantities of a signal is given by its r.m.s. value. One approach to this problem is to measure the *true r.m.s.* value, which gives strictly the value defined by equation (20.11). The other can be to make use of a *quasi-r.m.s.* responding method, wherein the square-law relationship used is only an approximation (Oliver and Cage, 1971; Coombs, 1972).

Measurement of the true r.m.s. value can be achieved with certain types of electromechanical instrument, and with the non-mechanical methods using a thermocouple and an analog multiplier. Electromechanical instrument methods are dealt with in Section 20.2.1.

The main properties of the thermocouple were outlined in connection

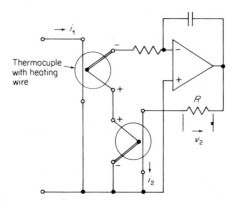

Figure   20.27   Thermocouple   r.m.s.
meter circuit

with Figure 20.22. The circuit shown in Figure 20.27 provides a d.c. current
i_2 which will be equal to the r.m.s. value of the alternating current i_1 if the
thermocouples have uniform properties. By this method influence errors
depending on the parameters of the environment can largely be eliminated.
Voltage v_2 shown in the figure will be proportional to the r.m.s. value of the
current i_1.

By applying operational amplifiers and analog multipliers in the circuit
shown in Figure 20.28 a device can be developed which provides true r.m.s.
value. However, it should again be noted that the frequency response of the
operational amplifiers restricts the upper frequency limit.

A quasi-r.m.s. responding meter can be developed by applying a circuit of
second-order characteristic (Chapter 17) approximated by breakpoints. Of
course, these methods provide only limited accuracy.

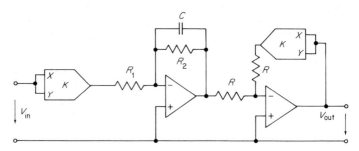

Figure 20.28   r.m.s.-meter employing operational amplifiers and
analog multipliers

Vector components of a sinusoid

Phase-sensitive rectification is employed when it is desired to find d.c. voltages proportional to the vector components of a sinusoidal signal that are directed toward and perpendicular (quadrature) to a reference direction. Phase-sensitive rectification can provide either the mean or the peak value of these components. It differs from the usual rectification in that the opening and closing of the diode (switch) is controlled by a separately supplied reference signal v_r, usually as a square-wave signal whose zero transition furnishes the phase reference. This reference signal is considerably larger than the v_m to be measured. It can be shown in a simplistic way that the output d.c. voltage V_{out} of the system is proportional to $V_{max} \cos \varphi$, where φ is the phase angle between the reference signal and the signal to be measured (Coombs, 1972; Blair and Sydenham, 1975). Figure 20.29 shows two schemes.

The cathode-ray oscilloscope (Stout, 1960) serves to display the time function of voltage. Its bandwidth ranges from zero to several hundred megahertz and with sampling front-ends, to several thousand megahertz.

The cathode-ray tube is a high-vacuum tube in which a focused electron beam, travelling at high speed from the cathode to the screen, is produced according to Figure 20.30. It traverses the space between the vertical y plates and horizontal x plates wherein the voltage applied to them deflects the beam in the direction of the field. The electron beam hits the screen which is covered with a phosphorous compound that generates visible light. The figure also shows the screen in front-view at the moment when a sinusoidal wave is connected to the y plates. In order to make the time function (x-axis expansion) of this wave appear on the screen a sawtooth

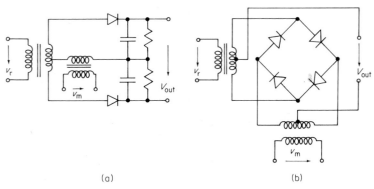

(a)   (b)

Figure 20.29   Phase-sensitive rectifiers: (a) Walter's circuit; (b) ring modulator

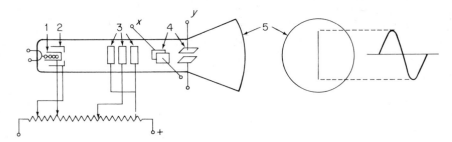

Figure 20.30   Construction of the cathode-ray tube: 1, cathode; 2, grid; 3, anode
and focusing system; 4, plate pairs producing horizontal deflection x and vertical
deflection y; 5, screen

voltage, produced in the device itself, is connected to the x plates which,
when uniformly increasing, deflects the beam on the screen from left to right
at uniform speed. Then, within a very short time during which the beam is
normally blanked out, it is returned to the starting point. In this way the
time function of the signal can be continuously displayed.

In the case where a periodic signal is to be displayed and measured, this
sweep signal must be triggered each time at the same relative phase point of
the periodic signal. This internal *trigger source* provides a stable display of
the signal applied to the y plates. A trigger source may also be derived from
an *external* source, this often providing a more stable display.

It is usual to characterize the sensitivity of an oscilloscope by the input
voltage magnitude which produces a 1 cm deviation of the beam. This
ranges from 0.1 mV/cm to 20 V/cm depending on the unit. Devices with low
upper frequency limit are more sensitive than broadband types. The time
function appearing on the screen supplies quantitative information with
uncertainty of 1 to 5 per cent of screen size. The input impedance of most
oscilloscopes is 1 MΩ shunted by a capacitance of 10 to 40 pF. This can be
increased by the use of a range of probes.

A storage oscilloscope can be constructed using a special variable persistence CRO tube or by the use of digital storage techniques combined with
D/A and A/D converters. The screen, in such instruments, can display, for
some minutes, the waveform of rapid transient *one-shot* phenomena: they
can store values for much longer periods.

Measuring transformers

There are two reasons for using measuring transformers (IEC, 1966). If the
voltage or current to be measured exceeds the measuring range of the
meters available, they can be reduced by a defined magnitude ratio with a
measuring transformer. Application is also justified in the case when an

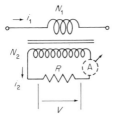

Figure 20.31   Schematic circuit of a current transformer

exactly defined ratio of voltages or currents has to be established in two circuits isolated from each other.

Measuring transformers can be characterized by the ratio of their primary and secondary quantities. The ratio of a theoretical *ideal* measuring transformer is a rational number (there is no phase error), the value of which is independent of the magnitude of the signal examined and also of the load of the secondary side. Practically these conditions cannot be fulfilled, and there exist ratio and phase errors which depend on the value of the current or voltage to be measured and on the load of the secondary winding.

The voltage transformer behaves similarly to a common voltage transformer designed in a manner to approximate the ideal conditions mentioned above. It is commonly used for measuring voltages of several times 10 or 100 kV in high-voltage distribution networks.

A current i_1 flowing in the primary winding (with a low number of turns) of a current transformer produces in the secondary winding a current i_2 determined by the excitation equilibrium (Figure 20.31):

$$i_1 N_1 = i_2 N_2$$

therefore

$$i_2 = i_1 \frac{N_1}{N_2}$$

This can be measured either by an ammeter inserted in the secondary circuit or by a voltmeter connected to the loading resistor R. For the mains frequency (50 or 60 Hz) the requirements for measuring transformers are strict and their limits of error are fixed in standards (Miljanich *et al.*, 1962). Using suitable cores, such as ferrites, their frequency limit can be extended to several MHz.

20.4   MEASUREMENT OF POWER AND ENERGY

20.4.1   Methods for Measuring Power

The *instantaneous* value of power represented by a voltage v and a current i is expressed by the equation

$$p = vi$$

and the *average* value of power for time T is given by

$$P = \frac{1}{T} \int_0^T p \, dt = \frac{1}{T} \int_0^T vi \, dt \tag{20.12}$$

In a *d.c. network* V is equivalent to v and I to i thus

$$p = P = VI$$

For a *sinusoidal* signal $i = \sqrt{2} \, I \sin \omega t$, $v = \sqrt{2} \, V \sin (\omega t + \phi)$ and $\omega = 2\pi f$ (I and V being r.m.s. values). From these the instantaneous value of the power is seen to be

$$p = vi = [\sqrt{2} \, V \sin (\omega t + \phi)][\sqrt{2} \, I \sin \omega t]$$

Simple transformation yields

$$p = [VI \cos \phi(1 - \cos 4\pi ft)] + [VI \sin \phi \sin 4\pi ft] \tag{20.13}$$

where f is the frequency of the signal. After averaging according to equation (20.12) the power is given by

$$P = \frac{1}{T} \int_0^T p \, dt = VI \cos \phi$$

As equation (20.13) shows, the instantaneous value of the power (first part of the equation) varies about the average P with a frequency $2f$, while $VI \sin \phi$ expresses the peak value of the reactive power, which similarly varies with a frequency $2f$, but with an average value of zero.

As any periodical signal can be composed from sinusoidal signals (Chapter 4), its power equals the sum of the powers of the sinusoidal components. To this the simply provable statement must be added that power can be produced only by currents and voltages of identical frequency.

Power measurements are of importance primarily in the networks of energy supply. They are, however, often required in high-frequency and low-power networks.

Methods suitable for measuring power are presented in Figure 20.32. The schemes demonstrate the measurement of the power dissipated on impedances R_x and Z_x, respectively. The relationships stated are only correct if the internal resistances of the voltage measuring devices, or if R_v, are so high that the power in them is negligible compared with the power to be measured. Some remarks must be added about the methods presented.

In a d.c. circuit power can be measured by measuring the voltage and the current. In the schemes shown in the figure the product of the values indicated by the meters gives the sum of the powers generated by resistance R_x and by the voltmeter. If the internal resistance of the voltmeter is not considerably higher than R_x, the result obtained must be corrected.

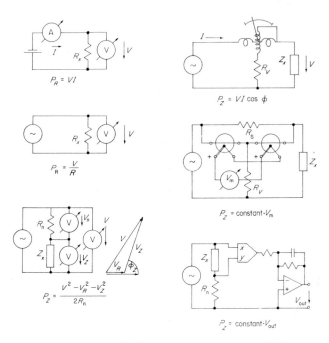

Figure 20.32   Power measurement methods

The power dissipated by a known value resistor R_x can be determined by measuring the voltage across the terminals of the resistor; power can be obtained from the relationship $P = V^2/R$.

In the case of a sinusoidal voltage V, the power of impedance Z_x can be calculated from the readings of voltmeters connected between three terminal pairs of nodes if the resistance R_n is known. The relative uncertainty of the calculated P_Z will become more dependent on the uncertainties of the voltage measurements, the closer the phase angle ϕ_z of impedance Z_x is to $\pi/2$.

For mains frequencies (up to a few hundred or a thousand Hz) the *electrodynamic* wattmeter is generally used. The power consumption of its voltage coil is of the order of watts, therefore the value indicated by the instrument must be corrected if necessary.

A possible instrument for measuring the power in high-frequency circuits is the *thermocouple* wattmeter (Coombs, 1972). It can be demonstrated that the difference between the d.c. voltages produced by two thermocouples is fairly proportional to the power in Z_x. In similar frequency ranges power can also be measured by exploiting the Hall effect (Drechsler, 1975) as well as by methods based on the measurement of the thermal energy produced by electric power: *calorimetric* methods (Coombs, 1972).

One possible type of electronic wattmeter (Friedl *et al.*, 1971) consists of operational circuits that perform the operations of multiplying and averaging.

20.4.2 Measurement of Energy

Energy W is the time integral of power:

$$W = \int_0^t P \, dt$$

The measurement of energy is primarily important in energy distribution networks, since the energy consumed is paid for on this basis. The *induction* type watt-hour meter is generally used for measuring energy. Its operation is based on the principle of the two-flux induction type instrument shown in Figure 20.4 (Golding, 1949).

The connection of a watt-hour meter into a network is similar to that of a wattmeter. It is to be noted that the accuracy for watt-hour meters is prescribed by standards.

Electric energy can also be measured by electronic methods. These are employed where special accuracy requirements exist.

20.5 MEASUREMENT OF FREQUENCY AND PHASE ANGLE

20.5.1 Methods of Frequency Measurement

The frequency f of a periodical signal is defined as the number of cycles of the signal occurring during a 1 second interval. If the period of the signal is T, then $f = 1/T$. In practice (see Section 20.1), the uncertainty of reproducing the unit of time is the smallest compared with the other basic units. Owing to the relationship between frequency and time interval, the same applies to the uncertainty of frequency measurement.

Frequency can be measured by two fundamental methods. The first possibility is by comparison of the unknown frequency, or period, with a known frequency, or period (this may be considered as frequency measurement of high accuracy). The alternative method is based upon calculation of the unknown frequency from the value of impedances, taking advantage of the phenomenon of resonance. This method is less accurate than the previous one.

Generation of a reference frequency

The basic SI unit of time (McCoubrey, 1966) *atomic time*, is defined by the frequency of the oscillation generated in the caesium atom. (Astronomical time is not equivalent.) A caesium beam resonator is commercially available in a portable construction (Heger *et al.*, 1976) having an uncertainty of 10^{-10} to 10^{-11}. For precision measurements in laboratories, however, it is, in most cases, the *quartz frequency standard* that provides reference frequency. As a result of the piezoelectric effect existing in certain crystals, particularly in quartz, an alternating polarity voltage impressed across the crystal produces mechanical dimensional changes; the force arising from such motions produces electrical charge. With appropriate feedback an oscillator can be made, the frequency of which is equal to the mechanical resonance frequency of the quartz crystal (Kartaschoff, 1978). A quartz oscillator placed in a thermostatted enclosure can yield a short-term frequency stability of about 10^{-10}.

Quartz oscillators can provide a reference base frequency from which *frequency synthesizers* (Bagley, 1967) are able to produce virtually continuously adjustable frequency values derived from this standard frequency. In this method the relative uncertainty of any derived frequency agrees closely with the relative uncertainty of the standard. The precision frequency source obtained in this way is a very useful laboratory device for conducting frequency measurements.

Measurement of frequency by comparison

An *analog comparison* can be obtained by means of the *Lissajous* patterns displayed on a CRT oscilloscope. Figure 20.33a shows the Lissajous pattern appearing on the screen. It is produced by applying a.c. voltages of frequency f_x and f_N to the horizontal and vertical plates respectively. It is

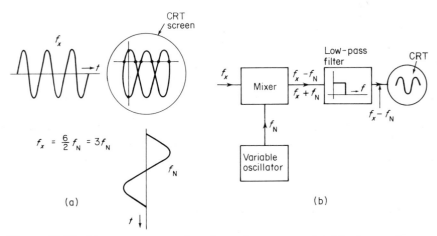

Figure 20.33   Frequency comparison in analog manner: (a) Lissajous patterns; (b) heterodyne method

practically expedient to have f_N adjustable. When the pattern appearing on the screen is motionless, the ratio of the frequencies f_x and f_N equals the ratio of the number of points produced by the intersection of the Lissajous pattern with a horizontal and a vertical straight line respectively. These points are shown dotted in the figure. If the pattern *rotates* slowly the frequency ratio obtained in this way must be modified by a correction term derived from the frequency of rotation.

The other kind of analog comparison uses the *heterodyne* method, the principle of which is shown in Figure 20.33b. The frequency f_x to be measured is mixed with a usually variable frequency f_N. By applying a low-pass filter to the mixer output the frequency of the signal obtained will be $f_x - f_N$; it can easily be observed, for instance, on the screen of an oscilloscope. At the state of *zero beat* $f_x = f_N$.

The methods outlined above permit comparisons of high accuracy, but they can be slow to use.

The method of *digital comparison* offers an ideal solution for frequency measurements. The *electronic counter* is an essential element of this method (Kartaschoff, 1978). The counter fulfils two important functions. It is able to directly count the number of the cycles of a periodic signal or it can be used to produce a pulse after a number of counts have occurred as determined by the user in advance (frequency division). It can define time intervals with accuracy corresponding to that of the frequency of the input signal.

Figure 20.34 illustrates the principle of measurement using a counter. The known frequency f_N is divided by n and the signal obtained in this way is used to close and open gate G. If, say, a 1 MHz quartz oscillator is used and a division by $n = 10^6$ performed, a 1 Hz signal will be obtained which means that the gate is closed for 1 s. In the closed state the counter counts the periods of f_x. If the counter indicates a decimal number N, then

$$f_x = \frac{N}{n} f_N.$$

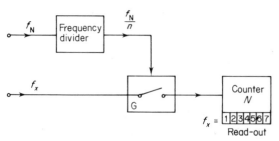

Figure 20.34   Digital method of frequency comparison

The uncertainty of the measurement is determined by the uncertainty of f_N to which the digital error of counting must be added; this is ± 1 l.s.d. (*least significant digit*) of the number indicated.

If the inputs f_x and f_N are mutually interchanged in the figure, the circuit will be suitable for measuring *period*. In this case the indication N of the counter will be proportional to the T_x period time of f_x:

$$T_x = N \frac{1}{n f_N}$$

A similar principle can be applied to construct instruments suitable for measuring time interval elapsed between specific points of an electrical waveform.

20.5.2 Measurement of Phase Angle

If the zero transitions of two periodic signals, having identical period T, are shifted from each other by time τ, the *phase angle* ϕ between them is given by

$$\phi = \frac{\tau}{T} 2\pi$$

At mains frequencies the phase angle between the current and voltage of a load is of importance because, for the case of given fixed power consumption, network losses increase with increasing phase angle. Phase angle, or $\cos \phi$, can be measured by such means as electrodynamic *quotient* meters (*crossed-coil* instruments) (see Kinnard, 1956).

A measurement having good demonstration features, but yielding poor accuracy, can be accomplished by measuring phase by means of the *Lissajous pattern*, using the cathode ray oscilloscope.

If a voltage $v_1 = V_{1max} \sin \omega t$ is applied to the vertical plates and a voltage $v_2 = V_{2max} \sin(\omega t + \phi)$ to the horizontal plates, a straight line at $45°$ or at $135°$ will appear on the CRO screen when $\phi = 0°$ or $180°$ respectively. A circle results when $\phi = 90°$ (provided that $V_{1max} = V_{2max}$ on the screen). Generally if $0° \leq \phi \leq 180°$, an ellipse will appear and ϕ can be determined from the ratio of its principal axis to its minor axis (Stout, 1960). Figure 20.35 illustrates another possibility, where

$$\sin \phi = v_2' / V_{2max}$$

An approximate measurement is possible with the use of a dual-beam oscilloscope by simultaneously displaying the two signals and measuring the magnitude of the phase shift. Care is needed to establish the internal phase difference between the two channels.

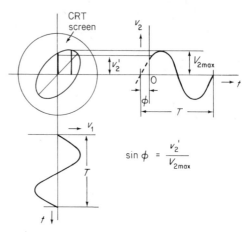

Figure 20.35   Phase measurement by means
of the Lissajous pattern

Measurement of phase shift by *digital methods* offers a significantly higher accuracy than the previous procedures. A possible solution of this is illustrated in Figure 20.36.

The accuracy is determined by the known standard frequency f_N furnished by a built-in quartz oscillator. Gate G_1 is controlled by voltages v_1 and v_2, being closed by the positive zero transitions of v_1 and opened by the subsequent positive zero transitions of v_2, in this manner being closed for an interval τ, in each period. This gate is logically AND connected with gate G_2, which is closed for an interval $t = n/f_N$. Accordingly during time t of the closed state of gate G_2, t/T signal bursts of frequency f_N, each lasting for period τ feed the counter. Each burst increases the content of the counter N

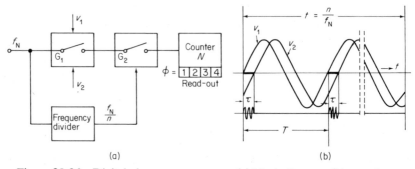

Figure 20.36   Digital phase measurement: (a) block diagram; (b) waveforms

by an amount $N_0 = \tau f_N$ giving

$$N = N_0 \frac{t}{T} = n \frac{\tau}{T} = n \frac{\phi}{2\pi}$$

Thus, the indication of the instrument is proportional to phase angle ϕ.
Generally, the interval t is not (and need not be) an integer multiple of period T. Truncation of a period of the signal passing gate G_2 gives rise to an error, the relative magnitude of which is smaller, the larger is the value of t compared with T.

20.6 IMPEDANCE MEASUREMENT

20.6.1 Characteristic Properties of Impedances

Resistance, inductance, and capacitance, in their theoretical singular senses, are termed *pure impedances*. The value of impedance describes the extent of properties of a component.

Pure impedance never appears alone; it contains, besides the *main* component, *additional* components. The impedances of this basic kind are termed *simple impedances*. The additional components, although usually producing undesirable effects, are often important qualifying parameters of simple impedance.

Impedance measurements may need to meet the following requirements.

Accuracy

This is the confidence interval of the main and additional components being measured. Accuracy depends on:

(a) Absolute accuracy of the reference, standards, elements applied (see Section 20.1);
(b) accuracy of the measuring instruments being used;
(c) sensitivity of the comparator circuit and the zero-detector;
(d) stray parameters arising as a consequence of the measuring arrangement.

Sensitivity of the actual measuring networks and the effect of the arising stray parameters will now be considered.

Span of measuring ranges

This is characterized by the set of ranges satisfying the accuracy prescriptions. It may be necessary to cover 10 to 15 orders of magnitude.

Frequency response

It is desirable that the impedance meter can be used over the widest possible frequency range with its accuracy remaining within permissible limits.

Measurement speed

This is proportional to the reciprocal of the time required to carry out the measurement and the evaluation of the measurement result. One advantageous feature of automatic measuring devices is the possibility of greatly increasing the measurement speed.

From the concept of simple impedance it follows that the *equivalent circuit* contains, besides the main component, additional components in numerous combinations. It is to be noted that some of the additional components are inherently built into the object to be measured and cannot be isolated, while others appear as stray parameters coming from the actual arrangement of the measuring circuit. Effort should be made to eliminate this latter effect in the course of the measurement.

The magnitude and position of additional components in the equivalent circuit affect the value and frequency response of the simple impedance. Frequency dependence, therefore, does not follow the ideal one. When stating accuracy, it is important to give the frequency range pertinent to the measurement.

The equivalent circuit of simple impedances is a *model* which always refers to an impedance that can be physically realized. When it is constructed it must be taken into account that the set of components causing additional effects always depends on the precision of the approximation of the model applied. Figure 20.37 presents the equivalent circuits of commonly met simple impedances along with their simplified versions.

It is important to realize that devices for the measurement of impedance provide the components of either the series or the parallel equivalent circuit. The correctness of the model can be evaluated by a comparison in the frequency range. It is, therefore, desirable that the impedances physically approximated by series equivalent circuits be measured by a method yielding the series equivalent circuit elements. Similar considerations apply to the cases of parallel circuits.

20.6.2   Typical Circuits for Impedance Measurement

Impedance measuring instruments can be classified as either *meter-type* or *null-type* devices. In meter-type devices voltage and current are measured, and from their values, the unknown impedance is calculable. A null-device compares the unknown impedance with known impedances in a network

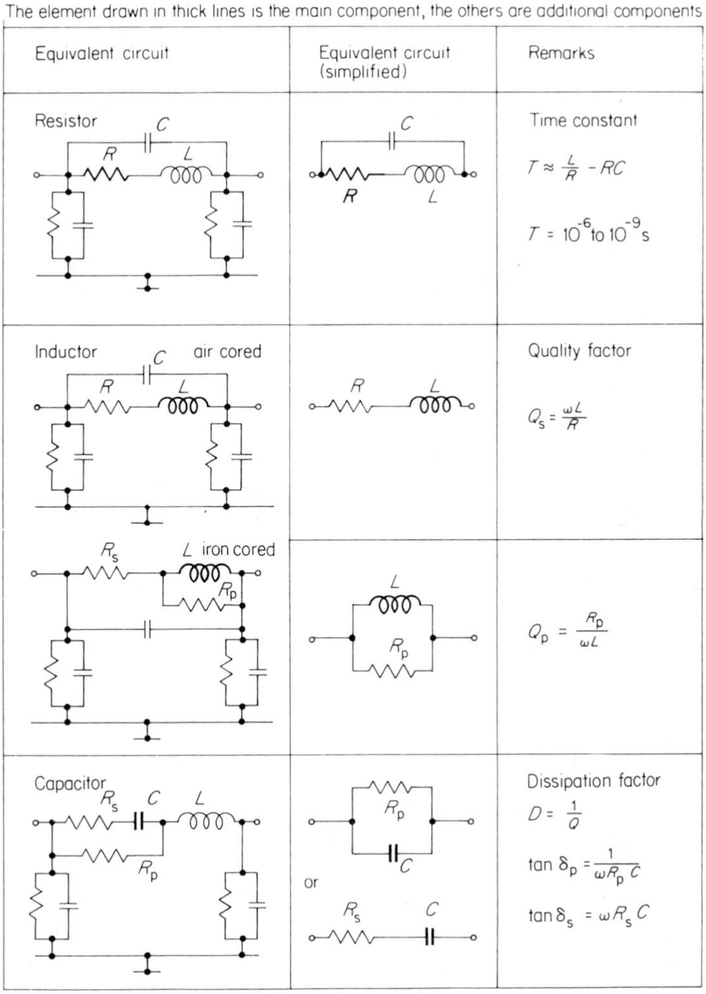

The element drawn in thick lines is the main component, the others are additional components

Equivalent circuit	Equivalent circuit (simplified)	Remarks
Resistor		Time constant $T \approx \frac{L}{R} - RC$ $T = 10^{-6} \text{ to } 10^{-9} \text{ s}$
Inductor — air cored		Quality factor $Q_s = \frac{\omega L}{R}$
iron cored		$Q_p = \frac{R_p}{\omega L}$
Capacitor	or	Dissipation factor $D = \frac{1}{Q}$ $\tan \delta_p = \frac{1}{\omega R_p C}$ $\tan \delta_s = \omega R_s C$

Figure 20.37   Equivalent circuits of simple impedances

having a known relationship between its elements when it is adjusted to give a zero output signal. Some meter-type methods are faster than null-type procedures because they require no balance; but they are usually less accurate.

In many cases *sensitivity* is an important feature of an impedance measuring method. The sensitivity of a measuring circuit is the response of the output voltage to a small change of the impedance to be measured near

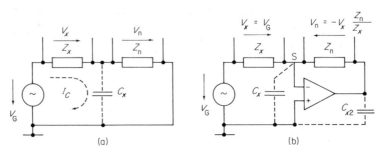

Figure 20.38   Simple comparison networks (a) series connection; (b) incorporating an operational amplifier

to the balance. Sensitivity always depends on the input voltage and on the values and arrangement of the components of the measuring circuit. It is determined by the properties of the measuring circuit and is termed *circuit sensitivity* (Hoyer and Wiessner, 1957). It can be expressed as follows:

$$H = \frac{dV_o/V_G}{dZ_1/Z_1} \quad \text{if} \quad Z_1 \rightarrow Z_{1o} \qquad (20.14)$$

where the value of Z_1 to be measured equals Z_{1o} in the balanced state. The deviation dZ_1 from Z_{1o} produces a voltage dV_0 on the output of the circuit. V_G is the voltage supplying the measuring network. The value of H is a suitable parameter for comparing various measuring networks.

Another important aspect is the effect of the *stray parameters* which will also be examined in the following treatment of impedance measuring circuits.

In Figure 20.38 stray *distributed* capacitances are substituted by concentrated *lumped equivalents* connected to nodes and drawn with dotted lines. Their effect in causing errors will be examined.

Simple comparison

The simplest way of comparing two impedances is to connect the known and the unknown impedances in series and measure the voltages arising. The scheme shown in Figure 20.38a represents a very accurate measurement method for measuring resistances, but it furnishes only the ratio of *absolute values* in the case of impedances.

Based on the figure, the following relationships can be written:

$$|Z_x| = |Z_n| \frac{V_x}{V_n} \qquad R_x = R_n \frac{V_x}{V_n} \qquad H_{max} = \tfrac{1}{4}$$

The current of stray capacitance C_x also flows through Z_x causing an error. In Figure 20.38b, owing to the presence of the operational amplifier, point S becomes a virtual ground and thus stray parameters C_{x1} and C_{x2} have no effect. By transferring voltages V_x and V_n through difference amplifiers into a comparator unit they can be compared for magnitude and phase (Hashimoto and Tamamura, 1976). In this way the components of Z_x can be determined (see also Figure 20.37).

Comparison by means of a ratio transformer

The circuit containing a ratio transformer is similar partly to the previous circuit, partly to the bridge network. The output voltages of a properly constructed transformer (inductive voltage divider) are proportional with very high accuracy (even in the loaded state) to the numbers of turns (Oliver and Cage, 1971). In the scheme (Figure 20.39) suitable values of N_1 and N_2 can yield the state where $V_o = 0$, in which case $Z_x = Z_n N_1/N_2$. The figure shows the stray capacitances pertaining to Z_x. These are in parallel connection partly with the coil having N_1 turns and partly with the null detector. In this connection their influence on the balance state is significantly reduced.

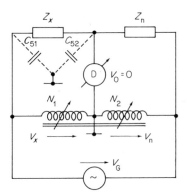

Figure 20.39 Impedance measurement with a ratio transformer

Bridge network

The bridge network is a widely used impedance-measuring network (Figure 20.40). Generally the equation yielding the balance condition, where $V_0 = 0$, has the form

$$Z_1 Z_4 = Z_2 Z_3$$

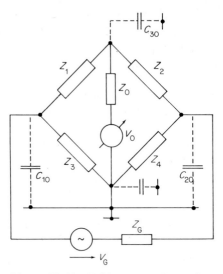

Figure 20.40   Bridge network consisting of impedance

Being complex this can be decomposed into two equations. From the
polar form of the impedances

$$|Z_1|\,|Z_4| = |Z_2|\,|Z_3|$$

and

$$\phi_1 + \phi_4 = \phi_2 + \phi_3$$

where ϕ_i is the phase angle of the corresponding impedance.
  In the case of resistances, where $Z = R$, the equation

$$R_1 R_4 = R_2 R_3$$

provides the condition of balance.
  Generally an electronic zero detector having $Z_o \rightarrow \infty$ and source having
small internal resistance of $Z_G \rightarrow 0$ are employed. With these assumptions

$$V_0 = V_G\left(\frac{Z_1}{Z_1 + Z_2} - \frac{Z_3}{Z_3 + Z_4}\right)$$

and

$$H = \frac{dV_0/V_G}{dZ_1/Z_1}\bigg|_{Z_1 \rightarrow Z_{1o}} = \frac{Z_2/Z_1}{(1 + Z_2/Z_1)^2} = \frac{F_0}{(1 + F_0)^2}$$

where $F_0 = Z_2/Z_1 = Z_4/Z_3$ is the ratio of the bridge arms connected to either terminal of the zero detector. F_0 is termed the *bridge ratio* (Hoyer and Wiessner, 1957).

In the case of a d.c. bridge H has its maximum value where $F_0 = 1$ and so

$$H_{max} = \tfrac{1}{4}$$

An interchange of the source and the zero detector has no influence upon the condition of balance, but it affects the sensitivity as illustrated in Figure 20.41.

For an a.c. bridge the value of H is usually a complex number: its absolute magnitude value will be considered in the following discussions:

$$|H| = \frac{1}{K + 2 \cos \phi}$$

where $K = |F_0| + 1/|F_0|$; $2 \leqslant K < \infty$, and ϕ is derived from $F_0 = |F_0| \, e^{j\phi}$.

Assuming $|F_0| = 1$, then at $\phi = 0°$, $\pm 90°$, $\pm 180°$ the maximum of the circuit sensitivity $H_{max} = \tfrac{1}{4}, \tfrac{1}{2}$ and ∞ respectively. The latter case is possible only for impedances truly free of loss. This cannot be achieved in practice.

As can be seen there exists, in general, an optimum for the connection nodes of the source and the zero detector.

Figure 20.40 indicates the stray capacitances reduced to the nodes of the bridge. Evidently, C_{30} has no effect on the balance, but C_{10} and C_{20} are coupled in parallel with impedances Z_3 and Z_4. Their effect can be reduced by choosing the grounding point so that the stray capacitances are coupled with the smallest possible bridge impedances.

The effect of stray parameters can be largely eliminated at the expense of making the network and balancing procedure more complicated, say, in the manner shown in Figure 20.42.

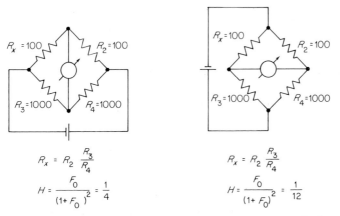

Figure 20.41 Effect of interchanging the source and the detector

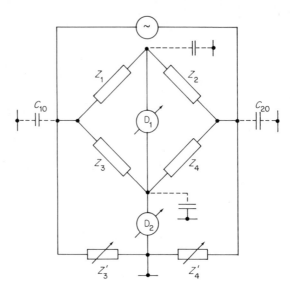

Figure 20.42   Elimination of stray parameters by
means of Wagner earth connection

When both detectors D_1 and D_2 indicate zero voltages, stray capacitances C_{10} and C_{20} are coupled to impedances Z_3' and Z_4' respectively, and do not, therefore, affect the elements of the main bridge.

This method can be automated using electronic elements.

Bridged-T circuit for impedance measurement

This circuit is considered balanced when $V_o = 0$ with a non-zero input voltage V_G (Figure 20.43a).

After star-to-delta transformation the network is as shown in Figure 20.43b for which

$$Y_{12} = \frac{Y_1 Y_2}{Y_1 + Y_2 + Y_3}$$

where Y_1, Y_2, and Y_3 are the admittances corresponding to impedances Z_1, Z_2, and Z_3 in Figure 20.43a.

V_0 becomes zero when there is zero admittance, that is when infinite impedance exists between points 1 and 2. From these conditions it follows that the balanced state occurs if

$$Y_{12} + Y_4 = 0 \quad \text{or} \quad Z_{12} + Z_4 = 0$$

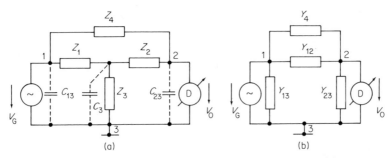

Figure 20.43 Bridged T-circuit for impedance measurement: (a) arrangement of the measuring network; (b) after star-to-delta conversion

The value of V_0, for the unbalanced state, is given by

$$V_0 = V_G \frac{Z_{23}}{Z_{23} + Z_4 Z_{12}}$$

From this

$$H = \frac{Z_{23}}{Z_{12}} = \frac{Z_3}{Z_1}$$

Theoretically this ratio can be arbitrarily high, that is, there is no upper limit to the sensitivity of the circuit. From the viewpoint of stray capacitances it is significant that, assuming lumped equivalent network elements, only capacitance C_3 coupled in parallel with Z_3, will cause error. Therefore, it is advisable to choose Z_3 as small as possible. *Twin-T circuits* can be treated similarly (Stout, 1960).

Measuring network with current comparator

In the circuit shown in Figure 20.44 the coils of N_1, N_2, and N_0 turns are wound on a common toroidal iron core having high initial permeability. When $V_0 = 0$ excitation of the core is zero. Given this ideal circumstance

$$I_x N_1 - I_n N_2 = 0$$

When there is no flux in the core, the impedances of coils N_1 and N_2 equal their resistances, and if these can be neglected compared with Z_x and Z_n then

$$Z_x = Z_n \frac{N_1}{N_2}$$

The number of turns can be varied including adjustment of a fractional

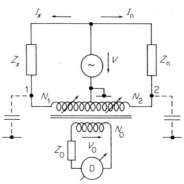

Figure   20.44   Current   comparator network

number of turns. If it is assumed that $Z_0 \to \infty$ and $Y_x = 1/Z_x$ is the admittance to be measured, the sensitivity of the circuit is

$$|H| = \omega \Lambda N_1 N_0 Y_x$$

where Λ is the *magnetic conductivity* of the iron core. Figure 20.44 shows the effect of stray capacitances which, in the balanced state where nodes 1 and 2 are practically at earth potential, can usually be neglected. This is more favourable than for the case of a bridge circuit. Measuring networks using a current comparator permit measurements to made that have very high accuracy in the lower frequency range (Miljanich *et al.*, 1962).

It is worth considering the *sensitivity* of the circuit when referred to the *main* and the *additional* components of the impedance to be measured.

Let the impedance represented by its series equivalent circuit $Z = R + jX$, in which either R or X may be the main component. Consider a balanced measuring network, then there will be a change in the value of Z by $\Delta Z = \Delta R + j\Delta X$. The relative magnitude of this change is

$$\left|\frac{\Delta Z}{Z}\right| = \left(\frac{\Delta R^2 + \Delta X^2}{R^2 + X^2}\right)^{1/2} = \left(\frac{\Delta R^2}{R^2}\frac{1}{1+Q^2} + \frac{\Delta X^2}{X^2}\frac{Q^2}{1+Q^2}\right)^{1/2}$$

where $Q = X/R$ is the *quality factor* of the impedance to be measured. Let us examine the following two cases:

(a)   Let $\Delta X = 0$ considering the sensitivity to a change ΔR. Using equation (20.14) this will be

$$\left|\frac{\Delta V_0}{V_G}\right| = |H| \left|\frac{\Delta Z}{Z}\right| = |H| \frac{\Delta R}{R} \left(\frac{1}{1+Q^2}\right)^{1/2}$$

(b)   Let $\Delta R = 0$ considering the sensitivity to a change ΔX,

$$\left|\frac{\Delta V_0}{V_G}\right| = |H|\frac{\Delta X}{X}\left(\frac{Q^2}{1+Q^2}\right)^{1/2}.$$

As can be seen from the foregoing, the sensitivity in the measurement of the main component is approximately Q times higher than the sensitivity referring to the additional component. This manifests itself in the performance specification of impedance measuring devices where less accuracy is generally stated for measurement of the additional component than for the main component.

20.6.3   Methods of Impedance Measurement

Methods for measuring resistances

Dominant measurement methods are summarized in Figure 20.45. In resistance measurements by the *volt–ammeter method*, R_x is determined by use of Ohm's law. This method is suitable for measuring both very small and very large resistances (milliohmmeter and megohmmeter, respectively). When the source can be considered as a constant current source the voltage on the terminals of R_x is directly proportional to the value of R_x. This is how measurements are made by *digital ohmmeters* (Oliver and Cage, 1971). The *series ohmmeter* method uses a meter of the d'Arsonval type; it is used generally only for indicative measurements, since it covers the range from 0 to ∞ and has limited accuracy. Ohmmeters working on the principle of *quotient meters* (cross-coiled instruments of d'Arsonval type, see Section 20.3.1) indicate the value of the resistor directly. The principle of *voltage comparison* offers very high accuracy, as voltages can be measured with high accuracy (see Figure 20.38). When small resistances ($R_x < 1$ to $10\,\Omega$) are measured, accuracy of measurement by the *Wheatstone bridge* method will be decreased because the resistance of the leads and the contact resistance of the connectors appear in the value of R_x being measured. In the measurement of high resistances ($R_x > 10^7\,\Omega$) the conductivity of the stray parameters becomes comparable with that of R_x, restricting accuracy.

A modified version of the Wheatstone bridge, the *Kelvin bridge*, is used to measure low resistances. For low resistances four terminals should be developed which separate the functions of current admission and voltage measurement.

In the scheme shown in Figure 20.45 the equality $R_3'/R_3 = R_4'/R_4$ must be ensured; this can be accomplished in the simplest way by making $R_3' = R_3$ and $R_4' = R_4$.

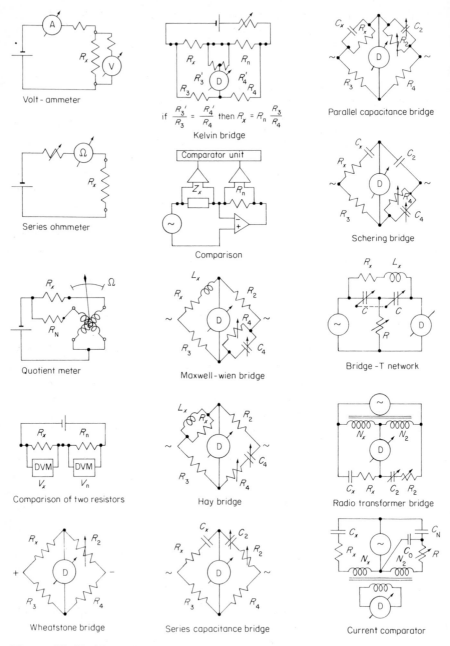

Volt-ammeter

Kelvin bridge

if $\dfrac{R_3'}{R_3} = \dfrac{R_4'}{R_4}$ then $R_x = R_n \dfrac{R_3}{R_4}$

Parallel capacitance bridge

Series ohmmeter

Comparison

Schering bridge

Quotient meter

Maxwell-wien bridge

Bridge-T network

Comparison of two resistors

Hay bridge

Radio transformer bridge

Wheatstone bridge

Series capacitance bridge

Current comparator

Figure 20.45 Networks suitable for measuring resistance, capacitance and inductance

The Kelvin bridge can be used with sufficient accuracy for values ranging down to $10^{-7}\,\Omega$ (Oliver and Cage, 1971).

When high resistances, up to $R > 10^7\,\Omega$, are to be measured, a properly *guarded* Wheatstone bridge should be used (Stout, 1960).

Methods for measuring inductances and capacitances

The most important measurement methods for this purpose are summarized in Figure 20.45. The method of *voltage comparison* is based on the *three-voltmeter principle*. If the impedance Z_x to be measured is connected in series with a known resistance R_n, the three voltages measured between the terminals determine the components of Z_x. The figure shows the schematic circuit diagram of an universal impedance meter (Hashimoto and Tamamura, 1976).

There exists a great variety of bridge circuits (Oliver and Cage, 1971). In the figures, bridge circuits are presented, as examples, for measuring series and parallel components. In addition to the impedance components indicated, the Q (quality factor) and the dissipation factor $1/Q = D$ can also be measured by the bridges.

Owing to the disturbing effect of stray parameters the bridge circuit can exhibit significant inaccuracy at higher frequencies, for instance, above 100 kHz. This error can be considerably reduced in some cases by the application of the substitution method, see Section 20.1.2.

Bridged-T and twin-T circuits are more suitable for measurements in the high-frequency ranges. Figure 20.45 includes, as an example, an inductance meter using a bridged-T network. Twin-T circuits are capable of measuring impedances with satisfactory accuracy even at frequencies of the order of 100 MHz (Stout, 1960).

The accuracy of measuring networks using ratio transformers and current comparators is limited by the properties of the iron. The accuracy obtained with them, however, is higher than that of bridge circuits since the uncertainty in the number of turns can be smaller than that of any other variable element.

20.7 MAGNETIC MEASUREMENTS

20.7.1 Magnetic Units

Magnetic flux (ϕ) and *magnetizing force* (H) are basic quantities for magnetic measurements.

From these can be derived other quantities, as follows, by means of appropriate physical relationships:

$$\text{\textit{Magnetic permeability}} \quad \mu = \mu_r \mu_0 = B/H$$

where $\mu_0 = 4\pi \times 10^{-7}$ H/m is the permeability of free space and μ_r is the relative permeability.

$$\textit{Magnetic flux density} \quad B = \phi/A$$

where A is the surface area crossed by the magnetic flux.

$$\textit{Magnetomotive force} \quad \Theta = IN = Hl$$

where I is the current passing through a coil of N turns, and l is the magnetic path length.

$$\textit{Magnetic conductivity} \quad \Lambda = \mu A/l$$

Magnetic loss is the dissipated power in the iron core caused by the changing magnetic flux.

Magnetic measurements are more difficult to make, and less accurate than electrical ones because magnetic flux must be considered as a distributed rather than a lumped parameter. Electric current is closely confined within an electrical conductor because the conductivity of the conductor is very large compared with that of insulation. In the case of magnetic flux it is difficult to totally confine the flux to a localized area because the magnetic conductivities of all of the possible paths for the flux are comparatively similar.

20.7.2   Measurement of Magnetic Flux

In Section 20.1 the definition of weber (W) as the unit of magnetic flux was given. The definition is based on a well known physical phenomenon. Whenever a magnetic flux linking with a coil is changed, a voltage is induced in the circuit, the magnitude of which is proportional to the rate of change of flux according to the relationship $v = d\psi/dt$. For measurement of a steady value magnetic flux two methods can be adopted.

The *fluxmeter*, Section 20.2, is able to measure the change of flux (Stout, 1960). If a *search coil* with known parameters (number of turns N, area of cross section A) connected to a fluxmeter is moved from a zero magnetic flux location to the area of the flux, then the deviation of the fluxmeter will be proportional to $N\phi$ where ϕ is the flux which crosses the area of the coil in its final position (Figure 20.46a). Obviously this method can also be used for the measurement of flux changes of any kind.

An alternative is the *Hall effect* method. Some semiconductors, for example indium arsenide, exhibit a phenomenon known as the Hall effect in the presence of steady magnetic field. If the square-shaped semiconductor plate is placed in a transverse magnetic field and a constant current I passes through the plate as shown in Figure 20.47, a voltage V_H appears which is proportional to the flux density B. The sensor plate can be made in

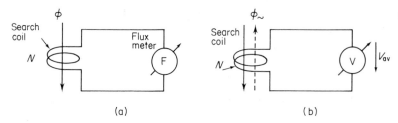

Figure 20.46   Measurement of magnetic flux: (a) fluxmeter; (b) average
value responding voltmeter

millimetre sizes. As this transducer converts the magnetic flux density into a
proportional voltage, all characteristic parameters of flux densities changing
in time can also be measured.

In the case of periodically changing magnetic flux the following relationship can be written for the search coil method:

$$v_i = d\psi/dt$$

and

$$V_{av} = \frac{2}{T} \int_0^{T/2} v_i \, dt = \frac{2}{T} \int_{-\psi_{max}}^{+\psi_{max}} d\psi = \frac{4}{T} \psi_{max}$$

where v_i is the instantaneous value of the voltage in the search coil placed in
the magnetic field under test, T is the time period of the change of flux and
$\psi_{max} = N\phi$ is the maximum value of the flux linking with the coil having N
turns. Figure 20.46b shows the arrangement of the measurement where the
true average value of the voltage measured is proportional to the maximum
value of the flux.

Each method mentioned above is applicable for the measurement of
magnetic flux density.

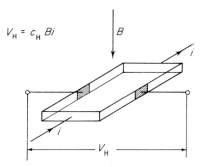

Figure 20.47   Parameters of the
Hall effect

20.7.3 Measurement of Magnetizing Force

In the case of a ring-shaped specimen the magnetizing force H can be calculated with good accuracy from the relationship $H = IN/l_{av}$ where N is the number of turns conducting the exciting current I and l_{av} is the mean circumference of the ring.

For specimens with shapes differing from rings, such as bars, the magnetizing force H can be determined by measuring the flux density in air adjacent to the specimen with a search coil which is close to its surface but which does not surround it. This method is based on the assumption that the tangential component of H at a common surface between a magnetic and non-magnetic medium is not altered when going from one medium to the other. From the value of the tangential component of flux density measured in air at a point adjacent to the surface of the magnetized specimen the magnetizing force, which is effective in producing the flux density B in the specimen, can be computed at that point in accordance with the relationship $B = \mu H$ (Harris, 1952).

20.7.4 Iron Loss Measurement

When iron is subjected to an alternating magnetic field a loss of power occurs due to the *magnetic hysteresis* effect and to *eddy currents*. In magnetically conductive materials that are also electrically conductive, eddy currents are induced as a result of the changing magnetic flux. Generally the goal of loss measurement is to determine the sum of these two losses.

For measuring the losses in iron cores two basic methods are now described.

The *wattmeter method* facilitates the measurement of magnetic loss with a wattmeter as Figure 20.48 shows. Consider a transformer the iron loss of which is to be measured. If the voltage coil of the wattmeter were supplied with the primary input voltage, the reading of the device would include not only the iron loss but also the copper loss of the primary winding. For this

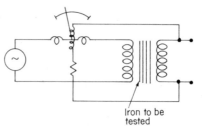

Iron to be
tested

Figure 20.48   Measurement of iron
loss with power meter

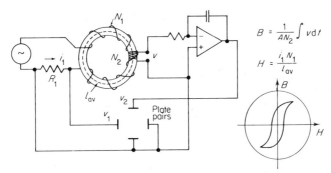

Figure 20.49   Demonstration of hysteresis loop with cathode ray oscilloscope

reason the voltage coil is connected to the secondary coil. The ratio of the primary and secondary voltages must be taken into account when calculating the iron loss from the reading of the power meter (Golding, 1949).

The other method for loss measurement was treated in connection with inductance measurement (see Section 20.6). The result of each inductance measurement method contains a resistance R_x which represents all losses of the coil. If the voltage on the terminals of an inductance having a parallel resistance R_{xp} is V, the total loss of the coil can be calculated as $P = V^2/R_{xp}$. When the current I passing through the coil is known then the loss, applying the series equivalent circuit with R_{xs}, is $P = I^2 R_{xs}$.

Figure 20.49 shows the demonstration of the hysteresis loop on the screen of a CRT oscilloscope. The magnetizing force H is proportional to i_1, which produces a voltage v_1 on resistor R_1. The vertical plates of the CRT are supplied by voltage v_1 which is proportional to H and the horizontal plates by v_2, which is proportional to the flux density B in the specimen.

REFERENCES

Bagley, A. S. (1967). 'Frequency and time measurements', in *Encyclopedia of Physics*, Vol. XXIII, Springer-Verlag, Berlin, pp. 350–60.

Bendat, I. (1971). *Random Data: Analysis and Measurement Procedures*, Wiley, New York.

Blair, D. P. and Sydenham, P. H. (1975). 'Phase sensitive detection as a means to recover signals buried in noise', *J. Phys. E: Sci. Instrum.*, **8**, 621–7. (Reprinted in 1982, B. E. Jones (Ed.), *Instrument Science and Technology*, Vol. 1, Adam Hilger, Bristol, pp. 134–41).

Coombs, C. F. (1972). *Basic Electronic Instrument Handbook*, McGraw-Hill, New York

Drechsler, R. (1975). *Messung elektrischer Energie*, VEB Verlag Technik, Berlin.

Dudley, R. L. and Laing, V. L. (1973). 'A self-contained, hand-held digital multimeter—a new concept in instrument utility', *Hewlett-Packard J.*, **25,** No. 3, 2–9.

Elnekavé, N. (1975). *Les Unités Électriques. Evolution des Procédés Utilisés pour la Définition et la Conservation des Étalons Foundamentaux*, Bureau National de Métrologie, April.

Friedl, R., Lange, W., and Seyfried, P. (1971). 'Electronic, three-phase four-wire power–frequency converter with high accuracy over a wide range', *IEEE Trans. Instrum. Meas.*, **IM-20,** 308–12.

Golding, E. W. (1949). *Electrical Measurements and Measuring Instruments*, Pitman, London (also 1968).

Harris, F. K. (1952). *Electrical Measurements*, Wiley, New York (also 1966).

Hashimoto, S. and Tamamura, T. (1976). 'An automatic wide-range digital *LCR* meter', *Hewlett-Packard J.*, **28,** No. 1, 9–16.

Heger, C. E., Hyatt, R. C., and Seavey, G. A. (1976). 'A cesium beam frequency reference for severe environments', *Hewlett-Packard J.*, **27,** No. 7, 2–10.

Herpy, M. (1980). *Analog Integrated Circuits*, Wiley, New York.

Hoyer, H. and Wiessner, W. (1957). 'Zur Genauigkeit der kapazitäts- und Verlust-faktorbestimmung mit Wechselstrommessbrücken', *Arch. Elektrotech.*, **43,** 169–77.

IEC (1966). *Publ. 185: Current Transformers*; *Publ. 186: Voltage Transformers*, International Electrotechnical Commission, Geneva.

IEC (1973). *Direct Acting Indicating Electrical Measuring Instruments and their Accessories:* Publ. 51, International Electrotechnical Commission, Geneva.

Josephson, B. D. (1962). 'Possible new effects in superconductive tunnelling', *Phys. Lett.*, **1,** No. 7, 251–3.

Kartaschoff, P. (1978). *Frequency and Time: Monographs in Physical Measurement*, Academic Press, New York.

Kinnard, I. F. (1956). *Applied Electrical Measurements* Wiley, New York and Chapman & Hall, London.

McCoubrey, A. O. (1966). 'A survey of atomic frequency standards', *Proc. IEEE*, **54,** 116–35.

Melchert, F. (1979). 'Darstellung der Spannungseinheit mit Hilfe des Josephson-Effektes', *Tech. Messen*, 59–64.

Miljanich, P. N., Kusters, L., and Moore, W. J. M. (1962). 'The development of the current comparator, a high-accuracy a-c measuring device', *Trans. AIEE (Commun. Electron.)*, **81,** 359–68.

Oliver, B. M. and Cage, J. M. (1971). *Electronic Measurements and Instrumentation*, McGraw-Hill, New York.

Stout, M. B. (1960). *Basic Electrical Measurements*, Prentice-Hall, Englewood Cliffs, NJ.

Taylor, B. N. (1967). 'On the use of the AC Josephson effect to maintain standards of electromotive force', *Metrologia*, **3,** 89.

Thompson, A. M. (1959). 'The cylindrical cross capacitor as a calculatable standard', *Proc. IEE*, **106B,** 307–10.

Handbook of Measurement Science, Volume 2
Edited by P. H. Sydenham
© 1983 John Wiley & Sons Ltd

Chapter

21 W. TRYLINSKI

Mechanical Regime of Measuring Instruments

Editorial introduction

By the 1940's design and implementation of fine mechanism concerned with information processing and low level energy use had reached a high level of maturity, its capability having been developed over centuries of experience (Sydenham, 1979).

Most likely because of the then rise to prominence of electronic alternatives, interest in the design of fine mechanics, especially in the English speaking western world, fell away to a marked extent. As an indication of this lack of interest it seems, apart from several valuable booklets by Geary of the Sira Institute in England, not one book on the design of fine mechanism was published by an English speaking author. Even today the sole and fortunately authoritative English language text, is that by Trylinski, which was translated from his native Polish language (Trylinski, 1971). In an attempt to ensure we do not lose the vast amount of published literature of the past, and to bring it to the attention of today's instrument designers who still need to have access to the elements of good fine instrument mechanics, a series of articles was compiled (Sydenham, 1980) with a view to eventual publication as a book.

This chapter provides a review of the mechanical design aspects of an instrument, the reader being guided to Sydenham (1980) and Trylinski (1971) for greater depth of available specialized literature and general design principles (in the former) and detailed design of elements (in the latter).

21.1 MATERIALS OF CONSTRUCTION

21.1.1 Ferrous Metals

In this, and subsequent subsections, materials are discussed (Davidson, 1971; Trylinski, 1971; Trylinski, 1978; Sydenham, 1980).

Carbon Steels

Mild and very mild steels, those with carbon content below 0.2%, are used for cold-worked parts such as headed screws with rolled threads. They are also used for carburized or cyanided parts needing high surface hardness, or for elements of some magnetic circuits not designed to meet high requirements.

Carbon steels with higher carbon content are used when higher strength, without recourse to heat treatment, is required, for example fine pitch gear wheels.

Free cutting, low carbon steel, of average carbon content 0.1% and with raised phosphorus content, is made to be machined on automatic lathes; it is brittle. It is not suitable for cold-worked parts. Screws, shafts, arbors, and leafed pinions are made from it.

Heat-treated carbon tool steel, carbon content over 0.6%, is used for parts that are exposed to wear. Special kinds of such steel, made to be machined on automatic lathes, are available. They are only slightly deformed by heat treatment when quenched in oil.

Mild carbon steel sheets for forming are used as strips or sheets with a very smooth surface and also for punched parts. Non-cleaned sheets should not be used because of the high cost of smoothing the surface.

Steel wire for the manufacture of springs is of two types: *cold-drawn* without previous heat treatment—for springs not designed to meet high requirements; and *piano wire*, heat-treated before final drawing. Springs are cold-coiled and those made of piano wire are tempered after coiling at a temperature of 210–220 °C. This raises the elastic limit of the wire.

Alloy steels

Alloy steels of the toughening type, those containing up to several per cent of Cr, Ni, Mn, of the low-carbon type for carburizing and of the high-carbon type which can be toughness-hardened, are often used not because of their strength or hardness but because they undergo only slight deformation as the result of heat treatment. With these alloys, accurate elements can be made without needing machining after heat treatment.

Nitriding steels are used for high-precision parts requiring very hard surfaces since nitriding causes almost no deformation.

Stainless steels of 12–14% Cr are corrosion-resistant, such as is needed in an oceanic or tropical climate. Their rust-proofing improves with surface quality and polishing.

Austenitic stainless steels of 18% Cr, Ni, are resistant to acids and caustic chemicals. These are used for membrane diaphragms and deep drawn parts.

Thin steel plate springs are made from toughness-hardened and cold-rolled ribbons. Thick springs of silicon–manganese steel are toughness-hardened after manufacture.

Special, magnetically soft, electrical steels, of 0.0–0.4%C content, are used for magnetic circuits needing very small coercive force and also for parts undergoing great plastic deformation in the production process.

21.1.2 Non-Ferrous Metals

Brasses

Leaded brasses (Cu 58%, Pb 2% or Cu 63%, Pb 1.5%) are very machinable. Hard sheets are used for mechanism plates needing bearing holes in them and for fine pitch gear wheels. They should not be bent in fabrication as they fracture easily.

Brass (Cu 63%, Zn 37%) is a ductile material, suitable for the plastic working of bending and drawing operations. It should not be used for fine pitch gear wheels because large burrs can arise in milling. Cast brass (Cu 60%, Pb 1%, Zn 39%) is used to make mechanism frames and levers.

Bronzes

Phosphor bronze containing 7% of tin is used in the form of bands, sheets, and wires for elastic elements such as flat springs, hairsprings, and membranes. Its Young's modulus ($E = 105 \times 10^4$ MPa) is lower than that of steel, permitting greater elastic deformation. Heavily loaded elastic elements are made of beryllium copper (Be $\approx 2\%$), a material which also has high electrical conductivity.

Nickel and its alloys

Commercially pure nickel, being highly corrosion-resistant, is used for moving parts of mechanisms working in corrosive media. Nickel silver (Ni 10 to 30%; Cu 55 to 63%; Zn, the rest: also known as *German silver*) because of its high corrosion resistance, low Young's modulus, and high strength, is used for contact springs; it contains no silver. Invar (Ni 36%, Fe 64%) has a coefficient of thermal expansion practically equal to 0 and is used for measuring instrument elements whose dimensions should not change appreciably with temperature.

Aluminium and its alloys

Housings, indicating dials, and pointers are often made of commercially pure aluminium because of its softness and ductility.

Aluminium casting alloys are suitable for thin-walled, intricate pressure castings, frames, housings, brackets, and levers.

Wrought aluminium alloys can have the strength associated with mild steel. However, their Young's modulus is only one third of that of steel ($E = 75 \times 10^4$ MPa) and thus, under equal loads, they will undergo three times the deformation that would result in steel.

Aluminium and all its alloys are used primarily because of their low density (2.8 g cm^{-3}).

Magnesium alloys

Magnesium alloys, both of the casting and of the wrought type, have low strength (50 to 120 MPa) and are very liable to corrosion. They are used in cases when light weight is of prime importance (density $\approx 1.8 \, \text{g cm}^{-3}$).

Zinc alloys

Zinc alloys are used for making very thin-walled, pressure-cast, parts of intricate shape. Surface-treated castings show considerable corrosion resistance, but are heavy (density $6\text{–}6.7 \, \text{g cm}^{-3}$), brittle and of very low strength. Inserts made of stronger materials, for example brass, are easily placed in the die so that the zinc alloy is poured around them. In this way ready for use, intricate parts are obtained.

21.1.3   Plastics

Thermoplastic and *thermosetting plastics* are distinguishable classes. Their properties are quite different from those of metals and this should be taken into account in the design of measuring instruments. Thermoplastics have low heat resistance (from 70 °C to some tens over a hundred degrees centigrade) while thermosetting plastics can withstand considerably higher temperatures—over 250 °C for some kinds.

The Young's modulus of plastics is but a fraction of that of metals (100, 300 or more times lower than that of steel). Consequently their deformations can be considerable, even under small loads. They show *elastic after-effect* and *strain relaxation*. Thermal conductivity is poor—a momentary local heating may cause destruction. Most plastics absorb moisture which causes them to swell. They are soft and, therefore, easily scratched. Almost all plastics are good electric insulators, some of them being very good. The high coefficient of thermal expansion (more than ten times higher than that of metals) and excessive casting contraction of moulding shrinkage result in a large scatter of dimensions of plastic parts. This, combined with moisture absorption (swelling) and seasoning, particularly of thermosetting plastics, results in instability of dimensions with time. There exists a wide variety of plastics which enables, despite the many shortcomings, the selection of appropriate plastic for almost any working conditions.

21.1.4   Minerals

Mineral bushes are used for *bearings* and *guides* when very high wear resistance is required.

Corundum is crystalline aluminium oxide (Al_2O_3). It occurs as almost colourless leucosapphire, blue-tinted sapphire or red-tinted ruby, the latter two being now almost exclusively synthetic products. It is very hard (about 23×10^3 MPa), wear resistant, and has a high Young's modulus ($E = 4.5 \times 10^5$ MPa). It is most often used for bearing bushes.

Diamond–crystalline carbon—is the hardest of all known materials (80×10^3 MPa). It is seldom used because its machining is expensive.

Chalcedony, in the form of bloodstone, garnet, and most often agate having a main component SiO_2, is softer (18×10^3 MPa). It is used mainly for *knife-edge bearings*, where the comparatively low Young's modulus is of importance ($E = 0.75 \times 10^5$ MPa).

21.2 STRENGTH IN INSTRUMENT DESIGN

Strength problems in fine mechanisms, particularly in precision instruments, have a different character from those occurring in machinery design. In the latter, the factor deciding part dimensions is usually the *stress level*, while in the design of fine mechanism and precision instruments it is the *elastic deformations* that are most important.

Useful accounts are Budynas (1977), Sydenham (1980), and Trylinski (1971). In measuring instruments accuracy often depends upon the accuracy of displacements of their kinetic system elements. Consequently, the rigidity of the mechanism frame must be very high so that deformations, caused by varying forces acting within the mechanism or by random external forces, will be very small, perhaps of the order of micrometres. Even a thick steel plate is liable to deformations of that order when loaded by apparently insignificant forces! Therefore, for example, in precision instruments for length measurements, instrument plates are made rigid, thick, with a high moment of inertia of the cross section obtained by ribbing. Instrument columns are often made in the shape of large-diameter tubes. Parts having very small dimensions are very flexible.

The criterion generally deciding the dimensions of a part of a measuring instrument mechanism is the acceptable *elastic deformation*—since it is decisive for the functioning accuracy, not the limit of elasticity. Forces transmitted by the parts of measuring instrument mechanisms are often very small. Such mechanisms actually serve to transmit *displacements* representing *information* (transmission of motion) and not forces. In some situations the required dimensions of parts, even if sized to fulfil the condition not to exceed permissible deformations with regard to mechanism functioning accuracy, are so small that the parts would be liable to damage during assembly or servicing. To avoid this, part dimensions are often determined such that they possess sufficient rigidity and strength to withstand assembly and servicing.

In certain cases, however, stresses do determine the dimensions of small mechanism parts. An example is the situation where point contact of a convex ball occurs with a concave spherical surface of a considerably greater radius or with another convex ball, as in centre bearings or in ball bearings (Trylinski, 1971).

Another example is the contact of a cylinder with a plane or with a concave cylindrical surface of a considerably greater radius, as in knife-edge bearings or in V-block bearings (see Section 21.5.1 and Trylinski, 1971). Necessary dimensions of these elements are determined by stresses calculated from the *Hertz formulae*. Another case occurs for flat or helical springs for which there is very little space. In such springs the occurrence of permanent deformations is sometimes permitted in order to increase their load capacity. Spring dimensions are determined by both permissible stresses and elastic deformations (Trylinski, 1971).

21.3 ACCURACY OF INSTRUMENT MECHANISMS

21.3.1 Accuracy of Ideal Mechanisms

This section deals with the concepts of ideal rigid links and joints without clearance and friction. References relevant here are Korotkov and Taic (1978) and Sydenham (1980).

Positioning function of a mechanism

This is a characteristic parameter of a mechanism. It is the relation between the input signal x_i, which can be a displacement or an angle of rotation (also called *displacement*) and the output signal x_e of the displacement:

$$x_e = f(x_i) \tag{21.1}$$

The *instantaneous mechanism ratio* in measuring instruments gives the *sensitivity*:

$$i = \mathrm{d}x_e / \mathrm{d}x_i \tag{21.2}$$

For involute toothed gears, flexible connector drives with constant wheel radii and for friction gears where the gear ratio is a constant

$$x_e = i x_i \quad \text{where } i = \text{constant} \tag{21.3}$$

These components can provide large angles of rotation or continuous rotary motion with high accuracy (see Sections 21.5.4 and 21.5.5).

Errors of systems containing non-linear stages

In the case of small angles of rotation or displacements, lever mechanisms can be applied, mechanism *play* being eliminated by means of springs, with

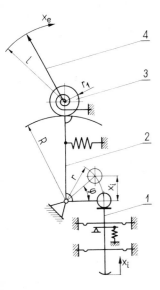

Figure 21.1 Sine lever and tooth mechanism with backlash eliminated by means of springs, $x_e = R_1/r_i$ arc sin (x_i/r). 1, plunger; 2, toothed sector; 3, pinion; 4, pointer

levers supported on springs (Section 21.5.1.3), on bearings without play, for instance, of the V-block type (Section 21.5.1.1), or on knife-edges (Section 21.5.1.2). Mechanisms, such as that system portrayed schematically in Figure 21.1, possess non-linear input–output relationships for their lever stages (Figure 21.2, curve 2), this being so even when an overall linear relationship is needed as shown in Figure 21.2, curve 1.

Design can be arranged in the following way so that the total additive effect of system error, due to such causes, closely provides the desired overall characteristic (Figure 21.2, curve 1) leaving only small residual errors:

(a) In the case of small displacements of the mechanism this error can often be reduced by introducing small changes into one or more of the selected dimensions of mechanism elements (by such means as plastic deformation or by relative translation of parts and their fixing positions—called *regulation* or *adjustment*) (Figure 21.2, curve 4). They aim to minimize the mean square error over the given ranges of mechanism motion.

(b) This source of error can also be reduced by applying a compensating mechanism or electrical transducer with a compensating characteristic (Figure 21.2, curve 3) at the output. However, compensating mechanisms are best avoided where possible: they may elongate the kinematic chain of the mechanism augmenting the influence of dimension errors (Section 21.3.2) and they may add random errors resulting from play, friction, and deformations.

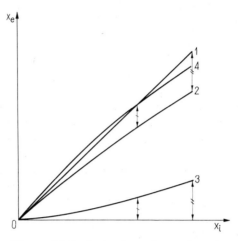

Figure 21.2 Reduction of non-linearity error in a system containing a non-linear stage: x_i, input signal; x_e, output signal; 1, demanded linear positioning function of a system; 2, non-linear positioning function evoked by a non-linear stage (for instance, by a lever mechanism as shown in Figure 21.1); 3, characteristic of compensating mechanism, or transducer; 4, positioning function after regulation

21.3.2 Accuracy of Real Mechanisms Under Static Conditions

This subsection deals with deformable links and joints with clearance and friction, as occur in practice, moving very slowly with negligible dynamic effects.

Errors caused by dimension and shape deviations

The dimensions and shape of real mechanism elements deviate from theoretical values resulting in *mechanism error* occurring at the output.

If the position output function x_e of a theoretical mechanism, for a definite value of x_i at the input, is given by

$$x_e = f(x_1, x_2, \ldots, x_n)_{x_i} \tag{21.4}$$

where $x_1, x_2, \ldots, x_n$ are linear and angular dimensions of fixed elements, frames, and mobile elements considered as independent variables, then, on the condition that deviations of these variables are small compared with their magnitudes, the resultant error e_r is given by

$$e_r = \frac{\partial f}{\partial x_1} \Delta x_1 + \frac{\partial f}{\partial x_2} \Delta x_2 + \cdots + \frac{\partial f}{\partial x_n} \Delta x_n \qquad (21.5)$$

The partial derivatives $\partial f/\partial x_1$, $\partial f/\partial x_2$, ..., $\partial f/\partial x_n$ are *influence coefficients* relating the influence of particular element dimension deviations upon the position error for definite individual input displacement values x_i: the influence of deviations of particular dimensions upon the position error of the output element may be different.

Where influence coefficients vary insignificantly with changes of displacement they can be regarded as constants. Dimensions having the highest influence coefficients should be kept within the narrowest tolerances, while conversely other elements of the mechanism may be made with reduced precision, without adverse effect on functioning accuracy.

In a given mechanism deviations of dimensions result in a systematic position error at the output as is expressed in equation (21.5). These can be reduced by adjustment discussed above.

Displacement error e_c (occurring in measuring instruments intended for comparative measurements) is error arising at the change of mechanism position from a position 1 to a position 2, that is

$$e_c = e_2 - e_1$$

where e_2 and e_1 are position errors in positions 2 and 1, respectively.

Random and hysteresis errors: limits of detection

Clearances in supports and guides, run-out of the inner ring in relation to the outer ring in low friction bearings and the resulting displacement of the centre of rotation, clearances and machining inaccuracies of toothed gears, all result in random errors of mechanism position at the output. These vary during motion, particularly when changes of direction occur.

Loading of the mechanism by a *quasi-statically* varying force during motion at the output and a change of direction of the mechanism motion evoke frictional forces, having changing values and direction, in the mechanism supports and guides. These forces, in turn, cause varying elastic deformations of mechanism elements, these being larger the more compliant is the structure. Elastic deformations, as for displacements of the centre of rotation in supports, result in mechanism position errors of both random and systematic character.

In reciprocal motion of a mechanism this error appears as *positional hysteresis* e_h (Figure 21.3). This is the difference between the position $x_{er,1}$ of the output element when the input element moves in one direction to

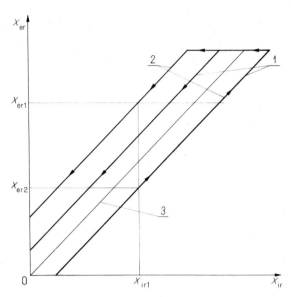

Figure 21.3   Function $x_{er} = f(x_{ir})$ for a real mechanism: 1, play eliminated by means of springs; 2, play not eliminated; 3, theoretical mechanism with no hysteresis

position $x_{ir,1}$ and the position $x_{er,2}$ of the output element when the input element moves in reverse direction to the same position $x_{ir,1}$. Where positional hysteresis occurs, to begin to move the output element from rest it is first necessary for the input element to perform a displacement constituting the *limit of detection* of the mechanism, its value depending on the former positions of the mechanism, as shown in Figure 21.3. Hysteresis and limits of detection can be reduced by the following means:

(a)   Springs can be applied in mechanisms to eliminate backlash in toothed gears (Figure 21.3) and, partially, clearance in supports.

(b)   Clearances can be reduced in slide bearings and rolling contact bearings.

(c)   The continuous rotation bearings mentioned above might possibly be replaced by spring supports (Section 21.5.1.3), or by supports without clearance such as the V-block type (Section 21.5.1.1), and knife-edge type (Section 21.5.1.2).

(d)   Mechanism elements can be made more rigid and of lower mass, the latter to reduce gravity forces and dynamic loads (Section 21.3.3), using materials of high Young's modulus and lower density in more suitable configurations.

21.3.3 Accuracy of Real Mechanisms Under Dynamic Conditions

The dynamic behaviour of mechanisms is generally covered, with respect to general rather than to fine mechanism, in such works as Harris and Credo (1976) and Mabie and Ocvirk (1958).

Influence of vibration and impact

Parts of a mechanism whose elements are not balanced in relation to stationary axes of rotation, in conditions of medium vibration and impact, may cause vibration of the moving system of the mechanism. Flexible connector drives and toothed gears, in which the *centre of gravity* is usually situated on the *axis of rotation*, can easily be made such that they do not excite *translational vibration:* elimination of *torsional vibrations* is more complex. Lever mechanisms may cause severe mechanism vibrations. Vibrations can be reduced by the following means:

(a) Unbalanced mechanism elements can be provided with suitable *counterbalances:* use of balancing lever systems may be applicable.
(b) Precise and accurate measuring instruments can be supported by elastic element (metallic springs, rubber elements, air bags) suspension systems in which the frequency of free vibrations of the instrument-support system is considerably lower than the forcing frequency of the medium.

Addition of *dither* or *jitter* to the mechanism, that is, imposing vibrations (often at the frequency of the easily available public power network) of small amplitude is a method sometimes used to significantly reduce the motion resistance of mechanisms, particularly at the moment of starting from rest. This can reduce hysteresis and limits of detection.

Influence of rapidly varying displacements

Rapidly varying displacements at the mechanism input may cause vibration of intermediary elements and of the output element, such as a pointer. To prevent *mechanical resonance*, the natural frequency of vibrations of the mechanism system should differ considerably from the forcing frequency. If the latter exceeds the critical frequency, the mechanism should be safeguarded against damage (see later Figure 21.17 for an example) from resonance.

In the case of rapid transient inputs, for example, a step change of displacement, the resulting vibrations should be adequately damped. The theory is given in Chapter 17. Damping is often deliberately provided as support friction, as vibration dampers of the friction or viscous type (consisting, for example, of a fixed bush containing a mobile bush with the gap between them filled with oil), or of the eddy current or pneumatic type.

Because of the difficulty of performing sufficiently exact calculations of the frequency of free vibration of a moving system (the rigidity of elements being hard to determine), the simplest way of establishing it may be to experiment on a model.

Consequences of vibration

Vibration of the output element of the mechanism evoked by one of the causes discussed above may result in dynamic error of the mechanism position. In pointer instruments it may make reading impossible. Vibration can bring about premature wear of mechanism gear wheels and supports and fatigue failure of elements.

21.4   INFLUENCE OF TEMPERATURE UPON THE FUNCTIONING AND ACCURACY OF INSTRUMENT MECHANISMS

21.4.1   Basic Considerations

Review accounts of the influence of temperature changes on instrument performance are available in Sydenham (1980) and Trylinski (1971). Variation of temperature affects several error mechanisms.

Variation of Young's modulus

This relation is not linear. Roughly approximated, the change to this modulus for a given temperature change is,

$$E_T = E_0(1 + \delta_1 \Delta T + \delta_2 \Delta T^2) = E_0[1 + (\delta_1 + \delta_2 \Delta T)\Delta T] = E_0(1 + \gamma \Delta T)$$
$$(21.6)$$

where ΔT is the temperature increase from the initial temperature T_0 to final T_1, E_0 and E_T being the Young's moduli at those temperatures. δ_1 and δ_2 are coefficients characteristic of a given material. Within narrow temperature ranges γ is usually assumed to be constant. As a rule γ is negative, meaning springs become more elastic as the temperature rises. This is called the *thermoelastic effect*.

Variation of dimension

This relation also is not linear. Roughly approximated, it can be written as

$$l_T = l_0(1 + \beta_1 \Delta T + \beta_2 \Delta T^2) = l_0[1 + (\beta_1 + \beta_2 \Delta T)\Delta T] = l_0(1 + \alpha \Delta T)$$
$$(21.7)$$

where l_T and l_0 are the lengths of a straight bar at temperatures T and T_0, respectively, and β_1 and β_2 are empirical coefficients characteristic of a given material. Within narrow temperature ranges α is usually assumed to be constant. As a rule α is positive. Materials used in instruments are compared in Sydenham (1980) and Trylinski (1971).

Variation of lubricant viscosity

As temperature rises, lubricant viscosity reduces resulting in the lubricant being pressed out of slide bearings which may then seize. Reduction of temperature can cause solidification of lubricants, resulting in increased resistance to motion, and even immobilization of mating parts. To prevent this, special lubricants are used (such as silicone oils) or the mechanisms are designed to run unlubricated.

Permanent deformation of elements made of plastics

This occurs when the temperature corresponding to their heat resistance is exceeded.

21.4.2 Influence of Temperature Upon Elastic Elements

In measuring instruments in which the deformation of elastic elements (for instance, hairsprings, membranes, and Sylphon bellows) is a measure of the variation of the magnitude of the sensed variable, a rise in temperature causes changes in their rigidity as a result of the decrease of Young's modulus (equation (21.6)), and a slight increase of rigidity because of the increase of element dimensions (equation (21.7)). In such cases the latter variation can often be neglected, the former being the dominant cause of thermal error of instruments. To reduce thermal errors elastic elements are made of materials having a low thermal coefficient of Young's modulus, such as Invar (Section 21.1.2), which, however, has a low limit of elasticity, or of other special multicomponent materials. Despite the availability of numerous materials instrument design often requires a mixture of properties that cannot be obtained simultaneously. For further discussion see Sydenham (1980).

Mechanical compensation of thermal error may be obtained by building elements having dimensions changing with temperature, for example, thermo-bimetals, into the kinetic system of the mechanism; in this way they compensate for the influence of temperature variations upon the elastic measuring element (Figure 21.4). In electrical and electronic instruments compensation can be obtained by building temperature-sensitive elements into the electric system. As a guide it is usually preferable to correct the mechanical system before resorting to other forms of post-sensing compensation.

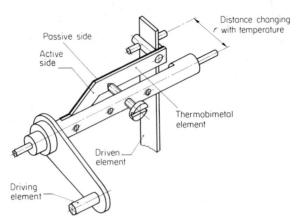

Figure 21.4 Temperature compensation by means of thermobimetal element. Reproduced by permission of Wydawnictwa Naukowo-Techniczne

21.4.3 Influence of Thermal Expansion upon Instrument Mechanisms

Where the ambient temperature field of an instrument changes, use of materials with different thermal expansion coefficients, for example, for mechanism frame and parts, may cause excessive increase of clearances or their reduction with resulting seizure (Figure 21.5). Similarly, in a toothed gear a rise of temperature may cause an inadmissible increase of backlash, and temperature reduction, gear seizure. To prevent this, it suffices in the example shown in Figure 21.5, to place the locating ring into the position marked with a dotted line, thus considerably shortening the element length causing the reduction of clearance. Alternatively the frame and mechanism

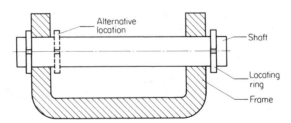

Figure 21.5 Reduction of the influence of thermal expansion, obtained by appropriate design in which locating ring is placed at a more effective position where thermal expansion of shaft is smaller. Reproduced by permission of Wydawnictwa Naukowo-Techniczne

could be made of the same material. This approach holds well when it can be assumed that the whole of the system is able to follow ambient changes such that temperature remains uniform throughout the system. The situation is less favourable under dynamic temperature changes for then the housing and the frame, having different thermal time constants, respond at different rates providing varying magnitudes of dimensional error. In this case making the entire instrument of the same material will not bring about the desired result. Appropriate design methods, such as that shown in Figure 21.5, may overcome such dynamic thermally induced dimensional errors.

21.5 MOTION TRANSFER ELEMENTS

21.5.1 Supports

This section deals with the supports used to allow rotary, and sometimes translational, motion between two links of a mechanism (Trylinski, 1971; 1978).

21.5.1.1 Sliding bearings

Bushes

Cylindrical sliding bearings can be of the *machine* type (Figure 21.6); of the *clock* type (Figures 21.7, 21.8); of *sintered* metals (Figure 21.9); or be made of one of many kinds of *plastic*. The friction couple M of a bearing is given by

$$M_f = \tfrac{1}{2}d\mu P \qquad (21.8)$$

where μ is the coefficient of friction, d the journal diameter, and P the load as shown in Figure 21.7. The coefficient of friction, for non-lubricated machine-type and clock-type bearings with brass or bronze bushes, in the first phase of work—during the running-in of the bushes—is $\mu = 0.15$ to 0.25. After the running-in process is completed this can increase to as high as $\mu = 70$. For bearings with mineral bushes $\mu = 0.14$ to 0.20. Although

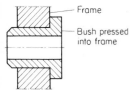

Frame

Bush pressed into frame

Figure 21.6 Machine-type support with bronze bush. Reproduced by permission of Wydawnictwa Naukowo-Techniczne

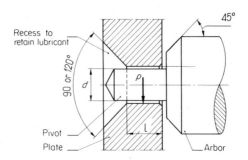

Figure 21.7 Clock-type support. Reproduced by permission of Wydawnictwa
Naukowo-Techniczne

lubrication considerably reduces the friction resistance of such bearings, loss
of the lubricant should be taken into account for this produces conditions
similar to those of non-lubricated bearings. For non-lubricated plastic bearings the coefficient of friction μ varies from 0.2 to 0.4 (depending on the
type of plastic and working conditions). Lubricating plastic bearings can
decrease the coefficient to 0.05. Porous bearings, usually made of sintered
iron, are impregnated with lubricant and do not require lubrication in
normal use. Their coefficient of friction, depending on material, sliding
speed, and unit pressure, can vary within wide limits, from 0.01 to 0.3.

The radial clearance c in small cylindrical bearings (Figure 21.10) is quite
large in relation to pivot diameter d. For example, for $d \leqslant 0.3$ mm values of
$c = 0.1d$ to $0.2d$ are typical. In reciprocating motion the pivot rolls inside
the bush in a variable direction (Figure 21.10). In vibratory conditions the
pivot axis may acquire random positions on the surface of a circle with
diameter equal to clearance c. This causes random error of the pivot axis
position and *lost motion* of the mechanism.

The bush, made of a softer material than the steel pivot, wears more
rapidly causing an increase of clearance. To avoid excessive clearance, or
bearing destruction by seizure, the contact pressure p where $p = P/ld$ must
be permissible for the bush material used and for the working conditions.
The same applies to the permissible product pv, where v is the sliding speed.

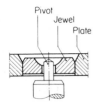

Figure 21.8 Support with bearing jewel. Reproduced by permission of Wydawnictwa Naukowo-Techniczne

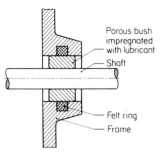

Figure 21.9 Support with sintered metal porous bush. Reproduced by permission of Wydawnictwa Naukowo-Techniczne

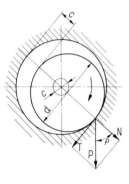

Figure 21.10 Pivot rolling up the bush: P is the load, T the force of friction, N the normal force and ρ the angle of friction

Centre supports

In an axially loaded centre support (Figure 21.11), a pivot with a spherical end rests in a block recess, which is generally conical in shape with a concave spherical centre portion, with $R \gg r$. Contact pressure, load capacity, and friction couple are calculated according to Hertz formulae (Trylinski, 1971). The pivot is usually made of steel. In a ruby block (relatively high Young's modulus) the pivot contact surface and its radius r_0

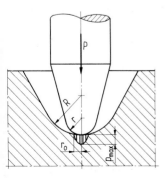

Figure 21.11 Centre support, axially loaded

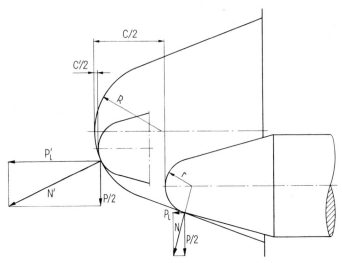

Figure 21.12   Centre support, trasversely loaded, $C'/2 < C/2$,
$N' \gg N_1$

are very small. As a result of this the friction couple is very small but contact pressures p_{max} are high and the load capacity is low. The use of steel (medium Young's modulus), agate or glass (low Young's modulus) for the block increases the load capacity but also increases the friction couple. For a steel pivot $p_{max} \leqslant 3000$ MPa.

In a transversely loaded centre support (Figure 21.12), the pivot should rest on a conical surface to eliminate excessive contact pressure. It should, therefore, have considerable axial clearance c and radial clearance. This results in considerable random error of axis position. In a bearing loaded this way the use of a material with low Young's modulus (steel, agate, glass) as the block material, instead of ruby, increases the load capacity and reduces the contact pressure; the friction couple remains the same. In bearings subject to vibration, blocks are made of materials with a low Young's modulus and pivots of special shock-resistant materials.

V-block supports

Figure 21.13 shows a knife-edge pivot where a cylindrical working surface rests on a V-block. The pivot can provide only swinging motion. Such bearings have no clearance, which ensures very accurate guiding of the pivot. With a small pivot radius r a small friction couple can be obtained.

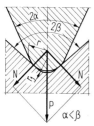

Figure 21.13   V-block support. Reproduced by permission of Wydawnictwa Naukowo-Techniczne

21.5.1.2   Rolling contact supports

Knife-edge bearings

A knife-edge shaped pivot with a cylindrical working surface, when providing swinging motion, *rolls* over a concave or flat block. Both the pivot and block are usually made of agate or of steel. Resistance to motion is very small. The axis of rotation changes as the pivot inclines.

Ball bearings

In precision instruments radial ball bearings of small dimensions (outer diameter $D \geqslant 2.5$ mm) or miniature angular bearings ($D \geqslant 1$ mm, Figure 21.14) can be used. The friction couple of a fully loaded bearing can be found from equation (21.8) where d is the shaft diameter. For radial ball bearings the friction coefficient, determined experimentally, is typically $\mu = 0.0015$ when the bearing is rolling. The *break-away* friction couple (that at the moment of starting from rest) for bearings lubricated with solid grease that is not warmed up by friction, can be up to four times greater.

When estimating the friction couple for lightly to zero loaded rolling contact bearings as used in precision instruments, the friction coefficient should be assumed to be several times greater—of the order of $\mu = 0.025$ to 0.1.

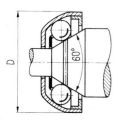

Figure 21.14   Miniature self-aligning angular bearing. Reproduced by permission of Wydawnictwa Naukowo-Techniczne

21.5.1.3 Supports on elastic elements

A flat spring (Figure 21.15), axially loaded by a tensile force P and by a deflecting couple M, is often used as a limited rotation pivot. For small deflection angles it can be assumed that the centre of rotation is situated approximately one third of the spring length l from the fixing point and that it is stable. Random transverse forces, such as are caused by vibrations, can result in displacement of the centre of rotation from that position.

Supports on *crossed-strip springs* (Figure 21.16), are used when random transverse forces are acting and when it is important to preserve the stability of the centre of rotation c. For symmetrical crossed straight springs the centre of rotation is situated at their crossing point. Provided the rotation magnitude is small the centre of rotation can be assumed to be at a fixed point.

With both flat- and crossed-spring supports small deflections, of the order of a few degrees, have a finite restoring moment that is approximately linear with deflection angle. For larger deflection angles this relation is not linear and the centre of rotation changes position with deflection. *Stressed suspension springs* (Figure 21.17) are often used to suspend a moving element of a measuring instrument.

The angle of deflection, even for considerable deflection, is linearly related to the restoring moment. A suspension spring of this kind generally has the shape of a thin band of rectangular section. Usually $n = b/t = 10$ to 15 where b is the width and t the thickness of the spring. In electrical measuring instruments the suspension springs often also serve as conductors supplying current to the moving element. Shock-absorbing devices in the

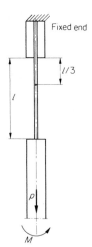

Figure 21.15 Support on a flat spring

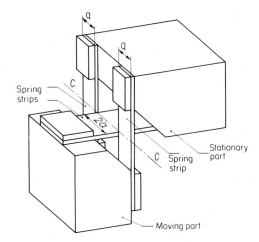

Figure 21.16 Support on crossed springs. Reproduced by permission of Wydawnictwa Naukowo-Techniczne

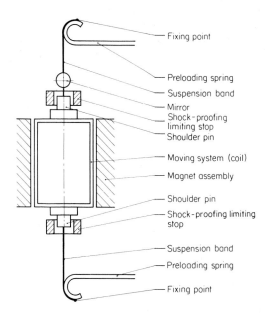

Figure 21.17 Support on stressed suspension springs

form of motion-limiting rings (Figure 21.17), prevent damage to the suspension spring which might result from vibrations and impacts. A general review of elastic elements is presented in Sydenham (1980), whilst Trylinski (1971) provides design considerations.

21.5.1.4   Aerostatic supports

These are radial or thrust sliding supports in which air is introduced under pressure into an interspace between the pivot and bush or block. The *breakaway torque* is almost equal to zero and the friction couple is negligible. Load capacity is considerably lower than that of plain slide bearings of the same dimensions. Manufacturing accuracy must be high. These supports are used when small resistance to motion is of importance, particularly at the start from rest or when the rotational speed is very high (tens to hundreds of thousands of revolutions per minute). At extreme speeds they are inclined to vibrate. Further information is available in Trylinski (1971, 1978).

21.5.1.5   Magnetic supports

It is impossible to suspend a supported element in the magnetic field of permanent magnets (Earnshaw's theorem) in a state of stable equilibrium. Permanent magnets, therefore, are used to provide support in the direction of action of the main force (for example, opposing gravitation). In the remaining directions of translation freedom other additional supports are used. Suspension on servo-controlled electromagnets can supply a stable support providing free suspension of a moving system in space. Magnetic supports based on permanent magnets are very compliant—the axis of rotation is easily displaced as a result of load action. Magnetic suspensions are reviewed in Geary (1964).

21.5.2   Guides

Sliding guides can be prismatic, with a cross section consisting of straight lines and arcs. Alternatively, an easier to manufacture section is the cylinder (Figure 21.18). Friction resistance of sliding guides can be high as they are difficult to keep clean. A coefficient of friction equal to 0.3 or more provides a major design constraint. The length (Figure 21.18) of the guiding element l should be sufficient in relation to width b to prevent seizure or excessive increase of resistance to motion. As shown in Figure 21.18 the line of action of force P should cross the doubly shaded area. Stick–slip action makes accurate positioning difficult with guides of this type. Aerostatic guides are

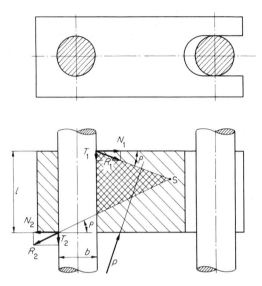

Figure 21.18   Cylindrical sliding guide. Reproduced by permission of Wydawnictwa
Naukowo-Techniczne

sliding guides in which air is forced into the guide contact area interspaces
acting as lubricant. Resistance to motion of such guides is negligible and the
stick–slip effect does not occur.

Rolling contact guides are also available in which the contact faces join
through rolling action of balls or bearing rollers. These have considerably
lower resistance to motion than sliding guides.

Guides using elastic elements can be applied where small displacements of
the guided element are satisfactory. Either flat springs (Figure 21.19), or

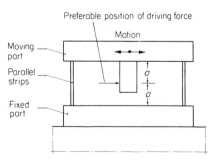

Figure   21.19   Guiding   by   flat
springs

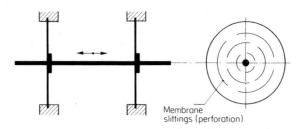

Figure 21.20 Membrane guiding system. Reproduced by permission of Wydawnictwa Naukowo-Techniczne

membranes (Figure 21.20), can be used. In order to obtain increased guided element travel or more compliant guides the plain disk can be slitted (Figure 21.20). Elastic element guides exert a restoring force on the guided element which, for small deflections, is approximately linear with displacement.

Friction losses and the mechanical hysteresis of elastic supports and guides are very small, they are caused by the material's internal friction and by friction at the fixing points. They can be reduced to very small amounts by appropriate design. Stick–slip effect does not occur and the hysteresis error is very small. Since deflections are usually very small and the loads operate the support at a point considerably below the elastic limit, service life is practically unlimited. The cause of destruction may be corrosion, or, in the case of suspension springs, electrical overload. Excessive vibration can, of course, fatigue the material. The design of guides is covered in Trylinski (1971).

21.5.3   Lubricants and Lubrication of Supports and Guides

Sliding supports and guides of precision mechanisms are primarily lubricated to protect them from corrosion and to reduce wear. Lubrication usually, although not always, also reduces friction resistance. Lubricated elements should be protected from ingress of dust which accelerates their wear and increases friction resistance.

Slide bearings for which a small friction couple is not expected, for instance, the cylindrical bearing block of Figure 21.6, that operate at low sliding speeds, are generally not provided with lubricant feed devices. They are lubricated with solid mineral greases to prevent the lubricant from leaking out.

Slide bearings transmitting power, for example, those of electric motors, usually use wick lubrication to which thick machine oil is applied. Alternatively, porous bushes are employed that are made of sintered metals impregnated with oil (Figure 21.9).

Bearings for which a small friction couple is expected—cylindrical clock-type bearings (Figures 21.7 and 21.8)—and particularly those using mineral bushes, are lubricated with fluid oils of low viscosity. A drop of oil is placed in the lubrication recess on assembly or after careful cleaning of the mechanism. Mineral oils, although they deteriorate slowly, are unsuitable in this case because they spread. On the other hand fatty unsaturated oils, such as bone oil, although they do not spread, quickly deteriorate by oxidation (especially in the presence of contaminations) and thicken. As a compromise *neutral oils* are used, being a mixture of mineral and bone oil that is saturated with an inert gas, oxidation inhibitors being added. Coating with a 0.01 to 0.1 μm thick *epilam film* prevents spreading. Synthetic oils are also produced, with characteristics adapted to the given support working conditions.

Guides often must be exposed to dust. If they cannot be suitably protected, non-lubricated guides or plastic bearing surfaces with graphite or molybdenum disulphite filler, are adopted.

Supports working at low temperatures (below −20 °C, and particularly below −40 °C) are lubricated with silicone oils which do not solidify or thicken quickly at such temperatures. However, silicone oils have poorer lubricating characteristics than those discussed above and they do spread.

Rolling contact bearings are lubricated to protect them from corrosion and to reduce wear. Frequently solid greases are used. These considerably increase the friction resistance, particularly at starting when the bearing is still cool. High-speed bearings (from ten-odd thousand up to several hundred thousand revolutions per minute) are sparingly lubricated with oil mist to prevent overheating of the bearing.

Choice of an appropriate lubricant for bearings determines the mechanism service life. Application of an inappropriate lubricant can cause destruction of the mechanism.

21.5.4 Toothed Gearing

Gears are used extensively in instruments. Their design, in general, is covered in numerous works. Specifically relevant are Michalec (1966), Sydenham (1980), and Trylinski (1971).

Kinematic accuracy of gearing

In kinematic chains of measuring instruments toothed gearing is used to transmit the measuring signal in the form of rotation angle. Kinematic accuracy is consequently required of such gearing. Most often involute gears are used, ensuring a constant velocity ratio. One type of such gears adapted to the specific working conditions prevailing in small mechanisms are gears

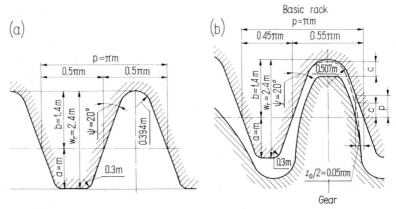

Figure 21.21   Basic rack of the BS 978/1968 involute profile: (a) designed for work with minimal backlash; (b) designed for work with considerable backlash. Reproduced by permission of Wydawnictwa Naukowo-Techniczne

made according to *BS 978/1968 Gears for instrument and clockwork mechanisms. Part 1* (Figure 21.21). In the case when backlash is eliminated in some way by, say, a bias spring and in other cases where the teeth always mate with the same side, and generally where considerable backlash is permissible, gears are constructed to the design of Figure 21.21b. This tooth form prevents seizing of the gearing caused by *run-out* of gears and permits easier to obtain gear machining accuracy of the outer diameter, pitch diameter, run-out, and accuracy of the bearing hole distances (if not adjustable), thus facilitating the production of fine pitch gears without any deteriorating influence existing upon the precision of their functioning; gear ratio does not change with working depth of tooth. In the case of unidirectional action the maximum deviation of driven gear position, expressed in radian angular units, is given by

$$\Delta\phi_{at} = \frac{f_{at,1} + f_{at,2}}{1000 R_2} \text{ radians} \tag{21.9}$$

where R_2 is the radius of the driven gear in millimetres, $f_{at,1}$ and $f_{at,2}$ are the total position errors for gears 1 and 2 respectively. These errors are measured by means of an instrument for the measurement of position error on the *pitch circle* (in μm), comprising the *run-out* (double eccentricity) of the axis. Where an instrument for such measurement is not available, it may be assumed that

$$f_{at} = f_{ct} \tan \psi \tag{21.10}$$

where f_{ct} is the *total composite error* in micrometres and, ψ is the *pressure angle*.

In the case of needing to work in both directions of rotation, when backlash is not eliminated by a preloading spring and importance is attached to the reduction of the positioning deviation of the gearing caused by lost motion (resulting in hysteresis error), and if the adjustment of centre distance is impossible, maximum positioning deviation of the driven gear in swinging motion is given by

$$\Delta\phi_{ct} = \frac{2(f_{at,1} + f_{at,2}) + f_d}{1000 R_2} \text{ radians} \qquad (21.11)$$

where f_d is the lost motion of the gearing on the pitch circle in micrometres. In this case gears of *basic rack without backlash* type (Figure 21.21a) are applied. Equations (21.9) and (21.11) are based on the assumption that mating gears might be mounted in any relative position including the most unfavourable. The deviation of a pair of mating gears can be reduced by mounting them in such a relative position that the maximum sum of position errors, $|f_{a1} + f_{a2}|_{max}$, should be as small as possible.

In gear trains the effect of deviations in particular pairs of mating gears upon the transmission deviation of the output shaft varies. Consider a gear train with n stages (Figure 21.22). Let shaft 1 be the driving shaft and n the driven, output shaft. The transmission deviation of all the stages reduced to the driven shaft of nth stage is given by

$$\Delta\phi = \Delta\phi_n + \sum_{k=1}^{k=n-1} \frac{\Delta\phi_k}{q_{k+1}q_{k+2}\cdots + q_{n-1}q_n} \text{ radians} \qquad (21.12)$$

where $\Delta\phi_k$ is the actual transmission deviation of the driven element with respect to the driving element for the train stage k (k is a whole number and $1 < k < n$) and where q_k is the speed ratio of the stage where $q_k = t_k/t_{k-1}$ in which t is the respective number of teeth. In gear trains which do not need to provide an accurately constant *instantaneous velocity* ratio, and whose module is very small $m \leqslant 0.3$ mm ($m = d_p/t = 25.4$ mm/P where d_p is the

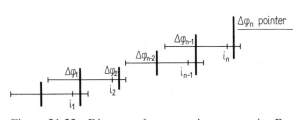

Figure 21.22   Diagram of a measuring gear train. Reproduced by permission of Wydawnictwa Naukowo-Techniczne

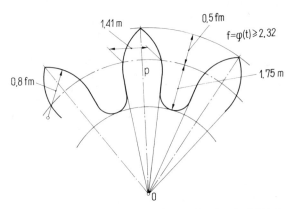

Figure 21.23 Swiss clockwork profile (NHS-56704).
The wheel may constitute the driving element the
pinion being driven, or vice versa. Reproduced by
permission of Wydawnictwa Naukowo-Techniczne

pitch diameter in mm, t the number of teeth and P the diametral pitch),
clockwork gears with long sharp pinion leaves and wheel teeth may be
applied. Typical examples of such gears are Swiss clockwork gears NHS-
56704 shown in Figure 21.23. Flank profiles consist of radial lines and face
profiles of circular arcs with their centres lying within the pitch circle. These
gears have a ratio that is variable within one pitch length. Due to the use of
very long teeth the tolerances of centre distances and wheel diameters may
be more than twice those of involute gears. This has particular importance in
the case of gear wheels made of plastic where great scatter of moulding
contraction exists. In situations where multiplying toothed gearing is combined with a driving source that is elastically compliant (for example
membranes), motion resistance combined with *wind-up* are key factors
causing hysteresis and, consequently, *reversal error* (lower indications for
higher resistance and vice versa). Resistance is not constant but varies as the
teeth work along the path of contact, being greatest at the beginning and at
the end of contact. For example, in multiplying use of involute gears having
a small number of leaves of the driven pinion ($t_2 = 8$ to 12) the instantaneous
torque transmitted by the wheel to the pinion at the moment of moving
from rest (friction at rest being greater than friction of motion) may
decrease at the beginning of contact to as little as 0.7 of the maximum
torque value. In multistage multiplying gears this can cause seizure of the
gearing. The following methods are used to reduce the oscillations of the
instantaneous torque of multiplying gears:

(a) In involute spur gears the number of leaves of the driven pinion should
     not be less than 14.

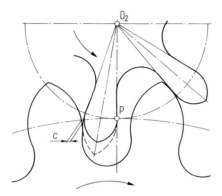

Figure 21.24 Swiss clockwork gears (NHS-56702). The dashed line shows the cycloidal profile of a pinion leaf end. The pinion constitutes the driven element. Reproduced by permission of Wydawnictwa Naukowo-Techniczne

(b) Swiss clockwork gears NHS-56702 (modified, cycloidal) can be used (Figure 21.24), for which the torque at the beginning of contact does not fall below 0.9 of the maximum torque. The mean friction resistance is smaller than for involute gears. A disadvantage of this gearing is the possibly significant variability of its ratio within one pitch length.

(c) Helical involute spur gears with considerably lowered tooth tips can be adopted, as can special gearing having a profile resulting in contact occurring only at the pitch point or in its vicinity. In this case the condition must be fulfilled that the sum of axial contact ratio and transverse contact ratio must be greater than 1.

Reduction gears

Reduction gears of high ratio are often used to reduce the speed of the output shaft of an electric motor or to increase the available torque. Multistage gears, of the involute type, are usually used. The gear train may operate in only one direction with considerable backlash, as an example this might be designed to BS 978/1968 (see Figure 21.21b). It may be required to work in reciprocating motion in which case lost motion is undesirable: a backlash-free design such as that shown in Figure 21.21a could be used.

In systems of backlash-free gearing used in feedback control systems (Truxal, 1958; Michalec, 1966), very high accelerations occur at the moment of a rapid change of direction. In those cases the total ratio should be divided into well chosen stages and the gear wheels designed so that the

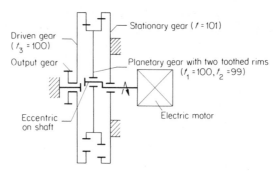

Figure 21.25 Epicyclic compound train with internal meshing. Total ratio is 10,000:1. Reproduced by permission of Wydawnictwa Naukowo-Techniczne

reduced moment of inertia at the motor shaft is as small as possible. An incorrectly designed gear train can have such a great value of the reduced moment of inertia, that very strong forces, appearing during the change of the direction of motion, can cause destruction of the motor shaft pinion and of its mating wheel.

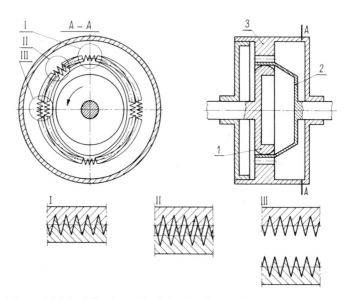

Figure 21.26 Working principle of a harmonic drive: 1, oval generator deforming the flexible ring; 2, 3, internal meshing of the stationary housing; I, II, and III, successive phases of mating. Reproduced by permission of Wydawnictwa Naukowo-Techniczne

In order to miniaturize reduction gears, multistage gearing can be replaced by *planetary gears* (Figure 21.25), or a *harmonic drive* (Figure 21.26) (see Dudley, 1962; Trylinski, 1978). In a single stage of such gears a ratio of 200, or even 300, can be obtained.

Gearing with minimal or no backlash

Gearing with very small backlash is obtained by the following means:

(a) By making gears with as small a run-out as possible and supporting them in a way enabling centre distance adjustment.
(b) By making the tooth thickness of mating wheels, having profiles without backlash (Figure 21.21a), exceed the nominal value, then running in mating wheels until the centre distance equals the bearing hole distance required.
(c) By making the spur gear wheels slightly conical, and displacing them axially until backlash is eliminated.

Reduction of backlash by the above methods does not increase the resistance to motion of the gearing.

Alternatively, backlash can be eliminated by means of a spring-loaded split gear. Elimination of backlash is obtained (Figure 21.27) by dividing the driving wheel into two wheels (in the figure: 1, fixed; 2, relatively rotating)

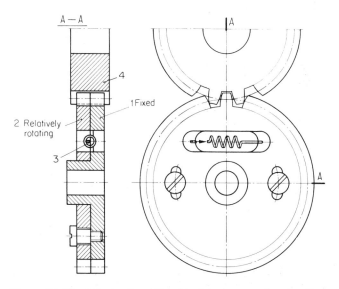

Figure 21.27   Elimination of backlash by means of spring 3:
1, 2, and 4, mating gears. Reproduced by permission of
Wydawnictwa Naukowo-Techniczne

biasing one against the other by means of springs (labelled 3). Another method is to press the teeth of the driving wheel into the driven one thereby closing up the mesh: support of one of the wheels must be mobile. Backlash elimination by means of springs considerably increases the resistance to motion of the gearing and its wear. Such gearing cannot be used to transmit power, or work at high rotational speeds.

The harmonic drive shown in Figure 21.26 (Dudley, 1962), has no backlash and minimal hysteresis error. Provided with a generator using roller contacts, it has resistance to motion similar to that of a gearing designed with backlash and it is suitable for transmitting power and great torques.

21.5.5 Flexible Connector Drives

Materials

Flexible connectors are made of organic materials (natural or synthetic) or of metals. Organic connectors are used in the form of strings, threads or cords, less often as woven bands. They are extensible, have considerable elastic retardation and hysteresis. Metal connectors are made of steel or bronze in the form of single or multiple strand wires and thin bands. They have a high Young's modulus and strength. To yield long life it is necessary to ensure that the outer member stress, due to bending, does not exceed the material's elastic limit. Chains of many forms can aiso be considered as flexible connector drives.

Friction flexible connectors

These are usually made of organic materials and transmit motion by virtue of friction (Figure 21.28). In such drives, slippage occurs, increasing with load to the point of total slippage.

A flexible connector drive of the friction type can have a constant velocity ratio if the torque transmitted is constant.

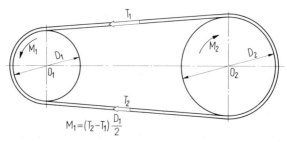

$$M_1 = (T_2 - T_1)\frac{D_1}{2}$$

Figure 21.28   Flexible connector drive, friction type

Toothed flexible connector drives

Flexible connector drives of the tooth type have been devised to maintain synchronism.

Fixed connector drives

These drives (Figure 21.29) serve to transmit motion within a limited angular range. Because of its unavoidable rigidity to bending the connector does not constitute a straight line tangent to both pulleys, but springs back slightly. A change to the torque loading the drive causes both elastic elongation of the connector and reduction of the *spring-back* causing position error of the driven wheel. The most accurate fixed connector drives would need to be made of metals (best of all is steel because of its high Young's modulus) in the form of very thin bands (0.05 to 0.15 mm thick). These can provide very accurate mechanical drives, more accurate than toothed gearing. For the velocity ratio to remain constant, the run-out of pulleys and bearing clearances should be kept to a minimum. Fixed connector drives formed as connectors made from thin metal bands are very durable and will not fatigue if the composite stress caused by the tensile force and by the bending evoked by winding the strip around the pulleys is small compared with permissible values. Such drives should be made of corrosion-resistant materials, otherwise corrosion may cause premature destruction. The design of instrument flexible connectors is covered in Trylinski (1971).

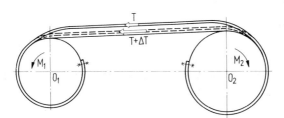

Figure 21.29   Fixed connector drive

21.6 SPECIAL ELASTIC ELEMENTS

21.6.1   Pressure-controlled Elastic Elements

There exist many special elastic elements suited to measuring instrument use. They are discussed in some detail in Trylinski (1971).

Properties

Pressure-controlled elastic measuring elements are made of metals such as brass, bronze, steel, and German silver. In the case of sealing elements, plastics are often used.

The characteristic of a pressure-controlled elastic element is represented by the relationship between effective pressure p (the difference of pressures on both sides of the element) and the deflection f of a selected point of the element (Figure 21.30). In Bourdon tubes (Figure 21.31) this point is situated on the free end of the tube; in membranes and bellows it is on the axis of symmetry. The ratio $S_p = df/dp = \tan \alpha$ defines the sensitivity of the element. Bourdon tubes (Figure 21.31) and bellows (Figure 21.34) have a rectilinear characteristic ($S_p = $ constant). Membrane (Figure 21.32) and capsule (Figure 21.33) characteristics are mostly curvilinear with variable sensitivity.

The phenomena of hysteresis and elastic retardation occur in pressure-controlled elastic elements. In measuring elements elastic retardation results in a dynamic, short-term, difference between the characteristic for the rising pressure and that for the falling pressure and thus this results in different deflection values occurring for a given value of pressure. The result of these phenomena will be instrument indication error called the *error of hysteresis*. The *hysteresis index* $= (\Delta f_h/f_{max})100\%$ where Δf_h is the maximum value of hysteresis and f_{max} is the greatest excursion of the element. Stabilized, pressure-controlled, elastic measuring elements show a hysteresis index ranging from 0.3 to 2%. Additional deflection Δf_T of the elastic element, caused by temperature variation of Young's modulus

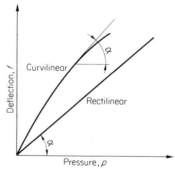

Figure 21.30   Characteristic of a pressure-controlled elastic element. Reproduced by permission of Wydawnictwa Naukowo-Techniczne

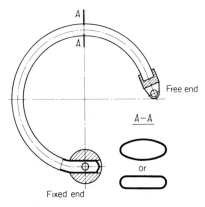

Figure 21.31 Bourdon tube. Reproduced by permission of Wydawnictwa Naukowo-Techniczne

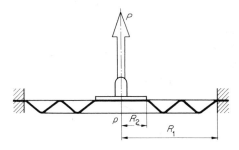

Figure 21.32 Corrugated membrane. Reproduced by permission of Wydawnictwa Naukowo-Techniczne

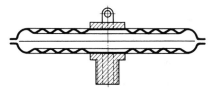

Figure 21.33 Closed capsule (aneroid). Dotted lines show the orifice of the differential capsule form. Reproduced by permission of Wydawnictwa Naukowo-Techniczne

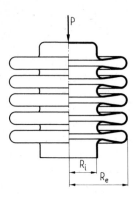

Figure 21.34   Sylphon bellows. Reproduced by permission
of Wydawnictwa Naukowo-Techniczne

E, results in thermal error of the instrument. In the case of a linear characteristic

$$\Delta f_T = f_0 \frac{-\gamma \Delta T}{1 + \gamma \Delta T}$$

where f_0 is the deflection at the temperature T_0 at which the characteristic was established, ΔT the temperature variation and γ the thermal (thermoelastic) coefficient of Young's modulus variation.

Bourdon tubes

A *Bourdon tube* is a sealed-end tube that is coiled into a circle and has an oval or flat-sided cross section (Figure 21.31). The free end of the tube is closed by an end-piece. End deflections are transmitted to the mechanism of the instrument. The working range of the tube p_{max} should be shorter than the proportionality range p_{pr}, so that coefficient $k = p_{pr}/p_{max} > 1$. A numerical value of coefficient k is adopted as $k = 2$ in the case of slow pressure variation, $k = 2.5$ at rapidly varying pressures, and $k = 3$ for tubes working at temperatures exceeding $+50\,°C$.

Membranes and capsules

Because of their small deflections and sharply curved characteristics, flat membranes are used less often than the corrugated variety. The latter have concentric, pressed-in, corrugation rings (Figure 21.32). The shape of the radial cross section of a membrane is called its *profile*. The proper choice of profile shape and sheet metal thickness enables variation of the characteristic shape within wide limits including a rectilinear characteristic.

Joining of two membranes along their edges by soldering, resistance welding or curling and crimping creates a capsule whose deflection is the sum of the deflections of the component membranes (Figure 21.33). Capsules intended for measuring the difference between pressures inside and outside the capsule are called *differential* or *open* capsules. Those intended for measuring absolute pressure have air removed from their interior so that the reference pressure is close to zero (inner pressure of the order of 13 to 40 Pa); this enables pressure measurement at varying ambient temperatures.

The effective area of the membrane is equal to the area of a piston which would be kept in equilibrium by the effective pressure p on one side and the concentrated force P on the other, the latter being equal to a force needed to counterbalance forces appearing as a result of the action of pressure p upon the membrane. Effective area $A_{eff} = P/p$ (Figure 21.32) is approximately, for small deflections, given by

$$A_{eff} = \tfrac{1}{3}\pi(R_1^2 + R_1 R_2 + R_2^2) \tag{21.13}$$

Sylphon bellows

Sylphon bellows (Figure 21.34) are used as measuring elements in cases where considerable deflection with rectilinear characteristics is required. The effective area (as defined above), is

$$A_{eff} = \tfrac{1}{4}\pi(R_i + R_e)^2 \tag{21.14}$$

Elastic bellows are also applied as sealing elements for the transmission of motion to a completely isolated space (for example, filled with aggressive or poisonous gases or liquids) or as flexible, play-free, couplings having great rigidity in the direction of rotation (torque) of the coupled shafts.

21.6.2   Thermal Bimetal Elements

Bimetal elements are elastic elements made in the form of plates or ribbons, consisting of two layers made of metals of different thermal expansion coefficients and joined by resistant welding or by soldering. Heating of the whole bimetal element causes greater thermal extension of the layer made of material with the higher coefficient of thermal expansion (called the *active side*) and smaller thermal extension of the layer with the lower coefficient (called the *passive side*). In consequence the bimetal element bends towards the passive side. The angle of deflection of a bimetal element not loaded by any force (Figure 21.35) is given by

$$\phi = k_T L \Delta T \tag{21.15}$$

and its deflection by

$$f = k_T \Delta T L^2 / 2 \tag{21.16}$$

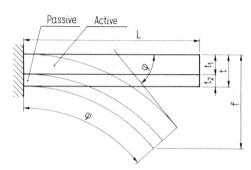

Figure 21.35 Thermobimetal element. Reproduced by permission of Wydawnictwa
Naukowo-Techniczne

where ΔT is the temperature rise of the element and k_T the bimetal element
sensitivity. The bimetal element has greatest sensitivity when

$$\frac{t_1}{t_2} = \left(\frac{E_2}{E_1}\right)^{1/2} \tag{2.17}$$

where E_1 and E_2 are the Young's moduli of the materials of which the
bimetal layers are made. Three limiting temperatures can be discerned.
They correspond to the following:

(a)   That causing permanent deformation due to heating of the free bimetal
element.

(b)   That causing permanent deformation due to heating of the thermal
bimetal element propped at the end, that is exerting a pressure when
heated. In this case the temperature is lower than in the former one.

(c)   That bringing about a decrease of the sensitivity coefficient k_T (loss of
linearity of temperature response).

Generally the passive layer is made of Invar, the active one being formed of
highly alloyed stainless steel with nickel present as the main alloying
component.

The Young's moduli of both metals are practically identical and the layer
thicknesses may be equal (equation (21.17)). Bimetal elements are used for
thermal compensation of measuring instruments (Figure 21.4) or for closing
or breaking electrical circuits according to temperature changes (Figure
21.36). Indirectly, bimetal elements may serve for electrical measurement of
various quantities such as pressure and temperature.

The *thermal inertia* of bimetal elements depends upon the rate of temperature rise, upon the conditions of heat supply to the element, and upon the
latter's thermal capacity. However, since the thermal capacity of the bimetal

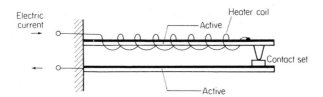

Figure 21.36   Indirectly heated thermobimetal contact system with thermal compensation. Reproduced by permission of Wydawnictwa Naukowo-Techniczne

element is very small, and its thermal conductivity is good, the error caused by this phenomenon may be kept, by skilful design, within narrow limits of the order of ±0.05 °C.

21.7   INTERFACE BETWEEN ELECTRICAL AND MECHANICAL REGIMES

21.7.1   General Remarks

In precision instruments the elements constituting the interface between electrical and mechanical regimes are often electromagnets for producing displacements not exceeding a few millimetres. These are often fed by direct current, rarely by alternating current. Electric motors are used to provide unlimited angular motion from large linear displacements down to micro-displacements.

21.7.2   Driving Mechanisms by Means of Electromagnets

Mechanism is often driven by d.c. electromagnets (Kallenbach, 1969), the armature being in the form of a mobile plate as shown in Figure 21.37. The

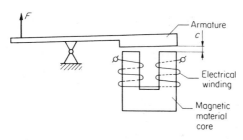

Figure   21.37   Electromagnet   actuator

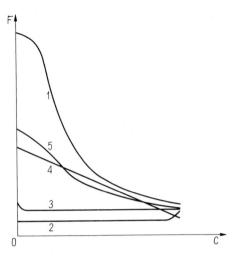

Figure 21.38   Characteristics of: 1, electromagnet (static); 2, electromagnet driven mechanism; 3, spring driven mechanism; 4, spring driving the latter; 5, electromagnet (dynamic)

static characteristic of such an electromagnet (the force F acting upon the
armature as a function of the gap C measured for a stationary armature and
a constant supply voltage) has a hyperbolic form (Figure 21.38, curve 1).
The force needed to move the mechanism from rest (overcoming friction
resistance and accelerating the mass) is greater than in the case of constant
velocity motion of the mechanism (curve 2). Consequently the mechanism
will be accelerated, the armature striking the electromagnet's face at a great
velocity. This simple solution for use is disadvantageous from both energy
and mechanical design considerations.

A more satisfactory method is to drive the mechanism by means of a
series spring put under tension by the armature (curve 3), the spring
characteristic being shown as curve 4. This method is acceptable where the
electromagnet does not need to work fast or when its task is only to attract
the armature and to hold it in this position. Where rapid functioning of the
electromagnet is required, dynamic phenomena should be taken into account. The curve illustrating the increase of current intensity I of the mobile
armature (Figure 21.39) as a function of time t, shows a characteristic saddle
resulting from the decrease of current intensity caused by armature movement and the resulting reduction of magnetic circuit reluctance. To shorten
the *time constant* of the electromagnet, the latter is supplied with current
(*overdriven*) of great intensity (instantaneous current density in a coil equal

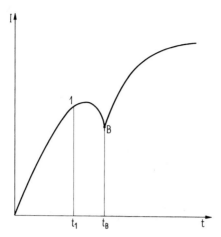

Figure 21.39 Current intensity I in the electromagnet as function of time t

to 100 A mm^{-2} or higher). To protect the electromagnet from overheating, the supply is stopped after a time $t_1 < t_B$, where t_B is the time taken to attract the armature. After the supply is cut off, the armature continues to move toward the electromagnet face due to both the stored kinetic energy of the armature and the mechanism and to the current circulating in the coil that is induced by the collapsing magnetic field. The static characteristic of the electromagnet depends only on its design and on the supply voltage. It can be calculated analytically by means of complicated calculations. The dynamic characteristic (Figure 21.38, curve 5) also further depends upon the dynamic properties of the driven mechanism. Its curve is always situated below that of the static characteristic. To date no sufficiently accurate modelling method for the theoretical determination of the dynamic characteristic has been developed. Electromagnets of the *solenoidal* plunger type give greater actuator displacement, and those of the rotating type (*torquers*) give angles of rotation up to several tens of degrees.

21.7.3 Electric Motor Drive of Mechanisms

Electric motor drives are discussed in Electro-Craft (1973), Kuo (1970), and Veinott (1970). The available diversity of small size electric motors enables a choice of the appropriate motor for the given working conditions; electronic control widens these possibilities still further.

If high starting torque is not required, continuously moving mechanisms are usually driven by single-phase induction motors such as the split-phase or capacitor-start motors. Shaded-pole motors, because of their reliability,

simple design, low cost, long life, and freedom from sparking r.f. interference caused by commutators and brushes, are particularly relevant. However, they have a slippage (*loss of speed*) depending upon the load. If a
constant rotational speed is of importance, then synchronous motors of the
permanent magnet, hysteresis or stepper kind (see below) can be used. If the
mechanism has to move at a slow rotational speed, then reduction gears are
built into the motors (see Section 21.5.4). For driving mechanisms having a
high input inertia, moving coil motors with a strong starting torque, very
short time of starting and braking are used (disc or shell, printed or woven
armature winding) in which the inertia of the armature is very low and the
power rate high.

Electronic control of many of the motors enables speed adjustment within
a very wide range from a few to several thousands of revolutions per minute.
Best control is obtained using a coupled tachogenerator or means of digital
measurement of speed to generate a velocity error signal for a closed-loop
system. A very short starting and braking time can also be obtained by
employing a continuously rotating shaft, driven by a motor with a large
moment of inertia of the rotor, coupling the driving mechanism with this
shaft, or by braking it, by means of rapidly acting electromagnetic clutches
of the friction or powder type (Lomnicky, 1973).

Control motors for use in feedback control systems for d.c., as well as a.c.,
operation of various systems are characterized by having a high speed of
response, do not overheat even when stalled, and have a linear characteristic
for torque as a function of voltage, at least to a high torque magnitude. A
special problem is constituted by designs needing exact angular or linear
positioning. Stepper motors or moving-coil motors are then employed, the
latter being controlled by means of photoelectric, magnetic or capacitative
feedback sensors which can yield as an inaccuracy as little as 10^{-5} of a
revolution.

Linear positioning with an inaccuracy of a few hundredths of a millimetre
within the range of several tens of millimetres in a time of several tens of
milliseconds can be obtained by means of linear d.c. motors (Wolf, 1975)
that are electronically controlled by speed and position sensors. Positioning
with an inaccuracy of nanometres within the range of several tens of
micrometres is possible by means of piles of plates made of a piezoelectric
(piezoceramic) material or by means of magnetostrictive elements (Spanner
and Dietrich, 1979). Vibrating movements can be produced by means of
piezoceramic or magnetostrictive elements or else by a.c. electromagnets.

REFERENCES

Budynas, R. G. (1977). *Advanced Strength and Applied Stress Analysis*, McGraw-
Hill, New York.

Davidson, A. (1971). *Handbook of Precision Engineering*, Vol. 2, McGraw-Hill, New York.

Dudley, D. W. (1962). *Gear handbook*. McGraw-Hill, New York.

Electro-Craft Corp. (1973). *CD-Motors, Speed Controls, Servo Systems*, Hopkins, Minnesota.

Geary, P. J. (1964). *Magnetic and Electric Suspensions*, Sira Institute, Chislehurst.

Harris, C. M. and Credo, C. E. (1976). *Shock and Vibration Handbook*, McGraw-Hill, New York.

Kallenbach, E. (1969). *Der Gleichstrommagnet*, Akademische Verlagsgesellschaft, Leipzig.

Korotkov, V. P. and Taic, B. A. (1978). *Osnovy Metrologii i Teorii Tochnosti Izmeritelnych Ustroistv (Basic Metrology and Theory of Accuracy of Measuring Instruments)*, Izdatelstvo Standartov, Moscow.

Kuo, B. C. (1970). *Theory and Application of Step Motors*, West Publishing Co., St Paul.

Lomnicky, H. (1973). 'Eine Magnetpulverkupplung und Bremse für Magnetbandgeräte der Datenverarbeitung', *Feinwerktechnik*, **4**, 157–62.

Mabie, H. H. and Ocvirk, F. W. (1958). *Mechanisms and Dynamics of Machinery*, Wiley, New York.

Michalec, G. W. (1966). *Precision Gearing: Theory and Practice*, Wiley, New York.

Spanner, K. and Dietrich, L. (1979). 'Feinste Positionierungen mit Piezo-Antrieben', *Feinwerktechnik Messtechnik*, **4**, 181–3.

Sydenham, P. H. (1979). *Measuring Instruments: Tools of Knowledge and Control*, Peter Peregrinus, London.

Sydenham, P. H. (1980). 'Mechanical design of instruments', *Measurement and Control*, **13**, series beginning October.

Truxal, J. G. (1958). *Control Engineers Handbook*, McGraw-Hill, New York.

Trylinski, W. (1971). *Fine Mechanisms and Precision Instruments: Principles of Design*, Pergamon Press, Oxford; Wydawnictwa Naukowo-Techniczne, Warsaw.

Trylinski, W. (1978). *Drobne mechanizmy i przyrzady precyzyjne. Podstawy Konstrukcji (Fine Mechanisms and Precision Instruments: Principles of Design)*, Wydawnictwa Naukowo-Techniczne; Warsaw, 3rd Edn.

Veinott, C. G. (1970). *Fractional and Subfractional-Horsepower Electric Motors*, McGraw-Hill, New York.

Wolf, A. (1975), Elektromechanische Linearmotoren für die Gerätetechnik, eine Übersicht für den Anwender, *Feingerätetechnik*, **3**, 100–4.

Handbook of Measurement Science, Volume 2
Edited by P. H. Sydenham
© 1983 John Wiley & Sons Ltd

Chapter

22 P. H. SYDENHAM

Electrical and Electronic Regime of Measuring Instruments

Editorial introduction

Few measurements systems today do not make use of the electrical form of signal. Solid-state micro-electronic circuitry can now provide almost any kind of information signal function required—sensing, conversion, conditioning, storage, display, and even low-level energy actuation.

This chapter introduces the literature, the basic building blocks, and the system design procedure concerned with the electronic part of an instrument.

22.1 INTRODUCTION

22.1.1 Relevance of Electrical and Electronic Technique

Although electricity was the most recent technological discipline to develop it was very rapidly found to be the most appropriate for many tasks required in a measuring system. Certain classes of instrument can be constructed without recourse to electrical principles, examples being the microscope, the micrometer, the direct recording water level gauge, and many more. However, if an application needs extensive information processing, signal transmission over long distances, uniformity of manufacture at low cost, ability to be extended and use systematic design procedures then electrical technique will usually (but not always) provide a better design solution than other alternatives.

This situation has only arisen in recent decades wherein solid-state integratable electronic circuitry has provided unprecedented ability to produce low cost, very powerful processing, and data handling ability.

This chapter provides an introduction to the capability and methodology of electrical and electronic techniques. It must be made clear that although electrical and electronic technique is so extensively used and is very attractive it is not always the optimum means of solving a measurement systems design need. Indeed, there are many applications where the use of electrical methods are either not applicable or they are not preferred, even if usable, for a variety of reasons. Reference to this chapter, the previous chapter on mechanical, and the following chapter on optical regimes should provide a background which will assist the choice of a suitable combination.

Despite the fact that electrical methods appear to be able to supply every measurement system data handling need it must be remembered that the natural world is usually the system that must be interfaced to man's inventions; that world does not possess much with the same form of physical manifestation as occurs in man-made electrical systems. Thus it is that the other energy regimes will always be important in measurement system design to assist formation of the input and output transduction interfaces. By itself an electronic system has no practical use—it must be interfaced to an application.

Electrical technique includes both the traditional electrical and the more recent electronic disciplines. Initially the two aspects could be reasonably well divided for purposes of exposition and administration. Prior to 1910 electronics was virtually unknown and electricity was, therefore, clearly defined.

Electrical methods were partially displaced by electronic methods (that grew out of the application of thermionic devices) but were taken up by electronics to provide the circuit theory needed, and initially much of the practice. As time progressed electronic methods were devised that could also handle most power needs previously only possible through the use of electrical devices. Today the distinction between electrical and electronic methodology is very blurred although many institutional mechanisms continue that might imply otherwise.

Until very recent times electronic technique was generally only suitable for stages following the sensor. In the 1970's decade, however, silicon integratable devices were developed to yield a new generation of electronically based sensors.

As time progresses electronic capability is steadily being extended.

It is certainly true that in any new application requiring a measuring instrument the designer must now consider the appropriateness of electrical methods for the specific task in question.

22.1.2 Historical Development of Electrical and Electronic Method

The origins of electricity and electronics (the distinction between the two is usually based on the premise that both are concerned with electron flow, the former being at a macro-level, the latter at the discrete electron level; this is somewhat artificial in modern terms of application) can both be traced to Ancient Greek times. Their contribution was, however, minimal compared with advances that began in the 17th century A.D. Furthermore their knowledge was rarely put to use and there was certainly no ordered use of principles of design.

Interest in, and application of, electricity started in earnest with work on electrostatic devices of the form of charge generators and detectors, this being in the century preceding 1800 A.D. Knowledge of charge, its quantity and storage, its transfer from place to place by the use of insulated wires, and its physical effects gave 18th century experimenters a tool for further experimentation. Out of their work gradually there grew useful applications. Electrical telegraphy first began in the late 18th century in a static electricity form. Static electricity led to knowledge about the quantity and potential of electricity and to the understanding that there exist two forms of electric charge: positive and negative. What did not develop so well at first was an appreciation of charge flowing as a current.

Experiments with static electricity, coupled with the improvement in vacuum technique, gradually provided knowledge of discharge effects in gases. These provided a place for the understanding of the behaviour of electrons, a maturity that took some two hundred years to reach the point where the electron was finally proven to exist as a discrete entity in J. J. Thomson's experiments of 1899.

Although there was much activity in seeking knowledge about the nature of electricity in the 18th century significant progress came only after the invention of the simple primary cell by Volta in 1800. Prior to that time experimenters had only very high voltage sources possessing very high output impedance. It was not exactly the preferred apparatus to use in investigation of the practical applications of electrical knowledge!

In 1800 the electric primary cell became generally available and, being simple to make, was rapidly adopted by experimenters. Its low output impedance, reasonably long duration of operation, low voltage, and ease of manufacture was admirably suited to the needs of the gentleman scientists of those times. Progress was steadily made in gaining understanding of the fundamental nature of practically useful electrical circuits and devices and in their everyday use.

During the 19th century the laws of circuits, both d.c. and a.c., were formulated through careful experimental and theoretical work. As new laws were published they were taken up to form technological devices. For example, the relationship between electricity and magnetism was established first as the motor effect in the 1820's. This enabled electrical indicators and electromagnetic actuators to be made. Better indicators of electrical quantities enabled more laws to be discovered and weaker effects to be applied.

The laws of electromagnetic radiation were also enunciated in the middle of the 19th century. By 1888 Hertz had experimentally proven that the theory was correct and that practical apparatus could be constructed that enabled it to be put to use. Marconi, in the last decade of the century, assembled the first practical radio system using simple, purely electrical, components and understanding.

By 1900 many of the basic circuitry laws and physical principles, which are today used to devise instruments, were known if only as scientific entities. The Victorian era produced much of the electrical groundwork needed in 20th century instrumentation.

A chance encounter occurred in the latter part of the 19th century, this being discovery of the *Edison effect*. It proved, experimentally, that a thermionic device could rectify alternating current. Edison did not, however, make use of this at the time. By 1900 designers of telegraph, telephone, and radio communications needed three important improvements in technological capability. Telecommunication needs provided strong commercial reasons to seek them. These needs were: how to amplify a weak electrical signal, how to rectify radiofrequency, alternating currents, and how to generate radio frequency currents. Each of these needs was being catered for at the time by very unsatisfactory methods.

The first decade of the 20th century saw these three basic needs being met by the invention of one basic device, the *thermionic valve*. It first emerged as the purposefully built diode in which the Edison effect was utilized. The diode was capable of rectification and signal generation. Soon after, the triode valve was devised; it could perform all functions of amplification, rectification, and generation. By the 1920's the thermionic device had evolved into many forms satisfying many needs, not only in telecommunications but also in the emerging instrumentation areas.

Electrical and electronic techniques began their partnership from the onset of the electronic discipline in the 1910's. Tuned circuits, special purpose devices built with valves as their basis developed, the bistable flip–flop and the d.c. amplifier being examples that initially fulfilled some special need.

The First World War proved that electronic devices were practical. Training in radio, given to military servicemen, helped the general public take an interest in electronics for wireless applications.

Gradually electronic technique spread into industrial usage. By the 1930's industrial electronics was clearly established, if not as well as it was in the teaching literature of the latter 1950's. It first began to find impact through the use of the *electric eye*, a photocell coupled to a suitable electronic amplifier.

Thermonic components, along with purely electrical devices such as the *magnetic* and *amplidyne amplifiers* were able to cope with most of the industrial needs that arose, but in general electronic method was, by modern standards, slow to be accepted. Basically valve-based electronic circuitry was then too expensive, too unreliable, and too sophisticated for the times to find the widespread use that we have come to accept for it today.

It is the field of computing that helped generate incentive to find better technological ways of carrying out the functions developed for valve operated apparatus. Digital computing, especially, provided impetus for activity for by the 1940's it was clear that very powerful computers could be built that far exceeded those economically possible with valves. A superior replacement for the valve was needed. Valves used too much power, they ran at too high a temperature, they were too large, they cost too much, and they were not reliable enough to be used in huge numbers.

In the 1940 decade the *transistor* was devised at the Bell Laboratories. It did not emerge quite as suddenly as it might seem but came about from gradual evolution of earlier work on the solid-state diode of the 1900 era and after. Valves, in fact, being able to satisfy many of the designers early needs, tended to slow down development of solid state devices.

From the invention of the practical transistor onward there has been great increase in the pace of electronic development. Integration made devices smaller, cheaper, and far less power consuming. They enabled the basic level of systems to be gradually, at an every quickening pace, extended in sophistication. As an example, in the 1960's a digital counter would normally have been assembled from discrete components using individually picked transistors. Today the whole multi-decade counter, with its display, would normally be purchased as a single basic commercial unit that is vastly smaller, far more reliable, and much less costly.

The discipline of electrical and electronic engineering has, for use in the design of instrumentation systems, passed from a stage where originally only relatively small-extent systems could be handled by the average person to one in which very complex and powerful systems are assembled with few technological constraints being placed upon the designer. It is now very much a procedure of being able to more fully express a designer's innovative powers and ability with extensive systems built from basic, very powerful, marketed units, doing this to the limit of his or her imaginative ability.

Space only allows a very short history to be given here. A detailed account is to be found in Sydenham (1979a), where reference to other historical studies is given in the general history of instrumentation presented there.

22.1.3 The Systematic Nature of Electrical and Electronic Engineering

Electrical engineering, as has been described above, developed with a well ordered basis of theoretical understanding. Indeed much of what resulted would not have emerged so rapidly without the theory being known. Unlike mechanical engineering where a large part of its output was able to emerge through the technological path of experimental experience and intuition, most electrical topics can only be understood and applied if the user has a good theoretical grounding in the principles involved. The physical nature of the subject material has allowed its constituent knowledge to be ordered systematically.

Electronic technique rests upon well understood, theoretically enunciated principles. When used at the level that most users assemble systems (from commercially marketed modules) it also requires a considerable degree of inventive skill. It can, however, be efficiently applied using reasonably straightforward and simplified theoretical aids, such as designing procedures and rule of thumb relationships. The nature of the range that electronic parameter values cover implies that the components, and their assembly, do not usually require very tight tolerances of design. People with very little training in the basics of electronics can build very worthwhile systems. Naturally specially-trained persons can handle the topic with greater ease.

This fact should not be interpreted as meaning that electronics does not somewhere require in-depth scientific understanding and application of sophisticated methodology. That has become today more the role of the component and subsystem designer, those people have provided a buffer for the end user and general systems designer.

Marketing forces have assisted simplification of electronic and electrical instrument design. The quest for sales, in large numbers, has enticed manufacturers to provide a truly useful and extensive application service. They have sought ways by which sophisticated devices can be used effectively by semi-trained persons.

This account is aimed at the semi-trained person. It is not intended for the specialist in some aspect of electronic or electrical engineering. It begins by presenting information about the most basic components used to form electronic systems. Whereas many of these may not be used as discrete devices very much in these times it is, nevertheless, important to have an understanding of the concepts and hardware entities that go into a marketed system module. These basics presented then allows discussion to rise to a higher level in the systems hierarchy wherein description can be given of the marketed building blocks that are commonly used to assemble the specific system needed in a given application. It is then feasible to rise yet further to the final design level in order to discuss how extensive systems are generated.

Other conceptual aspects of system design that need consideration are sources of noise in these systems and the power supplies that supply energy.

As an example of the systematic nature of electronic systems consider the diode-based electronic thermometer shown in Figure 22.1. It illustrates how a system is progressively broken down into subsystem blocks, each of which in turn can be further broken down, the process continuing until adequately basic levels of description are reached. In many systems there is no need to progress to the individual components' level as the necessary modules can often be assembled from inexpensive high-performance, units that only need wiring into the system in the appropriate manner.

The above procedure allows any system to be portrayed on paper at the level needed to adequately specify its functional architecture. Often the reverse occurs and the user may be initially confronted with the final subsystem details, not the primary stage block diagram. At first this may well overwhelm the beholder but the use of the systematic realization procedure to form the constituent, higher-level, blocks will reveal how the system operates as each circuit group is identified as a known and familiar building block.

22.1.4   Literature of Electronic and Electrical Engineering

It is of value to provide an overview of the extensive printed literature before continuing. There exist numerous texts on electronic and electrical engineering. They range, in marketing approach, from trivial expositions to esoteric academic theoretical works. Each has its usefulness. Although popular texts are often denigrated they have, over the years risen in quality, now providing excellent introductions for persons who wish to use and understand electrical and electronic technique at the level of practical use yet not become formally trained in this way. Academic works may often be found to be too advanced for middle-levels users. The reader is cautioned to study a text carefully to establish if it adequately matches the need.

The following selection of published works is presented to assist selection. It must, however, be stressed that many others are available. Electrical engineering *principles* are presented in Baitch (1972), Bureau of Navy (1970), and Smith (1976) as introductory courses. Similar material is also available in the many handbooks of electrical engineering, but there it is not usually presented in a didactic manner.

Electrical engineering overlaps electronic technique in the area of circuit networks. Linear networks are the easiest to design and theorize (see Leach, 1976; Van Valkenburg, 1974). Non-linear systems are covered in Fox (1957), Van der Ziel (1977), and Willson (1975). Systems of *circuitry* originally were developed by analysing given arrangements but a more positive design approach was developed through which the system can be synthesized to provide a prestated performance: Fox (1952) is but one text following this approach.

One particular class of network is the so-called *filter*. Modern design procedures for these use previously developed mathematical expressions to arrive at the desired performance using *feedback* placed around an *active*

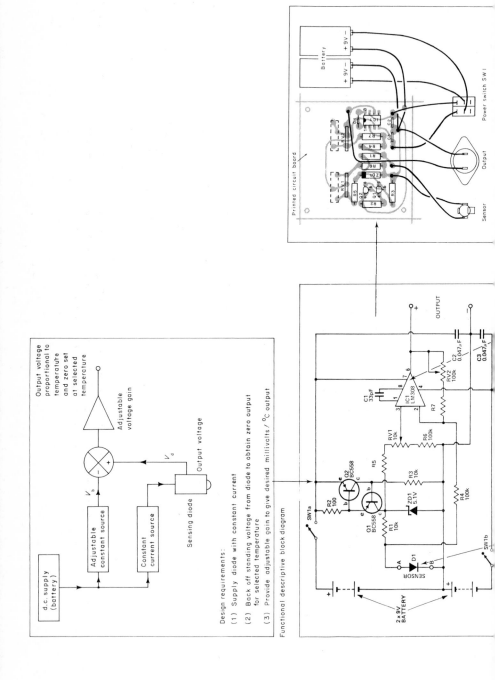

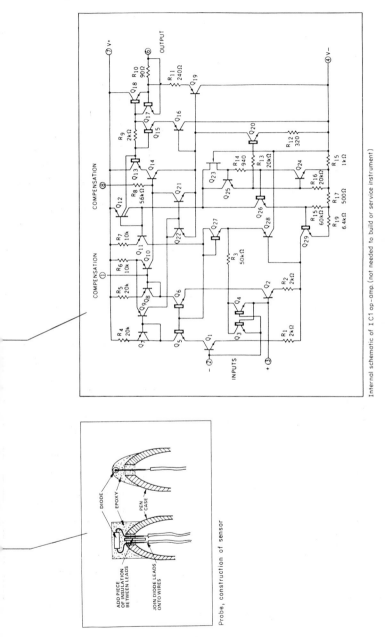

Figure 22.1 Breakdown of a solid-state, diode-based, temperature meter system is formed from basic components arranged in a systematic order. (Reproduced by permission of *Electronics Today International* and National Semiconductor Corporation)

amplifier element. Several texts exist on these forms of filter (refer to Burr-Brown, 1966; Hilburn and Johnson, 1973; Huelsman, 1977; Spence, 1970).

Computers have made simple circuitry design almost trivial and more complex systems possible. Director (1974) and Szentirmai (1974) describe computer aided design of circuits.

As in electrical engineering, but probably more so, there exist many works covering general electronic principles and method. A selection spanning the ranges of depth and of age (early texts contain useful conceptual information and theoretical derivations) is Brophy (1977), Carson (1961), Churchman (1971), Delany (1969), Langford-Smith (1955), Lowe (1972, 1977), Millman and Halkias (1976), Mitchell (1951), Pender and McIlwain (1950), Ruiter and Murphy (1962), Squires (1976), Starr (1959), Sydenham (1979b), and Waters (1978). Several of the electronic equipment and component suppliers offer packaged courses which often include aural recorded tapes and experimental apparatus. An example is Heathkit (1978).

Electronic circuits (these involve electrical circuits plus specialized components) are the subject of Jones (1978), Millman and Halkias (1967), Sands and Mackewroth (1975), and Senturia and Wedlock (1975). Lowe (1974) deals with drawing practice for electrical and electronic engineering: caution is, however, needed in using works on this aspect due to the existence of several different standards of practice.

Electronics, as did electrical engineering earlier, spawned numerous new words in its technical vocabulary. Dictionaries are often needed to understand terms used in written and spoken descriptions. Such definitions are given in Handel (1971), IEEE (1972), and Radio Shack (1975). There is a surprisingly close, but not identical, agreement on terminology across the world in this discipline but very few people appear to make general use of the standard terminologies that do exist and minor differences often occur.

It is difficult to separate electronic instrumentation from the total field of electrical and electronic engineering. For this reason many of the above mentioned general texts may well provide the information sought. There are, additional to these, many books available that indicate from their title that they relate to electronic instrumentation and testing, the main theme of this chapter. As with all works they must be consulted to establish their usefulness in a given situation. The following list refers to those specifically on instrumentation and testing: Banner (1958), Booth (1961), Diefenderfer (1979), Edwards (1971), Gregory (1973), Herrick (1972) IEE (1979), Kidwell (1969), Kinnard (1956), Lion (1975), Maeison (1974), Malvino (1967), Mansfield (1973), Norton (1969), Oliver (1971), Partridge (1958), Piller (1970), Prensky (1963), Regtien (1978), Soisson (1961), Studer (1963), Sydenham (1980), Terman and Pettit (1952), Thomas (1967), Turner (1959, 1963), Waller (1972a, 1972b), and Wedlock (1969).

Texts continuously appear that describe new devices as they arrive on the market. To the titles given here must be added, in any search for information, the enormous amount of application notes and internal publications provided by the component suppliers and marketing agencies. Electronic engineering is one of the areas of technology where tradition has given rise to sales backed by extensive published assistance about product use. Some of these publications are included in the following list on devices and components: Cleary (1969) and Thornton *et al.* (1966) on transistors; Lancaster (1976) on the transistor–transistor-logic assemblies, called TTL; and Lancaster (1977) on the next advance on TTL, but not its replacement, the alternative CMOS circuitry. Integrated circuits are covered in many books (see Millman, 1972; Rosenthal, 1975; Turner, 1977). Other components are dealt with in Sheingold (1972, 1974). Turner (1978) is on field effect transistors (FETs').

Thermionic valves are still used in a general way in some countries and will be found in older equipment; their operation is explained in the numerous electronic subject works compiled before 1960.

Often poorly considered when designing electronic aspects of a measurement systems is the need to supply adequately conditioned power. Texts on power supplies and on handling power include Csáki *et al.* (1975), Dewan and Straughen (1975), H-P (1973), Kepco (1966), Marston (1974), and Motorola (1961, 1966, 1967).

As will be explained in more detail below, one regime of electrical and electronic systems is those systems using continuously-varying signals that can take any level of voltage or current within defined limits. These are called analog signals and the components for use with such signals have become largely known as *linear* devices. The operational amplifier is a salient linear circuit unit. A selection of works on the linear regime includes Burr-Brown (1964), Clayton (1971), Connelly (1975), Faulkenberry (1977), Graeme and Tobey (1971), Gupta and Singh (1975), Melen and Garland (1978), NS Corp. (1976), and Signetics (1977).

The other kind of signal regime uses a signal form that can only exist at two, or sometimes three, levels the former being the more usual. This is known as a digital signal. As will be explained below several basic units exist for building digital systems. They are variously described as units or as combinations in Blakeslee (1975), Bouwens (1974), Breeze (1972), Helms *et al.* (1976), Kostopoulos (1975), Middleton (1977), Millman and Taub (1965), Namgoster (1977), Peled and Liu (1976), Rabiner and Roder (1973), and Wakerly (1976). General works on electronic method, mentioned earlier, also discuss the two regimes of signal use.

Manufacturing aspects of electronic systems using computer aids are covered in Cassell (1972). The works Dummer and Griffin (1962), Simpson (1976), and Waller (1972c) deal with testing and reliability of components and materials. Servicing electronic systems has become an important aspect due to the large size extent of equipment used and the responsible position held by it in the general integrity of a plant. Information about this topic is given in Garland and Stainer (1970), Sloot (1972), and Waner (1979).

Finally, in this introduction to the literature of electrical and electronic systems as used for measurement, mention is needed of the many books and booklets available from the commerical press and the marketing companies on the use of electronic instrumentation in specific fields. These often repeat some of the basic information in their introductions before moving into details of their specific topic. Two examples are Van Santen (1967) on weighing and Piller (1970) on electromedical instrumentation.

22.1.5   Signals in the Electrical and Electronic Domains

Measurement is the process of obtaining meaningful information about a topic. In the sense of relationship with the real physical world this is achieved by interpreting received modulations of energy or of mass transfer. The entity that conveys the information is known as a *signal*. Electrical and electronic systems make use of electrical energy for information transfer so their signals occur in some form of variation in the amplitude or phase relation of either voltage, current or impedance. These three entities are related according to *Ohm's law* in linearly operating circuits and through other relationships in non-linear systems. For example, the system may use a constant voltage-source impedance, variations in the, say, sensor causing current modulations to occur at the receiver. Alternatively the current may be held constant, impedance variations giving rise to voltage fluctuations. By suitable conversion, through the use of an impedance varying stage, a constant-voltage supply system can be used to supply varying voltage signals.

If the current flows only in one direction it is called a *direct current (d.c.)* system. If the current alternates backward and forward in the connecting link it is known as an *alternating current (a.c.)* system. Where the system has d.c. *bias* added to the a.c. signal, Figure 22.2, the current may, in fact, always move only in one direction. This is, however, still regarded as an a.c. system (possessing a d.c. level) because the a.c. signal can be recovered, as can the d.c. level. A d.c. level cannot carry information, unless it is modulated with an a.c. signal, but it is essential in many forms of electronic arrangement.

Signals varying in time (the same concept applies to those varying in space) can be broken down, see Chapter 4, into their *Fourier frequency* components. Sine wave signals are, therefore, the most basic signal form. Many of the concepts and explanations arising in electrical and electronic engineering rest upon the assumption that the a.c. signal in question is

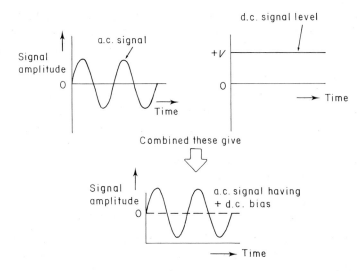

Figure 22.2   Amplitude–time graph of an electric a.c. signal
having d.c. bias

sinusoidal. If it is not then considerable error can occur in the inappropriate use of presented information.

Where no qualification is given as to the waveshape or to the averaging criteria used, it is usually safe to assume that the signal is indeed sinusoidal and that the *root mean square* (*r.m.s.*) value is that given.

As amplitudes of the signals and the frequencies that are used in instrumentation range over numerous decades the practice of compressing the scale from a linear expression is used extensively. The most usual form is the *decibel* method which provides compression of magnitude on a logarithm basis.

The range of frequencies involved in signals and systems explanation and operation begins at the cycle period of thousands of years (as arises say, in considerations of the movements of the orbiting planets) passing through the most commonly encountered region of around 10 Hz to 20 kHz (due to widespread interest in audio system) rising to much higher frequencies for communications and then into the lesser known and used areas above X-ray frequencies. Cosmic rays occur in the 10^{25} Hz region. The most dominant energy frequency spectra used are those of the *electromagnetic spectrum* and the *audio spectrum*. *Spatial frequency* systems, as arise in optical systems, closely follow many of the fundamentals expressed within the material contained in electrical and electronic method.

This account will now begin at the most fundamental system level at which most users of electronic and electrical systems will need to go, i.e. at the components level. Systems will then be considered at progressively higher levels of extent and sophistication. Due to the ubiquitous nature and flexible application within this instrument design regime further depth of detail will be found in other chapters of this work.

22.2 BASIC SYSTEM COMPONENTS

22.2.1 Materials used in Electrical and Electronic Systems

The features of materials and their use that places them in the electrical and electronic regime are their electrical characteristics. Suitable components and assemblies, however, must always be made of some form of substance and design will always rest on possession of knowledge about other properties of both electrical and non-electrical materials. For example, a transistor is formed of silicon, shaped and arranged with various electrical properties, arranged in various positions to which conducting leads are attached to make electrical connection. The whole is then mounted in a suitable container. Thermal and mechanical requirements of the transistor are as important as electrical needs.

Materials are all formed of combinations of the basic elements. Atoms form the smallest level of discreteness that generally concerns the electronic and electrical component designer. Atoms comprise relatively small mass *electrons* that are in orbit around a central, much larger mass, *nucleus* formed of *protons* and, perhaps, *neutrons.* The electron carries, by convention, the *negative electrical charge,* the nucleus possessing *positive charge* that retains one or more electrons attached to it.

Normally charges balance to neutrality and they usually attempt to establish this position in a material system. Removal of an electron leaves an atom positively charged; this is the *positive ion.* Addition of an electron forms a surplus of negative charge; this is the *negative ion.* Atoms are combined to form larger molecules and these can form myriads of compounds or mixtures which make up all solids, liquids, and gases.

When material is formed of charge-stable arrangements of similar, or different, elements it becomes unclear as to which atom is attached to which atom of the matrix. This (simplified explanation) results in a matrix formed of immovable atom nuclei between which there exist electrons moving around in what is known as the *sea of electrons.*

This primitive depiction of the make-up of materials is far from the complete, real situation but does provide for a superficial understanding of the classes of materials which find different use in electrical and electronic technique.

Conductors

A material in which the sea of electrons is very mobile, where any particular electron can easily be swept away and another attracted into the space resulting, is known as a *conducting* material. Typical conducting materials are the metals including silver, copper, aluminium, gold, and tin. It only requires a small energy level to cause a flow of electrons, the *current*, in these materials, hence the name *conductor*. Some liquids and gases can be good conductors under certain conditions.

Insulators

Some materials require very high energy potential to obtain electron flow. These are the *insulators* or *resistors*. Rubber, wood, glass, most plastics, and ceramics are typical insulating materials. Liquid and gaseous insulators also exist.

Conductors and insulators are at the two ends of the range of *resistivities* of materials. In reality there are different materials available with specific resistivities covering the range from one extreme to the other.

Electrical engineering, in the main is (or was so in the past) concerned with insulators and conductors and their application. The former are used in suitable arrangements to contain charge (see Figure 22.3) and its flow so that electric current flows in space where it is required. In a wire transmission system, for example, conducting wires are insulated to ensure that the electrical energy arrives at the desired destination with as little loss as possible. In a generator insulators force current to flow in conducting wires that are revolving in suitable placed magnetic fields.

Materials that form insulators and conductors generally obey the linear Ohm's law given by

$$V = RI$$

where V is voltage potential applied to exert forces upon the electrons, I is the electric current that flows, and R is the resistance to flow, a constant of proportionality for the material.

The electrical resistance between two points on a piece of material is decided by its specific resistance, the cross-sectional area of the material, and the distance between end connections.

Semiconductors

These form a class of material in which current flow is not necessarily proportional to impressed potential; and they have resistivity values just above those of metals. The physics of semiconductors can be most complex.

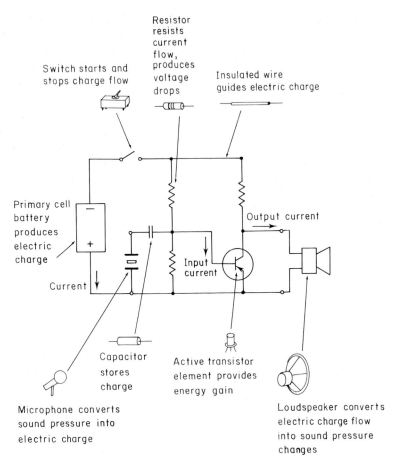

Figure 22.3   In a simple circuit (a sound amplifier is shown) selected materials provide electrical charge, controlling current flow into the speaker where it is converted back into acoustic energy at greater than input level

Fortunately electronic systems development is more likely to need only understanding of what semiconductors can do, than of why and how they actually provide their characteristics. For this reason it is not necessary to provide an in-depth study here. As time passes the electronic systems designer moves further away from such fundamental considerations. The only people being concerned being device developers and research workers seeking new devices and principles.

Semiconducting materials can usually be distinguished from conductors by their property that thermal energy will cause electron movement; which is not the case for conductors. This is usually (again simplistically) discussed in terms of the *energy gap* existing in the semiconducting material. If additional energy causes electrons to vacate their positions, causing current flow, the material is denoted as *n-type*. The spaces left are termed *holes*. A second form of current flow can occur in which adjacent valency electrons move to fill the holes. The materials in which this happens are termed *p-type*. As electrons move they leave holes thus giving the appearance that the holes migrate.

Suitable additions of impurity into extremely pure semiconducting materials (germanium was first used but silicon is now more generally the basic material in semiconducting devices) can enhance the n-type and p-type behaviour.

When a conductor is interfaced physically with a semiconductor, or two of the latter materials are so connected, a flow of electrons occurs across the boundary reaching some form of equilibrium situation. These connection regions are called *junctions*.

How suitable semiconducting material junctions are used for form rectifiers, amplifiers, and other devices is covered in Section 22.2.3.

22.2.2 Passive Devices

Subsystem components used to build energy conversion systems can be divided into *passive* and *active* kinds, these two being combined into a system to provide the required response.

Passive components are those devices that operate upon signals to provide conversion without being able to increase the output energy level beyond that of the input signal. The output signal energy is always less, due to losses, than the input signal energy. A combination of resistors, for example, can be used to *attenuate* a signal's energy but not to increase it. It is, however, possible to transform the energy product of voltage and current so that one is made larger than the original form at the expense of the other that is then made smaller. Examples of such transformations are the *voltage* step-up or step-down *transformer* and the *resonant* circuit.

The main passive components are the *resistor*, the *capacitor*, and the *inductor*. The resistor acts to *dissipate* energy and to resist energy flow by wasting what it does not allow to flow through it. The capacitor and the inductor, on the other hand, can *store* energy; they do possess losses but these are normally small compared with the storage energy capability. It is the energy storage property that enables signal frequency processing to be obtained. Purely resistive systems exhibit zero-order dynamic behaviour; capacitive and inductive systems can be used to realize higher-order dynamic systems as their number is increased in suitable arrangements. (For explanation of these responses refer to Chapter 17.) The behaviour of systems of passive components forms much of the material of introductory electrical engineering.

Calculation of circuit behaviour for resistive systems is relatively simple compared with those having the storage elements. The simpler resistive circuit theory can, however, often be applied to storage element systems provided the storage element is considered as a frequency-varying impedance.

The *capacitive reactance* (its apparent resistance) of a capacitor is given by

$$X_C = \frac{1}{2\pi fC} = \frac{1}{\omega C}$$

and the *inductive reactance* of an inductor by

$$X_L = 2\pi fL = \omega L$$

where f is the frequency of interest, in cycles per second (Hz), ω the angular frequency in radians per second (rad/s), C the capacitance in farad (F), and L the inductance in henry (H).

The time-transient behaviour of either a capacitor or an inductor combined with a resistor is a first-order response, that is, the rise or fall of signal level to a step input follows an exponential change given by the *time constant* τ of the combination in question; ($\tau = RC$ or R/L). Resistance acts to alter the time constant and it gives rise to losses in RC and LC combinations. Such losses are not necessarily of importance. Much of electronic engineering makes deliberate use of practical arrangements in which loss occurs, this being quite unlike power electrical engineering where reduction of losses is generally a prime design target. In recent times engineering of low energy level electronic information signal systems has also concentrated on less lossy arrangements as a means to increase the density of components and to reduce the size and cost of the necessary power supply.

Networks comprising only resistances can be converted into *equivalent networks* or be reduced to a single *equivalent resistance* value between two terminals. Inductors and capacitors can only be reduced to equivalent resistive values at a given frequency of operation and when one kind does not *interact* with others. Generally this means only series or parallel combinations of a like kind (of inductor or capacitor) are reducible. When the two forms interact to form a second-order or higher-order system reduction is more complex as it must take into account the vectorial, amplitude, and phase, nature of the two components.

When the two different forms of storage elements are combined second- or higher-order systems are formed. Losses arising from resistive elements can still be significant but they are, there, usually of secondary importance in basic theoretical considerations.

Figure 22.4 shows how inductive and capacitive reactance vary with

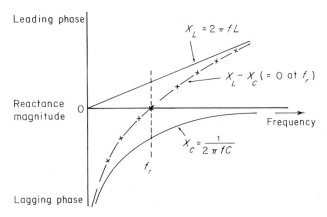

Figure 22.4 Behaviour of inductive and capacitive reactance with frequency. f_r is the resonant frequency

frequency. At a certain frequency, called the *resonant* frequency, where the two reactances X_L, X_C are equal in magnitude their vector addition results in a combined impedance of zero. This is because they each possess an opposite phase angle. In terms of the *complex algebra* description the imaginary components cancel leaving only the zero, real component to resist current flow.

Resonance, therefore, occurs at

$$\omega L = \omega C \quad f_r = \frac{1}{2\pi\sqrt{(LC)}}$$

where ω is the angular frequency, f_r the resonant frequency, L the inductance in henry, and C the capacitance in farad.

In practice no resonant system is free of loss; the real, resistive part has a finite value. This acts to limit the current in the resonant circuit at resonance: the less the resistive component the higher the series resonance current.

Resonance using an inductor and a capacitor can be established as a series or as a parallel connection of the two, and each of these can be used in a series or a shunt connection in a system. The availability of four options can lead to confusion about how a resonant stage influences the system. It is, therefore, necessary to verify which connection is used. In electronic method, where power loss is often permissible, shunting is commonly used to selectively remove signal frequencies that are not needed. This wasteful practice could not be tolerated in power electrical and electronic engineering.

As the windings of an electrical transformer form an inductor it is possible to resonate each winding with a capacitor. This forms the *tuned transformer*, each or one side only being used to increase the sharpness of the frequency selection process provided by resonance. Selectivity is quantified in terms of the *Q-factor*. The higher the *Q*-factor the sharper the resonance; its magnitude depends very much upon the resistance losses in the resonant system. Such losses can be made up by the use of active amplifying elements (Section 22.2.3).

Whilst the resistor, the capacitor, and the inductor are very common in electrical and electronic systems there do exist other passive elements. Some of these are the transformer, rectifier, temperature sensitive resistor (thermistor), light generating device, the panel indicator, and other transducers. These, however, can each be reduced, in theory, to combinations—the equivalent circuit—of resistance, inductance, and capacitance.

They are consequently, generally regarded in theoretical circuit considerations in an equivalent form. A thorough understanding, therefore, of the characteristics of resistors, capacitors, and inductors and their combination forms a most basic and important part of electrical and electronic method.

22.2.3 Active Devices

Networks formed only of passive devices can be devised that will provide much of the signal processing needed in electronic systems. However, their inability to maintain the signal energy level at, at least, that driving the system stage is a serious disadvantage for the signal progressively degrades down to the noise energy level (Section 22.5) and becomes lost. Furthermore the signal input energy may be insufficient to drive the required output transducer. As an example a signal can be mathematically integrated reasonably well with a simple *RC*, low-pass filter stage. To obtain accurate integration, however, it must also provide high attentuation. Some method must usually be added to the output to restore the signal level—this is called an *amplifier*. (Early circuit designs did just this but it is often more powerful to use the amplifier in an alternative less-obvious arrangement to obtain an improved overall performance.)

If the network has the ability to make up the circuit losses, or increase the energy at the output to more than occurs at the input, the arrangement is known as an *active* system.

Any arrangement that has active properties is capable of providing amplification. Amplifiers are able to provide larger energy signals at their output than occur at their input. At first sight it appears that the network provides the extra energy needed but in reality the amplifier acts as a control that allows flow of energy from a supply into the output in accordance with some relationship with the input signal. The output signal is often not even electrically connected to the input signal. Figure 22.5 shows the generalization of an amplifier.

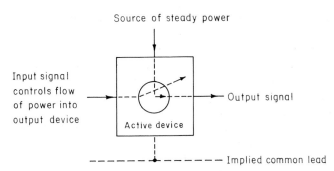

Figure 22.5   Generalization of a three-terminal amplifier
element

In electronic systems amplification, also called *gain*, is used in many ways
for many purposes. Each of the uses has a different terminology. For
example a *pre-amplifier* conditions a low-level signal ready for connection to
a following stage. A stage inserted to obtain a more optimum stage match is
called a *buffer amplifier*; as the name implies it buffers one stage from the
other so that they do not interact. At the output end of the cascaded system
there is often a need for a *power amplifier*; this raises the signal power
providing the specific matching needed to suit, say, the actuator coupled to
the output. Another is the *instrumentation amplifier*. These are vaguely
defined units that may carry out any, or all, of these functions: the term
generally implies that they are able to preserve signal purity and thus retain
the meaning of measurement signals passing through the system chain. The
operational amplifier is an instrumentation amplifier designed to use feedback in an analog computational mode: they are popularly called *op-amps*.

    Each of the above active amplifiers needs to be designed to suit the
application in question. Although there is a certain number of basic designs
in existence the specification of an amplifier can only be achieved properly
and satisfactorily when the task and the amplifier are well understood.

    It is the availability and use of active devices for processing information
that largely distinguishes electronics from electrics. Prior to the thermionic
valve era, that began around 1910, it was very difficult indeed to obtain gain
in a system. The thermionic valve overcame that prime need allowing
numerous active networks to be devised and implemented. It enabled
electronic technique to develop. Following the valve came the solid-state
semiconductor transistor element. This performs the same role as did valves,
operating it in quite similar ways, although not at the electron level of
understanding. Electrical method also has some active elements available.
The *magnetic amplifier*, the rotating electromechanical *amplidyne amplifier*,
and the *Ward–Leonard* motor-generator arrangement are methods for

obtaining gain. They are, however, not practical at very small power levels, the devices being too large compared with even valves for them to have been adopted in signal processing in the widespread manner that transistors have over the past years.

The most basic amplifier, in the general sense, is a three-terminal device having characteristics between the three terminals that can be expressed along the lines shown in Figure 22.6 for a representative electrical component. Transistors, valves, and any other amplifier can be described in this way. Gain occurs when signal input energy can be used to reproduce the same signal at a larger power level at the output. Several other forms of *characteristic describing curve* can be drawn.

The transistor is the most-used active electronic element today. It is capable, in different forms, of handling very small (picowatt) signals through to very large (kilowatt) signals. It is not practical to use exactly the same actual transistor assembly for all cases; the designer chooses that unit which is most suitable for the purpose in question. Since development of the transistor there have appeared many other active semiconductor devices that can provide gain. Each has its special virtues; devices include the silicon controlled rectifier (SCR), the field-effect transistor (FET), the thyristor, and many more. Sophisticated junctions have been devised using a variety of manufacturing methods, these including multiple junction devices. For most practical purposes it is only necessary to learn about the actual physics of operation within the device on rare occasions. Generally use of makers' published characteristic curves and numerical values suffices.

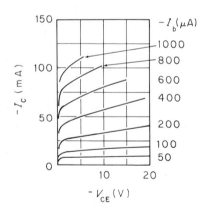

Figure 22.6 Characteristic curves for a typical, small-signal p–n–p, BC177–179 series, transistor. I_c is the collector current, I_b is the base current, and V_{ce} is the collector to emitter voltage

As it stands the transistor, as were valves, is not capable of amplifying a.c. signals centred at a zero d.c. voltage level. This is because the basic transistor p–n–p, or n–p–n, junctions can only pass current in one direction. Only one polarity of the alternating current signal waveform fed into the base connection will be amplified, the other being severely attenuated. It is, therefore, necessary to add extra passive components to the transistor to obtain practical operation. The design of transistor amplifiers used to be very important to electronic system builders but the introduction of low-cost, high-quality integrated-circuit amplifier modules enabled highly specialized amplifier designers to apply considerable effort to the design and to use many more active elements in the design than would have been used in former times. Thus it is common practice now to assemble systems using commercially available units chosen to suit the need. The user need know little about amplifier design; only the external characteristics.

The active element can be used to produce gain for use with *continuous linear signal analog* systems or it can be used to provide gain through a *switching* action for use in *digital* systems. In early times (1840's onward) the electromechanical relay was able to provide gain; it was a prime component in enabling telegraphy to be developed so successfully.

The design of amplifiers is covered in many texts. Of those already listed refer to Brophy (1977), Carson (1961), Clayton (1971), Cleary (1969), Csáki et al. (1975), Delany (1969), Faulkenberry (1977), Graeme and Tobey (1971), Heathkit (1978), Lowe (1977), Malvino (1967), Millman and Halkias (1976), Senturia and Wedlock (1975), Sydenham (1979b), and Thornton et al., (1966). The subject is taken up again in later sections.

22.2.4 Supplementary Devices

To the range of passive and active devices mentioned so far must be added a very wide range of *supplementary* components and devices. These are combined with the previously described active and passive elements to form electronic instrument (and other) systems. It is not practical, or necessary, to cover all of these as a description of a few will give the understanding needed here.

The variety available is most easily seen by reference to electronic component suppliers' catalogues. Electronics always has had a popular following. It began with radio in the 1920's, being added to by audio equipment and television interests. Popular interest then moved somewhat toward all manner of instrumentation for the domestic person's entertainment and use. At the time of writing the main thrust of commercial consumer component and module sales is for citizen-band radio, small computer systems, and general instrumentation. A typical suppliers' catalogue has around a hundred pages of listings.

Systems are generally built assembling these commonly available parts but design needs, where a very large market is assured, will allow new products to be considered. An example will be a host of new modules that will appear as the result of electronic technique being used to a greater extent in mass-produced motor vehicles.

Thermionic devices, although no longer used much for basic amplification, still find some places where they have not yet been replaced by superior solid-state equivalents. Examples are the cathode-ray tube used in television monitors and in the cathode-ray oscilloscope and the video camera. Thermionic devices are also used to obtain spectral emission in gas analysis equipment. The electron microscope also requires electron emission from a cathode.

Such supplementary devices are not normally built by the less specialized user but are obtained ready to use. If they fail they are generally replaced rather than being repaired.

Another example of a supplementary device is the number and letter *alpha-numeric* display device. Such devices again are purchased ready to wire into the circuit. Originally these were manufactured by the user and were a significant part of the cost of a system. Today they are often integrated into the complete, ready-to-use, subsystem module complete with any decoding and power drivers required.

Electronic systems need mechanical frames to support the various circuit boards, switches, variable controls, and connections to other stages. These are called *chassis* in older terminology and more recently *frames.* Most network systems are today formed in their final version on *printed-circuit boards.* These are boards in which all wiring connections between the commercially made components are made by selective etching of a copper coating to form conducting tracks. Holes are then drilled in the copper tracks, so formed, to take the leads of the components which are then soldered into the copper by hand or solder-flow methods. Wire-wrapped connections are also used, especially where greater reliability is called for. Printed wiring boards, called p.w.b. boards, are also used with wiring on both sides, selected holes being *plated through* to connect both sides where needed. The actual layout of components on a board more often is to suit manufacturing convenience and does not then follow the schematic block diagram layout.

Printed wiring boards are often connected into the complete system using plug and socket connectors. This makes for easy repair and fault finding. It also modularizes the system design.

A smaller circuit assembly can be made by printing suitable materials onto a usually ceramic substrate—called *printed circuitry.* These plus vacuum-deposited, conducting metal connecting tracks make up the passive part of *hybrid circuits.* To these are added un-encapsulated semiconductor active

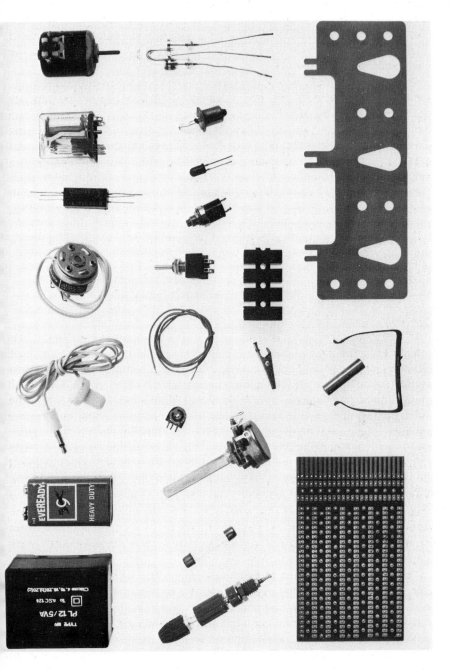

Figure 22.7   A range of supplementary devices used to form electronic systems

elements formed by the *monolithic process*. A complete hybrid circuit may then be totally encapsulated to provide environmental protection.

Still smaller assemblies are made by the monolithic methods using masks to lay down progressively conductors, junctions, and components onto the usually *silicon chip* base. A very large capital investment is needed to create these monolithic, *integrated circuits* (ICs'). The number of circuit elements possible on a single IC now ranges from the original fifteen or so to greater than one hundred thousand.

A myriad of major and minor supplementary components are also used (see Figure 22.7). Some have already been mentioned: the switch, plug and socket, printed wiring board. It is the ready, and relatively inexpensive, availability of electronic components that allows people with relatively little training to form quite sophisticated electronic instrumentation systems. In many instances the module required can be purchased in kit form ready for insertion of the components onto a ready etched and plated, printed wiring board. Very little skill is needed to assemble these the need for skill and understanding arises if they do not function properly!

22.3   SYSTEM BUILDING BLOCKS

22.3.1   Linear Units

As already mentioned, systems are built using a relatively small number of basic commercially available building block units. These divide into two groups, those for *analog* and those for *digital* use. Here we consider the former.

After many decades of research and development the satisfactory d.c. amplifier came into being in the 1950's for use in the first electronic computers: they used analog techniques. The demand for these was rather specialized and their cost and availability were, therefore, rather restricted. However, the introduction of integrated circuit semiconductor technology, in which the active and passive components of systems having many tens of elements are made by mass replication methods using photomechanical methods, led to the availability of a very cheap d.c. amplifier. These were taken up *en masse* because of the versatility that they offered when appropriate feedback connections are applied.

The satisfactory operational d.c. amplifier, the op-amp, must have *open-loop gain* (without feedback connections) of at least around 10,000, and draw relatively insignificant input current. It must also be reasonably stable with temperature and time, for any drift of its d.c. levels internally will provide a related output drift. Furthermore it should be able to handle both polarities of input signal. Integration allowed designers to package a design (Figure 22.8) with around twenty active junctions, plus some twelve passive

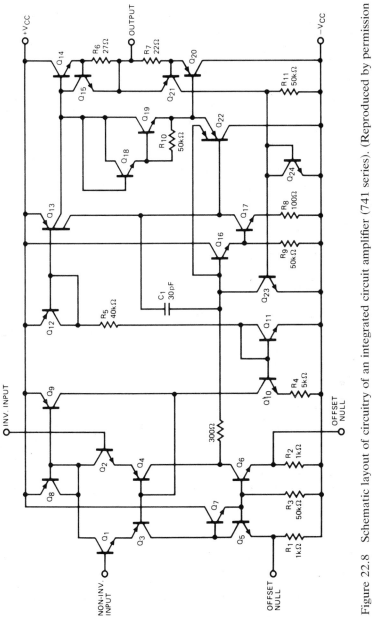

Figure 22.8 Schematic layout of circuitry of an integrated circuit amplifier (741 series). (Reproduced by permission of National Semiconductor Corporation)

resistors, and a few capacitors. Inductors are generally avoided in integrated circuits due to their relatively large size in integrated form at low frequencies of operation. The cost of the IC op-amp is so low as to now be almost insignificant compared with assembly costs to place it into use.

Naturally a *general purpose* op-amp may not be suitable for more exacting uses and more expensive designs are marketed for such cases. The selection of the op-amp must be made to suit the application. Guidance is generally given in texts on op-amp use and in makers' application notes.

To understand why the op-amp is so versatile consider the general case of an amplifier with a single feedback path and multiple inputs as shown in Figure 22.9. To simplify this explanation only resistors are considered but in practice the same derivation can be made for complex impedances in place of the resistors. It can be shown that the *transfer characteristic* between the inputs $V_1, \ldots, V_n$ and the output V_o, if the amplifier draws negligible current, is given by

$$V_o = -\left(\frac{R_o}{R_1} V_1 + \frac{R_o}{R_2} V_2 + \cdots + \frac{R_o}{R_n} V_n\right)\left[1 + \frac{1}{A}\left(1 + \frac{R_o}{R_1} + \frac{R_o}{R_2} + \cdots + \frac{R_o}{R_n}\right)\right]^{-1}$$

This simplifies greatly if the gain A of the amplifier, in the open-loop state, is very much larger than unity, say at least 10 000. Then the above becomes simply

$$V_o = -R_o \sum_{i=1}^{n} \frac{V_i}{R_i}$$

This expression is, therefore, independent of the amplifier gain and shows that the gain of the feedback connected op-amp is decided only by the ratio of the input and feedback resistors. These can be made very stable and accurate in value. As the gain of the op-amp is not a parameter, it can be seen that any variation of the gain does not alter the feedback connection gain, this leading to increased stability.

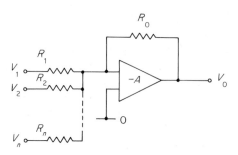

Figure 22.9   Generalized op-amp feedback connection

The gain feature is but one virtue of op-amp feedback usage. It can also be seen from the above expression that input voltages are summed according to a *multiplier factor* for each input channel. Furthermore a single input signal appears at the output with the opposite polarity. This is called an *inverting* arrangement. Modern op-amps provide both *inverting* and *non-inverting* inputs. Their output can swing around a zero voltage level by a typical ±10 V. They are rarely used in open-loop due to the existence then of extremely high gain (100 000 is typical) which has associated with it instability that rises in magnitude as the gain of the feedback system arrangements moves toward the open-loop gain value.

By suitable use of feedback connections the op-amp can be used to mathematically integrate, differentiate, multiply, divide, sum, subtract, convert from voltage to current source, provide buffer amplifiers of many kinds, and act as an analog signal filter (Chapter 9 concentrates upon multiple path feedback, active, filters using op-amps). The number of possible uses for linear circuit operations is enormous and grows continuously. Figure 22.10 shows a few examples to illustrate this versatility.

Op-amps are usually selected and designed according to application notes provided by their makers and to designs published in the, now many, texts on their use. Application notes are generally obtained on request from the op-amp marketing agencies. Suitable texts concentrating on their use are Burr-Brown (1964, 1966), Clayton (1971), Connelly (1975), Faulkenberry (1977), Graeme and Tobey (1971), Huelsman (1977), Melen and Garland (1978), and NS Corp. (1976). Most general electronic texts of origin more recently than around 1965 include chapters on op-amp use. Several manufacturing companies specialize in high performance op-amp production; they also provide relevant users guides and design assistance.

The actual manufacture of integrated circuits is discussed in most introductory texts about general electronics. It is continuously changing in refinement and a bewildering range of terms have been coined to describe the conceptual basis of the new technology. As it is not particularly important to the majority of electronic systems generation for measurement applications it will not be discussed here. The user should not be daunted by a lack of ability to understand just how an integrated circuit of any kind operates internally. Schematic diagrams are usually published for IC's (Figure 22.1 contains an example) but these are of passing interest to the user, becoming important only when the information provided elsewhere on the data sheet and application note fails to satisfy the need. On the very rare occasion it will be found that internal connections may give rise to unexpected operation when the IC is used in a system.

Op-amps and other amplifiers (buffers, servo, instrumentation, interface, converter) form only part of the range of linear devices. Some of the others are covered elsewhere: voltage and current regulators in Section 22.6,

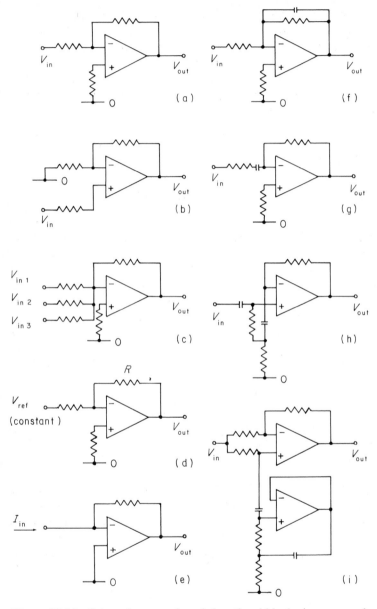

Figure 22.10   Selected op-amp based, functional blocks (power supply connections not shown) (a) voltage amplifier, inverting; (b) voltage amplifier, non-inverting; (c) voltage summation with amplification, inverting; (d) V_{out} proportional to R; (e) current to voltage converter; (f) integrator; (g) differentiator; (h) a.c. voltage amplifier; (i) notch frequency filter

analog-to-digital (A/D) and *digital-to-analog* (D/A) converter units plus *sample-and-hold* units are covered in Chapter 12. Other linears include (but here are less relevant) radio and audio, integrated circuits and video amplifiers for television use, arrays of transistors and diodes, timers, miscellaneous units made for use in industry and automobiles (special experimental runs are sold for original equipment manufactures OEM to develop large volume products), display drives, phase-lock loops and voltage level comparitors. As can be seen, some of these transcend the linear and digital boundary enabling conversion from one signal domain to the other, a very common practice.

22.3.2 Digital Units

Information can also be processed using digital signals by way of the application of logical principles. The foundation of these principles was laid by philosophers of the late 19th century who had little inkling that one day, their work would be applied to actual electrical hardware. *Boolean algebra* handles, in a mathematical sense, the interaction of logical situations. These are typified by the use of two state conditions the most general and basic being connective statements, such as AND and OR. For example, a switch must turn on if input A AND input B are energized but not for either one of them. This kind of thinking and problem solving was first developed to help philosophers conduct arguments on strictly objective logical lines. Their methodology was quite applicable to electrical switching networks and the first digital systems, as they were called, were formed using electromagnetic relays. These gave way to faster-acting thermionic valves and then to solid-state semiconductor devices such as the transistor and the diode.

The digital family of devices comprises a relatively few basic units, but as can be expected they are made in many forms. In the 1970's a considerable degree of standardization came about and two prime logic systems came to the fore. These are those using *transistor-transistor logic* (TTL) and those based upon *complimetary metal oxide semiconductor* (CMOS) logic families of device. TTL systems are always used with high and low logic levels of nominally +5 volt and +0.5 volt, respectively, CMOS, however, can be used at varying levels of supply. The range of digital devices also is often referred to as the *logic range*. Basic building units are now described.

Gates

Systems that perform logical AND, OR and the complimentary negated functions NAND, NOR plus some special cases, such as EXCLUSIVE OR, are called *gates*. As an example two electrical switches in series will provide transmission of a signal if both are closed, that is if switch A AND switch B

are closed. This is denoted $A \cdot B$. Placed in parallel, A or B will give transmission; denoted $A + B$. The negated NOT form results when the opposite switch state case is considered, the sense of the switches being reversed; denoted $\overline{A \cdot B}, \overline{A + B}, \bar{A}, \bar{B}$.

Modern practice no longer generally carries out gating functions with mechanical contact switches, the active semiconducting junctions being used as equivalent switches instead.

In use, logic IC's are provided with adequate buffering and suitable interfacing such that the user only has to first decide the logic gating needed and connect the units accordingly. The internal design of logic units is, again, of little importance to the user. The number of gate functions available in an IC is decided by the number of pins available for connection to external circuitry, two being needed to supply power to the IC. A 14-pin IC, a common unit, can provide twelve gate connections which can be supplied by the maker as either four, two-input/one-output gates; three, three-input/one-output gates; or six, one-input/one-output signal *inverters* (these negate the *logic polarity* but do not invert the the signal polarity in the same way as does a linear signal inverter). Logic IC's always contain only one kind of function on the IC chip: basic functions may be combined together to form a higher level function. The exception occurs when a much greater scale of integration (LSI) is used.

A basic theorem shows that all digital logic systems can be formed from NAND gates but in practice such methods are not cost effective and makers, therefore, supply a range of different IC's for appropriate selection. To make full use of its available logic elements, the gates on one physical IC unit are connected to different points of the circuit. Many of the functional units now to be described can be formed from gates but again it is generally more practical to use purpose-designed units.

Two active elements can be so interconnected that there exist two inputs and two outputs with a large degree of positive feedback between them. The interstage passive element coupling used between the two elements decides the function resulting. This family was originally described as the *multivibrators*. The three forms are the *astable*, *bistable*, and *monostable*. These provide three more basic digital system building blocks.

Astable

If the interstage coupling is capacitively coupled to provide a large amount of positive feedback from one active element to the other, the system will oscillate from one side in the low state, with the other at the high state, changing the states cyclically to form a square-wave output source. These units are also called *clocks* (because they are used to pace on a digital computational system) and *square-wave generators*.

Bistable

If the two stages are d.c. connected to provide positive feedback to the other, the system will rest with one side high and the other side low. A change of the state occurs if one input is fed with sufficient energy to *toggle* the system to the other state. Such a unit can divide pulses by two as each output state only repeats a state after two input transitions. In early bistable designs true short-duration pulses of energy were used; today it is the transition energy of a signal going from a high to a low state (or vice versa) that operates units. These are more commonly called *flip–flops*, *binaries*, *two-states*, and *memories*. They can also be used to store a binary state for a chosen output remains at a given state until toggled, or the power is disconnected. Several forms of flip–flop are available, some including gating into the inputs.

Monostable

If one intercoupling is made capacitive and the other a d.c. link (a combination of the astable and the bistable connections) the system will toggle over for an input transition returning back to the first state after a time period set by the circuit passive components. This is able to provide small time delays to digital signals and to restore the squareness of incoming signals as a repeater and signal restoration stage. These are called *one-shots* or *mono-stables.*

In each of the above multivibrator units there are two outputs, one being the *negated or complimentary* output of the other. Either output can be used depending on the logic sense needed. This often allows an inverter to be omitted.

Inverter

Already mentioned is another basic digital unit, the *inverter.* This is used to invert the logic sense of a signal. The need for this often arises in digital system design.

Inverters also can act to buffer one stage from those connected to it. Digital systems often require many connections to an output. The ability of a unit to drive a given number of other inputs is expressed by the *fan-out.* In TTL systems the number refers to standard TTL inputs, other uses have to be decided on their merits.

Schmidt trigger and comparitor

When a varying level analog signal needs to be quantized into two standard digital levels the Schmidt trigger can be used. This unit provides a rapid square-edge output transition from a high to a low, or vice versa, state at a given level of input of analog signal. The trigger level for a rising signal may be higher than for a falling signal, the difference is called *backlash*. Backlash is often deliberately enhanced to give the trigger system a *window* in which it will not keep retriggering until the signal level has changed its analog level by more than a given amount.

A similar element is the *comparitor*. In this two voltage levels are compared. If one exceeds the other the output *toggles*. By fixing one as a reference voltage a comparitor toggles at that level, as does a Schmidt trigger.

The above listed units form the basic building blocks of all digital systems. Figure 22.11 illustrates this point. Digital systems design rests on understanding their characteristics at the conceptual level. Where costs justify it systems formed of several of these units are manufactured as *large-scale integrated* (LSI) circuits. With time more and more basic functions are being made available as more sophisticated and more powerful, higher level systems units. They are the subject of the next section.

Digital systems convey their information primarily as the timings of the level transitions. The actual voltage, or current, levels of the square waveforms are not of such great significance as they are in analog systems. For this reason rise and fall times are kept as small as possible. Modern systems will make the transition in fractions of microseconds. Circuit connections and component layout can degrade the switching edge. For this reason high speed circuits must be carefully laid out and interconnected. Long interconnecting leads are especially bad unless designed appropriately. The fast rise times of the signals also leads to generation of stray signals. These can influence other parts of the circuit by electromagnetic induction. Furthermore the energy transients within a stage can cause influencing transients on the power supply line. This would effect other units in the system. For this reason it is usual practice to add *decoupling* filters to each small group of IC's.

The testing procedures for digital systems are quite different from those of analog circuits. First, information is only available about logic levels at the various positions in the circuit. Display, using a *logic analyser*, of numerous states is used to fault-find a digital system. Another feature is that the system can often be stepped on one step at a time at any speed that is convenient. Analog systems must act in real-time to provide proper operation of many of the circuit functions. Timing between stages is important in some digital designs and time delay may need to be considered in what is called a *race condition*.

The design of digital switching systems can be made very sophisticated using well established logical mathematical and graphical procedures. Design is largely aimed at reducing the *redundant* logic gates that generally

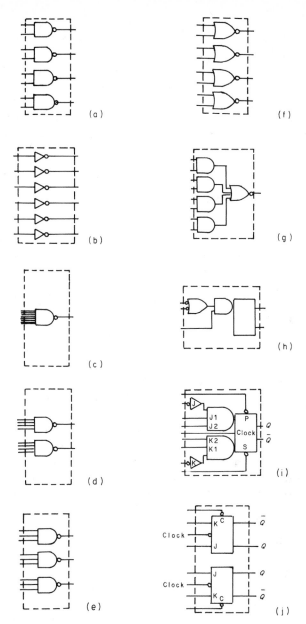

Figure 22.11 Schematic diagrams of some typical integrated digital circuits: (a) quadruple two-input positive NAND gates; (b) hex inverters; (c) eight-input positive NAND gate; (d) dual four-input positive NAND gates; (e) triple three-input positive NAND gates; (f) quadruple two-input positive NOR gates; (g) four-wide two-input AND–OR invert gates; (h) mono-stable multivibrator; (i) edge-triggered j–k flip–flops; (j) dual j–k master-slave flip–flops

occur in the first realized logic network. In many applications, however, it is often less expensive overall to leave redundancies in the system as elimination can be very time consuming and requires an in-depth understanding of switching theory.

Digital systems units are covered in Blakeslee (1975), Bouwens, (1974), Brophy (1977), Diefenderfer (1979), Heathkit (1978), Kostopoulos, (1975), Lancaster (1976, 1977), Middleton (1977), Millman and Taub (1965), Millman (1972), Namgoster (1977), Sydenham (1979b), and Wakerly (1976). Many of the other texts referred to earlier also contain sections of relevance. More detailed works exist on the various aspects, such as switching theory, Boolean algebra, and manufacture of digital systems.

22.4 SYSTEMS DESIGN

22.4.1 General Approach

Originally electronic instrumentation systems were devised using the basic active and passive elements, plus supplementary devices, to realize whatever basic functional units were needed. Almost all of design was carried out using *discrete* components. Design, therefore, first needed to form the functional building blocks which were then interconnected to obtain satisfactory performance. As an example, to build a fast reversible digital counter in the 1960's it was necessary to first develop a fast flip-flop using more active elements than were commonly used at the time for slower speed counters. A logical arrangement had then to be devised that would count as fast as possible providing the output digital code needed for the display. It was also necessary to develop the display for the numerical output as a distinct project facet.

Today the same task no longer exists: a complete counter like that mentioned above is now marketed with integral display. It used standard TTL signals and requires a very modest power supply. The most basic way to assemble the system is now to make use of standard integrated circuits driving an integrated digital display. The same unit is also available as a single LSI unit costing the equivalent of less than an hour of a technician's time. In 1960 it took several man-weeks of work plus significant expenditure on parts.

Systems design is now based on conceptual understanding of the task, this being committed to paper in the form of block diagrams that describe the characteristics needed. These are then refined until the first apparently workable system is ready to assemble. A prototype is then put together using temporary assembly with soldered or plugged joints. A number of proprietary assembly methods are available; many avoid the need to solder. When the prototype is thought to be satisfactory (from actual tests) the circuit is made in the printed wiring, printed circuit or hybrid integrated form.

It is generally the case that the first system design will not be entirely suitable. This occurs because design of any technological system relies on the designer making the correct assumptions about a system that he, or she, is not usually yet familiar enough with to make. The actual act of assembly and testing the prototype considerably improves understanding of the need. The degree of success in arriving at a suitable initial design relates to the complexity of the problem and the designer's experience. Where possible, proprietary modules are usually used to reduce the need for design knowledge. This eliminates many design decisions. If the modules are well tried and proven then the whole is more likely to be reliable.

Interaction between stages is probably the greatest source of prototype development difficulty. Stages may excessively load the preceding one, transient signals can influence other parts of the system. Race conditions, in logical systems, can also occur if some part of the system acts faster than others where timing is an important parameter. Many of these problems cannot be dealt with by rigorous mathematical methods and the designer's ingenuity is called upon to ensure that as many of these as possible are circumvented or allowed for in some way.

When a suitable prototype assembly has been developed it must then be tested for operation in a real situation. Temperature, vibration, mechanical strain, humidity, and many more influence quantities (Chapter 3) may prevent proper operation. In some cases it will be first necessary to construct an assembly that is very close to the actual construction method to be used because layout can be important. The prototype should never be destroyed or changed once the next model is being made. It will often be found that the second, supposedly superior, system exhibits faults that were not seen before. A check of the first prototype may well show that they were also in that model, thereby showing that the design is at fault, not the change in layout.

The design of integrated circuits is often more rigorously achieved but even there, where it is possible to devote more effort to a specific problem, similar methods are often utilized to produce a layout for integration. Prototyping of this level only occurs in specialized groups.

A very common error is to regard a single, apparently satisfactory, assembly as being necessarily typical of a batch production run. Too often production is begun on the strength of inadequate testing and tolerance-spread checking. It is wise to expand from one to ten, to, say, a hundred systems when very large numbers of an identical unit are needed.

The subject of overall systems development is taken up in Chapter 28. This chapter can only act to introduce electronic systems. In practice the electronic part of a measurement system is usually but a portion of the whole, each module being considered in turn, and then the whole, on such other grounds as reliability, production needs, and sales and service requirements.

22.4.2 Commonly Available Building Modules

Many modules are now available from which a system can be assembled. It is not always necessary to build systems using the basic linear and digital integrated circuits with, perhaps, a few discrete components. This section considers a few of these larger subsystems; the number constantly grows as designs and demand appear enabling integration to be used at a commercially viable level.

Counters

A cascaded chain of flip–flops will divide a time sequential string of digital logic levels according to binary division. By suitable arrangement the chain can be used to count up, or down, or be reversible according to an external direction control signal. If the signal passes from stage to stage in sequential order the pulse *ripples through* taking a considerable time (microseconds) to pass through many stages. This form of counter is called *asynchronous* for state changes happening in any stage of the chain are not necessarily made all at the same time. By suitable interconnection of the stages the ripple-through time can be greatly reduced by causing an input pulse transition to set each of the stages simultaneously. These are called *synchronous* counters. They are more complex to interconnect but as they are rarely built from discrete components that is not a problem for systems designers. Counters can easily count to around 10 MHz, the reversible kinds reversing in less than a microsecond.

Binary counters count in *binary* sequence. To change the contained binary code numerical value into the often more useful *decade* form they must be decoded by a suitable conversion logic network. It is sometimes more convenient not to count in pure binary but to use what is called a *binary coded decimal* (BCD) system instead. In this form four flip–flops are used to form a counting chain that passes through only ten, not sixteen, states. Feedback is used to achieve this. The actual coding used within the four-element system decides the form of decoding network needed to drive the chosen decade number display. The BCD units are then cascaded using simple connections to obtain the number of decades display needed. Many of these above design variables are removed by the use of ready-made, LSI counter/display modules. In many instances their price is so low as to make any thought of new development quite unrealistic. They are also well tried and proven.

Frequency counters and timers

Addition of a *clock source* (a square-wave generator of known frequency) and a suitable turn on, and off, gate system forms a method for measuring the frequency of a signal. The gate is, in one version, operated from the clock allowing the counter to accumulate one count per cycle of the signal being measured for a known period of time. This measures frequency of the signal. The count is displayed as frequency. If counts from the clock are measured for a time during which the gate is held open by the cycle time of the signal, period will be measured. Figure 22.12 shows the IC modules used to build such a unit. These instruments are described in some detail in Chapter 20.

Similar principles are used to count objects as part of a, say, batch control system.

Timing is also possible using linear modules in which an output is held, say, high for a preset duration. In linear methods mathematical integration is used to produce a time interval. Linear methods are not capable of the high precision afforded by properly designed digital counter/timers.

Registers

A register is a form of cascaded flip–flops. A digital binary number is passed into the digit stages from one end (*serial input*) or to each simultaneously (*parallel input*); each flip–flop then records a respective digit of the number. Action of the clock-pulse input causes the number to remain intact yet be stepped sideways in either direction. This is used as part of hardware of several mathematical operations on digital binary numbers.

Digital-to-analog converters

A digital number is often needed in an analog equivalent form, for example, as might arise when a electronic digital dalculation has been made and the resultant must operate an analog signal device. Several methods of conversion exist for this direction of conversion and for the reverse where analog signals are converted into digital forms. These functions are, also, rarely made any more from discrete components, the cost of ready-to-use modules being low enough to encourage selection for direct use in the larger system. The design of these is covered in Chapter 12. Although designers do not need to develop these converters it is very important that their internal operation is understood as the performance can vary greatly with application.

As an example of an electronic instrument system consider the apparently simple high quality multimeter. This instrument has undergone extensive change in internal design philosophy in recent years and exemplifies the general trend.

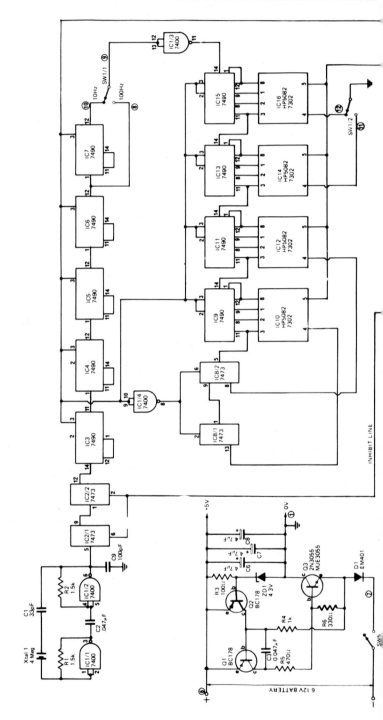

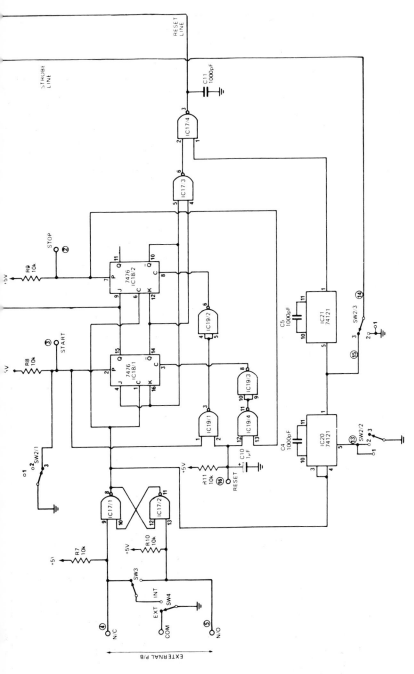

Figure 22.12   Schematic block diagram showing, as an example, how basic IC modules are used to form a general-purpose digital stop watch. (Reproduced by permission of *Electronics Today International*)

A multimeter has the purpose of allowing the user to measure at least voltage, current, and resistance of a system and its components. Originally these were always made from an electromagnetic meter indicating movement to which appropriate multipliers (series resistors—for voltage ranges), shunts (shunt resistors—for current ranges), and operational battery and series resistor (for resistance) were switched into circuit as needed by the operator (Chapter 20). A higher quality multimeter would normally have overload protection to safeguard the meter from accidental damage.

The entire unit used analog methods and may incorporate a linear amplifier in the circuit to raise the terminal resistance to 1 MΩ or more in order to remove loading effects in use.

The alternative design of more recent times and one that will no doubt eventually replace the earlier methodology has no moving parts, contains several hundred active elements and largely operates internally using digital, rather than analog, signals.

The user must still select which variable is to be measured but need not select the range; the systems are *auto-ranging*. The input signal is then converted into a digital equivalent using an analog-to-digital converter stage. The signal is then used with counters and digital display to show the value in digital format.

Another example is the use of digital methods to produce a very accurate, over long time, *sample-and-hold* module. In these the analog storage capacitor that stores the voltage level in the linear analog system has *internal leakage* that degrades the voltage level with time. If the capacitor is used to sample the level and the value so obtained is immediately converted to a digital equivalent form it can be stored for an indefinite time period. Recovery uses a digital-to-analog converter.

The relative cheapness, increased reliability and often enhanced accuracy of digital alternatives easily overcame the fact that internally their design is far more complex and uses many more elements than does the analog alternative.

22.4.3 The Microprocessor

The concept of a totally general-purpose basic electronic digital building block was taken a step further with the introduction of the *microprocessor*. This is an LSI circuit that contains the architectural features of a basic digital computer system. It can be set up internally, through the use of external *software* programming to perform an incredibly wide range of circuit functions. It operates internally as a digital computing system (Figure 22.13), but the interfaces used to connect it to the chosen application allow it to be connected to analog devices when needed. Its low cost enables it to be used as a most basic single component that is set by the user to perform specific tasks. It is set by software programming rather than by hard wiring.

In reality a considerable amount of peripheral equipment may be needed and the cost of this far exceeds the cost of the microprocessor itself (less than one man-hour of time). Even so the general-purpose nature of the module makes it a first choice when seeking to design a complex control system. Rather than make connections with wires the user writes programs. Once set, the microprocessor-based system can be reset reasonably easily. Use of microprocessors needs little understanding of the concepts, basic modules, and building blocks of electronics. It can be confidently expected that microprocessors will bring about a marked change in the design philosophy of electronic systems of the future. With them little electronic knowledge is needed: programming skills have been exchanged for network and system design expertise.

Despite this, peripherals still need to be designed and connected and in many situations, microprocessors and other marketed general-purpose computational designs will not match the performance possible from special-purpose designs built for a specific use.

In the same class as the microprocessor can be placed the *programmable logic controller* (PLC). These are systems in which the circuit wiring to provide the actual system task needed (basically switching networks) is again done by a form of direct programming into the computer-based sytem. Changes to the system operation are easily carried out without need to enter the electronic level of operation or to be expert at programming a computer.

The operational power that can be provided by one, or more, microprocessors has, as did valves in their more primitive way, allowed the designer still greater freedom of action in implementing conceptual schemes.

It is now quite realistic to consider building electronic systems that are self-checking, that can diagnose themselves, and make reconnections to overcome failures—these have been coined the *smart* systems. They can be made to maintain data on their past performance and, perhaps, to begin to design new systems with less human assistance. A certain level of artificial intelligence has already been achieved. The low cost, small space, and power needs of the microprocessor enable very extensive strategies to be implemented on reasonable budgets of finance and time.

Texts on the microprocessor have begun to appear. There has not however, occurred the degree of standardization in hardware that some though would result.

22.4.4 Interconnecting Stages

Whereas the modern electronic system designer has less to consider now about the internal design of building blocks and modules than in the past, it is still as important as ever that they be interconnected properly. If the connecting links are considered as additional stages when they are likely to

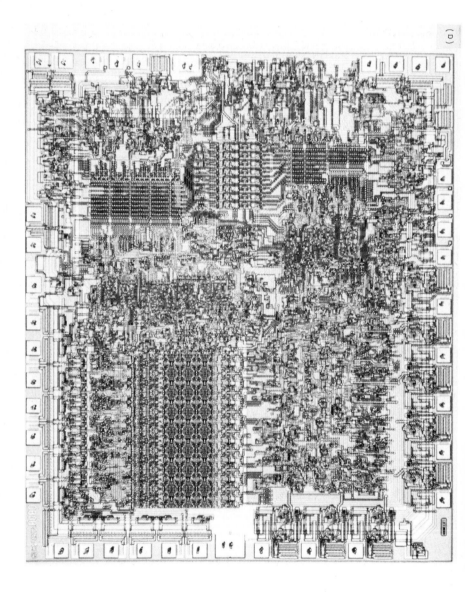

(a)

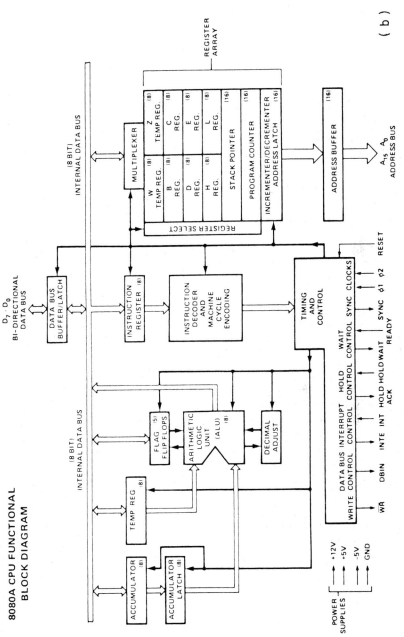

Figure 22.13 Internal arrangement of a typical microprocessor IC (Intel 8080 series): (a) physical construction; (b) schematic diagram of operation (Courtesy Intel Corporation)

receive the attention that they must be afforded. Even when modules are wired in close proximity, they must be connected appropriately.

It is necessary to consider the connections carefully for the link itself may act as frequency filter, alter the amplitude of signals, pick up unwanted signals, introduce phase shifts, be mechanically unreliable, and, in general, cause the system to fail to operate satisfactorily in one or more regards.

Transmission over distance

Within closely assembled circuitry simple printed and insulated wiring will generally suffice (unless very high frequencies are involved). If the distance between units exceeds a few metres then it becomes necessary to select a more suitable form of signal transfer method. The *open-wire*, as seen on telephone poles, is usable to around 10 MHz, but beyond that the higher frequency signals need some other transmission method. They are very lossy and not particularly secure from damage.

An alternative is the *coaxial cable*. This is formed from conductors suitably placed with a separating dielectric. They are made in circular, flat, and strip sections each having its special application. They can convey signals to around 5000 MHz. Multicore insulated cables could be considered to be of this type but they are not usually designed for high frequency use, having excessive losses.

Above the usable range of signal frequency suited to coaxial cables it is necessary to make use of *waveguides* and open-space *electromagnetic* radio links. Laser-based *optical fibre* links are also capable of conveying these high bandwidths. Although only just coming into use (refer to Wolf (1979) for detail) at the time of writing these will certainly play a more major role in future as they are not prone to the same stray signal, pick-up, effects as are electronic transmission systems.

Whichever kind of electric line is used certain basic generalizations can be made. First, a short line will possess small values of self, shunting, capacitance and resistive losses. As the line length is increased the capacitive effect becomes more dominant, the line acting as a low-pass filter: this gives attentuation and phase shift.

The coaxial cable exhibits an interesting feature for reasonably high frequencies of use. There the impedance, seen looking into the end of an adequately lossless line of any length, can be regarded as being a fixed pure resistance. This is called the *characteristic impedance*, Z_0, of the line. Common values are 600 Ω, 100 Ω, and 75 Ω. Thus the actual length of the line matters little.

When high frequencies need to be sent over transmission links a serious effect occurs if the terminating impedance is not purely resistive and equal to the characteristic impedance of the line used to make the connection. If mismatched, *reflections* of signal energy occur and time-varying signals are generated in the links that feed on, and back, at each end of the link. At lower frequencies the mismatched line acts to alter the signal magnitude and phase; reflections produce distortion and loss of power transfer. It is, therefore, always important to properly match high frequency signal terminations. Chapter 13 deals with signal transmission in greater depth. The subject is covered in Johnson (1950) and in Sinnema (1979).

Grounding and shielding

Another aspect of stage coupling is the need to preserve the signal-to-noise ratio (SNR). Ideally a coupling should not degrade the system but in practice there will always be some reduction.

Analog signals are more prone to this for their information is conveyed as an amplitude. Coding and the use of digital signals are often better alternatives to use.

Several connection techniques can be used to reduce stray pick-up in wiring. The principles apply to both long and to even very short leads in low-level, detection systems.

A first possibility is to transfer the information over the link at some frequency away from that of the expected noise. It is then theoretically possible to filter out the signal at the receiver. This method can be satisfactory (see Chapter 11 in the discussion of signal-to-noise improvement) but all filters allow some level of signal through and the price paid for sharp highly discriminating filters may not be acceptable; for instance, they tend to *ring* to rapid transient changes of signal level.

A better approach is to reduce the noise level at its source. If that is not possible then the wiring can be arranged such that two parts of its pick up similar noise signals which are out of phase, by 180°, at the detection stage. They then cancel to leave only the unbalanced contribution. This is known as *common-mode rejection*. A twisted cable pair does this as each wire has an equal signal induced in it by radiation from the noise signal source, the two being induced in opposite directions in the detector input circuit. To obtain the best from common-mode rejection, connectors and the input stage of the active circuitry must also be differentially arranged in this manner.

Shielding is also a method for reducing the signal induced into connecting leads. Leads form antennae in which electromagnetic fields will induce voltages. Thus the shorter the leads, the lower their impedance and the better they are shielded, the smaller will be the induced currents. Shields, and their necessary earthing connection, should not themselves form a significant resistance closed electrical circuit; the *earth loop* produced will then produce currrents that reinduce voltages elsewhere.

Radio frequency energy can also be very bothersome. Often the problems are overcome by preventing it escaping from the generating source by careful scientific design of the enclosures. Alternatively the detecting apparatus is screened.

The subject of signal transmission is seen as sufficiently important for there to be a chapter devoted to it in this Handbook (see Chapter 13). A chapter on the topic appears in Sydenham (1979b). Several specific texts are available that provide detail on long line transmission systems, for example Johnson (1950). Low noise systems, and how to couple them, are the subject of Morrison (1977), Motchenbacher (1973), and Ott (1976). Chapter 11 discusses methods for improving the signal-to-noise ratio of a system.

22.5 LIMITS OF DETECTION IN ELECTRONIC SYSTEMS

22.5.1 Sources of Fundamental Physical Limit on Detection

As in all of the energy regimes, electric energy systems also possess certain physical processes that prevent sensitivity being increased to infinity. No matter how well contrived and constructed is the apparatus the designer must face the fact that these fundamental effects cannot be eliminated. They provide an ultimate limit to detection for a given methodology. Three fundamental noise sources, and other controllable noise sources, are shown pictorially in Figure 22.14.

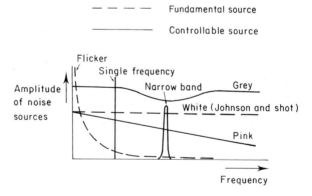

Figure 22.14 Frequency spectra of fundamental and other electronic noise sources. (Amplitudes of each type not shown to any relative scale)

Johnson noise

Within the resistive components of any electric component thermal energy causes the electrons to vibrate with increasing activity as the thermal energy increases. It is related to Brownian motion within matter. Thermal noise, also called *Johnson* noise after its prime investigator, has been quantified as

$$\bar{v}^2 = 4kTRB$$

where $\bar{v}^2$ is the mean square voltage generated, k is the Boltzmann constant, T is the absolute temperature of the resistance, R is the resistance in question (in ohms), and B is the frequency bandwidth of the signal in the resistor.

This can also be expressed in terms of noise current and noise power. Unlike Brownian motion noise of mechanical systems, which rarely is observed, this noise is relatively easily detected by electronic systems and is always a serious factor to consider when designing low level (below microvolt) detection stages. Clearly, from the equation, reduction of bandwidth, temperature of the resistor, and magnitude of the resistor reduces the Johnson noise level.

The frequency spectrum of this noise, in its purest form, is uniform, there being as much energy in a given bandwidth at any place in the frequency spectrum. If this condition applies the noise is said to be *white*. Grey and pink noise are of this general form but they possess different frequency spectra.

Shot noise

Wherever current flows the corpuscular nature of the electrons gives rise to discrete signal energy packets. If there are enough of them their sum averages to a steady current and the effect is not generally observable. If, however, each electron becomes an event, as it does when very low level signals are concerned, then the discrete randomness makes its presence detectable. This was first investigated for electron discharge in the thermionic valves with the assumption that the space-charge was not filling the space available. The theoretical expression derived to explain the effect quantitatively was thus an estimate for in most applications space-charge conditions alter the magnitude of the effect. The usual basic expression quoted is

$$\bar{i}^2 = 2eIB$$

where $\bar{i}^2$ is the mean square current generated, e is the electron charge, I is the average current passing, and B is the frequency bandwidth of the signal passing.

However, valves, transistors, and other active electronic devices each differ a little from this generalization. Suffice to say that this effect is also detectable by quite simple electronic apparatus and, therefore, must also be allowed for in detection stage design. Reduction in signal bandwidth and the through current level reduces the noise level. Choice of device also has an important bearing on the magnitude. This form of noise is also spectrally white.

In both of the above cases the formulations express the levels of noise signals generated within a component. The level that is received by the following stage depends upon the matching conditions.

Flicker noise

A third important fundamental noise source of electronic systems is called *flicker* noise. Since it varies approximately as the inverse of the frequency it is often referred to as $1/f$ or *hyperbolic* noise.

Despite considerable research it is still not possible to quantify this noise form on a general basis. As yet the physical reason for it in valves and in transistors is not proven. As a rule of thumb, an electronic systems' flicker noise begins to be significant compared with the other fundamental limits at around 400 Hz increasing below that frequency to infinity at absolute zero frequency. This is one reason for using carrier systems to provide good signal-to-noise ratios in detection equipment: the signal to be detected is modulated at generally well above 400 Hz.

Flicker noise has a direct counterpart in mechanical and thermal systems for the greater the integration used in an attempt to remove long-term, slowly changing, drift effects, the less the rate of improvement obtained.

The above mentioned noise sources are the three dominant sources that cannot be reduced beyond certain limits. The use of bandwidth reduction techniques reduces their level but at the usual expense of increase in response time. Cooling is often employed, including use of cryogenic stages. Electronic noise sources are well described in the literature (see Bennet, 1960; Connor 1976; Delaney, 1969; Usher, 1974; Van der Ziel, 1954; Whalen, 1971).

22.5.2 Sources of Noise Resulting from Imperfect Design

Assuming that fundamental limits do not restrict the design then the system must also be assembled, connected, and shielded such that additional noise pick-up is kept as small as possible. The common unit used to describe the degradation that always occurs from connection and from stage performance is *noise figure* (NF). It is given by:

$$NF = 10 \log \frac{SNR_{in}}{SNR_{out}} \, dB$$

A perfect stage adds 0 dB but in practice arond 2–3 dB is the best obtainable. Shielding, layout, wiring, device selection, connection modes, matching, and other factors must be considered as possible means to reduce the NF of a stage. The order that stages with differing NF are connected in cascade is also an important factor. As a guiding rule the stage with lowest NF should be used closest to the detection stage. Often an initial stage is inserted that provides little useful gain because its low NF assists the overall NF to be minimized. As practical measurement systems usually involve more than one energy regime the noise sources of each regime must be considered.

Noise reduction is the subject of Chapter 11. References to texts are given in Section 22.4.4.

22.6 POWER SUPPLIES

22.6.1 General Remarks

The various modules that make up an electronic system almost always (there are exceptions, the crystal radio set of early times being one) require a power source to operate them. Most systems operate basically from direct current supplies (alternating current signal sources are also used for other purposes—see Section 22.6.4) and it is not unusual for more than one voltage level to be needed. An operational amplifier, for example, commonly requires ±15 V with a zero supply line. Logic circuits operating with TTL circuitry need at least +5 and 0 V. These voltage supply lines are referred to as a *voltage rail, bus* or *line*. They are rarely drawn in circuitry diagrams, being indicated only as a line or simply as a voltage value.

In practice the size, cost, and performance of the power supply is a dominant feature of an electronic system.

Direct current power can be supplied: from primary, or secondary, electrochemical batteries and fuel cells; from rectified and smoothed a.c. mains supplies; from solar cells; from thermoelectric junctions that are heated; and from rotating electrical generators.

Whichever kind is used the important features of the supply will usually be: the method of generation of voltages supplied; stability of those voltages to changing load and changing input, and with time and temperature change; availability; life; capacity in terms of current and, hence, electrical power; size; weight and shape.

Past design tended to concentrate the whole power supply in one place generating each voltage needed at the desired total capacity at one point sending the energy to the respective modules over a cable or bus system. Current trends are now to distribute much more of the supply hardware over the whole system, supplying unfiltered and unstabilized energy for processing right at the place of use.

22.6.2 Obtaining d.c. Current

A wide range of types and size of primary electric battery are available. Each has features that suit some applications and not others. The cost penalty of primary cell supply and the changing needed are often lesser than that of additional weight and the need to connect to a power mains. In general the lower the weight for a given quantity of stored energy level the higher cost of the battery.

Secondary cells can be recharged making them a little more attractive but again they can become discharged at inconvenient moments and often they need constant maintenance to retain an adequate charge. Many primary cells are capable of being charged as secondary cells, the electrochemistry of the two being similar.

Electrochemical cells are discussed in Sydenham (1979b) and several texts are available on their design and use. Design of batteries is constantly being marginally improved; new cells are continuously being marketed.

A more acceptable alternative, and a less costly one overall, is possibly to rectify the a.c. mains, after suitable transformation, to obtain raw unfiltered and unsmoothed d.c. current. This is then filtered with RC low-pass filters (inductance designs are also used for better filtering) to yield reasonably smooth d.c. voltages.

The design of mains-derived power supplies is more complex than generally supposed. Good design aims to reduce losses of conversion, allows for adequate heat liberation in the rectifiers and transformer, and, above all, supplies the correct output voltage at the end of the conversion chain. This aspect is more relevant to electrical engineering than electronics but obviously it is one of the areas that transcends the two disciplines' boundaries.

Rotating generators must provide the correct voltages and, furthermore, be adequately free of cyclic transients (spikes, poor wave shape) produced by the non-ideal generators often used.

22.6.3 Stabilizing Voltages and Currents

If circuits possessed constant power level demand and if the basic supplies were unchanging then the d.c. energy supplied from the above methods could be used without need for further stages. However, in practice neither condition holds and it become necessary to add a serial stage between the source and the load module that causes the voltage (or current) supplied to remain within tight stable limits as supply and load vary. This stage is called a *regulator*.

The correct operation of many electronic circuits is based upon the assumption that the voltage supplied remains constant. (There is now, however, a detectable trend toward design of modules that can tolerate very wide variation.)

Ideally the perfect *voltage source* has an *internal impedance* of *zero*, meaning that any connected load current changes cannot cause changes in the internal voltage loss. The dual is the ideal *current source* which would have an *internal impedance* of *infinity*. In this case output load impedance changes cannot cause the current to alter in that load.

Thus stabilized supply design aims to approach these ideals by producing sources that appear to have a very low output impedance for a voltage supply and a very high output impedance for a current supply.

The simplest conceptual method in each case is to arrange a supply with an impedance such that the load sees the unit as an approximation to these ideals. Thus a large-current-capacity voltage supply, feeding a low-resistance shunt resistor, will appear to a higher resistance load, as a low impedance. Similarly an approximate current source can be made by use of a large resistance in series with a high voltage source driving a lower load. Such methods are very wasteful of power and are no longer normally used.

A second method is to make use of a non-linear device in which the impedance at the operating point is much lower at the operating point than elsewhere. The earlier used *barretter* tube did this. The modern *Zener* diode uses the same concept.

Zener diodes are (Motorola, 1967; Sydenham, 1979b) the simplest practical method of obtaining reasonably well regulated voltages. With a suitable reverse bias current level they exhibit a voltage characteristic that is largely constant over a very wide range of through current. To form a regulator they only need an additional resistor connected in series plus a suitable heat sink to liberate the thermal losses. Good temperature stability is obtained by the use of suitable combinations of Zener diodes. The degree of regulation of a Zener diode regulator depends upon the *dynamic resistance* around the operating point. It can reach low enough values to provide regulation within about 1 in 100 changes for a single Zener stage. In general whilst the Zener method played a most important general role in the past, the time has come where more sophisticated active regulators, made in integrated circuit form, can provide far superior performance at around the same cost. Zeners continue to provide reference voltages in circuitry where power levels are minimal.

Active regulators operate on the basic principle that a voltage sensor detects any change in output using a voltage comparitor circuit. This change is then used to alter the throughput of current to the load so that the voltage across the load is restored to the correct value. The Zener diode finds application here as a prime source of a stable reference voltage: it is used at very low power levels with substantially constant load. Active systems are, therefore, dynamic feedback control systems. Figure 22.15 is the schematic of an IC regulator.

A common element used to control the power flow is the power transistor. Base current is altered to control the series or shunt current of the stage feeding the load. Larger capacity systems make use of *silicon-controlled rectifier* (SCR) semiconductors (Motorola, 1966) and other components of the same family to control large power levels. High efficiency *switching regulator* supplies use variable time switching to control the power flowing.

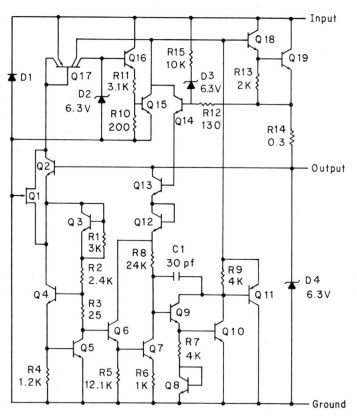

Figure 22.15 Internal schematic circuit of an active, feedback controlled, IC voltage regulator (LM309 series). (Reproduced by permission of the National Semiconductor Corporation)

High quality regulators first form a well regulated reference voltage in one section. This is then used to form the reference for the power control stage. Current control is obtained by conversion from voltage references into a current reference by the use of a stable resistor.

In use, the designer now selects an appropriate integrated circuit regulator from the linear device range adding any external passive and active components as desired (resistors to vary voltage output and perhaps a power transistor to handle a larger load current). A heat sink is often required to dissipate the losses.

Active supplies can generally hold the output constant to within 1 in 1000 of supply and output load variations; a ten percent variation in a 10 V d.c. source voltage will be reduced, by regulation, at the output to 1 mV variation. The variation is called *ripple*.

Obtaining regulation was the first design aim of electronic regulated supply designers. The next was to make the supply secure against overload from accidental short-circuit, transients, open connections, and overheating. Various additional circuits are now incorporated into the IC regulator to ensure that the unit does not fail in such instances.

The output terminal impedance of the IC voltage regulator is low enough for the resistance of connection leads to be large enough to significantly increase the value when seen at the load. Thus it is often better practice to place the regulator right at the circuit itself. This also helps to *decouple* the circuit transients that enter the supply lines through the regulator. This is one reason for the change in philosophy in power supply location; from concentrated toward distributed systems. It is now common practice to assemble a system with each printed wiring card carrying its own regulators, feeding the cards from a basic unregulated source.

22.6.4 Alternating Current Supplies: Signal Generators

Power to operate electronic systems is generally of d.c. form. Within systems the need often arises for signals of known waveform, amplitude, and frequency. Common waveforms needed are sinusoids, square waves, pulses, repetitive ramps called the sawtooth wave, plus special functions of many kinds including white and pink noise.

Modules that produce these signals are known as *signal generators*. The systematic repetitive waveform generator is basically some kind of *oscillator*. Signal generating modules may need to supply signal energies ranging from microwatts to kilowatts. For example, at each end of the range are the very low power, square-wave, clock source for digital computational circuit and the high power, radio-frequency sinusoidal waveform generator operating a microwave heater or dielectric welder.

Linear forms of waveshape are generally generated using linear oscillating circuits. Digital synthesis method are also available for producing analog waveforms.

Most oscillators comprise an amplifier in which the output is fed back into its input as positive feedback. Once set oscillating by an initial transient this arrangement produces a continuous wavetrain of the type the circuit is built to provide. Amplitude becomes limited by inherent features of the arrangement.

Numerous forms of oscillator exist. Design aims to obtain an adequately pure wave shape, stable amplitude, and frequency, plus sufficient output drive power. An op-amp can be connected with a second op-amp or some passive components to form an oscillator (Figure 22.16). A square wave can be generated using digital blocks. A square-wave source can also be obtained from a sine wave by the use of considerable amplification followed by a means to limit the amplitude.

Where complex wave shapes are needed special-purpose function generators are used. Proprietary *frequency synthesizers* are marketed in which a program can be entered to produce the signal needed.

The design of signal generators is covered in most general accounts of electronic technique. There is, however, a lack of texts specifically devoted to the subject.

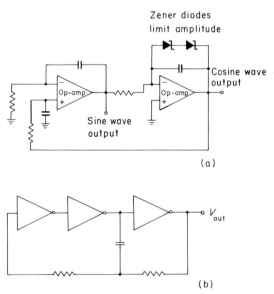

Figure 22.16 Integrated circuit modules connected to produce: (a) sinusoidal signals (using op-amp); and (b) square-wave signals (using digital IC's)

The electronic regime of a measuring instrument system must be designed, at the functional level, to interface with the remainder of the multiregime system. This chapter has outlined electronic and electrical components and procedures. Other chapters of this work provide more detail of other aspects of the design process.

REFERENCES

Baitch, T. (1972). *Electrical Technology*, Wiley, Sydney.

Banner, H. W. (1958). *Electronic Measuring Instruments*, Chapman and Hall, London.

Bennet, W. R. (1960). *Electrical Noise*, McGraw-Hill, New York.

Blakeslee, T. R. (1975). *Digital Design with Standard MSI and LSI*, Wiley, New York.

Booth, S. F. (1961). *Precision Measurement and Calibration; Selected Papers on Electricity and Electronics* (3 Vols), National Bureau of Standards, Washington DC.

Bouwens, A. J. (1974). *Digital Instruments Course* (several parts issued), N. V. Philips Gloeilampenfabrieken, Eindhoven, Netherlands.

Brophy, J. J. (1977). *Basic Electronics for Scientists*, McGraw-Hill–Kogakusha, Tokyo.

Breeze, E. G. (1972). *Digital Display Systems: Fairchild Application Note 212/1*, Fairchild Camera and Instrument Corp., Mountain View, USA.

Bureau of Navy (1970). *Basic Electricity*, Dover, New York.

Burr-Brown (1964). *Handbook of Operational Amplifier Applications*, Burr-Brown Research Corp., Tucson, USA.

Burr-Brown (1966). *Handbook of operational amplifier active R.C. Networks*, Burr-Brown Research Corp., Tucson, USA.

Carson, R. S. (1961). *Principles of Applied Electronics*, McGraw-Hill, New York.

Cassell, D. A. (1972). *Introduction to Computer-aided Manufacturing in Electronics*, Wiley, New York.

Churchman, L. W. (1971). *Survey of Electronics*, Rinehart Press, San Francisco.

Clayton, G. B. (1971). *Operational Amplifiers*, Butterworths, Sevenoaks, UK.

Cleary, J. F. (Ed.) (1969). *Transistor Manual*, General Electric Co., Chicago.

Connelly, J. A. (1975). *Analog Integrated Circuits*, Wiley, New York.

Connor, F. R. (1976). *Noise*, Edward Arnold, London.

Csáki, F., Ganszky, K., Ipsits, I., and Marti, S. (1975). *Power Electronics*, Akadémiai Kiadó, Budapest.

Delaney, C. F. G. (1969). *Electronics for the Physicist*, Penguin Books, Harmondsworth, UK.

Dewan, S. B. and Straughen, A. (1975). *Power Semiconductor Circuits*, Wiley, New York.

Diefenderfer, A. J. (1979). *Principles of Electronic Instrumentation*, Saunders, Philadelphia.

Director, S. W. (1974). *Computer-aided Circuit Design: Simulation and Optimization*, Dowden, Hutchinson & Ross, Stroudsburg, Pa.

Dummer, G. W. A. and Griffin, N. B. (1962). *Environmental Testing Techniques for Electronics and Materials: Series on Electronics and Testing*, Vol. 15, Pergamon, Oxford.

Edwards, D. F. A. (1971). *Electronic Measurement Techniques*, Butterworths, Sevenoaks, UK.

Faulkenberry, L. M. (1977). *An Introduction to Operational Amplifiers*, Wiley, New York.

Fox, J. (1952). *Modern Network Synthesis*, Wiley, New York.

Fox, J. (1957). *Nonlinear Circuit Analysis*, Wiley, New York.

Garland, D. J. and Stainer, F. W. (1970). *Modern Electronic Maintenance Principles*, Pergamon, London.

Graeme, J. G. and Tobey, G. E. (1971). *Operational Amplifiers—Design and Application*, McGraw-Hill, New York.

Gregory, B. A. (1973). *Electrical Instrumentation; an Introduction*, Macmillan, London.

Gupta, K. C. and Singh, A. (1975). *Microwave Integrated Circuits*, Wiley, New York.

Handel, S. (1971). *A Dictionary of Electronics*, Penguin Books, Harmondsworth, UK.

Heathkit (1978). *Electronic-Courses for Learn-at-Home*: Course 1 *DC Electronics*; Course 2 *AC Electronics*; Course 3 *Semiconductor Devices*; Course 4 *Electronic Circuits*; *Advanced Digital Technique Course*; *Microprocessor Course*. Available with experimental back-up and optional cassettes, Heath-Schlumberger, London.

Helms, H. D., Kaiser, J. F. and Rabiner, L. R. (1976). *Literature in Digital Signal Processing*, Wiley, New York.

Herrick, C. N. (1972). *Instruments and Measurements for Electronics*, McGraw-Hill, New York.

Hilburn, J. L. and Johnson, D. E. (1973). *Manual of Active Filter Design*, McGraw-Hill, New York.

H-P (1973). *D.C. Power Supply Handbook*, Hewlett-Packard, Palo Alto, USA.

Huelsman, L. P. (1977). *Active R.C. Filters*, Wiley, New York.

IEE (1979). *Electronic Test and Measuring Instrumentation: Testmex 79*, IEE, London.

IEEE (1972). *IEEE Standard Dictionary of Electrical and Electronic Terms*, Wiley, New York.

Johnson, W. C. (1950). *Transmission Lines and Networks*, McGraw-Hill, New York.

Jones, M. H. (1978). *A Practical Introduction to Electronic Circuits*, Cambridge University Press, Cambridge.

Kepco (1966). *Power Supply Handbook*, Kepco Inc., Flushing, USA.

Kidwell, W. M. (1969). *Electrical Instruments and Measurements*, McGraw-Hill, New York.

Kinnard, I. F. (1956). *Applied Electrical Measurements*, Wiley, New York.

Kostopoulos, G. K. (1975). *Digital Engineering*, Wiley, New York.

Lancaster, D. (1976) *TTL Cookbook*, Howard W. Sams, Indianapolis, USA.

Lancaster, D. (1977). *CMOS Cookbook*, Howard W. Sams, Indianapolis, USA.

Langford-Smith, F. (1955). *Radiotron Designers Handbook*, AWV, Sydney (numerous editions).

Leach, D. P. (1976). *Basic Electric Circuits*, Wiley, New York.

Lion, K. S. (1975). *Elements of Electrical and Electronic Instrumentation*, McGraw-Hill, New York.

Lowe, J. F. (1972). *Experiments in Electronics*, McGraw-Hill, Sydney.

Lowe, J. F. (1974). *Electrical and Electronic Drawing*, McGraw-Hill, Sydney.

Lowe, J. F. (1977). *Electronics for Electrical Trades*, McGraw-Hill, Sydney.

Maeison, E. C. (1974). *Electrical Instruments in Hazardous Locations*, Instrument Society of America, Pittsburgh.

Malvino, A. P. (1967). *Electronic Instrumentation Fundamentals*, McGraw-Hill, New York.

Mansfield, P. H. (1973). *Electrical Transducers for Industrial Measurement*, Butterworths, Sevenoaks, UK.

Marston, R. M. (1974). *Thyristor Projects using SCRs and Triacs*, Butterworths, Sevenoaks, UK.

Melen, R. and Garland, H. (1978). *Understanding IC Operational Amplifiers*, H. W. Sams, Indianapolis, USA.

Middleton, R. (1977). *Digital Equipment Servicing Guide*, H. W. Sams, Indianapolis, USA.

Millman, J. and Taub, H. (1965). *Pulse Digital and Switching Waveforms*, McGraw-Hill, New YOrk.

Millman, J. and Halkias, C. C. (1967). *Electronic Devices and Circuits*, McGraw-Hill, New York.

Millman, J. and Halkias, C. C. (1976). *Electronic Fundamentals and Applications*, McGraw-Hill, New York.

Millman, J. (1972). *Integrated Circuits: Analogue and Digital and Systems*, McGraw-Hill, New York.

Mitchell, F. H. (1951). *Fundamentals of Electronics*, Addison-Wesley, Cambridge, USA.

Morrison, R. (1977). *Grounding and Shielding in Instrumentation*, Wiley, New York.

Motchenbacher, C. D. (1973). *Low Noise Electronic Design*, Wiley, New York.

Motorola (1961). *Power Transistor Handbook*, Motorola Inc., Phoenix, USA.

Motorola (1966). *Silicon Rectifier Handbook*, Motorola Inc., Phoenix, USA.

Motorola (1967). *Zener Diode Handbook*, Motorola Inc., Phoenix, USA.

Namgoster, M. (1977). *Digital Equipment Trouble Shooting*, Reston, Reston, USA.

Norton, H. N. (1969). *Transducers for Electronic Measuring Systems*, Prentice-Hall, Englewood Cliffs, New Jersey.

NS Corp. (1976). *Linear Applications Handbook*, (Vol. I, 1973; Vol. 2, 1976), National Semiconductor Corp., Santa Clara, USA.

Oliver, B. M. (1971). *Electronic Measurements and Instrumentation*, McGraw-Hill, New York.

Ott, H. W. (1976). *Noise Reduction Techniques in Electronic Systems*, Wiley, New York.

Partridge, G. R. (1958). *Principles of Electronic Measurements*, Prentice-Hall, Englewood Cliffs, New Jersey.

Peled, A. and Liu, B. (1976). *Digital Signal Processing*, Wiley, New York.

Pender, H. and McIlwain, K. (1950). *Electrical Engineers Handbook: Electric Communication and Electronics*, Wiley, New York.

Piller, L. W. (1970). *Electronic Instrumentation Theory of Cardiac Technology*, Staples Press, London.

Prensky, S. D. (1963). *Electronic Instrumentation*, Prentice-Hall, Englewood Cliffs, New Jersey.

Rabiner, L. B. and Roder, C. M. (1973). *Digital Signal Processing*, IEEE, New York.

Radio Shack (1975). *Dictionary of Electronics*, Radioshack (Tandy Electronics), USA.

Regtien, P. P. L. (1978). *Modern Electronic Measuring Systems*, Delft University Press, Netherlands.

Rosenthal, M. P. (1975). *Understanding Integrated Circuits*, Hayden, Rochelle Park, USA.

Ruiter, J. H. and Murphy, R. G. (1962). *Basic Industrial Electronic Controls*, Holt, Rinehart and Winston, New York.

Sands, L. G. and Mackewroth, D. (1975). *Encyclopaedia of Electronic Circuits*, Prentice-Hall, New Jersey.

Senturia S. O. and Wedlock, B. D. (1975). *Electronic Circuits and Applications*, Wiley, New York.

Sheingold, D. H. (1972). *Analog–Digital Conversion Handbook*, Analog Devices, Norwood, USA.

Sheingold, D. H. (1974). *Non-linear Circuits Handbook*, Analog Devices, Norwood, USA.

Signetics (1977). *Analog Data Manual*, available from N. V. Philips Gloeilampenfabrieken, Eindhoven, Netherlands.

Simpson, A. (1976). *Testing Methods and Reliability—Electronics*, Macmillan, London.

Sinnema, W. (1979). *Electronic Transmission Technology*, Prentice-Hall, London.

Sloot, W. (1972). *Solid-state Servicing*, H. W. Sams, Indianapolis, USA.

Smith, R. J. (1976). *Circuits, Devices and Systems: A First Course in Electrical Engineering*, Wiley, New York.

Soisson, H. E. (1961). *Electronic Measuring Instruments*, McGraw-Hill, New York.

Spence, R. (1970). *Linear Active Networks*, Wiley, Chichester.

Squires, T. L. (1967). *Beginner's Guide to Electronics*, Newnes–Butterworths, Sevenoaks, UK.

Starr, A. T. (1959). *Electronics*, Pitman, London.

Studer, J. J. (1963). *Electronic Circuits and Instrumentation System*, Wiley, New York.

Sydenham, P. H. (1979a). *Measuring Instruments: Tools of Knowledge and Control*, Peter Peregrinus, London.

Sydenham, P. H. (1979b). *Electronics—It's Easy* (3 Vols), Modern Magazines, Sydney.

Sydenham, P. H. (1980). *Transducers in Measurement and Control*, Adam Hilger, Bristol, UK.

Szentirmai, G. (1974). *Computer-aided Filter Design*, IEEE, New York.

Terman, F. E. and Pettit, J. M. (1952). *Electronic Measurements*, McGraw-Hill, New York.

Thomas, H. E. (1967). *Handbook of Electronic Instruments and Measurement Techniques*, Prentice-Hall, Englewood Cliffs, New Jersey.

Thornton, R. D., Linvill, J. G., Chenette, E. R., Boothroyd, A. R., Willis, J., Searle, C. L., Albin, H. L., and Harris, J. N. (1966). *Handbook of Basic Transistor Circuits and Measurements*, Wiley, New York.

Turner, R. P. (1959). *Basic Electronic Test Procedures*, Holt, Rinehart and Winston, New York.

Turner, R. P. (1963). *Basic Electronic Test Instruments*, Holt, Rinehart and Winston, New York.

Turner, R. P. (1977). *ABC's of Integrated Circuits*, H. W. Sams, Indianapolis, USA.

Turner, R. P. (1978). *ABC's of FET's*, H. W. Sams, Indianapolis, USA.

Usher, M. J. (1974). 'Noise and bandwidth', *J. Phys. E.: Sci. Instrum.*, **7**, 957–61. (Reprinted in 1982, *Instrument Science and Technology*, Vol. 1, B. E. Jones (Ed.), Adam Hilger, Bristol, pp. 110–15.

Van der Ziel, A. (1954). *Noise*, Prentice-Hall, Englewood Cliffs, New Jersey.

Van der Ziel, A. (1977). *Nonlinear Electronic Circuits*, Wiley, New York.

Van Santen, G. W. (1967). *Electronic Weighing and Process Control*, Philips Technical Library, N. V. Philips, Gloeilampenfabrieken, Eindhoven, Netherlands.

Van Valkenburg, M. E. (1974). *Circuit Theory: Foundations and Classical Contributions*, Dowden, Hutchinson and Ross, Stroudsburg, Pa.

Wakerly, J. F. (1976). *Logic Design Projects Using Standard Integrated Circuits*, Wiley, New York.

Waller, W. F. (1972a). *Electronic Measurements*, Macmillan, London.

Waller, W. F. (1972b). *Electronics Testing and Measurement*, Macmillan, London.

Waller, W. F. (1972c). *Electronic Component Testing*, Macmillan, London.

Waner, W. (1979). *Trouble Shooting Solid-state Circuits and Systems*, Reston (Prentice-Hall), Englewood Cliffs, New Jersey.

Waters, F. J. (1978). *ABC's of Electronics*, H. W. Sams, Indianapolis, USA.

Wedlock, B. D. (1969). *Electronic Components and Measurements*, Prentice-Hall, Englewood Cliffs, New Jersey.

Whalen, A. D. (1971). *Detection Signals in Noise*, Academic Press, London.

Willson, A. N. (1975). *Nonlinear Networks: Theory and Analysis*, Wiley, New York.

Wolf, H. F. (1979). *Handbook of Fibre Optics*, Garland STPM Press, New York.

Handbook of Measurement Science, Volume 2
Edited by P. H. Sydenham
© 1983 John Wiley & Sons Ltd

Chapter

23 J. PRASAD and G. MITRA

Optical Regime of Measuring Instruments

Editorial introduction

Optical technique has been important to measuring and observing instrument design, making its first impact as the telescope and microscope of Galilean times. Combined with fine mechanics a vast array of instruments has been continuously developed.

Optical methods made significant strides when machine computation reached a state where previously impracticable amounts of man-years of design effort presented lens system designers with a release from this constraint. A further boost came when the traditional optical–mechanical industries combined their expertise with 20th century electronics.

The finally realized laser source, in the 1960's, provided still more capability for optical methods to provide means for measurement.

This chapter aims to introduce the optical regime of instrument design, a regime that will undoubtedly find still widening use in the future of measurement systems design.

23.1 INTRODUCTION

Optical instruments have been extensively used for varied applications in diverse areas and thus have immensely contributed to the science of measurement. During the past two decades, the interest in this branch of physics has further been stimulated and sustained with the availability of high power coherent light sources and detectors. This era has thus also witnessed many original approaches towards developing sophisticated optical instruments and systems and associated technological advancements in meeting intricate measurement needs for different sectors of economy.

The present chapter is aimed primarily towards elaborating the fundamental issues connected with optical technology and understanding the basic principles underlying the construction of important optical measurement instruments and their applications.

The subject matter is developed under the sections optical materials, optical elements, light sources and detectors, geometrical optics and optical design, interferometry, optical transfer function, holography, and optical measuring instruments.

The last category is considered under the broad subdivisions of optical workshop instruments, industrial process and control instruments, opto-medical instruments, and laser-based optical instruments and systems.

23.2 OPTICAL MATERIALS

Raw materials used in the optical shop are considered under the categories, blank materials, abrasives, optical tooling, blocking materials, polishers, cleaning agents, and optical cements.

23.2.1 Blank Materials

Glass is the most commonly used *blank* material for the production of optical elements. It is a transparent dielectric substance, in which the molecular structure is characterized by an extended network lacking in periodicity and symmetry but with atomic forces comparable to those present in crystals. Glasses of optical grade (commonly known as *optical glass*) are used for quality optics.

Optical glass is extremely homogeneous, non-absorbing, with negligible *remanent strain, bubbles,* and *striae.* The specifications of A grade optical glass state that it should not possess any visible striae or cord, with birefringence less than 10 μm/cm that is, no colour beyond light grey is to be seen in the *crossed Nicols* test and it should be totally free from bubbles larger than 0.01 mm in diameter.

Optical glass is characterized by its *refractive index* and the *constringence* or *Abbe value* (or simply ν value) defined as $(N_d - 1)/(N_F - N_C)$, where N_F

Table 23.1   Percentage composition of some typical optical glasses

Glass type	SiO_2	B_2O_3	Na_2O	K_2O	CaO	BaO	ZnO	PbO	Al_2O_3	Fe_2O_3
Boro-silicate crown	71.0	14.0	10.0	—	—	—	—	—	5.0	—
Crown	74.6	—	9.0	11.0	5.0	—	—	—	—	—
Light flint	62.6	—	4.5	8.5	—	—	—	24.1	—	—
Barium flint	45.2	—	—	7.8	—	16.0	8.3	22.2	—	—

and N_C are the refractive indices of the material corresponding to the F and C lines of the spectrum of hydrogen respectively and N_d pertains to the helium yellow line.

The physical significance of ν is that it relates to the dispersion caused by the material: the higher the ν value, the lower is the dispersion. Glasses with higher ν are customarily called *crown* glass and those with lower ν as *flint* glass. Thus crown glasses have dispersion considerably less compared to flints.

Basically, optical glass contains silica, which when mixed with the oxides of potassium, calcium, barium, lead, and more in different proportions, yields optical glasses of different properties. Some typical compositions are given in Table 23.1.

A wide choice of optical glasses varying in refractive index and ν value is available commercially. The choice of appropriate material is governed by the end use and design considerations.

Testing of optical glass involves evaluation of refractive index, ν value, striae, strain, and bubble contents. The test methods are simple and can easily be adopted in actual practice by following standard specifications.

Some other blank materials employed in different systems are quartz (both crystalline and amorphous varieties), rock salt (NaCl), sylvine (KCl), fluoride (CaF_2), calcite, KBr, ZnS, calcspar, Iceland spar, solid solutions of thorium bromide with thorium iodide, Corning 7940 glass, Mylar, sapphire, and Irtran.

Curves illustrating transmission of electromagnetic radiation through atmosphere, as also of some selected materials, can be seen elsewhere in manufacturers' data sheets and in works such as Hardy and Perrin (1932) and Sydenham (1976). The spectral characteristics of optical glasses render them largely unsuitable for incorporation in infrared or ultraviolet imaging systems: other special materials are available for use at these wavelengths.

Plastics are also frequently used in optical technology, particularly for making inexpensive and not too critical optical elements. Polymethyl methacrylate (Perspex) has properties like crown glass and polystyrene is equivalent to flint type. The principal drawbacks in their usage are: the nonavailability of materials with wider choice of refractive index and ν value; lack of homogeneity; lower softening point; fragile nature; and proneness to scratching.

23.2.2 Abrasives

Glass is quite a hard substance and for its surface working special abrasives are used. Obviously, abrasive must be hard compared to the surface being ground. On Moh's scale, the hardness numbers of diamond, ruby, topaz, and quartz are, respectively, 10, 9, 8, and 7.

Carborundum (SiC), emery in natural as well as artificial forms (fused

Al_2O_3, aloxite, alundum), and boron carbide (B_4C) are some of the abrasives commonly used in optical workshop practice. These are mostly used in powder form suspended in a water medium. Diamond powder is available in almost all grades and is used for impregnation on working tools particularly when large stock removal is necessitated.

The average grain sizes of abrasives used in the optical shop range between 3–300 μm. Rough work requires a faster cutting rate which can be achieved by using coarser grains. Silicon carbide or carborundum powder is most suitable at this stage. Once the desired shape and size have been given to the blank, finer grades of abrasives are progressively employed to obtain better finish. Aluminium oxide or emery is normally used for this purpose.

Grain size grading of abrasives is made in several ways. Examples are as the elutriation time taken, in minutes, for the emery to settle through a column of water contained in a vessel 30 cm in diameter and 1 m high, using sieve or mesh size through which a particular grade can pass, or trade designations.

Table 23.2 gives the commonly used nomenclature of particle size grading and the corresponding average grain size.

The quality of uniformity in grain size is often indicated by numbers $0, 1, 2, 3, \ldots, 10$; 0 being perfectly graded and 10 badly graded.

For polishing operation, extremely fine grain hard powder free from impurities is used. Rouge (Fe_2O_3), cerium oxide (CeO_2), putty powder (SnO), chromium oxide, and diamond paste are some of the common polishing agents.

Tripoli, diatomaceous earth, chalk, talc, slate, and pumice powder are some other abrasive materials. Garnet optical powders have also been used for the purpose particularly on ophthalmic machines employing auto-feed slurry pumping systems.

Table 23.2   Abrasive gradation

Mesh size	Other designation	Elutriation time in minutes	Approx. average size (μm)
60	M 60	—	290
100	M 100	—	150
180	M 180	1	85
220	M 220	—	75
700	M 302	5	20
850	M $302\frac{1}{2}$	10	16
1000	M 303	20	13
1300	M $303\frac{1}{2}$	40	10
1800	M 304	60	5
3000	M 305	—	3

23.2.3 Tools for Optical Manufacture

Truing, smoothing, and polisher forming tools are normally made of cast iron. Molten iron treated with meehanite (CaSi) gives a fine structure free from porosity and other defects. Polisher-holder tools may be made of aluminium. Gun-metal has also been used.

A radius turning attachment is used for providing the desired curvature on the tool surface. As a turning process is quicker compared to lapping, it is desirable to produce as accurately a curved surface from the tool as is possible by turning.

Templates and gauges are generally made of brass, 2 mm thick, the edge reduced to about 0.5 mm. Collets for holding the job during curve generation on automatic machines may be made of brass (aluminium is also used). Clamping bells required in centring and edging operation are normally made of brass.

Test plates, test spheres, polygons, standard angles, optical flats, and the like are made to the required tolerance and durability requirements, from specified materials such as white plate glass, ophthalmic glass, optical glass, or low thermal expansion coefficient materials such as Pyrex, fused silica, CerVit and Corning ULE.

23.2.4 Blocking Materials

After smoothing, several optical elements are normally *blocked* over a tool and later operations performed upon them *in situ*. This requires a material which holds the optical elements rigidly during working. The blocking material should be hard enough to withstand the pressure and friction of the smoothing and polishing tool, yet should be soft enough for easy extraction of the secured optical elements whenever needed.

Pitch is the most common blocking material as it is soft when warm, but hardens on cooling. Normal tar can be hardened to the desired degree by mixing with it additives such as saw dust and cotton wool. Blocking material should be cloth filtered to eliminate undesired hard foreign particles. Beeswax, plaster of paris, and hydrated lime are used for blocking optical flats, prisms, and similar elements. Low melting point alloys are also used as blocking material. For the preparation of spot-blocks by the casting process, a self-polymerizing organic material commonly used for acrylic dentures and commercially available under the trade name *Stellon*, has been used experimentally (Prasad and Bande, 1973).

23.2.5 Polishers

A *polisher* is an accurately shaped cast iron tool upon which the polishing medium remains mounted. Polishing of optical surfaces is achieved by lapping the job with a preformed polisher of desired shape and size and applying polishing powder. The accuracy of the surface quality generated on the finished product largely depends upon the quality of the polisher. A polisher material should be hard enough to retain its form during polishing, yet it must be amenable for alteration as and when needed. Pitch, as stated earlier, possess these properties: a polisher is formed of warm pitch of the desired shape and size. On cooling, it becomes sufficiently hard to retain its character.

Polisher pitch is softer than blocking pitch. Mixing turpentine with hard pitch produces polisher pitch. Polishing grade pitch is commercially available. As the hardness of pitch depends upon the temperature, the viscosity and hardness of blocking and polishing pitch should be maintained in accordance with the ambient temperature of the shop floor.

Felt, cloth, and wax are some other materials used in preparation of polishers. Wax polishers cause less sleeks and scratches but are slow compared to pitch and do not so readily produce accurate surfaces. For hand polishing, wood pitch is considered best.

For better results, polisher pitch might be prepared to a consistency such that it is readily but not deeply indented by the thumbnail. Thus it should be neither too soft nor viscous. Loaded polishers are made by addition of wood flour, cotton wool, or yellow felt to the pitch. These help to maintain the shape of polishers. Addition of beeswax to pitch reduces its tendency to scratch the working surface.

Teflon has also been tried with success as a polisher. Different types of polishing pads are also available commercially; are mainly used for medium grade flat work.

23.2.6 Cleaning Agents

At various stages of the work, the optical surfaces are cleaned to remove grease, pitch, or wax. Teepol solution is very useful for degreasing. Methylated spirit, isopropyl, and anhydrous alcohol are commonly used for general purpose cleaning of optical surfaces. Benzene, benzol, and trichloroethylene dissolve pitch and are used during the *deblocking* operation. Kerosene dissolves beeswax efficiently. Acetone is an excellent solvent and is used for cleaning polymerized H.T. cements.

23.2.7 Optical Cements

Optical cement must be transparent, reasonably colour free and must not develop much strain on the cemented surfaces. These are normally non-adhesive, but join two glass surfaces after some physical or chemical operation has been performed upon them. Decementing may or may not be possible, depending upon the type of cement used. Some cements set on heating to a particular temperature followed by curing at another temperature.

Canada balsam, a natural oleo-resin with refractive index 1.52, is commonly used in optical cementing work. Its unique characteristic is that it has very little tendency for granulation or crystallization on drying from a solution. Cellulose caprate (refractive index 1.47–1.49) is another widely used, thermosetting cement.

Thermosetting cements readily release the cemented components upon heating. Many other types of commercially available optical cements are set by a polymerization process, on exposure to ultraviolet light. These are extremely fast drying, but once cemented, the re-opening process becomes quite tedious.

23.3   OPTICAL ELEMENTS

According to the ray concept, a beam of light travels in rectilinear path inside any homogeneous medium. The optical elements interposed in the light path cause a change in course of the original path by means of *refraction, reflection,* or *diffraction.*

On the above basis, the optical elements which can be regarded as the building blocks of any optical system can broadly be considered under three categories: refracting elements, reflecting elements, and diffraction gratings.

Many other miscellaneous optical elements, for example, graticules, filters, fibre optic elements, polarization components, and optical modulators, are also employed for varied purposes in the system design.

This section deals with basic understanding of the function of the various types of optical elements.

23.3.1   Refracting Elements

Refracting elements are made up of an optically transparent homogeneous medium bounded by two or more surfaces. When a ray of light crosses from one medium to another, the path of propagation of the ray is altered in accordance with the well known *Snell's law,* which essentially forms the fundamental basis for developing designs of the varied types of refracting elements.

The *lens* and the *prism* are two basic types of *refractors.*

The lens

The purpose of a lens is to form image of an object, that is, a lens is essentially an *image forming* element. The image may be *real, virtual,* or *formed at infinity.* A lens which forms a real image is called a *positive* lens; similarly a virtual image forming lens is called a *negative* lens. When a lens is used to form the image at infinity, it is called a *collimating lens,* since it transmits a parallel beam after refraction.

A single lens element is characterized by its two radii of curvature, axial thickness, diameter, and the glass type. The line joining the two centres of curvature is called the *optical axis* of the lens.

Associated with each lens there are three pairs of points called *cardinal points,* which are of immense value in analysing its behaviour. These are the *focal, principal,* and *nodal points* of the system.

When an incident parallel beam of light traverses through a lens it either *converges* to an axial point (in case of positive lens) or appears *diverging* from a point (in case of negative lens). This situation can be conceived both for bundle of rays travelling from left to right or vice versa. These convergence points of parallel incident rays are called *focal points.* The planes perpendicular to the optical axis passing through the first and second focal points are known as the *first* and *second focal planes,* respectively. Obviously, rays passing through focal points become parallel to the optical axis after refraction. The surface generated by the locus of the point of intersection of the incident and refracted rays is called the *principal plane.* Like focal planes, there are two principal planes and their axial intersection points are designated as *first* and *second principal points.*

The *effective focal length* (or *focal length*) is defined as the distance from the focal point to the corresponding principal point, while the distance from nearest lens surface to the focal point is called the *back focal length. Nodal planes* are a pair of conjugate planes corresponding to unit angular magnification. *Nodal points* are the axial intersection points of these planes. Nodal planes coincide with corresponding principal planes when the lens is immersed in air.

The prism

A prism is essentially a non-image forming optical element used for two main purposes, namely to deviate the light path to any desired direction and to create wavelength dispersion of incident light into its constituent colours.

The incidence angle corresponding to which the refraction angle becomes $90°$ is termed the *critical angle,* I_c. From Snell's law, it follows that $I_c = \sin^{-1}(N'/N)$, where N and N' are the refractive indices of the incident and refracting media respectively. This equation forms the basis of design of many deviating prisms.

The *angular deviation* β suffered by a ray traversing through a prism immersed in air, with vertex angle α, can be shown to be given by

$$\beta = I - \alpha + \sin^{-1}[(N'^2 - \sin^2 I)^{1/2} \sin \alpha - \cos \alpha \sin I]$$

where I is the incident angle, and N' is the refractive index of the prism material.

In the minimum deviation position,

$$N' = \sin [\tfrac{1}{2}(\alpha + \beta_m)]/\sin (\alpha/2)$$

which is a very suitable equation for refractive index measurement.

As the refractive index corresponding to each colour (wavelength) is different, the deviation suffered by each of them is different; this causes wavelength dispersion of the incident beam into the constituent colours after refraction. The *resolving power* of a prism is a measure of its capability for separating adjacent spectral colours.

For erection of an inverted image without angular deviation, the prisms commonly used are Leman, Goerz, Hensolt, Porro types 1 and 2, and Abbe (or Konig). The Abbe prism does not cause lateral displacement. The Schmidt prism erects an inverted image and deviates it by 135° (or 45°). Penta and Amici prisms cause a 90° deviation. A special feature of the penta prism is that the transmitted beam remains stationary even when the prism is turned. Hence, it is also called an *optical square*. A Dove prism inverts without deviation, and on rotation the transmitted image rotates synchronously.

The actual configurations of the various types of prisms can be seen in standard test books such as Jacobs (1943), Smith (1966), Levi (1968), and Smith (1978).

23.3.2   Reflecting Elements

Reflectors are those elements which reflect back the incident ray in the original medium; they are called *mirrors* or *reflectors*. A mirror may be flat, concave, or convex. Concave and convex mirrors act like positive and negative lenses respectively regarding their imaging properties.

In many specialized applications, particularly in astronomical telescopes, reflectors of parabolic, elliptical, or hyperbolic shape are used. Parabolic mirrors are also extensively employed in lighthouse and other projection systems.

Beam splitters which permit only a specified fraction of the incident light to pass through and maintain a known definite intensity ratio between the reflected and transmitted beams can also be considered as a class of reflectors.

Hot, cold, partial, pellicle, multilayer dielectric metal mirrors are some other useful reflecting components employed in optical system design.

23.3.3 Diffraction Gratings

Diffraction gratings essentially consist of a large number of closely spaced, equidistant, lines or rulings formed on plane or concave optical surfaces like a prism or filter, it causes spectral separation of the incident light beam.

The *efficiency* of a grating is determined by the amount of energy diffracted in any desired direction. This is enhanced by control of the groove shape. Sawtooth groove patterns of different step angle enable light concentration in the preferred direction (called *blazing*).

Linear gratings with straight apertures are produced by ruling precise parallel grooves, with a diamond tool, on a suitably worked blank. They can be replicated from a master ruled grating. These may be of *transmitting* or *reflecting* type. Concave reflection gratings are frequently used for efficient spectral separation. Spectroscopic grade gratings carry extremely fine grooves (of the order of 400 lines per millimetre).

The *resolving power* of a grating, that is, its ability to separate adjacent spectra, is given by mN, where m is the order of diffraction and N is the total number of lines.

In recent years holographic gratings have also become commercially available. Their development has greatly extended grating applications and performance.

Holographic gratings are produced by recording optical interference between two parallel laser beams, obliquely set relative to each other. Relative obliquity of the interfering beams enables control of line density. Compared with the ruled grating the holographic grating remains free from *periodic* error. With proper choice of the recording medium, blazed holographic gratings can also be made. Other advantages of holographic gratings reported include greater number of grooves possible, up to 6000 grooves per millimetre; large dimension, up to 400 mm diameter; no ghosting; very low level of stray light; recording is possible on concave, toroidal, and aspheric blanks; and a possibility of aberration correction.

A grating formed with concentric circular grooves (or obstructions) is called a *Fresnel zone plate*. It has the property of a lens but with several focal lengths.

Metrological gratings carry identical opening and obstruction widths and are usually much coarser (tens of lines per millimetre). The constructional details of a versatile radial pattern generating engine, suitable for the manufacture of medium accuracy masters of metrological patterns as scales, line gratings, absolute multi-track encoder disks, spirals, circles, zone plates, and variable density and variable width circular tracks has been reported (Jain and Sydenham, 1980).

23.3.4   Miscellaneous Elements

Graticules

Graticules are fine measuring scales or well defined patterns placed in the focal plane of an optical instrument. These are employed for determining size, distance, direction, position, or any other attribute of the object which is viewed coincident with them. Graticules are also often denoted by terms such as diaphragm and web, or in accordance with their use, examples being sighting scales, cross-lines, (optical) micrometers, stage micrometers, and bomb sights.

Since a graticule is usually required to be viewed in the instrument under high magnification, tolerances in the manufacture of graticule blanks are generally very severe, for example scratch width, dig or pit diameter should not exceed 0.01 and 0.05 mm respectively; surface accuracy should be within two fringes and sphericity 0.5 of a ring; thickness within 0.05 mm and decentration should not exceed 0.5 to 1.0 minute (Prasad and Singh, 1970).

Graticules are made by several different techniques, namely ruling and etching, ruling followed by vacuum deposition of chromium or kanthal, photographic processes, and photo-etching. Details of the procedures followed in graticule manufacture and compositions of some photo-resists suitable for this work are given elsewhere (Horne, 1974; Prasad and Jain, 1969, 1972; Jain and Prasad, 1969, 1973).

Filters

Filters can be considered under the three broad categories: absorption or colour filters, interference filters, and spatial filters.

Absorption filters derive their performance from bulk interaction between light and the filter media; the transmission generally being a smoothly decreasing function of thickness. Such filters can be made in a variety of base materials: gelatine, glass, and plastic are commonly used. Selective colour transmission is obtained primarily by ionic absorption or by selective scattering. Gelatine filters are low in cost and provide a wide colour choice. Glass filters are widely used in photography and colorimetry and also in the production of interference filters for blocking extraneous pass-bands. Plastic filters are available both in sharp-cutoff and intermediate bandwidth types.

Thin film *interference filters* operate in the same manner as a Fabry–Perot interferometer. These are usually designed for normal incidence, but may be constructed for specific non-normal applications. Non-normal incidence results in a shift of the pass-band towards shorter wavelengths. Two basic types of interference filters are available, metal–dielectric and all-dielectric. These are manufactured by thin-film, vacuum deposition techniques. The experimental techniques adopted are described in Horne (1974), Macleod (1969), and Chhabra and Prasad (1976).

Under spatial filters are included components, which in a functional sense, are used to modify the spatial characteristics of the incident light. A typical example is the fine pinhole (about 5–10 μm in diameter) used in a Fourier system for filtering out the unwanted portion of diffraction caused by the circular aperture of the microscope objective when used in a laser beam.

Fibre optic elements

Fibre optics is based on the ability of thin smooth strands of transparent materials to convey incident signal by the total internal reflection process without interference from neighbouring fibres. It is achieved by drawing thin (of the order of 10 to 100 μm), coated, optical fibres, having a core and cladding made of different refractive index materials.

Fibre optics elements can broadly be grouped into two types, *non-coherent* and *coherent*. Non-coherent fibre bundles are used for transporting light and are often called *light guides* or *light cables*. Coherent fibre bundles are used in applications requiring image transfer without loss of image information. The flexible coherent fibre bundle is the basic component of many endoscopic medical and industrial inspection instruments used in remote examination of internal cavities. The rigid form of coherent fibre bundle, for example a face plate, is the essential component of image intensifiers, enabling high optical efficiency in image transfer. These components are also used as windows of cathode-ray tubes built for direct recording of high speed computer print out or other data recording.

Low loss (attenuation approaching 1 dB km^{-1}) graded index fibres, for use in optical communication, have also been developed.

Polarization components

These components are used for the production, control, and analysis of light in special polarization states. Their operation depends on the properties of *birefringence* (double refraction), *dichroism* (a special case of birefringence), and *change in polarization characteristics* upon reflection (at, and near, Brewster's angle).

Retardation plates (quarter, half and full wave), the Savart polariscope, Soleil–Babinet compensator, dichroic polarizers (sheet type), laser-polarizing beam splitters, and Cornu depolarizer, are some of the important components.

Optical modulators

In communication, display, data recording, and measurement, it is often desirable to control the amplitude, phase, frequency, and state of polariza-

Table 23.3   Types of optical modulator

Type	Principle	Typical materials	Typical applications
Electro-optic	Induced birefringence on application of external electric field	ADP, KDP, $LiNbO_3$, ADA	Amplitude, phase, frequency and polarization modulation, high speed shutter, laser Q-switching, mode locking, cavity dumping
Magneto-optic	Induced optical activity on application of external magnetic field	Fused silica, dense glasses H_2O, CS_2, P	Deflection, scanning, spatial modulation, information processing
Acousto-optic	Diffraction and scattering of light waves by sound	Fused silica, GaAs, YAG, YIG, TiO_2	Scanning, optical delay line, heterodyning
Mechanical	Chopping of light beam by rotating sector disc	—	Signalling, IR detection

tion of the light beam. Optical modulators are the devices used for these purposes. They are generally based on electro-optic (Pockels, Kerr), magneto-optic (Faraday, Cotton–Mouton) or acousto-optic (Debye–Sears, Raman–Nath) effects.

With the advent of the laser source of radiation, optical modulators have assumed much importance (Hartfield and Thompson, 1978). Table 23.3 gives some basic information about the various types of optical modulators.

23.4   LIGHT SOURCES AND DETECTORS

This section deals with the basic characteristics and salient features of common types of radiation sources and detectors used in optical instruments.

23.4.1   Light Sources

These are considered under incandescent filament lamps, arc lamps, infrared sources, and lasers.

Incandescent filament lamps

Incandescent filament lamps are the most widely used sources of incoherent illumination. They are commercially available with power ratings from a watt to 10 kW. Efficacy of energy conversion to light increases with filament operating temperature, though lamp life expectancy decreases as filament temperature is increased. Filaments are usually *coiled-coil* to increase their emissivity, efficacy, and mean luminance. Ribbon filament lamps are also available which produce uniform luminance over larger area.

Tungsten–halogen lamps are a special class of incandescent filament lamps, with iodine or bromine compounds added to the normal filling gases. These maintain about 90% of their initial light output throughout their life compared to 50% with the conventional incandescent lamps.

Arc lamps

In arc lamps, the optical radiation is generated by the passage of an electric discharge through an ionized gas. Carbon arcs can be cited as an illustrative example; they are widely used as the high intensity source required in cinematographic projection. High intensity discharge lamps and compact-source arc lamps of different types and configuration are extensively used as illuminators in various optical systems.

Infrared sources

Nernst Glower and Globar are the most common general-purpose sources often found in infrared spectrometers. A Nernst Glower consists of a fragile cylinder made by sintering a mixture of zirconium, yttrium, thorium, and some oxide. Upon electrically heating, it emits radiations of wavelength ranging from 2 to 15 μm. A Globar source has a silicon carbide rod heated to about 1500 K.

Low-intensity carbon arcs have also been used as spectrometer sources. The high-intensity carbon arc has been used in solar simulators.

In the near-infrared region, the tungsten filament lamp and xenon arc lamp are useful.

Lasers

Laser is an acronym for *light amplification by stimulated emission of radiation.* Essentially, a laser consists of a cavity in which the electrons of the lasing material are continuously pumped to a higher energy level. For a short interval of time, a greater number of electrons exist in a higher energy state than those in lower energy levels. This is called *population inversion.* At this stage, electrons from the higher energy level fall back to a lower level

Table 23.4   Basic data of some typical lasers

Laser	Wavelength (μm)	Power (W) Typical	Power (W) High	Pulse energy (J) Typical	Pulse energy (J) High	Continuous wave (CW) or pulsed (P)
Ruby	0.6943	—	—	5	120	P
YAG–Nd	1.06	100[b]	—	1	100	P
YAG–Nd	1.06	10	1000	—	—	CW
Glass–Nd	1.06	—	—	10	300	P
GaAs	0.85–0.905	6×10^{-3} [b]	—	10^{-5}	3×10^{-3}	P
GaAs	0.85–0.905	0.02	1	—	—	CW
Dye[a]	0.36–0.65	0.01[b]	—	10^{-4}	1	P
Dye[a]	0.41–0.78	0.2	1	—	—	CW
He–Ne	0.6328	0.005	0.05	—	—	CW
Argon	0.4880, 0.5145 and others	2	15	—	—	CW
CO_2	10.6	50[b]	—	2.0	400	P

[a] Tunable.
[b] Average power (W).

emitting an highly coherent, directional, monochromatic light beam. Due to these unique properties the laser source has been found to be extremely useful in such applications as interferometry, holography, communication, and ranging.

Lasing materials can be in solid, liquid or gas form. Common examples in each class respectively are ruby, yttrium aluminium garnet–neodymium, glass–Nd and gallium–arsenide lasers; dye lasers (tunable wavelength); and He–Ne, argon ion, and CO_2 lasers.

Lasers are now available with varying power output, as *pulse* or *continuous wave* (CW) devices, in different (but not all) spectral regions. Table 23.4 gives basic data of some commercially available lasers (Ready, 1978).

23.4.2   Detectors

The human eye is the basic detector for use with optical sighting instruments. Photographic film is another widely used light detection and recording medium. Since these conventional detectors are well understood described here are only the important categories of electro-optical devices.

Photoemissive devices

Amongst this class, the photomultiplier tube possesses the broadest range of applications, including use in low light level systems. It consists basically of a

Table 23.5   Basic characteristics of typical photocathodes

Designation	Photo-sensitive material	Spectral response range (μm)	Conversion factor (lm/W)	Luminous sensitivity (μA/lm)	Radiant sensitivity (mA/W)	Quantum efficiency
S-1	Ag–O–Cs	0.4–1.2	94.0	25	2.35	0.36
S-4	Cs–Sb	0.3–0.65	1036	40	41.5	12
S-11	Cs–Sb	0.32–0.65	803	60	48.2	14
S-20	Na–K–Cs–Sb[a]	0.3–0.82	429	150	64.2	18
—	Na–K–Cs–Sb[a]	0.4–0.96	800	85	68	21
—	Ga–As	0.2–0.94	148	250	37	13.5

[a] Multi-alkali. Abridged from RCA booklet on PM tubes.

photoemissive cathode (or photocathode) followed by an electron multiplier section.

The most commonly used photocathodes are S-1, S-4, S-11, multi-alkali, and solar blind. Another few types are also used to a lesser extent. The basic characteristics of some typical photocathodes are given in Table 23.5.

Photovoltaic devices

These devices generate electrical power upon exposure to light. Silicon and selenium cells are important examples.

The p–n silicon cells are made from single crystals or by monolithic electronic manufacture. Selenium cells are composed of an iron substrate coated with a polycrystalline layer of the hexagonal form of Si, followed by a thin layer of noble metal.

The Se cell is less expensive compared to the Si cell, but its efficiency is also lower.

Photoconductive devices

These devices are essentially a semiconductor whose electrical conductivity changes, when exposed to light, between the connecting electrodes. Typical examples are CdS and CdSe cells.

The characteristics of these devices vary considerably with temperature and aging.

Infrared detectors

These can broadly be considered under two groups, thermal and photon detectors. As the name implies, the design of thermal detectors is governed by the change in some property of the detecting element upon absorption of the infrared radiation, whereas in case of photon detectors energy is absorbed in such manner that atomic configuration is affected. It might be remarked that *photon detector* is a generic name; it covers the devices enumerated in earlier subsections.

The three basic types of thermal detectors are thermocouples and thermopiles, bolometers, and pneumatic detectors (Golay cell). The most used IR photon detectors are the intrinsic detectors prepared from materials such as Si, Ge, PbS, PbSe, InSb, and InAs. They are suitable for use in the spectral region 1 to 7 μm. Impurity activated germanium or InSb and HgCdTe cells are suitable for longer wavelength detection.

Image tubes

Image tubes are used for the detection and amplification of the optical image. These can broadly be classified as T.V. camera tubes (or pick-up tubes), image intensifier or converter tubes, cathode-ray tubes (CRT) and storage tubes.

The television camera tube permits sequential conversion of an incident image into electrical video signals via the processes of photodetection, spatial accumulation of charges, and signal reading. Image intensifier or converter tubes enable detection of an incident image and visualization of the corresponding secondary image. A CRT converts an electric input signal into visual information. Storage tubes, with electric or light input are

Table 23.6   Characteristics of typical classes of TV camera tubes. Abridged from catalogue GTE 038, Thomson-CSF

Type	Principle	Spectral response, λ_{max} (μm)	Modulation transfer function (MTF) (%)		Average sensitivity
			400 TV lines	600 TV lines	
Vidicons	Charge storage tube with photoconductive target	0.55	50–75	20	100 nA
Silicon target vidicons	Mosaic of photodiodes	0.4–1.1	40	10	500 nA
Super vidicons	Vidicon with image intensifier	UV–visible–near IR	25	10	100 nA
Pyricon	Pyroelectric target	8–14	15% at 100 TV lines		3 μA W^{-1}

designed to produce and store the charge patterns. The output can be either an electric signal or image display.

Table 23.6 gives the important characteristics of commonly used typical classes of TV camera tubes.

A more recent concept is detection and sensing by charge coupled devices (CCD). Basically, CCD is an analog shift register which accepts and transports information in the form of individual charge packets, generated by the input optical or electrical signal. Charge packet integration and their transfer is achieved by varying the potential level of the array of electrodes located on the surface of the semiconductor (Si) device at the rate of an external clock signal.

This method can be expected to receive great stimulus and increased application in imaging (particularly in high performance, low light level), memory, and information processing systems.

23.5   GEOMETRICAL OPTICS AND OPTICAL DESIGN

In geometrical optics, diffraction effects are neglected and the *ray aspect* of light is considered. Cophasal surfaces orthogonal to a set of rays are defined as geometrical wavefronts (or simply *wavefronts*).

An *ideal* lens produces an image exactly alike the object. However, an ideal lens is only a theoretical postulate since in reality there are certain parameters inherent in the refraction process which cause *aberrations* and deviate the image structure from its ideal shape. The purpose of optical system design is to work out technical specifications of each of its constituent elements, tailoring the geometry such that the entire system gives an adequately aberration-free acceptable image.

23.5.1   Aberrations

There are five *monochromatic* and two *chromatic* aberrations. *Spherical aberration, coma, astigmatism, petzval curvature,* and *distortion* fall in the first category, with *longitudinal* and *lateral* (or *transverse*) *chromatic aberrations* forming the second.

Spherical aberration arises due to the fact that rays originating from an axial object point and intersecting the refracting surface at different heights do not meet after refraction at a single point. In other words, the paraxial focus is different from the marginal focus; the axial distance between them is called *longitudinal spherical aberration* ($\Delta l'$).

When rays from an off-axial object point pass through the lens, the paraxial and marginal bundles again have different foci. A comet-like flare is then seen in the image plane, which is generated by the partial overlapping of circular light patches of increasing diameter. This off-axial aberration is called *coma* (OSC'). It causes asymmetry in the image.

Astigmatism occurs because rays from an off-axial object point lying in the meridional and sagittal planes focus at different points. The distance between these two foci is called *astigmatism*.

An image of a straight object perpendicular to the optical axis is formed on a curved surface instead of a plane. This is due to *petzval curvature*.

Owing to unequal lateral magnification throughout the field the appearance of an image changes from the original object, this occurring even when the above aberrations are not present. *Pin cushion distortion* means that the outer zones are more magnified compared to the inner; the reverse manifestation is called *barrel distortion*.

Any incident polychromatic beam suffers *dispersion* due to the wavelength dispersive nature of the material of the refracting element. Thus with incident white light, a point object has different image locations for different colours. The axial and transverse distances between the violet and red foci are called *longitudinal* and *lateral chromatic aberration* (CPD') respectively. General accounts of lens types and aberrations are available in Hardy and Perrin (1932), Jacobs (1943), and Levi (1968).

23.5.2 Elements of Lens Design

It is possible to minimize aberrations of an optical system by choosing suitable combinations of two or more lens elements. Consistent with the stipulated focal length, f number, image definition, and other requirements, the selection of proper lens combination, their glass types, radii of curvature, thickness, axial separation, and stop specifications falls within the purview of lens design. Detailed analyses on lens design are available in several standard treatises (Buchdahl, 1954; Conrady, 1957, 1960; Cox, 1964). The relationship between ray and wavefront aberrations and their evaluation has been extensively covered in Hopkins (1950) and Prasad and Narasimham (1973). The method of quasi-invariants is covered in Unvala (1962). Zoom lens are covered in Clark (1973).

The analysis of an optical system requires tracing a large number of rays through it and consequently determining the residual aberrations, which become the basis for further refinement. Depending upon the computational technique adopted, several ray tracing schemes have been reported, Feder (1951) and Smith (1966) being examples.

Ray tracing schemes require numerous calculations, of simple nature, the whole process being very time consuming and tedious by traditional methods. In recent times electronic computation has been applied to this with great effect.

In either case considerable experience is required to design even a simple corrected lens system. For this reason an instrument designer would normally call upon the services of a lens designer or make use of commercially available, proprietary lens systems when lens design is a critical feature of the whole system.

Stop and pupil

To control the illumination and field coverage, certain diaphragms are used in optical systems. An *aperture stop* determines the maximum cone of light which passes through the system; its location has a marked influence upon off-axial aberrations of the system, though spherical aberration and longitudinal chromatic aberration remain independent. A *field stop* is the diaphragm used to control the image size.

Entrance and *exit pupils* are defined as the images of the aperture stop formed by the elements placed on the left and right of the stop, respectively. The *ratio* of the *pupil diameters* represents system *magnification*.

Field angle

Field angle is defined as the angle subtended by the object at the entrance pupil; it specifies the maximum object size that can be imaged by the system. As the field angle increases, the effects due to off-axial aberrations become more pronounced, with consequent increased difficulties in the design process.

Effects of non-image forming elements

Non-image forming elements, such as plane parallel plates and prisms, also introduce aberrations when placed in non-collimated light beam (Prasad and Mitra, 1971, 1972, 1973; Prasad and Jain, 1973). Such elements are often employed as a *corrector* for compensation of residual aberrations of the system. In Prasad *et al.* (1975) the aberrations produced by a single plane parallel plate, immersed in air and inclined at an angle with the vertical are discussed. They provide equations for calculating the lateral displacement, tangential coma, and astigmatism of the system.

A parallel plate can also be used to convert large angular displacements into small translational equivalents. This has been used in metrology (Sydenham, 1969).

Performance analysis

Performance of the system may be expressed in terms of *residual aberration characteristic curves* corresponding to different field angles, or in terms of a *spot diagram* depicting the image plane intersection points of several rays passing through various zones of the system. The *resolving power test* is based on Rayleigh's criterion of resolution, the *Strehl intensity ratio*, and as is discussed later, the *optical transfer function* OTF technique, are other methods of image quality assessment.

23.5.3 Optical Modules

Illustrations of configurations of often met optical systems are given in Figure 23.1. Their design clearly becomes progressively more complex as the number of elements rises. Zoom lenses, for instance, were made economic by the advent of electronic computer methods of design.

23.6 INTERFEROMETRY

Interferometric measurements serve as a very powerful tool yielding a high degree of precision (within fractions of a wavelength of light).

Under suitable conditions, two light beams *interfere* with each other producing a resultant pattern consisting of alternate bright and dark bands (or *fringes*). Shape and location of these bands depends upon the difference in *total optical paths* travelled by the two interfering beams. Accordingly, length and phase variables may be measured directly in an interferometer, while refractive index, displacement, velocity, and more can be evaluated by converting them into corresponding *optical path difference* (OPD).

This section deals with fundamentals of interferometry followed by a brief description of common types of interferometers.

23.6.1 Conditions of Interference

Consider two light waves A_1 and A_2 of the same angular frequency ω, given by

$$A_1 = a_1 \sin(\omega t + \phi_1)$$
$$A_2 = a_2 \sin(\omega t + \phi_2)$$

where a_1, ϕ_1 and a_2, ϕ_2 are the corresponding amplitude and phase respectively, at any instant of time, t.

Straightforward mathematical analysis shows that superimposition of A_1 and A_2 produces a resultant amplitude A given by,

$$A = a \sin(\omega t + \theta)$$

where

$$a = [a_1^2 + a_2^2 + 2a_1 a_2 \cos(\phi_1 - \phi_2)]^{1/2}$$

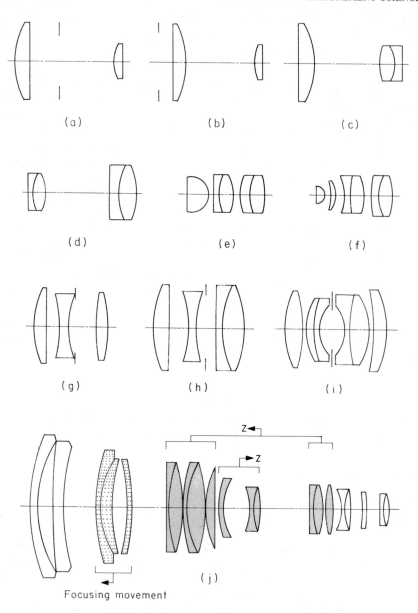

Figure 23.1 Configuration of some optical systems: (a) Huygen eyepiece;
(b) Ramsden eyepiece; (c) Kellner eyepiece; (d) Lister-type micro-
objective; (e) Amici-type micro-objective; (f) oil-immersion micro-
objective; (g) Cooke triplet; (h) Tessar lens; (i) double Gauss; (j) zoom
television lens-Taylor Hobson (Clark, 1973). Reproduced by permission of
A. D. Clark

and

$$\theta = \tan^{-1}\left(\frac{a_1 \sin \phi_1 + a_2 \sin \phi_2}{a_1 \cos \phi_1 + a_2 \cos \phi_2}\right)$$

From this it can be seen that the resultant intensity pattern is maximum (I_{max}) or minimum (I_{min}), when $\phi_1 - \phi_2 = 2n\pi$ or $(2n+1)\pi$ respectively, where n is zero or an integer.

Fringe *visibility* (or *contrast*) is defined as $(I_{max} - I_{min})/(I_{max} + I_{min})$. For best visibility (or contrast), a_1 and a_2 should be equal or nearly equal.

Summarizing the conditions of interference: the two interfering beams must be of the same wavelength (*monochromaticity*), the phase difference between them must remain closely constant over the period of observation (otherwise the fringe pattern would dance around causing a blurred patch—the *coherence condition*), and the state or plane of electromagnetic vibrations in both beams must be the same (Fresnel–Arago law).

To satisfy the above conditions, light sources used in interferometry generally have a high degree of monochromaticity and are highly coherent so that a large path difference can still produce a meaningful interference pattern.

Further, all interferometers, to a greater or lesser degree, remain sensitive to stray mechanical vibrations. Adequate precautions must be taken during their installation to minimize these effects.

23.6.2 Interferometers

The principle of interferometry is used in many measuring instruments. Such use is discussed in Dyson (1970) and in Sydenham (1976).

Michelson interferometer

The Michelson interferometer is used for the measurement of wavelength of light, slow air drift, calibration of a standard length, and other scientific applications.

Light from an extended source S (white light or monochromatic) is divided into two paths by a beam splitter B (Figure 23.2). The end mirrors M_1, M_2 reflect back the respective beams 1, 2, which recombine to form fringes in the viewing telescope T. In one of the arms, a glass plate P identical in construction to the beam splitter is introduced to compensate the inequality of glass path traversed by the interfering beams. In practice corner-cubes can be used instead of flat mirrors, this making adjustment considerably easier (Sydenham, 1976).

The Twyman–Green interferometer is a versatile instrument that is particularly suited to optical shop-floor applications. It is similar in construction

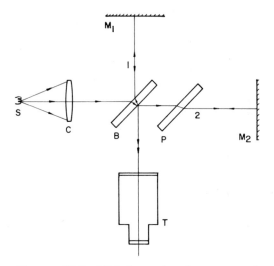

Figure 23.2 Michelson interferometer: S, source; C, collimating lens; B, beam splitter; P, compensator; M_1, M_2, mirrors; T, telescope

to the Michelson interferometer. Here the incident light is formed as a well collimated monochromatic beam and the compensator plate is dispensed with. The optical job under test is placed in one arm of the interferometer, while the other arm provides the reference beam. In cases of a lens, prism, and the like, the beam passes twice through the test piece; hence the fringe pattern depicts twice the amount of total error, which is inclusive of surface defects, material inhomogeneities, and aberrations. Proper interpretation, however, can lead to evaluation of desired defects (Rimmer et al., 1972).

The invention of the laser source enabled development of an inequal path Twyman–Green interferometer of different geometry that is suitable for testing large optics in a production line (Grigull and Rotten-Kolber, 1967; Buin et al., 1969; Bruning and Herriot, 1970).

Fizeau interferometer

A Fizeau interferometer is suitable for qualitative as well as quantitative assessment of surface quality. In its classical form it is used (Figure 23.3), for flat surface testing. A collimator C sends a parallel beam of light from source S onto the test surface T. A good quality plane parallel glass plate F whose lower surface is of high degree of flatness (master surface), is placed over the test piece with an air gap present. The collimator focuses the return beams onto the observation point O that is suitably located using a beam splitter B. The interference pattern is formed between the rays reflected

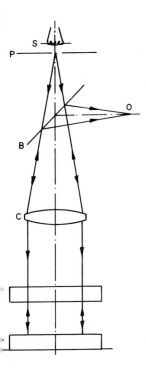

Figure 23.3   Fizeau interferometer: S, source, P, pinhole; B, beam splitter; C, collimating lens; F, master surface; T, test surface; O, observation point

from the master and test surfaces. Departure of linearity of the fringes express the surface error in terms of fractional fringe width (each fringe corresponds to $\lambda/2$ departure).

Numerous modifications of this classical instrument exist (Murty, 1978a). Concave and convex surfaces are tested in a diverging and converging beam respectively, the master flat then being replaced by the complimentary master surface. Using a laser source, a spherical surface may be tested against a master flat surface. In this interferometer, a converging beam is focused at the centre of curvature of the test surface. Another set of fringes is obtained when the beam is focused at the vertex of the test surface, and as such, the radius of curvature can also be measured.

Rayleigh interferometer

This is a very useful instrument for the measurement of the refractive indices of gases, liquids, and for the control of their composition. Light from a slit source is collimated and made to fall upon two tubes placed side by side, one containing the sample under test the other a standard sample. Fringes are viewed through a focusing lens and a magnifier. Fringe displacement is measured by tilting a compensator plate placed in the light path passing through one of the tubes. White light sources may be used enabling easy measurement of the fringe displacement.

The Michelson *stellar interferometer* employs two widely separated slits whose interdistance can be varied. Each slit selects different parts of the wavefront. From the change in visibility of the fringes seen through the telescope, stellar diameter may be evaluated.

Jamin interferometer

Used in refractometry, a Jamin interferometer has a thick plate which splits the incident beam into two using reflections at the front and back surfaces. An identical plate recombines the beam. The Mach–Zehnder interferometer is a modification of this. Using beam splitters for beam division and recombination, the separation between the interfering beams is made large. It is extensively used in hydrodynamic measurements.

Point-diffraction interferometer

Generation of an aberration-free reference wavefront without using high quality optics is achieved in a point-diffraction interferometer (Smartt and Strong, 1972). A pinhole on a transparent substrate is located slightly off-axis at the focal plane of the lens system under test. Diffracted wavefronts from the pinhole produce reference wavefronts. An absorption coating is placed over the pinhole substrate to enable matching of the intensities of the interfering beams.

Scatter fringe interferometer

Large concave mirrors, as used in astronomical telescopes, are effectively tested in a scatter fringe interferometer (Burch, 1962). A scatter plate placed perpendicular to the optical axis passing through the centre of curvature of test surface, splits a converging beam into two components. The specularly transmitted beam, after reflection through a small segment of the mirror, serves as the reference beam, while the scattered component fills the whole test surface and is the test beam. A semireflecting mirror is interposed to superimpose both the beams onto another scatter plate (identical to the first one). The instrument is comparatively less sensitive to vibrations, since the interfering beams are closely confined to each other.

Shearing interferometer

In a shearing interferometer, the test beam is split into two beams which are compared against each other thereby doing away with any reference beam

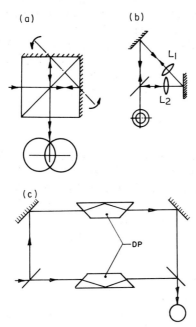

Figure 23.4   Shearing techniques:
(a) lateral shear; (b) radial shear;
(c) rotational shear produced by
counter rotating Dove prisms (DP)

as is needed in comparison type interferometers. Whilst superimposed the
beams are laterally, rotationally or radially sheared relative to each other, so
that different parts of the two identical wavefronts interfere. There exist
several techniques for achieving the desired shear. Figure 23.4 gives simple
examples. Employing birefringent crystal elements, such as the Wollaston
prism or Savart polariscope, shear is readily obtained. This type is generally
called a polarizing interferometer.

A shearing interferometer is less sensitive to vibrations (relative to
Twyman–Green interferometer) and is suitable for testing of a lens, wedge
angle, homogeneity of transparent samples, optical flatness, and the optical
transfer function (Prasad, 1963; Francon, 1966; Steel, 1967; Malacara,
1978; Murty, 1978b).

Multiple-beam interferometer

If the test surface and the inner surface of the master plate of a Fizeau
interferometer are partially reflecting, each ray splits into several components. Consequently a number of beams of varying path length take part in

forming a multiple-beam interference pattern, where high finesse fringes are formed. High precision measurement of thin film thickness, surface contour, and separation of neighbouring wavelengths are performed in a multiple-beam interferometer (Tolansky, 1948).

23.7 OPTICAL TRANSFER FUNCTION

Various criteria exist to express the performance characteristics of a lens system. Measurement of the individual aberrations gives clues to the designer in respect of its defects. A *standard interferogram* depicts the phase error across the pupil. The *resolving power test* indicates the final resolution limit attainable by the system. However, none of these attributes yields a completely objective assessment of the quality and image forming capabilities of the optical system under test.

The optical transfer function (OTF) has been used as a merit function which, to a large extent, serves this need even though it must be appreciated that this technique also is not the final solution to all lens evaluation problems.

23.7.1 OTF Representation

OTF is basically a resolving power test, but unlike the classical resolution test, it expresses the image contrast at different spatial frequencies of a sinusoidal object. It is, in this respect, similar to the time frequency response tests associated with electrical circuits (Coltman, 1954; Schade, 1964).

For elucidation of the OTF concept, consider two bright object points separated by a dark space. Theoretically the image of this object assembly should be two bright points separated by a dark region. In such a situation, the object is said to be well resolved and both the object and the image possess unity contrast. However, since any real system is afflicted with aberrations and associated diffraction effects, in actual practice some light would encroach into the dark space, with consequent reduction in image contrast. As the two object points are brought closer, the contrast decreases until it becomes zero; this is taken as the *resolution limit*. The function relating image contrast to the corresponding object spacing is called the modulation transfer function (MTF).

According to Fourier mathematics, any periodic function can be expressed in the form of a series containing sinusoidal terms.

Upon combining the above two concepts the OTF can be taken as a functional representation of the image contrast and location corresponding to various spacings of sinusoidally varying objects.

The OTF and the point spread function are interrelated according to Fourier transformation rules, each being transformable into the other, the point spread function being the representation of intensity variations in the image of a point object. Applying Fourier transform rules it can further be shown that the OTF is the *autocorrelation* of the *pupil function*, where the pupil function is the intensity distribution of the lens system at the exit pupil and autocorrelation is a standard mathematical operation (Francon, 1963; O'Neill, 1963).

Mathematically, the OTF can be expressed in the form

$$O(\xi_x, \eta_y) = M(\xi_x, \eta_y) \exp[-ik\phi(\xi_x, \eta_y)]$$

where $O(\xi_x, \eta_y)$ is the OTF; ξ_x, η_y are frequency coordinates in the image plane; $M(\xi_x, \eta_y)$ is the modulation transfer function (MTF): and $\phi(\xi_x, \eta_y)$ is the phase transfer function (PTF).

It might be noted that the OTF comprises both the MTF and the PTF. The significance of the phase term is that the image does not remain located at the ideal conjugate point but is displaced by the amount given by the PTF.

The above equation further reveals that for complete evaluation of system performance, both the MTF and PTF are required. However, it has been seen that for a moderate range of object frequencies, the MTF gives sufficient information of practical interest, and is, therefore, the quantity most commonly evaluated.

23.7.2 OTF Evaluation

Scanning methods

This method (Murata, 1966) involves scanning the image intensity distribution of a known object, which may be a sinusoidal grating of variable frequency, square-wave grating, single slit, two square-wave gratings rotating in opposite direction (producing Moiré patterns of varying frequency), or of any other arbitrary shape.

The basic experimental set-up is given in Figure 23.5. When the object O is composed of sinusoidally varying intensity gratings of different frequencies, the response of P remains proportional to the corresponding intensity in the image. The image scan therefore directly provides I_{max} and I_{min}, and hence the contrast. In the case of square-wave gratings, an intermediate electronic subsystem is introduced which allows only the fundamental of each frequency to be passed thus indirectly achieving generation of sine wave targets.

Since for good performance the object contrast at all frequencies must be identical, it might be remarked, that the actual realization of variable frequency sinusoidal or square-wave gratings is in itself a cumbersome task. Furthermore, in both cases, since the object frequency range is limited with

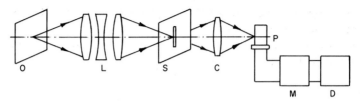

Figure 23.5   Scanning method of determining OTF: O, object; L,
test lens; S, scanning slit; C, condenser; P, photomultiplier tube; M
and D, detection and display

regard to generation of higher frequencies, a relay lens located between C
and L must be employed. This introduces its own defects into the final
result, and hence it should have much better corrections compared to those
of the test lens.

The aforesaid difficulties are met to the same extent by choosing object

Table 23.7   Comparison of test objects in OTF evaluation

Object shape	Advantages	Disadvantages
(1) Sine-wave grating	Direct reading	Difficult to make; limited range of spatial frequencies; intermediate relay lens needed for obtaining higher frequencies
(2) Square-wave grating	Direct reading	Same as in (1). Additional electronic subsystem needed for filtering fundamental of each frequency
(3) Two counter-rotating gratings	Rapid direct display	Same as in (2)
(4) Pinhole or single slit	Easy to make; wider range of spatial frequencies	Low light efficiency; computer needed for data conversion; correction factor needed for finite width
(5) Knife edge	Same as in (4). Also better light efficiency compared to pinhole or single slit	Computer needed for data conversion; stray light affects the OTF measurement to the largest degree; very low signal to noise ratio at high spatial frequencies

of simple geometry, like a single slit, a pin hole or a knife edge, but these have other disadvantages.

Table 23.7 summarizes the relative advantages and disadvantages of commonly used test objects.

Interferometric methods

The desired autocorrelation (or self-convolution) is achieved in lateral shearing interferometers; this provides the basis of an analog technique of OTF measurement (Hopkins, 1955). It is known that if the path difference between the two laterally sheared beams is changed, the total light flux across both of the sheared beams varies sinusoidally. Further, the shear magnitude represents the spatial frequency. Hence, by variation of the shear and measurement of the corresponding contrast, the OTF can be evaluated.

Polarizing type, lateral shear interferometers for OTF measurement have also been developed (Francon, 1966).

23.7.3   OTF Applications

OTF analysers have been applied with practical advantage in many interesting situations, for example, as a *go/no-go* gauge for final checking of optical elements, giving the cumulative effect of residual aberrations, diffraction, and inaccuracies caused during manufacturing processes; providing means of evaluating performance of complex cascaded electro-optical systems; effecting improvement in the quality of optical system at the design stage and many more.

As an illustration, qualitative representation of MTF curves is made in Figure 23.6. Curve A is the plot of a diffraction-limited (theoretical best)

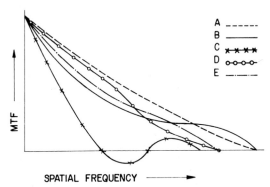

Figure 23.6   Qualitative representation of MTF curves: A, diffraction limited system; B, with central obstruction; C, defocusing effect; D, E, different designs of the same optical system

system; B and C show the effects of adding a central obstruction and of defocusing respectively; and D, E are representations of different practical designs of the same optical system.

It might be noted that while D has better performance at lower spatial frequencies; the quality of E is superior in the high frequency domain. Thus MTF curves gives a reliable quick answer and enable fast selection of a good lens for a particular requirement.

23.8 HOLOGRAPHY

In the conventional image recording process, the photographic emulsion records the time average square modulus of the complex amplitude function Hence, the phase information contained in the object wavefront is lost. In 1948, Gabor originated the basic ideas of wavefront reconstruction through which both the phase and amplitude information could be obtained. It involves two steps. First, a complex *interferogram* (called a *hologram*) formed by interference between the object and reference wavefronts, is recorded. Second, the hologram is suitably illuminated to reconstruct the original object beam in space.

Holography remained in a dormant stage until the early 1960's. Then highly coherent laser sources became available, and considerable improvements in the process were effected (Leith and Upatnieks, 1962). With the laser becoming an integral part of the practice of holography, this subject has since developed very fast into an active field that has provided new measurement techniques and solutions to a wide range of difficult problems.

23.8.1 Recording of a Hologram

In principle, holography is a kind of lens-less photography. The object is illuminated by laser light. The transmitted or scattered object beam is allowed to interfere with a reference beam derived from the same source. A suitably placed photographic plate records the hologram.

Any geometry employed in a comparison-type interferometer, with suitable modification to ensure that the reference beam subtends a certain minimum angle with the object beam at the plane of recording, can be used in making holograms.

The quality of a hologram is influenced by several factors, namely the various modes present in laser light, diffraction caused by the laser aperture speckles, and stray vibrations. Their effects are minimized to some extent by applying *spatial filtering* of the incident laser beam using a pinhole placed at the focus of a microscope objective, by properly attenuating the reference wavefront for intensity matching, by mounting the experimental set-up on a good vibration isolation table, and by employing high speed, high resolution photographic emulsions (non-conventional media are also used) for recording of the hologram.

When the interfering beams fall on the same side of the recording emulsion, a *transmission hologram* is obtained. To record a *reflection hologram* they are introduced from opposite sides of the emulsion, enabling the interference to take place within the depth of the photosensitive layer.

23.8.2 Wavefront Reconstruction

When a transmitting hologram is illuminated by a reference wavefront (spatially filtered as before), the object wavefront and its complex conjugate are reconstructed in space together with two more beams of light travelling near the transparency axis. The latter two beams consist of the directly transmitted reference beam and a beam whose intensity is proportional to the object intensity; these are of little consequence. The object wavefront produces a virtual image of the original object behind the hologram (as seen through it), while the complex conjugate component forms a real image in front. Angular separation between them, and from the axially travelling beams, depends upon the angle between the object and reference beams during recording. The reconstructing beams need not be identical to the recording reference beam. For example, use of a longer wavelength and more divergence give rise to image magnification.

In the case of a reflection hologram, reconstructed wavefronts are obtained on reflection under white light illumination. Holographic images are three dimensional in nature, retaining the parallax effect.

23.8.3 Applications of Holography

The basic concepts of holography were initially introduced to achieve very high magnification in microscopy (Gabor, 1948). Using a shorter wavelength, for example an electron or X-ray beam, for recording, carrying out reconstruction with a visible radiation wavelength, provides a very high order of magnification in addition to large depth of focus.

Holograms, because of the three-dimensional imaging properties, have been used for display and demonstration purposes. Three-dimensional motion and television pictures by holographic principle are very potential applications, though yet to be fully developed (Jacobson *et al.*, 1969; DeBitetto, 1970).

Holographic interferometry utilizes the interference patterns(s) formed by making multiple exposures on the same photographic plate corresponding to different stages of the object wavefront in a dynamic situation. Vibration analyses have thus been possible by making a continuous exposure of the vibrating object (Powell and Stetson, 1965). Phase change in the surrounding medium due to the passage of a high velocity missile, thermal gradients of objects, hot spots in electronic integrated circuits, mechanical strain, and particle size distribution are some of the applications of this technique. Contour generation of complex three-dimensional objects, achieved by changing the laser wavelength between two exposures, has also been reported (Hilderbrand and Haines, 1967).

The holographic technique has also been applied in the fabrication of diffraction gratings and simulated optical elements and in data storage and information processing (Caulfield and Lu, 1970).

23.9 OPTICAL MEASURING INSTRUMENTS

Optical measuring instruments have been extensively used for different extremely precise and reliable measurements in various disciplines of science and technology. The present section gives a brief description of some basic types of such instruments. They are arbitrarily grouped under the categories: optical workshop instruments; industrial process and control instruments; opto-medical instruments; and laser-based optical instruments and systems.

Interferometers, OTF measurement techniques, and holographic instruments, insofar as these have already been dealt with in previous sections, are not included in this section.

23.9.1 Optical Workshop Instruments

Refractometers

Refractometers are employed for refractive index and Abbe value measurement. Depending upon the precision needed, several instruments are available commercially.

Abbe and Pulfrich refractometers are the most commonly used instruments for precise measurement. The design of both these instruments is based upon the total internal reflection principle.

An Abbe refractometer consists of two 90°–60°–30° prisms, a pair of Amici prisms, and a telescope. One right-angled prism is used for illumination while the other serves as the test specimen holder which is placed over it with liquid of high refractive index at the contact area. The Amici prisms are rotated in opposite direction to compensate for the dispersion introduced by the test specimen and stage prism. The telescope can be swivelled along an arc to pick up the total internally reflected light and its position is read on a scale directly calibrated in terms of refractive index. The instrument also incorporates the provision for measuring the refractive index of liquid samples.

In the Pulfrich refractometer, the emergent angle I' of the grazing incident rays, as measured by the telescope, is used for the determination of the refractive index N of the test samples, according to the equation $N = (n^2 - \sin^2 I')^{1/2}$, where n is the refractive index of the supporting glass block.

Curvature measurement

Radius of curvature, r, measurement of optical surfaces is conventionally made employing mechanical or optical *spherometers.*

In mechanical spherometers, the sag of the test surface is obtained by locating the correct contact point with the help of a plunger. This quantity in conjunction, with other known parameters, enables calculation of r.

The optical spherometer is based on the autocollimation principle. The instrument is first focused on the test surface and then at its centre of curvature. The distance between the two image positions determines r.

The well known *Focault knife edge test* method is usually employed for testing long radius of curvature, concave, surfaces and assessing sphericity errors at different zones.

Autocollimators

Parallelism of a glass plate, prism and wedge angles, and surface flatness can be efficiently tested using an *autocollimator.* It is also frequently used for alignment purposes.

When the surface under test is not exactly *normal* to the incident rays, the reflected image is transversely displaced. This image displacement is measured (manually and photoelectrically) precisely yielding test surface inclination.

In the *angle Dekkor* method, the test specimen is placed horizontally upon a flat base, while the autocollimator can be swung in a vertical plane. A precision goniometer is essentially a spectrometer with an in-built auto-collimating telescope.

Focometers

The two standard techniques conventionally used for focal length measurement are the *focal collimator* and *nodal slide* methods. For better precision, an autocollimating microscope can be used.

23.9.2 Industrial Process and Control Instruments

A wide variety of optical instruments are available for various measurements in industry. Brief details of some common types of instruments are now discussed.

Microscopes

A *microscope* consists essentially of a condenser lens to illuminate the object, an objective lens for image formation, and an eyepiece optical system for viewing. Opaque objects are illuminated from the top either by a beam splitter placed below the objective or by fixing miniature lamps by the side of the objective. An industrial version of stereozoom microscope is shown in Figure 23.7.

A tool-room microscope eyepiece is fitted with special graticule patterns. Micro-threads, tool shape, and the like are checked by matching their images with respect to a corresponding ideal pattern contained in the graticule. Dimension measuring microscopes are mounted on a high precision translating rail. The travel distance is read from a scale using an optical vernier.

Figure 23.7 Stereozoom microscope. (Reproduced by permission of Bausch & Lomb)

In the case of metallurgical specimens (where it is important to study fine scratches in polished surfaces, faults, and inclusions which, due to glare, are usually not visible under normal illumination) a special objective, carrying a beam splitter at its back, is used.

When the object has only variation in phase or refractive index and no differential absorption, a *phase contrast* microscope is used. With the help of a phase shifting device, the variations in the object are converted into intensity variations in the image. On phase retardation, a denser medium appears brighter and vice versa in the case of phase advancement.

In *fluorescence microscopy*, the specimen (usually biological) is dyed with an appropriate dye and observed under ultraviolet light using filters.

Study of the crystals is done on a *polarizing microscope.* Plane polarized light, after passing through the specimen, changes its state of polarization and the output beam is studied by an analyser.

Interference objectives are basically Twyman–Green types of micro-interferometers which are coupled with a microscope to measure the micro-surface structure of polished reflecting surfaces.

Telescopes

Basically a *telescope* has an objective and an eyepiece. *Astronomical telescopes* of the reflecting type contain at least a large primary mirror and a small secondary mirror. *Terrestrial telescopes* are often of the refracting type fitted with additional inverting optics for image erection.

The *precision theodolite* and *sighting level* are another important class of telescopic instrument. They are extensively used in land surveying, geodesy, and tacheometry. Figures 23.8 and 23.9 illustrate, as examples, the optical configurations of modern versions of the micrometer theodolite and universal automatic level.

The technical data of the Wild T_1 micrometer theodolite shown is as follows. Telescope magnification with standard eyepiece is $30\times$. The field of view at 1000 m is 27 m with a shortest focusing distance of 1.7 m. Bubble sensitivity per 2 mm run, for the circular level is 8 minutes of arc and for the plate level 30 seconds of arc. Liquid compensator setting accuracy is ± 1 second of arc, and the working range ± 2 minutes of arc. Graduation intervals of the horizontal and vertical glass circles is $1°$ of arc with direct reading on the micrometer to 6 seconds of arc and by estimation 3 seconds of arc.

In the case of the Wild $NA_2(NAK_2)$ universal automatic level shown the compensator setting accuracy is enhanced to ± 0.3 seconds of arc and working range to ± 15 minutes of arc. For precise levelling a parallel plate micrometer, reading to 0.1 mm direct and 0.01 mm by estimation, is attached to the basic instrument.

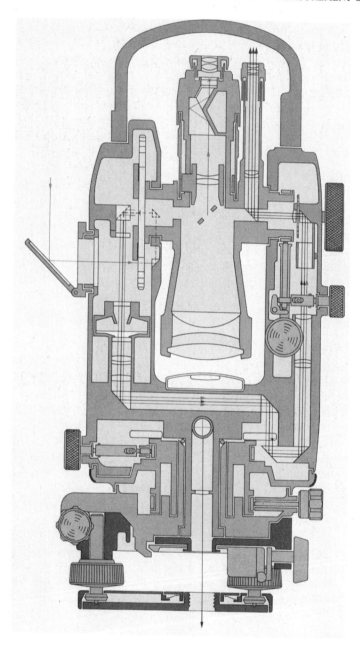

Figure 23.8   Micrometer theodolite. (Reproduced by permission
of Wild Heerbrugg Ltd)

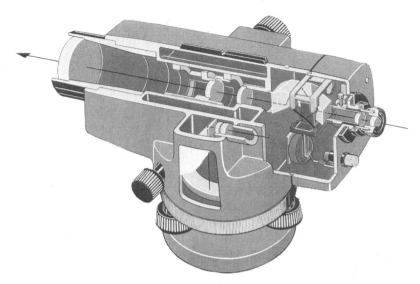

Figure 23.9   Universal automatic level. (Reproduced by permission of Wild Heerbrugg Ltd)

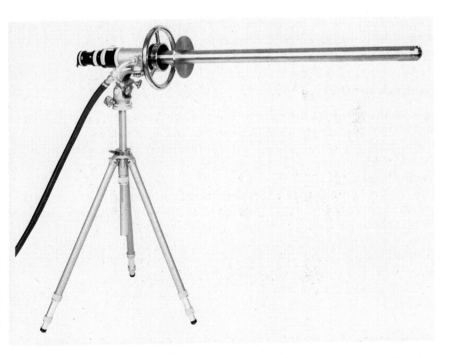

Figure 23.10   Portable endoscope. (Reproduced by permission of S. R. Clave)

Endoscopes are often employed in industry for observations inside high temperature enclosures or other inaccessible areas. Figure 23.10 shows a commercial model of a portable endoscope for photographic and visual examination of the inside walls of operating furnaces and boilers. Photo-visual endoscopes suitable for use in very-high pressure (25 kg cm^{-2}) chambers with corrosive atmospheres are also available.

Projectors

Projection equipment consists essentially of a high power lamp, a suitably designed condenser lens system and a well corrected projection objective lens. The condenser fully illuminates the entrance pupil of the objective which forms a magnified image on a distant screen. The *diascope* is used for the projection of transparencies (the *overhead projector*) and the *episcope* for opaque objects.

Profile projectors are made epidiascopic in design so that both opaque and transparent objects may be imaged. A set of projection objectives of different magnifications are mounted on a revolving turret. The test specimen is mounted on a glass platform provided with *X–Y* translational motion. The image is received on the rear of a ground glass screen. Reflecting mirrors are used to fold the optical path.

Optical micrometric readers are incorporated on several machine tools; these help in accurate positioning of the work piece. An *optical dividing head* is used for precise angular and displacement measurements.

Spectrophotometers

Spectrophotometers are used for a variety of applications in industry, including routine batch composition analyses and pollution control.

The principal component of the instrument is the *monochromator,* a source which provides a continuous range of wavelengths in the specified spectral region.

A commercial model of the instrument, together with the schematic layout of the optical system of the monochromator used is given in Figure 23.11. The specifications, as given by the manufacturer are, wavelength range 0.35–0.80 μm, source a tungsten lamp, dispersing element a holographic concave grating with 1200 grooves/mm, and wavelength accuracy better than 0.02 μm. The detector is a silicon photodiode.

A *colorimeter* basically employs optical filters instead of the dispersing element.

Miscellaneous

A large number of other instruments based on optical principles are used for quality control and other purposes in industry. A selection of examples,

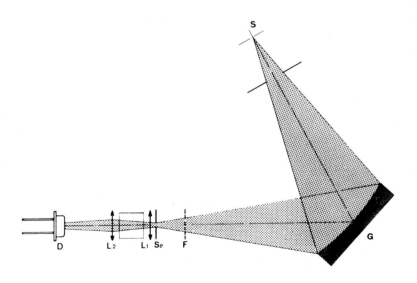

Figure 23.11 Spectrophotometer with schematic layout of monochromator comprising: S, source; G, holographic grating; F, removable filters; Se, exit slit; L_1, L_2, lens; and D, detector. (Reproduced by permission of Jobin Yvon)

which have not been dealt with already, includes the *infrared (IR) based instruments* such as the IR spectrophotometer, IR gas analysers, and non-contact type on-line IR moisture analyser, which are based upon the principle that depending upon its composition, each substance has its own characteristic IR *absorption band*(s), which serves as a finger print for its identification and quantitative evaluation.

Optical pyrometers are used for the temperature measurement of industrial furnaces wherein the flame luminosity is compared with that of an incandescent filament under varying electric power input conditions until complete photometrical matching is obtained. The *saccharimeter* is used to determine sucrose concentration in the sugar industry: in these the property of optical rotation is used. *Photometers* are used for determining the light output of various sources and illumination measurements. *Moiré fringe* methods are used extensively by the machine-tool industry for precise determination and control of linear and radial movements and in many aspects of gauge-room inspection. Lastly in these examples are the *liquid* and *gas chromatographs* used for rapid estimation of sample purity and for conducting chemical analyses.

23.9.3   Opto-medical Instruments

For diagnostic purposes, observation of different parts inside the human body is made by the *endoscopic class* of optical instruments. The *cystoscope, laparoscope, arthroscope, bronchoscope, gastroscope, colonoscope, auriscope* are some of the instruments belonging to this family. These vary mainly in their length, diameter, and other configuration parameters, to suit the particular application.

An *endoscope* is essentially a tele-microscopic system with a large number of intermediate image transfer relay lenses or coherent fibre bundles. Usually a low wattage miniature lamp, fitted at the viewing side, provides illumination. Use of a higher wattage light source and photography have become feasible in case of optical fibre based endoscopes.

Ophthalmic instruments are designed to examine and measure various parameters of the human eye. Retinal observation is done by an *ophthalmoscope*. It has a miniature low wattage light source which is focused, by a condenser, onto a tiny mirror, the light reflected from the mirror illuminates the retina. During observation, any of the number of small lenses of different dioptric power, mounted on a rotatable disc, can be brought into the return light path. A lens introduced in the correct focus position, gives the refraction error of the eye. A *retinoscope* measures the retinal curvature and is based on the Foucault knife-edge test principle.

Retinal photography is taken by a *fundus camera*. It focuses the filament of a lamp onto the pupil while the eye, to be examined, looks at a small steel

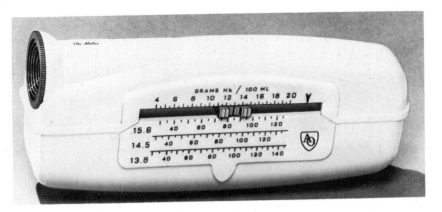

Figure 23.12   Haemoglobinometer. (Reproduced by permission of American Optical)

ball placed suitably in the light path. The same objective, used for illumination, focuses the image of the retina onto a photographic plate. After initial adjustment is made, a photograph is taken under electronic flash illumination.

A *cold light source*, which is basically a non-coherent fibre bundle transmitting only visible light (IR content is filtered out by thin film device) is often used in retinal surgery.

An *ophthalmometer* is used to measure corneal curvature; the size of the reflected image from the cornea is measured by a variable image-doubling technique.

An *Hbmeter* (Figure 23.12) determines the haemoglobin concentration in blood. The instrument compares the absorption of light by haemoglobin in a carefully defined depth of haemolysed blood with a standard glass wedge of similar absorption characteristics.

Several other opto-medical instruments not covered above are the ear microscope, slit lamp, trial sets and refractors, optometers, keratoscope, projection perimeter, Nagel anomaloscope, photocoagulators, focimeter, non-contact tonometer, and colony counters (Knoll, 1969).

23.9.4   Laser-based Instruments and Systems

The highly coherent, high power, well collimated beams obtainable from lasers have rendered them extremely useful in different sophisticated instruments. Both accuracy and range of many conventional systems are enhanced using the laser source. Shop floor, as well as field, alignment problems can easily be accomplished precisely with a laser beam (Sydenham, 1976).

As an example, the schematic diagram of a military tank, laser-based,

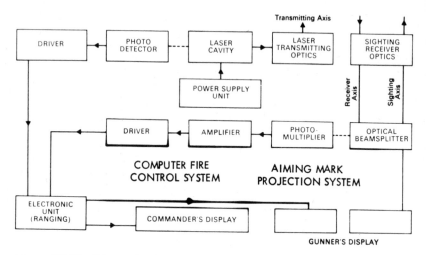

Figure 23.13   Schematic diagram of tank laser range-finder system. (Reproduced by permission of Barr & Stroud)

range-finder system is shown in Figure 23.13. The laser transmitter–receiver system and optical sight are housed in the tank. The line of sight in elevation is aligned to the axis of the gun by means of a precision parallel linkage and in azimuth by coincident turret mounting. Bore-sighting is achieved using controls on the unit with either muzzle boresight or muzzle reference system. Gunlaying is achieved through a ballistic graticule.

A laser *speckle interferometer* is used to measure small displacement, vibration, deformation, turbulence, contour generation of three-dimensional objects, and for thermal mapping; it is analogous to holographic interferometry. It is based on the principle that scattered laser light from a grainy surface (grain size larger than the wavelength of light employed) gives rise to an interference pattern (called *speckle*). Overlapping of two speckle patterns produces Moiré fringes.

Using a laser source and flexible optical fibre as a probe, it has been found possible to measure vibrations inside closed chamber or of other places where normal optical techniques may not be suitable. The fibre bundle carries the object beam forming one arm of a Twyman–Green type interferometric set-up (Cookson and Bandyopadhya, 1978). The laser interferometer has also been used in the measurement of refractive index variation (Raymond and Schlien, 1978), plasma density (Jackson and Call, 1978), and velocity (Hemsing, 1979).

A laser *Raman spectrophotometer* is based on the principle that a change in radiation frequency occurs due to molecular scattering. This has become a very powerful tool for studying vibrational and rotational energy levels of a

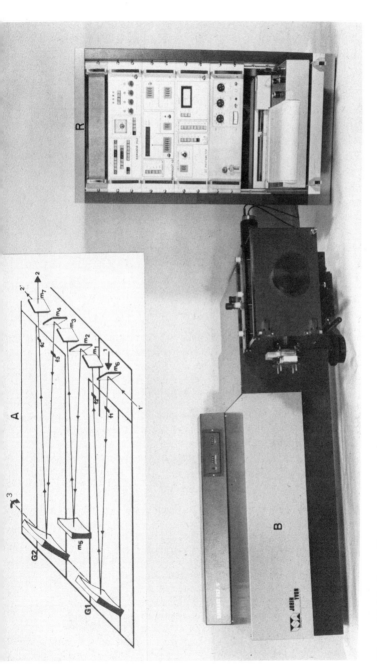

Figure 23.14 Laser Raman spectrophotometer: A, optical diagram; B, basic unit; R, detection system; 1, entrance beam path; 1', alternate entrance beam path when m_6 is present; f_1, f_2, f_3 and f_4, stepping-motor controlled slits; m_6, kinematically mounted plane mirror which permits use of 1' as entrance path; 2', alternate exit beam path when m_7 is present; 3, grating holders and shaft; G_1, G_2 holographic gratings; m_1, m_2, m_3, m_4, m_5, plane mirrors and concave mirror for imaging exit slit of first monochromator f_2 on entrance slit f_3 of second monochromator; and m_7, kinematically mounted plane mirror to permit use of 2' as exit path. (Reproduced by permission of Jobin Yvon)

test sample. A commercial model of a laser Raman spectrophotometer, together with its optical diagram, is shown in Figure 23.14. The instrument employs two concave, aberration corrected, holographic gratings with 2000 grooves/mm and offers a resolution better than $0.5\,\mathrm{cm}^{-1}$ at $0.5145\,\mu\mathrm{m}$.

ACKNOWLEDGEMENTS

We are grateful to Dr Harsh Vardhan, Director, Central Scientific Instruments Organisation, for his kind permission to undertake this work. We also wish to express our sincere thanks to the colleagues who have rendered assistance in preparation of the manuscript. Finally, we are indebted to various companies for sending us their product literature and permission to reproduce material.

REFERENCES

Bruning, J. H. and Herriot, D. R. (1970). 'A versatile laser interferometer', *Appl. Opt.*, **9**, 2180.

Buchdahl, H. A. (1954). *Optical Aberration Coefficients*, Oxford University Press, London.

Buin, A. P., Semenova, M. P., and Kiryukhina, L. A. (1969). 'Inspection of the surface quality of large scale optical components of an unequal arm interferometer', *Sov. J. Opt. Technol.*, **36**, 720.

Burch, J. M. (1962). 'Scatter-fringe interferometry', *J. Opt. Soc. Am.*, **52**, 600.

Caulfield, H. J. and Lu, S. (1970). *The Applications of Holography*, Wiley–Interscience, New York.

Chhabra, D. S. and Prasad, J. (1976). 'Experimental studies on the preparation of near infrared filters of Fabry–Perot construction', *CSIO Commun.*, **3**, 68.

Clark, A. D. (1973). *Zoom Lenses*, Adam Hilger, Bristol.

Coltman, J. W. (1954). 'The specifications of imaging properties by response to a sinewave input', *J. Opt. Soc. Am.*, **44**, 468.

Conrady, A. E. (1957 and 1960). *Applied Optics and Optical Design*, Parts I, II, Dover, New York.

Cookson, R. A. and Bandyopadhya, P. (1978). 'Mechanical vibration measurement using a fibre optic laser Doppler probe', *Optics Laser Tech.*, **10**, 33.

Cox, A. (1964). *A System of Optical Design*, Focal Press, London.

DeBitetto, D. J. (1970). 'A front-lighted 3-D holographic movie', *Appl. Opt.*, **9**, 498.

Dyson, J. (1970). *Interferometry as a Measuring Tool*, Machinery Publ. Co., Brighton.

Feder, D. (1951). 'Optical calculations with automatic computing machinery', *J. Opt. Soc. Am.*, **41**, 630.

Francon, M. (1963). *Modern Applications of Physical Optics*, Wiley, Chichester.

Francon, M. (1966). *Optical Interferometry*, Academic Press, New York.

Gabor, D. (1948). 'A new microscope principle', *Nature*, **161**, 777.

Grigull, V. and Rotten-Kolber, H. (1967). 'Two beam interferometer using laser', *J. Opt. Soc. Am.*, **57**, 149.

Hardy, A. C. and Perrin, F. H. (1932). *The Principles of Optics*, McGraw-Hill, New York.

Hartfield, E. and Thompson, B. J. (1978). 'Optical modulators', in *Handbook of Optics*, W. G. Driscoll and W. Vaughan (Eds), McGraw-Hill, New York.

Hemsing, W. F. (1979). 'Velocity sensing interferometer (VISAR) modification', *Rev. Sci. Instrum.*, **50**, 835.

Hilderbrand, B. P. and Haines, K. A. (1967). 'Multiple wavelength and multiple source holography applied to contour generation', *J. Opt. Soc. Am.*, **57**, 155.

Hopkins, H. H. (1950). *Wave Theory of Aberrations*, Oxford University Press, London.

Hopkins, H. H. (1955). 'Interferometeric methods for the study of diffraction images', *Opt. Acta*, **2**, 23.

Horne, D. F. (1974). *Dividing, Ruling and Mask-making*, Adam Hilger, Bristol.

Jacobs, D. H. (1943). *Fundamentals of Optical Engineering*, McGraw-Hill, New York.

Jacobson, A. D., Evtuthov, V., and Neeland, J. K. (1969). 'Motion picture holography', *Appl. Phys. Lett.*, **14**, 120.

Jacobson, A. R. and Call, D. L. (1978). 'Interferometric techniques for measuring the refractive index variation of a liquid with temperature', *Rev. Sci. Instrum.*, **49**, 318.

Jain, P. K. and Prasad, J. (1969). 'Graticule production', *Bull. Opt. Soc. India*, **3**, 28.

Jain, P. K. and Prasad, J. (1973). 'Suitability of polyvinyl cinnamate photo-resist for graticule production', *J. Optics*, **2**, 21.

Jain, P. K. and Sydenham, P. H. (1980). 'Radial metrological pattern generating engine', *J. Phys. E: Sci. Instrum.*, **13**, 461–6.

Knoll, H. A. (1969). 'Ophthalmic instruments', in *Applied Optics and Optical Engineering*, Vol. V, R. Kingslake (Ed.), Academic Press, New York.

Leith, E. N. and Upatnieks, J. (1962). 'Reconstructed wavefronts and communication theory', *J. Opt. Soc. Am.*, **52**, 1123.

Levi, L. (1968). *Applied Optics*, Vol. 1, Wiley, New York.

Macleod, H. A. (1969). *Thin Film Optical Filters*, American Elsevier Publishing Co., New York.

Malacara, D. (1978). 'Radial, rotational, and reversal shear interferometers', in *Optical Shop Testing*, D. Malacara (Ed.), Wiley, New York.

Murata, K. (1966). 'Instruments for the measuring of optical transfer functions', in *Progress in Optics*, Vol. V, E. Wolf (Ed.), North-Holland, Amsterdam.

Murty, M. V. R. K. (1978a). 'Newton, Fizeau and Haidinger interferometers', in *Optical Shop Testing*, D. Malacara (Ed.), Wiley, New York.

Murty, M. V. R. K. (1978b). 'Lateral shearing interferometers', in *Optical Shop Testing*, D. Malacara (Ed.), Wiley, New York.

O'Neill, E. L. (1963). *Introduction to Statistical Optics*, Addison-Wesley, Reading, Mass.

Powell, R. L. and Stetson, K. A. (1965). 'Interferometric vibration analysis by wavefront reconstruction', *J. Opt. Soc. Am.*, **55**, 1593.

Prasad, J. (1963). 'Comparative studies of optical path difference measurement with Michelson and polarisation interferometers', *Phys. Chem. Glasses*, **4**, 112.

Prasad, J. and Bande, V. (1973). 'A new method of making spot-block', *J. Optics*, **2**, 19.

Prasad, J. and Jain, P. K. (1969). 'Preparation of silver iodobromide photosensitive emulsions', *Res. Ind.*, **14**, 57.

Prasad, J. and Jain, P. K. (1972). 'Phenol-formaldehyde photo-resist', *Res. Ind.*, **17**, 4.

Prasad, J. and Jain, P. K. (1973). 'Spherical wave propagation through systems of inclined cascades with wedge substrates', *Opt. Acta*, **20**, 69.

Prasad, J. and Mitra, G. (1971). 'Refraction of spherical wavefronts through a cascade of optically homogenous non-absorbing media', *Opt. Acta*, **18,** 857.

Prasad, J. and Mitra, G. (1972). 'Refraction of spherical wavefront in stratified isotropic media', *Optik*, **35,** 134.

Prasad, J. and Mitra, G. (1973). 'Wavefront aberration caused by transverse planar layers placed in divergent light', *Proc. Ind. Nat. Sci. Acad.*, **39A,** 251.

Prasad, J., Mitra, G., and Jain, P. K. (1975). 'Aberration of a system of arbitrarily inclined planar surfaces placed in non-collimated light beam', *Nouv. Rev. Optique*, **6,** 345.

Prasad, J. and Narasimham, V. M. L. (1973). 'On the evaluation of wavefront aberration of optical systems. I. Spherical and plane refracting surfaces', *Proc. Ind. Nat. Sci. Acad.*, **39A,** 213.

Prasad, J. and Singh, R. (1970). 'Tolerances in the manufacture of precision optical components', *Res. Ind.*, **15,** 15.

Raymond, L. B. and Schlien, A. D. (1978). 'Novel interferometer for the measurement of plasma density', *Rev. Sci. Instrum.*, **49,** 861.

Ready, J. F. (1978). *Industrial Applications of Lasers*, Academic Press, New York.

Rimmer, M. P., King, D. M., and Fox, D. G. (1972). 'Computer program for the analysis of interferometric test data', *Appl. Opt.*, **11,** 2790.

Schade, O. H. (1964). 'An evaluation of photographic image quality and resolving power', *JSMPTE*, **73,** 81.

Smartt, R. N. and Strong, J. (1972). 'Point-diffraction interferometer', *J. Opt. Soc. Am.*, **62,** 737.

Smith, W. J. (1966). *Modern Optical Engineering*, McGraw-Hill, New York.

Smith, W. J. (1978). 'Image formation: geometrical and physical optics, in *Handbook of Optics*, W. G. Driscoll and W. Vaughan (Eds), McGraw-Hill, New York.

Steel, W. H. (1967). *Interferometry*, Cambridge University Press, London.

Sydenham, P. H. (1969). 'Position-sensitive photocells and their application to static and dynamic dimensional metrology', *Opt. Acta*, **16,** 377–89.

Sydenham, P. H. (1976). *Laser Gauging*, University of New England, Armidale, Australia.

Tolansky, S. (1948). *Multiple-Beam Interferometry*, Oxford University Press, Oxford.

Unvala, H. A. (1962). 'Quasi-invariants in ray-tracing', *Optik*, **19,** 551.

Handbook of Measurement Science, Volume 2
Edited by P. H. Sydenham
© 1983 John Wiley & Sons Ltd

Chapter

24 P. H. SYDENHAM

Transducer Practice: Displacement

Editorial introduction

Many texts covering measurement systems are more statements of application of principles than of the principles themselves. This Handbook, in contrast, predominantly concentrates on principles.

To provide appreciation of the application of principles now follow four chapters that review the commonly met measurands of displacement, flow, temperature, and chemical analysis. In each case the aim is to lay foundational understanding, not to provide yet more catalogues of specific practice. As will be seen each of these classes of measurand rests upon quite different physical principles yet makes use of common philosophies of transduction and signal processing.

This chapter discusses what is probably the most transduced physical variable, length, one that finds extensive use in indirect transductions.

24.1 INTRODUCTORY REMARKS

24.1.1 General Comment

Of all the physical measurands that have now been defined that concerning *displacement* (also called *length, movement, motion, dimension*) probably is the most measured system parameter. Not only is it used directly (such as to measure a gap), but it is also often used indirectly as transforms of other variables (for example, strain of a membrane caused by a pressure measurand exerted on that membrane).

It is probably impossible to estimate accurately the extent of length measurement usage. Some feeling for the breadth of use can be obtained from study of the results of the US study of its National Measurement System (Section 3.6.1 in Volume 1) and in particular the charts presented in (Norden, 1975).

Man has developed innumerable sensing systems that can collectively transduce over 30 decadic orders of length into electrical regime signals, these ranging from subatomic sizes (femtometres) to celestial distances (parsecs).

In each case the sensor must possess static and dynamic performance (Chapters 16, 17) suited to the task. It must also have appropriate mechanical size and shape features, these being set by the application and the limits of the principle used.

Numerous different principles have been employed to convert length into *mapped* signals. Although this chapter concentrates on detail of electrical output devices, the general information, such as standards and classification of ranges, is relevant to all systems.

This chapter provides discussion of the nature of the length measurand, classification problems, derived measurements based on length determination, standards of length, basics of principles used, and angle transduction.

24.1.2 General Literature

Numerous books have been published on length metrology since routine mechanical measurements entered industry at the late end of the 19th century. Books predating *c.* 1940 included very little description of electrical length transduction systems. The evolution can be traced in chapters of Sydenham (1979).

Gauging metrology

An extensive list of books devoted to what has become known as *gauging* or *tool-room metrology* is provided in Sydenham (1980a), an extensive bibliography of books on measurement science and technology. A selection of these is Batson and Hyde (1931), Hume (1970), Miller (1962), Moore (1971), Parsons (1970), Rolt (1929), Sharp (1970), and Thomas (1974). The general mechanical design aspects of instruments—see Chapter 21 and (Sydenham, 1980b)—are also particularly relevant to mechanical gauging as early length metrology relied heavily on ingenious mechanism and innovative fine-mechanics to realize the mechanical sensitivities sought. This problem has now been greatly relaxed by the availability of electronic signal amplification. In larger industrial size gauging measurement *optical tooling* is the name used to describe techniques; Kissam (1962) is a text on these procedures.

General instrumentation works

As the IMEKO bibliography revealed (see Chapter 32) there exist many books on measurement that catalogue the principles and practice of the commonly used variables. Such general works invariably contain a chapter, or more, on length and measurands derived from it. Examples are Jones (1974), Mansfield (1973), Norton (1969), Sydenham (1980c), and Woolvet (1977).

Also possibly relevant to an information search are Herceg (1972) in which preliminary chapters describe the use of inductive length sensors; Neubert (1976) in which general principles are described with examples of applications, many of which are length related; and Roughton and Jones (1979) which includes significant material on length sensing.

Other works will be introduced in this chapter as their topics arise in the sections following.

24.2 THE LENGTH DIMENSION MEASURAND

24.2.1 Nature of the Variable Length

The concept of the physical parameter we call *length* is so self-evident that few people have even see need to question it. It was intuitively defined as a measurement parameter by earliest civilized man (Sydenham, 1979).

In reality very little is understood about its nature and it is axiomatically accepted as a parameter because over the centuries it has, in the main, always provided satisfactory results.

It is an *extensive* variable (see Chapter 1); two lengths can be added, end to end and in line, to form a length equal to the arithmetic sum of their individual lengths.

Extending the concept of length to describe the size and shape of objects in space poses still more philosophical questions. Again no absolute proof can be furnished that proves that ordinary *space* is only three-dimensional, not four, five or more. This matter has been considered in depth by philosophers such as Bertrand Russel, the conclusions again being that measurement practitioners can accept the three-dimensional nature of space as an axiomatic feature. As put by (Freeman, 1969): 'The statement that ordinary space has just three dimensions is usually taken to be an empirical truth and is commonly connected with the fact that three numbers suffice to locate a point uniquely'. This rationale is then justified in that paper for the appearance of dimensions in bio-topology, planetary orbits, stability of atoms, and wave propagation. The *Oxford Universal Dictionary* somewhat sidesteps definition defining length as the 'quality of being long or the linear measurement of'.

Thus practitioners who need to use length measurements can be assured that although we do not appreciate the philosophical nature of length it is a well proven and stable parameter and that three such measurements (some may be converted into angles—see Section 24.2.3) will be needed to define position in a space.

This simplistic situation contrasts with more complex definitions of other variables, such as temperature discussed in Chapter 26. It is appropriate here to distinguish between *absolute* length and length *change* as they arise in practice.

By definition (see Section 24.3), the unit of length is defined by physical apparatus. This enables everyone to measure, and quote in numerical form, the extent of a distance in a uniform manner. Thus a piece of steel can be said to be *x* metres long.

In many applications the absolute length is not of prime concern, what is important is change to that absolute length—often called the *gauge interval*—for example, the expansion of a piece of steel as it is heated. In this case the change in length needs careful measurement. As a ratio concept (a relative determination) is involved it can be argued that the gauge interval does not need to be known in an absolute sense to the same order of precision as a length in isolation. Such changes are often also referred to as *displacements*. Hence the synonomous use of the terms length and displacement.

24.2.2 Range Classification of Displacement/Length Transducers

The question of how to classify length transducers on the basis of absolute range capability is addressed in Sydenham (1972). There it was stated, using a 'field of application' basis,

> As yet no universally accepted classification of the regions comprising this spectrum exists but it does group crudely into five ranges
>
> —microdisplacement (up to a micrometre)
> —industrial (a micrometre to ten metres)
> —surveying (ten metres to tens of kilometres)
> —navigation (kilometres to hundreds of kilometres)
> —celestial (hundreds of kilometres upward).
>
> There are, however, no clear-cut boundaries between each.

This proposal apparently satisfied the US study mentioned in Section 24.1.1 for they adopted it without change (Norden, 1975).

In practice a short-range device may well be applied in what might be at first thought to be a long-range measurement; for example, a microdisplacement transducer is used in a 100 m long gauge interval rock strain meter.

Range of a transducer principle can often be extended by making use of the extensive nature of the length measurand. Several short-range sensors can be mechanically cascaded using additional sensing to measure which sensor is in use and then the position within the range of the individual sensors. Capacitive methods are particularly suited to this arrangement, as are optical scales. Such cascading can provide, in principle, an infinite-range device having the finest discrimination. Clearly a statement of a length measuring system's discrimination alone can be most misleading. Often the discrimination of a system is confused with precision and accuracy features (see Chapter 3).

Considering the range of single sensor systems their discrimination capabilities will vary from as little as 1 in 1 (in a go/no-go inspection gauge), through 1 in 100 (for simpler electronic analog sensors) to an upper limit of smaller than 1 in 10^{10} (for laser interferometer and other modulated transit wave systems).

Several comprehensive reviews of length transducers in the various ranges have been published. In the microdisplacement region see Garratt (1979), Sydenham (1969, 1972); in the industrial, surveying, and celestial regions see Sydenham (1968, 1971). Sydenham (1980c) contains more than the usual proportion of attention to length measurements. Although the electronic circuitry aspects of the systems involved is changing rapidly in their implementation the principles on which transduction is based in the above sources remain current.

24.2.3   Derived Measurands

The dimension length is related to the measurement of numerous other measurands. Table 24.1 provides a list of commonly encountered measurands involving length. Combination of length and the other base SI units, when used to provide derived units, is shown graphically in Figure 24.1.

These relationships can provide the basis of suitable sensor systems. For example a volume meter can be made by combining the outputs of three suitably placed length transducers (three direct measurements). Pressure can be measured by first using the force of the pressure exerted on a given area to deflect the constraining member that defines the area. The deflection is then sensed with a displacement sensor.

Often a measurement need that involves a quite complex dimensional conversion relationship can be easily obtained using a suitable single-stage practical conversion element. An example is one form of relative humidity %RH meter design that uses change in %RH (a reasonably complex dimensional quantity) to change the length of a slightly tensioned membrane material: a length sensor converts the change into electrical signals. Although often economic to devise, these methods can be difficult to calibrate on a traceable basis because of the difficulty of establishing a traceable path, to the base SI units, through declared and agreed steps.

Table 24.1 Some common measurands involving the length variable

Measurand	Derivation showing units involved
Length	m
Relative length change	m/m
Area	m^2
Volume	m^3
Angle	m/m
Velocity	$m\,s^{-1}$
Acceleration	$m\,s^{-2}$
Jerk	$m\,s^{-3}$
Density	$kg\,m^{-3}$
Angular acceleration	$rad\,s^{-2}$
Current density	$A\,m^{-2}$
Mass flow rate	$kg\,s^{-1}$
Mass per unit length	$kg\,m^{-1}$
Pressure	$kg\,m^{-1}\,s^{-2}$
Force	$kg\,m\,s^{-2}$
Magnetic flux	$kg\,m^2\,A^{-1}\,s^{-2}$
Illuminance	$cd\,sr\,m^{-2}$

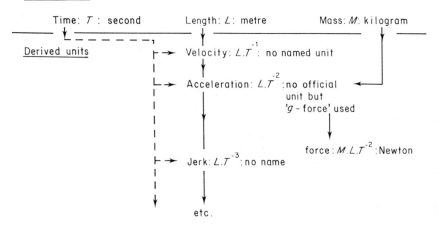

Figure 24.1 Length related measurands and their relationship to the base SI units. --- division; —— multiplication

Position definition

Discussion is now concentrated on the measurands involving multiple use of lengths, that is, area, volume, and position.

Position of a theoretical *point* in a space can be uniquely defined by three separate length dimension parameters (distances and, see below, some alternative use of angles), the parameters being each referred to a common system of spatial axes.

Predominant reference systems used are (see Figure 24.2) the rectangular *Cartesian*, the *polar*, and the *triangular* forms. In each case position P can be

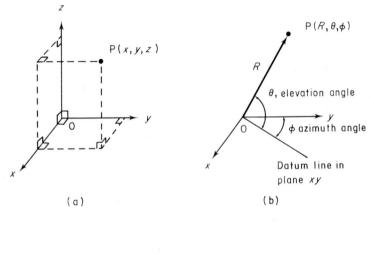

(a)

(b)

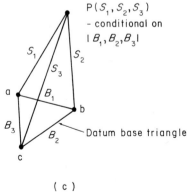

(c)

Figure 24.2   Commonly used reference systems for multidimensional definition: (a) Cartesian; (b) polar; and (c) triangular

defined uniquely by three measurements: in (a) and (c) three lengths are needed and in (b) one length and two angles. Note that each system also relies upon well defined datum positional relationships. In the Cartesian system the spatial quadrature and linearity of the x, y, z reference axes and method of projection from P to the axes form a collectively large source of errors in defining position of part P. In the triangular system errors will arise in definition of the base lengths B_1, B_2, B_3. This method has the feature that provided all measurements of P are made with respect to the base triangle, orientation of the triangle in space is not critical. Furthermore, in practice it is easier to define a triangle mechanically as the structure is kinematically pure.

In the polar system problems of direct length measurement are replaced, in two definitions, by those of angle determination which, in each case, is traceable to two lengths each which are only required as ratios.

Thus each system has practical sources of error. In the 1970's considerable interest arose in calibrating three-dimensional inspection and machining machines: for comment and citations see Scarr (1979).

As a generalization, positional systems with linear axes distances less than tens of metres usually use the Cartesian system. Its attractive practical feature is that motions of P can be produced that do not, to first order at least, interact with other axes. It is easy to comprehend how to direct or observe position in such systems. In the other two systems mentioned only quite complex motional paths of P exist that will not cause another measurand to alter. The availability of inexpensive, powerful, digital computing largely reduces the problem of interaction for computation can be used to reduce interactive difficulties.

As range increases the costs to produce an accurate Cartesian reference framework become prohibitive. Furthermore, large sizes often dictate that the measurement system be taken to the task in which case the mass of the framework prevents portability. The optical tooling method produces a Cartesian framework using light translation slides and optical sighting devices (Kissam, 1962). Automation of the triangular coordination system is discussed in Sydenham (1980c).

Point-object positioning systems lie in four groups, according to the number of length dimensions that are constrained. For full spatial control three dimensions, of which at least one must be a length (the others may be angles), need measurement. By appropriate constraint two measurements will define *plane* motion. One measurement will define distance along a straight line. Zero-dimensional control is where the point is controlled to have no departure from a defined line—alignment to a line or a line in a plane.

Where an extended object exists then two, not coincident, points will define position and orientation of the object. In this case there then occur six degrees of freedom. Definition of position of an extended object within a space, therefore, requires six distinctly separate measurements, these being some combination of lengths and, or angles.

24.3   STANDARDS OF LENGTH

24.3.1   SI Definition

The base SI unit for length, at the time of writing, is defined using an optical interferometry apparatus in which 1,650,763.73 wavelengths in vacuum of radiation from a krypton-86 source equal one metre.

In early 1982 recommendations were made that the metre be redefined in terms of the time standard not as a length; the key recommendation follows:

'the metre is the length of the path travelled by light in vacuum in the fraction (1/299,792,458) of a second'.

This is a translation of the legal French language definition and, as such, one should interpret 'light' to mean a spectrum of electromagnetic radiation that is broader than visible light. It is to be expected that this definition will be adopted in 1983.

In earlier times, before the krypton definition method, the metre was legally defined at its base level by the mechanical interval between two lines engraved on a platinum–iridium bar, called the International Prototype Metre.

Apparatus that realizes the base SI unit is, in turn, used to calibrate line standards which are used to calibrate scales, gauge blocks, end bars and, for longer lengths, surveying tapes. A relatively recent addition for the standardization of industrial length measuring instruments is the laser interferometer. Practical length calibrations having uncertainties around 1 part in 10^8 are routinely obtainable using a proprietary system (Hewlett–Packard). The National Bureau of Standards, NBS of the USA, has officially declared that individual units need no traceable calibration tests for them to be used as standards.

At the base level errors of defining the accuracy of length are as small as a part in 10^{10}. Successive, traceable stages, from that apparatus through to shop floor use, do, however, degrade the uncertainty. As an indication, inspection level, mechanical standards (blocks, scales) of length can provide around 1 part in 10^6 with the interferometer capability being more than an order better.

Length standards have been reviewed many times in the literature, (see

Puttock, 1958 and Scarr, 1979). The methodology of intercomparison of line standards is discussed in Pettavel, (1963).

24.3.2 The 1982 Recommendations

Early in the 1970s it was demonstrated that the wavelength stability of, continuous-wave, laser radiation could be improved by use of a method that alters the cavity length to keep the radiation locked into a specific absorption line in certain gaseous vapours (such as iodine). The method has been proven to be stable to parts in 10^{10} using relatively easily defined apparatus operating under conveniently procured ambient conditions.

Proposals were made in consequence of this work for adoption of this method as a legal replacement of the krypton procedure.

Adoption of a new method for defining a legal SI base unit requires exhaustive study of the proposal. For this reason alone the krypton method is still in use in 1982 as the legal procedure despite the existence of superior apparatus in several standards laboratories.

A second reason for not readily adopting the laser, direct length, definition method as the legal procedure of definition of the metre is the existence of a quite different calibration path, one that is more fundamental.

The velocity of light (dimension of metres per second) relies on the definition of both the metre (length) and upon the second (time) units for its specific numerical value. The unit, second, is definable, using an atomic transition method, to an uncertainty of parts in 10^{14}.

It is, therefore, feasible to define length in terms of the time standard multiplied by an adopted numerical constant for the velocity of light.

Practical reasons made this concept somewhat difficult to implement but standards scientists, by the early 1980s, were able to develop and produce adequately, suitable apparatus for transferring time into traceably calibrated length.

In 1982 general agreement was reached, by the countries contributing to legal adoption of the metre, that a new definition, that given in Section 24.3.1, be adopted in the future and that the krypton method would then no longer be the legal method for defining the base SI unit of the metre.

24.3.3 Traceability

Where traceability of a length measurement is needed the shop or field standard must itself be calibrated against another, higher level of length standard. It is usual to hold reference length standards in an organization against which the shop-floor measuring tools are calibrated at chosen intervals. These reference standards are certified against others held by accredited calibration laboratories or by the national laboratory responsible for measurement standards.

24.4 BASIC PRINCIPLES USED IN LENGTH TRANSDUCTION

24.4.1 General comment

Innumerable principles have been adopted to transduce length into electrical signals. Space does not permit an exhaustive list. In general practice, however, certain methods have found greatest favour.

This section discusses these, classifying them in terms of the physical principle adopted—namely electrical resistance, inductance and capacitance change, plus optical methods. A small miscellaneous subsection is included to illustrate the diversity of approaches that have been utilized. Variously applicable to discussion of principles are Garratt (1979), Neubert (1976), and Roughton and Jones (1979).

24.4.2 Electrical Resistance Change

A relatively easy way to transduce length change into an electrical signal is to use the displacement to vary mechanically the properties of an electrical resistance.

This can be realized either by using the mechanical movement to alter the tapping point, as in a potentiometer, or to alter the bulk properties of an electrical resistance element. These two alternatives are depicted in Figure 24.3.

The variable tapping point method can be devised to provide a wide range of transfer characteristics (logarithmic and linear; linear and rotary potentiometers are common place) and the dynamic range can be tailored to suit the need (units with several metres range have been made). The main disadvantage is that the necessary electromechanical contact can become unreliable and generate noise.

Resistance displacement sensors using bulk characteristic changes are based on the equation

$$R = l\rho/A$$

where R is electrical resistance, l is the length of the element, ρ is the coefficient of electrical resistance, and A is the cross-sectional area of the element.

Strain, a change in length, causes variations in l, ρ, and A. A given change in length Δl will cause a corresponding change in resistance ΔR giving the ratio

$$G = \frac{\Delta R/R}{\Delta l/l}$$

where G is the *gauge* factor of the element.

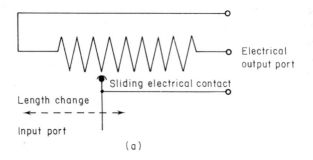

(a)

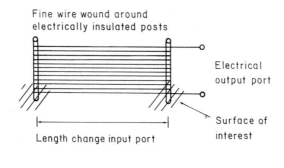

(b)

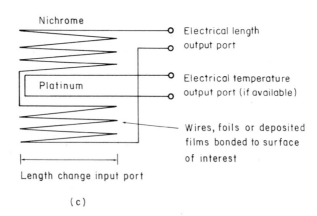

(c)

Figure 24.3   Principles of electrical resistance displacement transducers: (a) potentiometer; (b) unbonded strain gauge; and (c) bonded strain gauge

These types are termed *resistance strain gauges* and, as shown in Figure 24.3, can be of the unbonded or bonded form. Semiconductor strain gauges can offer higher gauge factors compared with metal elements (nominally 2.1 compared with 10 or more) but they exhibit greater error of linearity and temperature effect error. Typical ohmic values used are 350 Ω and 120 Ω.

Well devised resistance displacement sensors can follow dynamic responses as high as the mechanical structures bearing them. Such systems can generally be regarded as zero order (see Chapter 17) and thus are capable of providing electrical output that follows mechanical input without phase shift, delay or oscillation.

Additions of extra elements (see Roughton and Jones, 1979), can greatly reduce the effect of unwanted thermal, common-mode, noise pick-up, and non-linear characteristics.

Resistance gauges are discussed in Garratt (1979), Mansfield (1973), Neubert, (1967, 1976), Norton (1969), Roughton and Jones (1979), and Sydenham (1972, 1980c).

24.4.3   Inductance Change

Appropriate dynamic relationships between a current-carrying electrical wire (or wires), a magnetic field, and a force form the basis of several kinds of inductive displacement sensor. In general they adhere to some variation of the parameters of the basic inductance equation:

$$L = N\frac{d\phi}{di}$$

where L is the inductance, N the number of turns cut by the flux linkages, and ϕ the magnetic flux.

Displacement sensors can be formed by the following methods.

a)   Variable reluctance

Varying the magnetic circuit reluctance so as to alter the self-inductance of an inductor: usually by changing the gap distance in the magnetic loop or by altering the permeability of the circuit. Figure 24.4 gives examples.

b)   Variable mutual inductance

These make use of changes in mutual inductance of a coupled circuit of some kind. Numerous forms have been reported: shorting ring proximity, series-coupled coils, and the, predominantly used, linear variable differential transformer (LVDT) in which a moving magnetic element alters the inductance of two coils connected in a differential bridge configuration. Often a

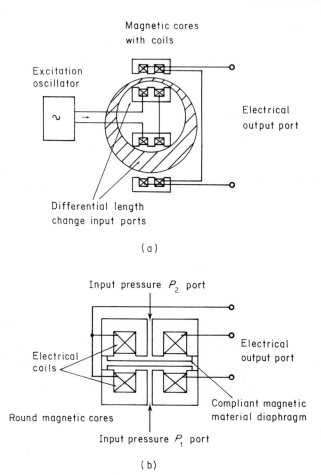

Figure 24.4   Examples of variable reluctance inductive displacement sensors: (a) steel tube wall thickness gauge; (b) E cores in differential pressure gauge

third coil is added to introduce the excitation field to the system without a d.c. connection. Figure 24.5 shows two differential configurations that are commonly used in practice.

(c)   Generating systems

Faraday's law covers voltage generation by a conductor moving relative to a magnetic field. For this the equation is as follows:

$$e = -N\frac{d\phi}{dt}$$

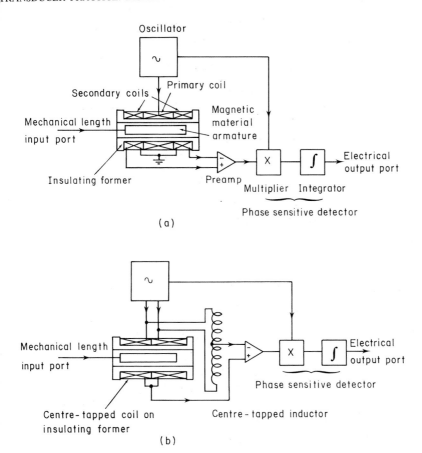

(a)

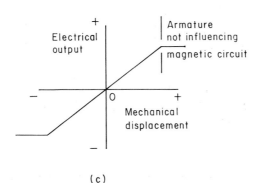

(c)

Figure 24.5 Commonly used inductive differential transformer displacement sensors: (a) linear variable differential transformer (LVDT); (b) centre-tapped solenoidal coil; and (c) transfer characteristics of above devices

where e is the voltage generated, N the number of turns of a coil, ϕ th magnetic flux, and t time.

This class of length-related sensor is fundamentally a velocity, not length sensing system. One integration of the signal will, however, yield displace ment: this has practical application especially where velocity is also o interest.

In generating systems a permanent or electromagnetic d.c., magnetic field is supplied, the signal output appearing ready to use as a voltage propor tional to velocity.

In the variations of inductance methods an a.c. source of excitation i needed in order to interpret the inductance changes. The resultant outpu signal is a varying-amplitude, approximately sinusoidal, waveform tha changes phase by 180° at the null point of the differential connections.

Rectification alone of the sinusoid will yield a d.c. signal proportional t displacement but will not give directional information indicating at whic side of the null the system rests.

In practice carrier demodulation is used as shown, schematically, in Figur 24.5. This technique has many other names such as phase sensitive detec tion, synchronous detection, and lock-in detection (see Chapter 11 of Vol. 1 Blair and Sydenham (1975), and Regtien (1978) for detailed explanation) The necessary circuitry for modulation and demodulation has progressivel reduced from extensive racks of modules to an integrated circuit package.

Inductive methods are discussed in Garratt (1979), Herceg (1972), Mans field (1973), Neubert (1976), Norton (1969), Roughton and Jones (1979) and Sydenham (1972, 1980c, 1981).

In recent years excellent mathematical models of inductive systems hav been developed (see, for instance, Abdullah, 1977; Hugill, 1978, 1981).

Inductive sensors can provide reliable performance with good linearity Performance is a little affected by moisture and other contaminants but ca be significantly perturbed by ambient magnetic fields. The moving element the *armature*, can be made to have a small mass and can be supporte without mechanical contact with the sensor unit. Long stroke units, usuall of the LVDT kind, need to be approximately twice the dynamic stroke i length. Units to a metre in length have been made but it is more usual t cascade the coil elements (using flat coil layouts) to achieve longer trave Such methods are commonly used to transduce the axial travel of numerica controlled tools (Sydenham, 1980c).

24.4.4 Capacitance Change

The *duality* of magnetic and electric field systems suggests that for eac inductive sensor arrangement there are capacitance counterparts. This i indeed the case.

Capacitance displacement sensors use variation of some parameter of the basic equation

$$C = \frac{\varepsilon A}{d}$$

where C is capacitance, ε the dielectric constant of the gapping medium, A the area of overlapping plates, and d the gap between the plates. Numerous forms have been reported for both linear and rotary position transduction.

In the simplest displacement sensing arrangements one capacitance only is utilized, variations of capacitance being caused by gap or area changes. Figure 24.6 shows some capacitance sensor configurations. The capacitance change resulting can be used to alter the frequency of a local oscillator or to alter the balance of an a.c. bridge configuration.

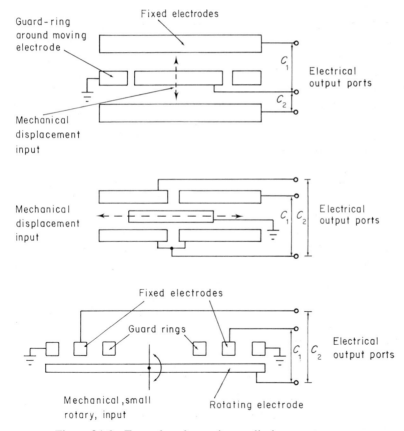

Figure 24.6 Examples of capacitance displacement sensors

Differential methods are to be recommended because there dielectric constant variations, due to change in humidity or contaminants, will alter both units equally. For the same reasons as the differential use of inductive gauges, common-mode noise pick-up, excitation supply drift, and stray reactance effects can be minimized.

Capacitance systems also need a.c. excitation and here, as with inductive systems, phase sensitive detection is usually employed.

A comparison of the performance of capacitive and inductive sensors and their necessary electrical termination conditions into transformer ratio bridges have been published by Hugill (1982). This study generally supports the popular assertion that capacitive systems are preferable for fine discrimination applications to inductive alternatives.

The basic capacitive systems transducer can be made tubular or pancake in shape. Units have been cascaded to provide metre length ranges using digital, individual unit identification, and analog interpretation with each unit.

Further detail of capacitance sensors is available in Garratt (1979) Mansfield (1973), Neubert (1976), Norton (1969), Roughton and Jones (1979), Sydenham (1972, 1980c).

Reactive systems require excitation for operation. The modulation frequency must be at least ten to twenty times higher than the highest frequency component sought in the displacement signal spectrum. Typical systems are modulated in the one to tens of kilohertz region.

24.4.5 Optical Methods

Methods of displacement sensing using a beam of radiation fall into two distinct groups according to the displacement's motion relative to the optical–electronic detector. It can be transverse or longitudinal to the direction of the beam.

The electromagnetic radiation used is often in the optical region but the methodologies discussed below have been applied using all groups ranging from coherent microwave to incoherent ionizing radiation.

Transverse to beam methods

A beam of optical or near-optical radiation having uniform spatial intensity across it can be used to monitor position as shown in Figure 24.7.

As can be seen several forms of this kind of non-contact sensor have been reported. These were reviewed in Sydenham (1969) and Sydenham (1972) Improvements, further development using the lateral effect cell and more recent citations, are to be found in Day and Marples (1971), Woltring (1974, 1975, 1977), and Noorlag and Middelhoek (1979).

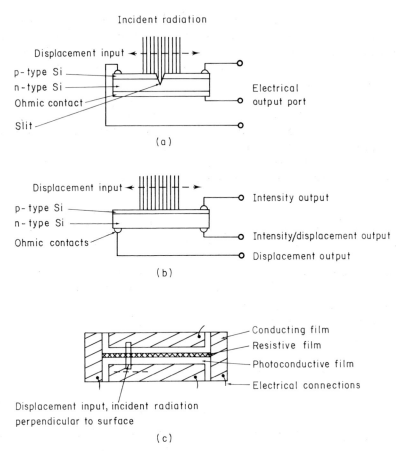

Figure 24.7 Transverse motion optical position sensitive photocells: (a) dual p–n silicon photodetectors; (b) lateral effect detector; and (c) thick-film, photo-potentiometer

The sensitivity and dynamic range of the cells are largely decided by the physical size of sensor system. Conversion to other ranges, or to rotary motion input, is possible using suitable optical elements (see Sydenham, 1969).

Position sensitive detectors can provide high dynamic frequency performance (to typically 1 MHz), their upper frequency limit usually being decided by the electronic characteristics of the photodetector and its following amplifier.

Similar methods have been used to detect position using an ionizing radiation source. A simple method here is to count the radiation received by a detector from a shuttered (collimated) ionizing radiation source for which

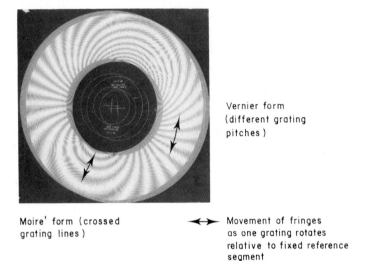

Vernier form
(different grating
pitches)

Moiré form (crossed   ◄—► Movement of fringes
grating lines)   as one grating rotates
                                       relative to fixed reference
                                       segment

Figure 24.8   Demonstration of two forms of fringes formed
using two identical radial gratings placed together. Arrows
indicate direction of movement of fringes

the overlap between two apertures changes as one moves with respect to the other. One report is Noltingk and O'Neill (1969). Such methods lack the discrimination of the optical forms which can provide, as a guide, micrometre discrimination over millimetre ranges.

Another method that makes use of transverse vignetting of a source of radiation falling onto a detector is the *Moiré fringe* concept. In these bar-space transmission, or reflection, optical gratings are used to form, by simple geometrical shuttering, fringes that are much larger than the fine grids forming them (see Figure 24.8).

Explanation of the geometry involved in the fringe size amplification is provided in Sydenham (1967). Moiré methods are used for linear and rotary measurement. Short pieces have also been used for strain investigations of mechanical components.

A collimated, coherent, radiation source beamed across a fine wire (or through a fine slit) will be diffracted to produce a spatial line of points of radiation. The spacing, distance to the aperture, and wavelength relate to the size of the aperture. As an example, this method is used commercially to control drawn-wire size, the source being a continuous-wave helium–neon laser, the detector a linear array of sensitive elements.

Longitudinal methods

If a beam of radiation has spatial or temporal modulation imposed along its length it can be used to detect distance by counting the discrete number of whole cycles and monitoring analog phase within a cycle.

The method, especially when based on optical wavelengths (around 600 nm), can yield less than 10 nm discrimination for high velocities of motion.

In this method wavelengths beyond the microwave region must be detected first by the use of interference of the energy with itself (for example, use of optical interference) for as yet no detectors have been developed that can follow the amplitude and phase variation of the cyclic energy variations of the carrier. For this to be performed satisfactorily the radiation must be adequately coherent. In this class are optical interferometers (see Figure 24.9).

For signal frequencies at and below the microwave region, radiation detectors can first convert the energy modulations into phase coherent electronic signals which can then be phase processed in the electronic signal domain. Microwave interferometers, such as the National Physical Laboratory (NPL) 'Teramet', and ultrasound systems have been devised for measuring length.

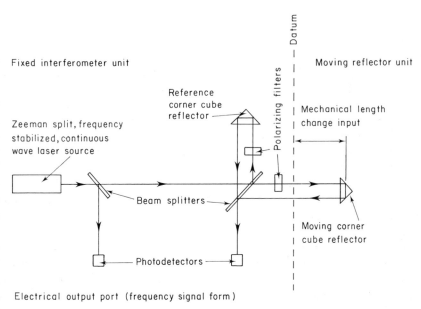

Figure 24.9   Principle of advanced laser-based optical interferometer as used in the Hewlett–Packard system

Optical length sensors based upon laser sources are reviewed in Sydenham (1976) where an extensive bibliography is provided (see also Sydenham, 1968, 1972). Interferometry applied to measurement is covered in Dyson (1970).

In many cases, however, the radiation beam is modulated at frequencies much lower than the functional frequency of the beam. For example, some electromagnetic distance measuring (EDM) meters use an optical beam (10^{14} Hz carrier) modulated at microwave frequencies (10^9 Hz). In this case the beam acts only as a continuous d.c. carrier for the modulation.

Time of flight methods fall into this class. The simplest form relies on timing the interval of time required for a pulse of radiation to travel to and from a target. A derivative of these is the method whereby the receipt of a return pulse initiates transmission of the next: a characteristic frequency results. Travel time then decides frequency of retransmission which, in turn, yields distance in terms of output frequency.

All time of flight methods require accurate enough knowledge of the velocity of the radiation in the travel medium. Optical wavelengths require correction for ambient flight path parameters such as moisture content, temperature, and pressure if uncertainties less than parts in 10^5 are required. Acoustic methods suffer from medium velocity variations in a more severe manner: typical uncertainties lie in the parts in 10^4 order.

Correction for the influence parameter errors on the radiation path's travel velocity are possible using additional sensors. Use of differential methods often enables reduction of such errors. For example double wavelength optical EDM systems can be corrected to have error less than 1 in 10^6.

Holography is a procedure wherein the three-length-dimensional features, plus radiation characteristics, of an object are recorded on a two-length-dimensional medium (usually a photographic plate) called the *hologram*. The object can then be reconstructed as an apparent three-dimensional object by suitable viewing of the hologram (or of any piece of the hologram).

Although the mathematical explanation is complex the practice of making a hologram is quite straightforward. Figure 24.10a shows the principles of the arrangement for forming a hologram, Figure 24.10b is the basic procedure for viewing the reconstructed image, and Figure 24.10c shows what a transmission hologram looks like when viewed in the normal manner with incoherent light. This plate only records the information about the object and is, by itself, usually not a measuring tool.

By visually superimposing the image, on a second, real, similar object it is possible to directly view relatively large shape differences: parts can be inspected against a standard for geometrical factors in this way.

Consider the hologram plate exposed, as shown in Figure 24.10a, but not developed. Let the object now be distorted in shape (as would a metal part

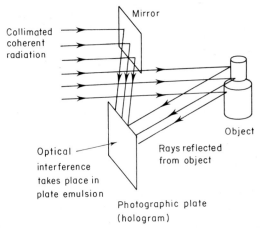

Mirror

Collimated
coherent
radiation

Object

Optical
interference
takes place in
plate emulsion

Rays reflected
from object

Photographic plate
(hologram)

(a)

Viewing position

Collimated
coherent
radiation
of suitable
wavelength

Virtual
image

Hologram

Real
image

(b)

(c)

Figure 24.10 Basic practice of holography: (a) making a hologram; (b) viewing a hologram; and (c) appearance, in incoherent light, of a time lapse hologram of a circular disk having a radial crack

being heated or a car tyre being slightly deflated) and re-expose the hologram plate. After development the plate will reconstruct an apparent object having fringes across its surface. The relative pitch and shape of the fringes relate to the distortions that occurred in the object. This method, called *time-lapse* holography, can detect distortions as small as a fraction of the wavelength of the radiation used.

Time-lapse holography is also possible using a pulsed radiation source. This alternative enables viewing of distortions in high speed, moving objects such as turbine blades under power.

A further longitudinal optical measuring method class consists of what have come to be called *non-contact probes*. These operate on a wide range of principles in which changing geometry of triangular beam transmission geometry is used to produce a transverse displacement from a longitudinal movement; an example is to be found in Sydenham (1969). The advantage of such probes is that they can detect small longitudinal displacements from relatively large stand-off distances.

Although the radiation methods discussed above largely make use of the visible, optical frequencies the principle has been applied to wavelengths ranging from microwave to X-rays.

24.4.6   Miscellaneous Methods

Many other principles of measuring length variables have been reported. Each has its particular features that may make it more suitable than the concepts explained above. Some examples now follow; others are to be found in Garratt (1979), Roughton and Jones (1979), and Sydenham (1968, 1972, 1980c).

Thickness monitoring

Vacuum deposition of thin films requires measurement of the film thickness (fractions of wavelengths of light). Methods used include sensing radiation transmission changes, direct weighing, and frequency-pulling of a quartz crystal resonance frequency as a test film increases alongside that required.

Ultrasonics

A sound wave is directed through a material. The transit time to a receiver is a measure of distance. An alternative method uses the travel time of a surface wave launched along the object's surface.

Mössbauer effect

Recoil-free emission and resonant absorption of nuclear gamma-rays have been used to monitor displacements as small as 0.1 nm.

24.5 ANGLE TRANSDUCTION

24.5.1 The Need for Angular Measurement

Measurement of distance can be obtained using a suitable mechanical conversion element—a roller, rack and pinion, wire around a drum—to transform the length variable into a equivalent rotary form.

In multi-axis position determination angles can often be monitored *in lieu* of distances.

Alignment error along a line, or from a flat surface, can also be measured in terms of angle.

In each of the above cases a length needs to be defined to provide absolute size to the transformation.

Angle transducers are reviewed in Sydenham (1968, 1980c).

24.5.2 Standards of Angle

In principle standard definition of angle can be achieved using length standards because angle is defined as a ratio of two lengths.

In practice it is convenient to use standardized angles formed between the surfaces of glass, or metal, cubes and polygons. Divided radial scales are also employed.

A uniformly rotating body can also be used to define angle if it possesses a datum position that synchronizes angular position with a temporal source that drives the body around. These, *chronometric*, angular devices are capable of great precision of circular motion division.

24.5.3 Small Angular Excursions

As with length sensing there is generally a distinct difference between the technique that may be used to measure small excursions of less than a few degrees of arc and those providing near, or more than, full circle transduction; in general the smaller the angular dynamic range capability of the sensor the smaller the attainable discrimination.

Angles can be sensed, if the range is small, to as little as 10^{-10} radians. For large-range devices the range excursion can be infinite.

The simplest small angle method is to sense the small linear displacements occurring at the end of a rotating radius arm. Figure 24.11a shows this approximating concept. In its use in *optical levers* and *autocollimators* (Figure 24.11b) a reflecting surface acts to rotate the optical beam producing a small linear, non-contact, displacement on a position sensitive detector of the kind shown in Figure 24.7. In these cases sensitivity of angular transduction is proportional to the distance of the detector from the mirror. Further-

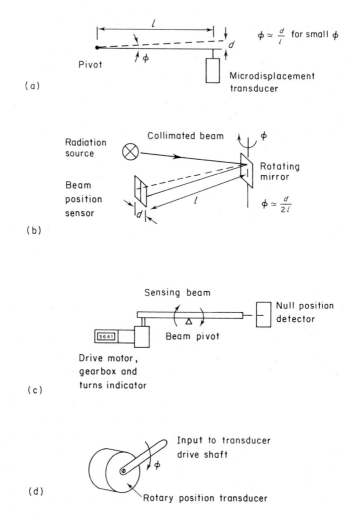

Figure 24.11   Methods of sensing small angular changes: (a) linear sensor at end of radius arm; (b) optical lever, autocollimator; (c) null following displacement balance; and (d) direct drive

more, each reflection doubles the angular excursion produced, giving additional sensitivity.

Another method is to use a null position detector to sense when a measuring arm (Figure 24.11c) has reached a chosen datum angle, the arm being used to drive, or being driven by, some other kind of angular indicator having large angular range. In this application the null detector need not have linear sensing characteristics away from the null point but should have a stable, adequately sensitive null point.

In some applications it is possible to make use of direct drive of an angular sensor of the kinds described in the Section 24.5.4 below.

The Michelson form of optical interferometer, described above in Section 24.4.5, can be arranged to measure small angular excursion.

These are but a few examples: more are to be found in Dyson (1970), Mansfield (1973), Norton (1969), and Sydenham, (1968, 1976, 1980c).

24.5.4 Large Angular Excursions

The small excursion methods described above are generally unsuited to large excursions—other methods are used.

Divided circle scales (Figure 24.12a), formed on metal or glass discs, are used where manual reading of the angle suffices. These can be read to as small as 0.1 seconds of arc (12,960,000 of these increments exist in one 360° circle) where electrical output is acceptable, or needed, a much wider range of options exists.

The simplest is to use a single or multiple-turn resistance potentiometer in the same manner as a linear potentiometer (Figure 24.12b). Practical difficulties are the finite angular range, relatively poor accuracy (<0.1%), contact noise, sometimes restricting discretization error due to wire size, and possibly the contact friction which causes a noticeable torque requirement at the shaft.

The reactive linear displacement sensor principles—inductive and capacitive arrangements—can also be used to transduce angle. The 1940's concept of the *synchro* is a rotary inductive transformer that produces three-phase signals that can be used to position a corresponding actuator unit. Vernier action (Figure 24.12c) has been added to these to improve the discrimination to less than one second of arc.

Optical scales, of various kinds, have also been used to provide electrical angular position signals. Basically these divide into: first, those providing information about movement from a chosen datum by counting lines (Figure 24.12d) as they pass (called *incremental*); and second, those (Figure 24.12e) that read position from a unique code value at any position (called *absolute*).

The distinction between incremental and absolute also applies to a number of reactive sensor designs.

In most of the above systems centring error of the driven element of the sensor can cause reading error. If multiple coincident readings can be taken at diametrically opposite positions the errors can be reduced by averaging the readings.

The synchro device shown in Figure 24.12c makes use of simultaneous multipole, determination, hence gaining accuracy by *spatial* averaging. If the

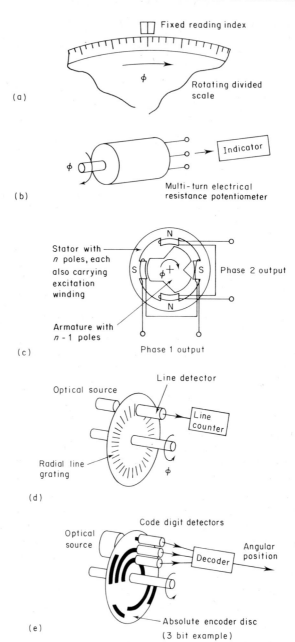

Figure 24.12   Passive sensing of large angular excursions: (a) divided circle; (b) multi-turn resistance potentiometer; (c) inductive vernier; (d) optical, incremental; and (e) optical, absolute

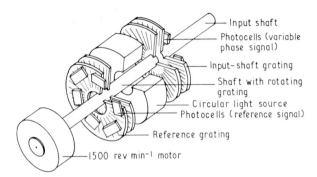

Figure 24.13 Dynamic input angle transducer incorporating spatial and temporal averaging. (Reproduced by permission of The Institute of Physics)

errors are random in nature the gain will rise as $\sqrt{n_s}$ where n_s is the number of *spatial* averaging processes.

Some systems use multiple reading heads (or multipoles) to improve accuracy in this way.

If the scale can be rapidly and continuously rotated with respect to a datum scale a system can be devised that provides a larger number of determinations in a given period of time. If these are averaged over time—*temporal averaging*—a further improvement in accuracy is obtained, again as $\sqrt{n_t}$ where n_t is the number of temporal averages available in a measurement averaging period. This feature enables speed of output response to be exchanged for improvement in accuracy.

Combination of both spatial and temporal averaging of a rotary 'scale' system can provide significant (over ×100) improvement in accuracy and, in some cases, reduce the necessary bearing tolerances. An optical example of this dual improvement is depicted in Figure 24.13. In use the two photocell circles provide spatial averaging across the discs, their a.c. signals providing a phase relationship each time a disc line crosses another to shutter the light source from the detector. Thus, in one rotation (of say, 20 ms duration), a disc with 10,000 lines will provide opportunity to make 10,000 phase comparisons. Some additional means is needed to provide the coarse angle position—counters or separate encoders.

Large excursion methods are reviewed in Mansfield (1973), Norton (1969), and Sydenham (1968, 1976, 1980c).

REFERENCES

Abdullah, F. (1977). 'The application of mathematical models in the evaluation and design of electro-mechanical instrument transducers', *J. Appl. Sci. Eng. A*, **2**, 3–26.

Batson, R. G. and Hyde, J. H. (1931). *Mechanical Testing*, Vol. 1, *Testing of Materials of Construction*, Chapman and Hall, London.

Blair, D. P. and Sydenham, P. H. (1975). Phase sensitive detection as a means to recover signals buried in noise', *J. Phys. E.: Sci. Instrum.*, **8**, 621–7 (Reprinted in 1982, *Instrument Science and Technology*, Vol. 1, B. E. Jones (Ed.), Adam Hilger, Bristol, pp. 134–41).

Day, P. E. and Marples, V. (1971). 'Phase calibration of displacement sensitive vibration transducer systems', *J. Phys. E.: Sci. Instrum.*, **4**, 137–8.

Dyson, J. (1970). *Interferometry as a Measuring Tool*, Machinery Publ. Co., London.

Freeman, I. M. (1969). 'Why is space three dimensional?', *Am. J. Phys.*, **37**, No. 12, 1222–4.

Garratt, J. D. (1979). 'Survey of displacement transducers below 50 mm', *J. Phys. E.: Sci. Instrum.*, **12**, 563–73.

Hardway, E. V. (1971). 'Position sensor combines low cost with high accuracy and reliability', *Electronics*, **44**, 86–8.

Herceg, E. E. (1972). *Handbook of Measurement and Control*, HB-72, Schaevitz Engineering, Pennsauken, US. (Revised 1976).

Hugill, A. L. (1978). 'Synthesis of inductive displacement—measuring system using computer aided design', *Proc. IEE*, **125**, No. 5, 417–21.

Hugill, A. L. (1981). 'Probable flux path modelling of an inductive displacement sensor', *J. Phys. E.: Sci. Instrum.*, **14**, 860–4.

Hugill, A. L. (1982). 'Displacement transducers based on reactive sensors in transformer ratio bridge circuits', *J. Phys. E.: J. Sci. Instrum.*, **15**, 597–606.

Hume, K. J. (1970). *Engineering Metrology*, Macdonald, London.

Jones, E. B. (1974). *Instrument Technology*, Vol. 1, *Pressure, Level, Flow and Temperature*, Newnes–Butterworths, Sevenoaks, UK.

Kissam, P. (1962). *Optical Tooling for Precise Manufacture and Alignment*, McGraw-Hill, New York.

Mansfield, P. H. (1973). *Electrical Transducers for Industrial Measurement*, Butterworths, Sevenoaks, UK.

Miller, L. (1962). *Engineering Dimensional Metrology*, Arnold, London.

Moore, W. R. (1971). *Foundation of Mechanical Accuracy*, MIT Press, Cambridge, Mass., USA.

Neubert, H. K. P. (1967). *Strain Gauges*, Macmillan, London.

Neubert, H. K. P. (1976). *Instrument transducers*, Clarendon Press, Oxford.

Noltingk, B. E. and O'Neill, P. C. (1969). 'Remote measurement of displacement by use of X-ray source', *J. Phys. E.: Sci. Instrum.*, **2**, 198–200.

Norlag, D. J. W. and Middelhoek, S. (1979). 'Two dimensional position sensitive photodetector with high linearity made with standard i.c. technology', *Solid St. Electron Dev.*, **3**, No. 3, 75–82.

Norden, B. N. (1975). 'National Measurement System. Length and related dimensional measurements—a micro study', *NCSL Newsletter*, **15**, No. 2, 21–26.

Norton, H. N. (1969). *Handbook of Transducers for Electronic Measuring Systems*, Prentice-Hall, Englewood Cliffs, NJ.

Parsons, S. A. J. (1970). *Metrology and Gauging*, Chapman and Hall, London.

Pettavel, J. (1963). 'A survey of the development of line standard metrology', Paper 1282, *Societe Genevoise D'Instruments de Physique, Geneva*.

Puttock, M. J. (1958). 'Standards of length', *Cartography*, **2**, 92–6.

Regtien, P. P. L. (Ed.) (1978). *Modern Electronic Measuring Systems*, Delft University Press, Delft.

Rolt, F. H. (1929). *Gauges and Fine Measurements*, Macmillan, London, 2 Vols.

Roughton, J. E. and Jones, W. S. (1979). 'Electro-mechanical transducers in hostile environments', *Proc. IEE*, **126**, IIR. Nov., 1029–52.

Scarr, A. J. (1979). 'Measurement of length', *Measurement and Control*, **12**, 265–9.

Sharp, K. W. B. (1970). *Practical Engineering Metrology*, Pitman, London.

Sydenham, P. H. (1967). 'An optical incremental shaft resolver using plastic radial gratings', *J. Phys. E.: Sci. Instrum.*, **44**, 146–50.

Sydenham, P. H. (1968). 'Linear and angular transducers for positional control in the decametre range', *Proc. IEE*, **115**, No. 7, 1056–66.

Sydenham, P. H. (1969). 'Position-sensitive photocells and their application to static and dimensional metrology', *Opt. Acta*, **16**, No. 3, 377–89.

Sydenham, P. H. (1971). 'Review of geophysical strain measurement', *Bull., N.Z., Soc. Earthquake Engng*, **4**, No. 1, 2–14.

Sydenham, P. H. (1972). 'Micro-displacement transducers', *J. Phys. E.: Sci. Instrum.*, **5**, 721–33.

Sydenham, P. H. (1976). *Laser Gauging*, Dept. of Continuing Education, University of New England, Armidale, NSW.

Sydenham, P. H. (1979). *Measuring Instruments—Tools of Knowledge and Control*, Peter Peregrinus, London.

Sydenham, P. H. (Ed.) (1980a). 'A working list of books published on measurement science and technology', International Measurement Confederation IMEKO (Delft: Applied Physics Dept., Technische Hogeschool Delft).

Sydenham, P. H. (1980b). 'Mechanical design of instruments', *Measurement and Control*, **13**, 365–72 and subsequent issues.

Sydenham, P. H. (1980c). *Transducers in Measurement and Control*, Adam Hilger, Bristol.

Sydenham, P. H. (1981). 'Survey of displacement and vibration transducers', in *Machine Condition Monitoring*, P. Wells (Ed.), Caulfield Institute of Technology, Caulfield, Australia.

Thomas, G. G. (1974). *Engineering Metrology*, Butterworths, Sevenoaks, UK.

Woltring, H. J. (1974). 'New possibilities for human motion studies by real-time light spot position measurement', *Biotelemetry*, **2**, 132–46.

Woltring, H. J. (1975). 'Single and dual-axis lateral photo-detectors of rectangular shape', *IEEE Trans. Electron. Dev.*, **ED-22**, 581–90.

Woltring, H. J. (1977). 'Measurement and control of human movement', *Ph.D. Thesis*, Catholic University of Nijmegen, Netherlands.

Woolvet, G. A. (1977). *Transducers in Digital Systems*, Peter Peregrinus, London.

Handbook of Measurement Science, Volume 2
Edited by P. H. Sydenham
© 1983 John Wiley & Sons Ltd

Chapter

25 T. J. S. BRAIN and K. A. BLAKE

Transducer Practice: Flow

Editorial introduction

Without doubt the measurement of flow of a material is one of the measurements greatly in demand in both scientific and industrial enterprise. As can be seen by this review of the salient methods, the measurement of flow rests heavily on rigorous understanding of the static and dynamic behaviour of fluids. This account purposefully concentrates (as an example of methodology) on the mathematical derivation of the principles used, this being a class of measurand where the mathematical modelling approach is vital to a proper understanding of the process being used to map flow into an equivalent signal. Practical details of construction are to be found in the many texts on flow measurement (IMEKO, 1980), in addition to those references cited here.

25.1 INTRODUCTION

Fluid flow rate is a measure of the mass or volume of fluid passing a point in a measured time, and in many fluid flow systems flow rate has to be determined. In this chapter a survey is presented of flow measurement methods suitable for use in *closed conduit systems*. The flowmeters dealt with can be divided into two main groups:

(a) *rate meters*, which measure the rate of flow or flow velocity at a given instant; and

(b) *quantity meters*, which measure the quantity of fluid passed in a given time.

While rate and quantity can be obtained from meters of both types, and integrated flow rate is frequently obtained from rate meters, quantity meters are seldom used to measure *instantaneous* flow rate accurately. All of the volumetric displacement meters dealt with are basically quantity meters, while the remaining meters measure either instantaneous flow rate or flow velocity. *Full flow* velocity meters measure the *mean flow* velocity in a pipeline, but in *point velocity* methods the actual velocity at a point in the flow is measured and several point velocity determinations must be taken across a cross-section of the flowing fluid stream before mean velocity can be determined, for instance as in Chapter 6 of Ower and Pankhurst (1977).

The ranges of flow and velocity as well as the pressure, temperature, accuracy, and rangeability given for each method are meant only to act as guides. While the methods experience widest use within these ranges, in most cases their use is not strictly limited to the given ranges. A useful list of references is given at the end of this chapter, but it is worthwhile noting that Dowden (1972) gives an extensive bibliography on fluid flow measurement. It should also be noted that much of the information put forward here was obtained from Brain and MacDonald (1977). General works on, or including accounts of, flow measurement are Benedict (1969), Linford (1961), Norton (1969), Sira (1978), and Sydenham (1980).

For the best accuracy all of the devices dealt with in this chapter, except perhaps the laser velocimeter, require calibration. Some attention is given, in Section 25.6, to primary standard methods for evaluating the performances of flowmeters.

25.2   VOLUMETRIC FLOW MEASUREMENT METHODS

This section deals with the three main methods in current use for measuring the volumetric flow of fluids, namely, *volumetric displacement*; *variable area*; and *pressure difference*. The flowmeters employed in each method have been extensively used over the years. Meters incorporated in variable area and pressure difference methods are basically *rate meters*, while meters used in volumetric displacement methods measure *totalized flow*.

25.2.1   Volumetric Displacement

Volumetric displacement meters (positive displacement meters) measure the actual volume of fluid passed at line conditions in a given time by dividing the flow into discrete volumes and summing the volumes as they pass through the meter. A large number of ways of dividing and integrating the flow have been devised since the first positive displacement meters appeared and numerous different types of meters have been developed. While almost all of the remaining positive displacement meters can be considered under the headings *rotary displacement meters* and *piston meters*, apart from rotating vane meters and diaphragm meters, it is clearly not appropriate to describe here in detail all of the available displacement meters, and only brief outlines are presented to give the principles of operation of the more important methods. Further details of the metering devices dealt with here may be found in Linford (1961) and ASME (1971).

For volumetric displacement meters very good accuracy in totalized flow can be achieved and direct readout can readily be obtained without expensive integrating systems. These meters are, however, best suited for the measurement of clean fluids since even small amounts of grit or other suspended matter can cause significant wear or damage to their moving parts. It is also true that, in general, these mechanical meters tend to require more maintenance than the other volumetric devices considered in Sections 25.2.2 and 25.2.3 of this chapter.

Piston meters

Piston meters are only used to meter liquid flows. The main types of meters in current use are *semi-rotary piston meters* and *reciprocating piston meters*.

A semi-rotary piston meter is outlined in Figure 25.1 where four stages of its operating cycle are shown. This meter consists of a hollow piston mounted eccentrically inside a cylinder. The cylinder and the piston have

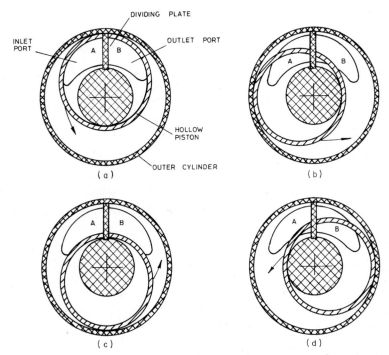

Figure 25.1   Cross-section of semi-rotary piston meter at four stages, (a)–(d), of operation

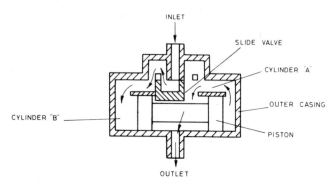

Figure 25.2   Reciprocating piston meter

the same length, but as can be seen from Figure 25.1 the piston has a much smaller diameter than the cylinder. The piston is split and its motion is such that when operating it slides along the dividing plate which separates the incoming from the outgoing liquid. At the start of a cycle liquid enters the meter through the inlet port A and flows into and, depending on the piston position, around the piston. The pressure of the incoming fluid forces the piston to move around the cylinder in the direction shown in Figure 25.1 until the liquid in front of the piston has been forced through the outlet port B and the piston itself is also emptied, position (d). The incoming fluid then moves the piston back to position (a) ready to begin another cycle.

The principles of operation of reciprocating piston meters may be described by reference to Figure 25.2 where a meter with two cylinders and a double-ended piston is shown. For this meter, in the position shown, cylinder B is being filled with liquid while cylinder A is discharging its contents. As the liquid fills cylinder B, it forces out the liquid in cylinder A and moves the slide valve to the right, closing the inlet to cylinder B and opening the inlet to cylinder A. When the inlet to B is fully closed the inlet to A is open and cylinder A then charges, causing cylinder B to discharge its contents. Various designs of reciprocating meter are available and meters with as many as five pistons are often used.

Piston-type meters are suitable for measuring low liquid flow rates in the range 10^{-6} to $10^{-2}\,\mathrm{m}^3\,\mathrm{s}^{-1}$. They have been used to meter flows at pressures up to 10^6 Pa and temperatures as high as 150 °C. They usually have rangeabilities between 10:1 and 20:1 and can attain accuracies of ±1 per cent of output reading.

Nutating disc meters

Nutating disc meters are also used for liquid flow measurements, and in these devices the meter parts are arranged as shown in Figure 25.3. The disc

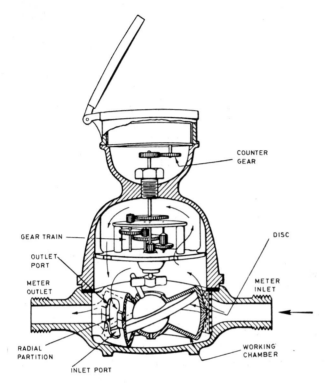

COUNTER GEAR

GEAR TRAIN

DISC

OUTLET PORT

METER OUTLET

METER INLET

RADIAL PARTITION

WORKING CHAMBER

INLET PORT

Figure 25.3   Nutating disc meter

which is mounted on a ball is enclosed in a specially shaped chamber. The top and bottom of the chamber are conical extending inwards and the side walls of the chamber are spherical. A radial partition prevents the disc from rotating about its own axis during operation. When the meter is functioning liquid enters the chamber alternately above and below the disc as it passes to the meter outlet. This motion causes the disc to nutate and for each nutation cycle a discrete volume of liquid flows through the meter. The cyclic motion is transmitted by a shaft to the readout system and totalized flow is determined by summing the number of cycles.

This type of meter is simpler to manufacture and requires less maintenance than other mechanical positive displacement meters and hence it is a commonly used device. Nutating disc meters experience most use at liquid flows in the range 10^{-6} to $10^{-1}\,m^3\,s^{-1}$. They usually have a rangeability of around $10:1$ and can be used to meter flows to within ± 1 per cent of reading. It is claimed that they can be used at temperatures as high as 120 °C and pressures up to 3×10^6 Pa.

Rotary displacement meters

Rotating *vane* and *gear-type* meters are the most commonly used rotary type of displacement meters for liquids. Here we will consider only rotating vane meters. Details of gear-type meters may be found in Linford (1961) and ASME (1971).

A rotary sliding vane meter is shown in Figure 25.4. This meter consists of a drum and a set of vanes arranged, as shown, inside a meter casing. The drum is driven in the direction shown by the liquid being metered. As the drum rotates each compartment which is enclosed by two vanes and the casing walls is filled at the inlet and then in turn emptied through the meter outlet. Totalized flow is obtained by summing the number of cycles of the drum. Various designs of rotating vane meters have appeared over the years and in the meter shown the vanes can slide in and out of their recesses to give a sealing arrangement between the vanes and the casing walls which helps to keep any leakage of the liquid across the vanes at an acceptable value.

Liquids usually metered using this type of device are petroleum products at flows between 10^{-3} and $0.5 \, m^3 \, s^{-1}$. Pressures and temperatures are generally below 10^6 Pa and 100 °C respectively. Under very carefully controlled conditions these meters can be used to measure totalized flow to within ±0.2 per cent of reading.

For gases, lobed rotary displacement meters have been one of the most popular devices used for gas flows up to about $2 \, m^3 \, s^{-1}$ at line conditions, while for low flow rates of the order of 10^{-3} to $10^{-6} \, m^3 \, s^{-1}$, at pressures and temperatures close to ambient, wet gas meters have experienced wide usage

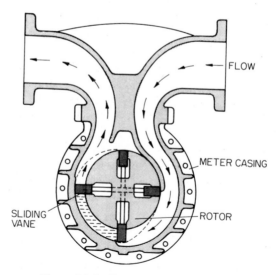

Figure 25.4   Rotary sliding vane meter

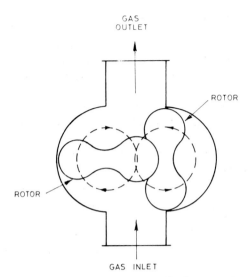

GAS
OUTLET

ROTOR

ROTOR

GAS INLET

Figure 25.5 Lobed rotary displacement
meter

as secondary standard flow measurement devices. Here we will consider only the lobed rotary displacement (*Roots-type*) meter. Details of other positive displacement meters for gases may be found in ASME (1971) and Walker (1966).

A lobed rotary displacement meter is outlined in Figure 25.5. This meter incorporates two *figures of eight* lobed rotors located inside a shaped meter casing. When the meter is operating the rotors are driven in the direction shown by the flowing gas such that for each rotation cycle a calibrated volume is swept out. As with other rotary displacement meters totalized flow is obtained by summing the number of rotor cycles. For this type of meter, to ensure that leakage across the rotors is kept within acceptable values it is important that the spaces between the rotors and the casing walls should be kept very small. It is also important that the resistance of the rotors to the flowing gas should be kept low or excessive pressure differences will occur across the meter.

Roots-type meters are used at pressures up to 80 bar, but temperatures do not generally exceed 60 °C. Rangeabilities of up to 25 : 1 can be achieved and flow rates have covered ranges from $2 \times 10^{-3}\,\mathrm{m^3\,s^{-1}}$ to $2\,\mathrm{m^3\,s^{-1}}$ at line conditions. Accuracies of better than ±0.5 per cent of totalized flow are attainable with clean gases.

Diaphragm meters

Diaphragm meters measure the flow of gases and are commonly used by gas-producing companies for the sale of natural gas to both domestic and commercial consumers. These meters experience wide usage since they can be manufactured cheaply and measure volume directly. They are not, however, suitable for the measurement of corrosive gases or large gas flow rates.

A meter of this kind is shown in Figure 25.6 at four stages in its operating cycle. The device consists of a rigid housing containing four chambers. Flow to, and from, the chambers is controlled by slide valves. The walls of chambers B and C are flexible and expand and contract as they are filled and exhausted respectively. The movement of the flexible chambers is transmitted through a linkage arrangement to a mechanical readout system which counts the number of displacements to give totalized gas flow.

For carefully calibrated and maintained meters accuracies, in totalized flow, of within ± 1 per cent can be achieved. These meters experience most use in measuring gas flows in the range 10^{-4} to $10^{-1}\,\mathrm{m}^3\,\mathrm{s}^{-1}$ at pressures and temperatures close to ambient.

25.2.2   Pressure Difference

Over the years numerous different types of pressure difference devices have been introduced for both liquid and gas flow measurements. In this section, however, only the most widely known pressure difference flowmeters, *orifice plates*, *nozzles*, and *venturi tubes*, will be considered. Considerable data have been obtained on the performances of these meters and standard documents concerning their use have been produced in several countries (BSI, 1964; ISO, 1979; VDI, 1969). If the specifications laid down within these documents are closely followed flows can be measured to better than ± 2 per cent without calibrating the flowmeter. It should be noted, however, that the performances of these devices can be markedly affected by changes in upstream pipework configurations and when compared with other meters their rangeability is low, being within $4:1$ in most cases.

Pressure difference flowmeters introduce a constriction to the flow which causes the velocity of the fluid to increase until it reaches a maximum at the area of minimum contraction. This increase in velocity causes the static pressure to decrease and a pressure difference is created between the meter upstream pressure and the pressure at the contraction. The flow rate can be determined by measuring this pressure difference as shown in the following section.

Derivation of the equation for mass flow of a compressible fluid through a construction

There now follows the derivation, from physical principles applied in the fluid dynamic situation, of the equation for mass flow of a compressible fluid through a constriction.

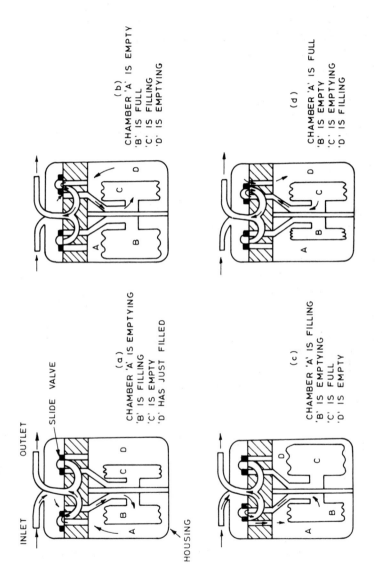

Figure 25.6 Diaphragm meter – states of operation

Notation

A	Cross-sectional area of fluid flow
H	Enthalpy per unit mass
K, K_a	Constants
p	Absolute pressure
r	Pressure ratio, p_2/p_1
T	Absolute temperature
U	Fluid velocity
W	Fluid mass flow rate
γ	Isentropic exponent
ρ	Fluid density

Subscripts

1	Upstream plane (see Figure 25.7)
2	Downstream plane (see Figure 25.7)
t	Ideal theoretical

Consider the subsonic flow of a perfect gas flowing horizontally through a constriction, Figure 25.7, according to the isentropic process law

$$\frac{p}{\rho^\gamma} = K \tag{25.1}$$

As shown in Figure 25.7 the gas expands from conditions p_1 and T_1 at area A_1 to conditions p_2 and T_2 at the area of minimum contraction, A_2.

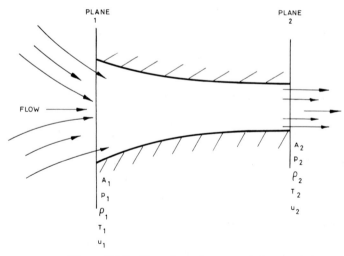

Figure 25.7   Flow through a constriction

From the conservation of energy equation we can write

$$H_1 + \frac{U_1^2}{2} = H_2 + \frac{U_2^2}{2}$$

or

$$\frac{U_2^2 - U_1^2}{2} = H_1 - H_2 \tag{25.2}$$

For an isentropic process it can be shown that

$$H_1 - H_2 = K_a \int_{p_1}^{p_2} p^{1/\gamma}\, dp$$

$$= K_a \left(\frac{\gamma}{\gamma - 1}\right) p_1^{(\gamma-1)/\gamma}(1 - r^{(\gamma-1)/\gamma}) \tag{25.3}$$

But

$$K_a = \frac{p_1^{1/\gamma}}{\rho_1}$$

and equation (25.2) can be written as

$$\frac{U_2^2 - U_1^2}{2} = \frac{p_1}{\rho_1}\left(\frac{\gamma}{\gamma - 1}\right)(1 - r^{(\gamma-1)/\gamma}) \tag{25.4}$$

For the mass continuity equation we can write

$$W_t = \rho_1 A_1 U_1 = \rho_2 A_2 U_2 \tag{25.5}$$

From equation (25.1)

$$\frac{\rho_2}{\rho_1} = \left(\frac{p_2}{p_1}\right)^{1/\gamma} = r^{1/\gamma} \tag{25.6}$$

so from equation (25.5)

$$U_1 = \left(\frac{p_2}{p_1}\right)^{1/\gamma} \frac{A_2}{A_1} U_2 \tag{25.7}$$

Inserting this expression for U_1 in equation (25.4) we obtain

$$U_2 = \left\{ 2\left(\frac{\gamma}{\gamma - 1}\right)\frac{p_1}{\rho_1}(1 - r^{\gamma-1/\gamma}) \Big/ \left[1 - \left(\frac{A_2}{A_1}\right)^2 r^{2/\gamma}\right]\right\}^{1/2} \tag{25.8}$$

Using the value of U_2 in equation (25.5) and noting from equation (25.6) that

$$\rho_2 = \rho_1 r^{1/\gamma}$$

we can write for the gas mass flow

$$W_t = A_2\left\{ 2\left(\frac{\gamma}{\gamma - 1}\right)p_1\rho_1 r^{2/\gamma}(1 - r^{(\gamma-1)/\gamma}) \Big/ \left[1 - \left(\frac{A_2}{A_1}\right)^2 r^{2/\gamma}\right]\right\}^{1/2} \tag{25.9}$$

The above formulations form the basis for design of differential pressure flow devices of which several forms of the many are now discussed.

Nozzles

A nozzle is a pressure difference device which has a shaped convergent entry. This entry is often followed by a short cylindrical throat as shown in Figure 25.8 where a standard ISA 1932 nozzle is illustrated.

To obtain equation (25.9) in the same form as that given in standard documents for nozzles (BSI, 1969; ISO, 1979; VDI, 1969) we note that, at the nozzle inlet, the inlet diameter is d_1 and the area of minimum contraction is at the nozzle throat which has a diameter d_2. We can write that

$$\beta^2 = \frac{d_2^2}{d_1^2} = \frac{A_2}{A_1} \tag{25.10}$$

and noting that

$$p_1 = \frac{p_1}{p_1 - p_2}(p_1 - p_2) = \frac{p_1 - p_2}{1 - r} \tag{25.11}$$

we can rearrange equation (25.9) as follows:

$$W_t = A_2 \frac{1}{(1-\beta^4)^{1/2}} \left(\frac{(1-\beta^4)\gamma(1-r^{(\gamma-1)/\gamma})r^{2/\gamma}}{(\gamma-1)(1-r)(1-\beta^4 r^{2/\gamma})} \right)^{1/2} [2\rho_1(p_1 - p_2)]^{1/2} \tag{25.12}$$

For standard nozzles where the flow from inlet to throat can be considered

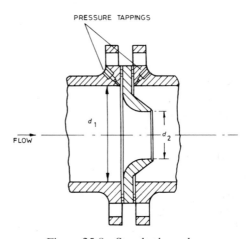

PRESSURE TAPPINGS

FLOW

d_1

d_2

Figure 25.8   Standard nozzle

close to ideal, the term ε, the *expansibility factor*, is given by

$$\varepsilon = \left(\frac{(1-\beta^4)\gamma(1-r^{\gamma-1/\gamma})r^{2/\gamma}}{(\gamma-1)(1-r)(1-\beta^4 r^{2/\gamma})} \right)^{1/2}. \tag{25.13}$$

For both nozzles and orifice plates the *velocity of approach factor* is defined as

$$E = \frac{1}{(1-\beta^4)^{1/2}}. \tag{25.14}$$

Equation (25.12) can then be written as

$$W_t = A_2 E\varepsilon[2\rho_1(p_1-p_2)]^{1/2}. \tag{25.15}$$

To account for the fact that, due mainly to fluid friction effects and density gradients, actual fluid flows do not comply exactly with the assumptions made in the ideal gas mass flow derivation given above, a factor known as the *discharge coefficient* is introduced. The discharge coefficient, C_d, is defined as

$$C_d = \frac{\text{actual rate of flow}}{\text{theoretical rate of flow}}$$

Hence the actual mass flow rate of fluid through a nozzle can be obtained from

$$W = C_{d,n} E\varepsilon \left(\frac{\pi}{4} d_2^2 \right)[2\rho_1(p_1-p_2)]^{1/2} \tag{25.16}$$

where $C_{d,n}$ is the *nozzle discharge coefficient* which is obtained either from standard documents or directly by nozzle calibration. If the flow is incompressible, $\varepsilon = 1$, equation (25.16) is considerably simplified.

Nozzles tend to be more suitable than orifice plates (see next section for description) for metering dirty fluids since they are less affected by wear. Also nozzles have higher discharge coefficients than orifice plates. Orifice plates are, however, cheaper to manufacture than nozzles and can be more readily installed into pipelines. Standard nozzles are used to meter liquids and gases and are most useful at *pipe Reynolds numbers* between 10^4 and 10^7, pipe diameters being greater than 50 mm.

Orifice plates

A *sharp-edged orifice* plate, which is essentially a thin flat circular plate with a cylindrical hole through its centre, is shown in Figure 25.9. In this illustration are shown the various standard pressure tapping arrangements which can be used to measure flow (BSI, 1964; ISO, 1979).

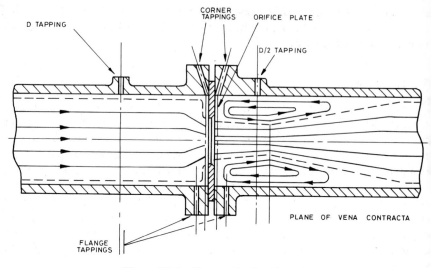

Figure 25.9   Sharp-edged orifice plate

Since orifice plates have a simple geometry and can be manufactured reliably and cheaply, they are easily the most popular pressure difference devices. Due to their wide usage their performance characteristics are well known and their behaviour can be reliably predicted for a range of different installation conditions. Standard plates experience use at pipe Reynolds numbers in the range 10^4–10^7 in pipes with diameters greater than 50 mm.

One of the main disadvantages of orifice plates, when they are compared with nozzles, is that their inlet edge tends to wear in use and this can give rise to significant errors. Also their discharge coefficients are considerably lower than those of nozzles. For nozzles discharge coefficients are usually between 0.9 and 0.99, while for orifice plates discharge coefficients are approximately 0.6. The main reason for this difference may be noted by examining Figures 25.8 and 25.9. It can be seen that in the case of the sharp-edged thin plate orifice the area of minimum contraction, the *vena contracta*, occurs downstream of the plate. The equation used to determine flow through the orifice plate is

$$W = C_{d,o} E \varepsilon \left(\frac{\pi}{4} d^2\right) [2\rho_1(p_1 - p_2)]^{1/2} \qquad (25.17)$$

where d is the orifice diameter. Clearly the discharge coefficient $C_{d,o}$ must take into account the additional contraction which occurs between the plate and the downstream tapping plane. For a nozzle the area of minimum contraction occurs at the throat. It should also be noted that the isentropic flow assumption used to determine ε for nozzles cannot be used for orifice plates, and hence the expansibility factors used for orifice plates are determined by experiment.

Venturi tubes

A *standard venturi* tube is shown in Figure 25.10 where it can be noted that the pressure difference is determined using tappings at the inlet and throat. Since this device has a diffuser section at its exit it allows the fluid pressure to recover smoothly and hence gives excellent pressure recovery when compared with orifice plates or nozzles which cause substantial pressure head loss. Like the nozzle it is more suitable than orifice plates for metering dirty liquids, but of the three basic pressure difference measuring devices dealt with here it is the most expensive to manufacture. Standard meters experience most use at pipe Reynolds numbers in the range 10^5–10^6 and, depending on construction, their discharge coefficients vary from about 0.98 to 0.995.

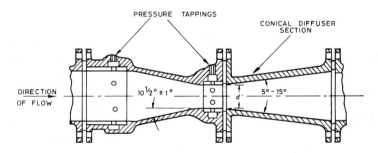

Figure 25.10   Standard venturi tube

25.2.3   Variable Area

The most widely used form of *variable area* flowmeter is the tapered tube and float meter and only this type of device will be dealt with here. Details of other forms of variable area meter may be found in ASME (1971). In the *tapered tube and float* meter flow is in the vertical plane, as shown in Figure 25.11, where it can be seen that the fluid flows upwards in a tapered vertical tube of circular cross-section which has its narrow end at the meter inlet. The flow supports a float which, in certain designs, is provided with slots cut slantwise as shown. These slots cause the float to rotate giving central stability. Since the tube is tapered, for a given flow rate the velocity of the fluid varies along the length of the tube and the float achieves an equilibrium position at a height where the fluid velocity is sufficient to support the float.

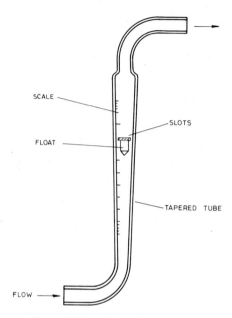

SCALE

SLOTS

FLOAT

TAPERED TUBE

FLOW

Figure 25.11   Variable area meter

When flow is increased to a new steady value the float support velocity occurs at a greater cross-sectional area higher up the tube and the float rises to assume this new equilibrium position. The height of the float can, therefore, be used to give a measure of the flow velocity or flow rate provided that the various other factors which govern the float position, such as the density of the metered fluid and the pressure difference across the float, remain constant. Derivation of the theoretical flow equation is given at the end of this section.

For measurements at pressures and temperatures close to ambient the tapered tube is usually made of transparent glass and the float position can be observed directly against a scale which is often scribed on the tube wall and graduated in rate-of-flow units. In meters designed for high working pressures the metering tube is made of metal and such means as magnetic couplings can be used to give an indication of the float position.

Variable area meters give a direct indication of the rate of flow at the instant of observation and while they cannot cover such high flows as pressure difference meters they usually cost less (considering cost of a total unit) than most orifice plate, nozzle or venturi tube metering systems. The accuracy which can be obtained with variable area meters, without special calibration, is generally no better than ±2 per cent of full-scale deflection and hence they tend not to be used in applications where high accuracy is needed.

Variable area instruments are manufactured to cover flow rates which range from near zero to $0.5\,\mathrm{m^3\,s^{-1}}$ for gases and $0.1\,\mathrm{m^2\,s^{-1}}$ for liquids. Pressures and temperatures of the metered fluid are generally close to ambient, but special instruments suitable for pressures up to $3.5 \times 10^6\,\mathrm{Pa}$ and temperatures up to $350\,^{\circ}\mathrm{C}$ can be obtained. Rangeabilities of $10:1$ can usually be achieved.

Derivation of the theoretical flow equation for compressible fluids in variable area meters

As with flowmeters using a pressure dropping constriction (Section 25.2.2) it is possible, for the variable area meters, to derive a theoretical equation to provide a model for instrument development.

Notation

A Area of tube corresponding to diameter D (see Figure 25.12)
A_f Area $\pi/4(D_f^2)$ (see Figure 25.12)
a Annular area between float and tube at plane 2 (see Figure 25.12)
D_f Diameter of float
F Buoyed weight of float
p Absolute pressure
r Pressure ratio, p_2/p_1
U Fluid velocity
W_t Theoretical mass flowrate
α The ratio $(A/A_f)^{1/2}$
γ Isentropic exponent
ρ Fluid density

Subscripts

1 Upstream plane (see Figure 25.12)
2 Downstream plane (see Figure 25.12)

If we consider the metered fluid to be an *ideal gas*, flow through the annular area at plane 2, Figure 25.12, can be derived using the same method as that already used in Section 25.2.2 for flow through a constriction. If any *potential energy differences* between planes 1 and 2 are *neglected*, equation (25.9) for the configuration shown in Figure 25.12 can be written, after squaring both sides, as

$$W_t^2 = a^2 \left\{ 2\left(\frac{\gamma}{\gamma - 1}\right) p_1 \rho_1 r^{2/\gamma} (1 - r^{(\gamma-1)/\gamma}) \middle/ \left[1 - \left(\frac{a}{A}\right)^2 r^{2/\gamma}\right] \right\}. \quad (25.18)$$

Since the taper angle of the tube is small it is neglected in this analysis.

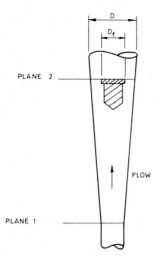

Figure 25.12   Basic elements of tapered tube and
float meter

By noting that

$$\alpha^2 = \frac{A}{A_f} \quad \text{and} \quad \frac{a}{A} = \frac{\alpha^2 - 1}{\alpha^2}$$

and putting

$$p_1 = \frac{p_1 - p_2}{1 - r}$$

we obtain

$$W_t^2 = 2(p_1 - p_2)A^2 \rho_1 \left(\frac{\gamma}{\gamma - 1}\right)\left(\frac{1 - r^{(\gamma - 1)/\gamma}}{1 - r}\right)\left(\frac{1}{[\alpha^4/(\alpha^2 - 1)^2 r^{2/\gamma}] - 1}\right)$$

$$(25.19)$$

If we now consider the *force balance* at the float, then, neglecting any
pressure difference due to elevation, we can write

$$A(p_1 - p_2) = F + W_t(U_2 - U_1) \qquad (25.20)$$

From *mass flow continuity* we obtain

$$U_1 = \frac{W_t}{A\rho_1} \quad \text{and} \quad U_2 = \frac{W_t}{a\rho_2} = \frac{\alpha^2 W_t}{(\alpha^2 - 1)A\rho_1 r^{1/\gamma}}$$

Inserting these values for U_1 and U_2 in equation (25.20) gives

$$A(p_1 - p_2) = F + \frac{W_t^2}{A\rho_1}\left(\frac{\alpha^2}{(\alpha^2 - 1)r^{1/\gamma}} - 1\right) \qquad (25.21)$$

Using equation (25.21) to eliminate $(p_1 - p_2)$ from equation (25.19), taking the square root of both sides of equation (25.19) and rearranging the terms, we obtain

$$W_t = (2AF\rho_1)^{1/2} \left\{ \frac{1}{\left(\dfrac{\alpha^2}{(\alpha^2-1)r^{1/\gamma}}-1\right)\left[\left(\dfrac{\alpha^2}{(\alpha^2-1)r^{1/\gamma}}+1\right)\left(\dfrac{\gamma-1}{\gamma}\right)\left(\dfrac{1-r}{1-r^{(\gamma-1)/\gamma}}\right)-2\right]} \right\}^{1/2}$$

(25.22)

Equation (25.22) is usually written in the form

$$W_t = (\alpha^2 - 1) Y \sqrt{(2AF\rho_1)}$$

(25.23)

where

$$Y = \frac{1}{(\alpha^2-1)} \left\{ \frac{1}{\left(\dfrac{\alpha^2}{(\alpha^2-1)r^{1/\gamma}}-1\right)\left[\left(\dfrac{\alpha^2}{(\alpha^2-1)r^{1/\gamma}}+1\right)\left(\dfrac{\gamma-1}{\gamma}\right)\left(\dfrac{1-\gamma}{1-r^{(\gamma-1)/\gamma}}\right)-2\right]} \right\}^{1/2}$$

(25.24)

For *incompressible* fluids the flow equation can be derived in a similar manner to that given above using the *energy equation, mass continuity* and the *force balance* relationships at the float. It should be noted, however, that in the case of liquids where *density* is much *greater* than that of gases the *differences* in *elevation* between planes 1 and 2 must be taken into account (ASME, 1971). For an incompressible fluid the flow equation is usually written as

$$W_t = (\alpha^2 - 1)\sqrt{(2AF\rho)}.$$

(25.25)

25.3 MASS FLOW MEASUREMENT METHODS

In many industrial fluid flow systems the mass flow and not the volumetric flow of the fluid must be measured. In the chemical engineering industry mass flow rates are often needed to determine energy balances in process plants, and accurate mass flow rate measurements are also clearly required when a fluid product is sold on a weight basis. In the aerospace industry the mass rate of fuel consumption of, say, a jet aircraft or a rocket, is frequently needed since to determine the thrust exerted the mass of the fuel propellant delivered must be known. Indeed, in any combustion engine work output is directly related to the mass flow of fuel delivered and mass flow information is usually of greater value than volumetric flow data.

Mass flow measurement methods can be divided under two main headings: *true* mass *flow* methods and *inferential* mass *flow* methods. A true mass flowmeter is one in which the reaction of the basic sensing element is dependent on the mass flow of fluid through the device. In inferential mass flowmeters volumetric fluid flow rate and density at line conditions are measured and, by multiplying flow rate by density, mass flow rate is obtained.

The methods considered in Sections 25.3.1 and 25.3.3 deal with attempts to produce true mass flowmeters, while in Section 25.3.5 inferential mass flow systems are reported. The critical flow nozzles dealt with in Section 25.3.4 are used to measure gas flow rates and for these devices, in many cases, measurements of only the nozzle upstream pressure and temperature are sufficient to enable the gas mass flow to be determined. At present, generally, when mass flow is needed inferential mass flow systems are used while true mass meters tend to experience more use in specialized applications.

25.3.1 Pressure Difference

As with the previously described methods, mathematical modelling leads to a rigorous understanding of the principle used.

Notation

C_1, C_2 Constants

dm/dt Mass flow rate

Δp_A Pressure difference between inlet and throat of venturi A (see Figure 25.13)

Δp_{AB} Pressure difference between the throat of venturi A and the throat of venturi B (see Figure 25.13)

Δp_B Pressure difference between inlet and throat of venturi B (see Figure 25.13)

v_{01} Velocity imparted to fluid by pump

ρ Fluid density,

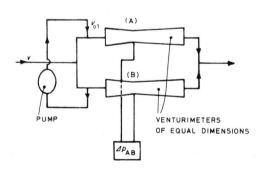

Figure 25.13   Twin venturi mass flowmeter

A flow measurement method in which a differential pressure signal gives a measure of the fluid mass flow was one of the earliest suggestions put forward for a mass flowmeter. As such well known and well tested flow measurement devices as venturi tubes, orifice plates, and pitot tubes can be used in this method various developments of the basic idea have taken place (Scanes, 1962). In most of the meters developed, however, serious problems were encountered and at present mass meters, which use a differential pressure signal to meter flow, experience very limited usage.

To illustrate the principle of operation a system which utilizes two venturi tubes is shown in Figure 25.13. Flow is divided equally between these identical venturi tubes. Assuming that the flow is *incompressible* and that *Bernoulli's equation* may be applied, along with the *mass continuity equation*, it can be shown that

$$\Delta p_A = C_1 \rho \left(\frac{v}{2}\right)^2 \tag{25.26}$$

$$\Delta p_B = C_1 \rho \left(\frac{v}{2}\right)^2 \tag{25.27}$$

If the pump is switched on and some fluid is extracted from one tube and added to the other at a constant rate, v_{01},

$$\Delta p_A = C_1 \rho \left(\frac{v}{2} + v_{01}\right)^2 \tag{25.28}$$

$$\Delta p_B = C_1 \rho \left(\frac{v}{2} - v_{01}\right)^2 \tag{25.29}$$

Provided the upstream pressure is not disturbed by the fluid extraction and injection the pressure difference between the throats of the two venturi meters, Δp, will be proportional to the mass flow rate since

$$\Delta p_{AB} = \Delta p_A - \Delta p_B$$
$$= 2C_1 \rho v v_{01} \tag{25.30}$$

and

$$\rho v \propto \frac{dm}{dt}$$

Therefore

$$\frac{dm}{dt} = C_2 \Delta p_{AB} \tag{25.31}$$

It should be noted that not only does the production of a constant injection velocity prove difficult in a system such as that described above but also the flow conditions in the venturi tubes are markedly disturbed by both the injection and withdrawal of the fluid.

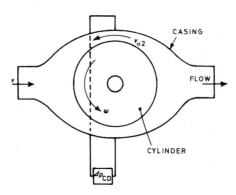

Figure 25.14   Meter of Brand and Ginsel

The problems encountered in extracting and injecting fluid at a constant rate were to some extent overcome in one design (Brand and Ginsel, 1951). In this meter the components are arranged as shown in Figure 25.14. The flow is symmetrically divided by means of a cylinder, the ends of which form a close seal with the walls of the flow enclosure. This central cylinder is driven at constant speed, ω, by an electric motor. The rotation of the cylinder produces a velocity of circulation equivalent to v_{02}. As in the case of the twin venturi tubes it is necessary that v_{02} should be less than $v/2$ or reverse flow will occur in one of the limbs. While this meter was a distinct improvement on earlier inventions, even Brand and Ginsel concluded that it would prove very expensive to manufacture and that its use would, therefore, be restricted.

25.3.2   Fluid Momentum

Various meters which utilize a fluid momentum signal to measure mass flow have been reported in review papers (Brain, 1969; Katys, 1964). Coriolis acceleration methods devised by Li and Lee (1953) and more recently by Tullis and Smith (1979), and a vibrating gyroscope method developed by Decker (1960), are rather interesting methods which have been described, but meters which utilize an axial flow transverse momentum principle have become the most commonly used true mass flowmeters which operate on a momentum principle.

The basic principle of the axial flow transverse momentum type of flowmeter, first introduced by Orlando and Jennings (1954), can be described with the assistance of Figure 25.15. The annular space in the impeller is fitted with axial vanes and the impeller is rotated at constant speed, ω, by means of a driving motor. The fluid enters the annular space, is

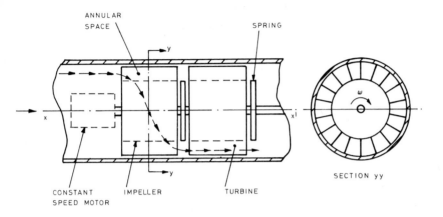

Figure 25.15   Meter of Orlando and Jennings

caused to swirl, and leaves the impeller with an angular velocity equal to that of the impeller. The fluid then enters the turbine, which is similar to the impeller and free to rotate about the same axis but is constrained against rotation by a calibrated spring. Since the meter is designed such that on leaving the turbine all angular velocity has been removed from the fluid, losses and secondary flow effects being negligibly small, the torque produced on the turbine is proportional to the mass flow. The angle through which the turbine is turned against the control spring may be measured and transmitted to a remote indicator and, since this angle is proportional to the torque, this measurement gives a measure of the mass flow. A simplified analysis of the meter's operating principles is given at the end of this section.

Although the principles of operation of this mass flowmeter are simple, certain factors, such as adequate vane length, accurate control of the constant-speed motor, variations in the radius of gyration due to centrifugal force effects, and fluid coupling effects between the impeller and turbine, must be considered in the design of the flowmeter.

Meters of the type described above give a direct indication of mass flow rate and flow totalization can also be readily achieved. Provided the meters are operated within the density ranges specified by the manufacturers, changes in density do not affect the meter performance. The meters are, however, expensive and require rotating seals as well as an accurate constant-speed driving motor. They can be used to meter both gases and liquid to within ±1 per cent over a 10:1 flow range, and meters for liquid flows in the range 0.25–$40\ \mathrm{kg\,s^{-1}}$ and gas flows in the range 3×10^{-2} to $7\ \mathrm{kg\,s^{-1}}$ are commercially available.

Analysis of the operating principles of a momentum meter

Notation

A_1 Vane cross-sectional area
C Constant
dH Angular momentum of fluid mass, dm, about xx'
dI Moment of inertia of fluid mass, dm, about xx'
dm Mass of fluid leaving impeller in time dt
dm/dt Mass flow rate
K Constant
k Radius of gyration of fluid mass, dm, about xx'
n Number of axial vanes in impeller
v Fluid velocity
X Torque on turbine
θ Angle through which turbine has turned
ρ Fluid density
ω Angular velocity

The torque, X, produced on the turbine by the fluid may be calculated using Newton's second law:

$$X = \frac{dH}{dt} \qquad (25.32)$$

Consider a mass of fluid, dm, leaving the impeller in time, dt, where

$$dm = \rho n A_1 v \, dt \qquad (25.33)$$

The angular momentum of this mass of fluid, dH, about XX' is

$$dH = \omega \, dI \qquad (25.34)$$

and

$$dI = k^2 \, dm \qquad (25.35)$$

Therefore

$$dH = \omega \, dm k^2 \qquad (25.36)$$

$$\frac{dH}{dt} = \omega k^2 \frac{dm}{dt} \qquad (25.37)$$

and

$$\frac{dm}{dt} = \frac{X}{\omega k^2} = CX \qquad (25.38)$$

$$= K\theta, \quad \text{Since } X \propto \theta$$

25.3.3 Thermal

Surveys of thermal methods of flow measurement have been given by various workers such as Benson (1971). While thermal mass flow methods can be used to measure flows over large ranges of flow rates, thermal mass flowmeters have experienced most popular use at low flows since the heat to be added to the metered fluid often becomes excessive when the flow rate is large. While these meters are often treated as true mass flowmeters, it is important to note that their output depends not only on the fluid mass flow but also on variations in the properties of the metered fluid, which can prove to be a serious limitation in many cases.

A well tried device suited for determining low gas mass flows is outlined in Figure 25.16a. This instrument consists of a tube which is uniformly heated so that the power input to the heated tube is constant. Thermocouples TC-1 and TC-2, connected to the meter readout system, are located as shown. When the gas is stationary the temperature distribution along the tube is symmetrical (see Figure 25.16b) and there is a null readout at the meter which measures the differential between TC-1 and TC-2. When gas is allowed to flow through the tube the temperature becomes asymmetrical, TC-2 becoming hotter than TC-1. The difference in thermocouple readings is affected by three main factors.

(a)   the *heat capacity* of the gas,
(b)   the *heat transfer* from the heated tube to the surroundings, and
(c)   the *gas mass flow rate*.

Provided (a) and (b) are held constant the meter output gives an indication of mass flow rate.

In these instruments the flow-sensing element is external to the flow and hence they can be used to measure toxic or hazardous gases. In addition, since they have no moving parts they are not greatly affected by wear. The basic instrument is, however, only suited for small flows and different

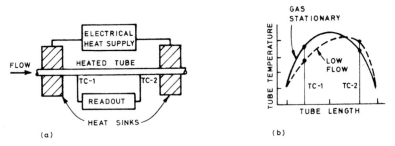

(a)   (b)

Figure 26.16   Thermal mass flowmeter

calibrations are required for different gases. These flowmeters can be used to measure gas flows within the range 2.5×10^{-10} to 5×10^{-3} kg s^{-1} and each meter has an accuracy estimated to be within ±2 per cent of full-scale deflection.

25.3.4 Critical Flow

Critical flow nozzles are used to measure the mass flow of gases. When a nozzle is operating critically the pressure drop between the nozzle inlet and throat is such that mass flow through the device for a given upstream pressure is a maximum and sonic velocity exists at the throat. At this pressure drop mass flow through the meter is only a function of the nozzle upstream pressure, temperature, the physical properties, of the gas, and the ratio of the nozzle throat area to its inlet area. It is independent of downstream pressure. A derivation of the nozzle flow equation is given at the end of this section.

Various nozzle shapes which can be used in critical flowmeters have been considered by Arnberg (1962). It is likely that toroidal inlet venturi nozzles of the type initially proposed in Stratford (1964) and recently evaluated by Brain and Reid (1978), Figure 25.17, will eventually be adopted as ISO standard devices since the performances of these nozzles have been thoroughly evaluated and nozzle discharge coefficients can be reliably predicted. These nozzles are designed to experience widest use at throat Reynolds numbers above 10^5, in pipes with diameters between 0.025 and 0.3 m. Upstream pressures are generally greater than 10^5 Pa and nozzle inlet temperatures close to ambient. It is estimated that if the nozzles are very carefully manufactured, installed, and used, flow rates can be determined to better than ±0.5 per cent.

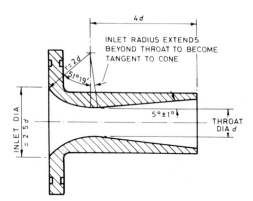

Figure 25.17   Critical flow venturi nozzle
tested by Brain and Reid

The ranges covered by critical flow nozzles are mainly limited by the minimum pressure ratio required to establish sonic flow and also by high pressure losses incurred in many installations. However, critical flow venturi nozzles such as those described in Brain and Reid (1978) have now been developed in which head losses across the device can be less than some 10 per cent of the upstream pressure. It should also be noted that since critical flowmeters are unaffected by downstream pressure variations they make excellent flow controllers, and in many applications flow rate varies almost linearly with nozzle upstream pressure. Another advantage which is worth considering is the fact that critical flow nozzles are generally less affected by changes in upstream pipework configurations than subsonic pressure difference devices.

Derivation of critical flow equation

Notation

A   Cross-sectional area of fluid flow
C^* Critical flow function
C_d Nozzle discharge coefficient
p   Absolute pressure
r   Pressure ratio, p_2/p_1
T   Absolute temperature
W   Fluid mass flow rate
γ Isentropic exponent
ρ Fluid density

Subscripts

1   Upstream plane (see Figure 25.7)
2   Downstream plane (see Figure 25.7)
t   Ideal theoretical

If we consider a gas expanding through a constriction as shown in Figure 25.7 then from equation (25.9) we can write

$$W_t = A_2 \left\{ 2\left(\frac{\gamma}{\gamma-1}\right) p_1 \rho_1 r^{2/\gamma} (1 - r^{(\gamma-1)/\gamma}) \middle/ \left[1 - \left(\frac{A_2}{A_1}\right)^2 r^{2/\gamma} \right] \right\}^{1/2}$$

For critical flow nozzles the ratio A_2/A_1 is usually very small and the gas at the upstream plane can be considered to be at rest. Putting

$$\frac{A_2}{A_1} = 0$$

we obtain

$$W_t = A_2 \left(\frac{2\gamma}{(\gamma - 1)} p_1 \rho_1 r^{2/\gamma} (1 - r^{(\gamma - 1)/\gamma}) \right)^{1/2} \tag{25.39}$$

For a given p_1 and ρ_1, W_t will be a maximum when the term

$$Y = r^{2/\gamma} (1 - r^{\gamma - 1/\gamma})$$

is a maximum, that is, when

$$\frac{dY}{dr} = 0.$$

Differentiating Y with respect to r and putting the resulting expression equal to zero we obtain

$$r = \left(\frac{2}{\gamma + 1} \right)^{\gamma/(\gamma - 1)} \tag{25.40}$$

Inserting this value for r in equation (25.39) and noting that for an ideal gas

$$\rho_1 = \frac{p_1}{RT_1}$$

we obtain

$$W_t = A_2 \frac{p_1}{\sqrt{T_1}} \sqrt{\frac{\gamma}{R}} \left(\frac{2}{\gamma + 1} \right)^{(\gamma + 1)/2(\gamma - 1)} \tag{25.41}$$

The *critical flow function* C^* for an ideal gas is defined as

$$C^* = \sqrt{\gamma} \left(\frac{2}{\gamma + 1} \right)^{(\gamma + 1)/2(\gamma - 1)} \tag{25.42}$$

and equation (25.42) is written as

$$W_t = C^* A_2 \frac{p_1}{\sqrt{(T_1 R)}} \tag{25.43}$$

As with the subsonic devices considered in Section 25.2, for real flow conditions a discharge coefficient must be used to obtain the working equation

$$W = C_d C^* A_2 \frac{p_1}{\sqrt{(RT_1)}} \tag{25.44}$$

For more details on the derivation of the critical flow equation see Arnberg (1962). For several real gases C^* functions to satisfy the above equation have been derived in Johnson (1965).

25.3.5 Inferential

As already mentioned at the start of Section 25.3, inferential mass flowmeters, volumetric fluid flow rate, and density at line conditions are determined separately, and by multiplying the flow rate signal by the density signal, mass flow rate is obtained. These density compensation methods are at present the most commonly used methods for determining fluid mass flow, mainly since well tried and tested volumetric or velocity meters can be incorporated to measure the flow. Also, with the rapid developments which have taken place in electronic instruments the various output signals from inferential mass meters can be processed effectively and reliably and the costs of these electronic processing systems have fallen sharply in recent years.

Two different types of inferential mass flow metering systems are shown in Figure 25.18. The principles of operation of orifice plates and turbine meters are dealt with in Sections 25.2.2 and 25.4.1 respectively.

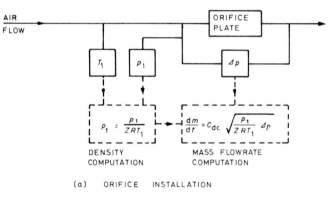

(a) ORIFICE INSTALLATION

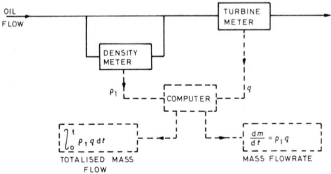

(b) TURBINE METER INSTALLATION

Figure 25.18   Inferential mass flow systems: (a) orifice installation; (b) turbine meter installation

Figure 25.18a shows an orifice plate installation for measuring the flow of dry air. In this case measurements of the upstream temperature and pressure are used to measure the fluid density and mass flow is determined from the equation

$$\frac{dm}{dt} = C_{dc}\left(\frac{p_1}{ZRT_1}\Delta p\right)^{1/2} \tag{25.45}$$

where dm/dt is the mass flow rate, C_{dc} is the orifice calibration coefficient, p_1 and T_1 are the orifice upstream pressure and temperature respectively, Z is the gas compressibility factor, R is the gas constant, and Δp is the pressure difference across the orifice plate.

It is worthwhile noting here that in certain systems, particularly in the measurement of liquid hydrocarbons, a pressure, volume, temperature databank is stored in a small computer. Pressure and temperature signals are fed to the computer, the fluid density is calculated and then used along with the volumetric flow rate to determine the mass flow.

In Figure 25.18b a density meter installed outside of the flow line is used along with a turbine meter to measure a flow of oil. The signal from the density meter is proportional to ρ_1, and the signal from the turbine meter is proportional to q, the volumetric flow of the oil. These signals are fed to a computer which gives two outputs, the mass flow rate $\rho_1 q$ or totalized flow

$$\int_0^t \rho_1 q \, dt$$

where ρ_1 is the density of the metered fluid at the density meter tapping and q is the volumetric flow rate at the turbine flowmeter.

The ranges and accuracies of volumetric flowmeters are dealt with in those sections of this chapter dealing with these devices. For density meters, various types are manufactured and these include vibrating element, buoyant beam, and spinner-type devices (Dowden, 1972; Newcombe, 1969). Standard meters are available to measure liquids at pressures up to 7×10^6 N m^{-2} and temperatures in the range -30 to $100\,°C$ with an accuracy estimated to be within ± 0.01 per cent. For gases the temperature range is -10 to $85\,°C$ for standard meters and measurements can be made at pressures up to 1.4×10^7 N m^{-2}. Accuracies of some 0.1 per cent of reading are claimed for these instruments.

Density-measuring elements can be installed inside or outside the pipeline, but while it usually preferable that they be installed inside the line this can prove inconvenient for flowmeters such as orifice plates or turbine meters where calibration is affected by changes in the upstream flow pattern and this must be carefully considered before installing an element in the flow line.

25.4 FULL FLOW VELOCITY METHODS

This section deals with four common instruments for full flow velocity measurement: *turbine, electromagnetic, ultrasonic,* and *vortex* flowmeters. All operate using recently developed techniques and the latter two are not yet fully established.

All of these methods are susceptible to the flow pattern varying from ideal. *Asymmetry* or *swirl* may cause errors of several per cent in some cases. The meters should be installed in a situation where the velocity profile is close to fully developed. If this is not possible some form of flow-straightening device should be incorporated upstream.

25.4.1 Turbine

The development of turbine flowmeters began during World War II and improvements in performance and in range of applicability have continued steadily to the present day. A large number of instruments are available commercially, many of which may appear to be very similar. However, the bearings and rotor must be matched very closely to the fluid properties and flow range to be handled if the meter is to give accurate and lasting service, so that choice of instrument is even more important here than for most flowmeters.

The construction of a typical meter is shown in Figure 25.19. A rotor with a number of angled blades is mounted coaxially on the meter body centre

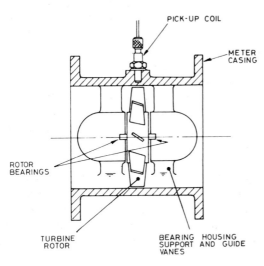

Figure 25.19   Turbine flowmeter

line. It is supported upstream and generally downstream by a spider arrangement which often incorporates flow-straightening vanes. Several types of bearing design and materials are available and much research is aimed at producing improved types. In most meters the fluid plays an active role in lubricating the bearings. The blades are angled so that the rotor rotates as fluid passes through the rotor. Over the working range the rate of rotation is, to a first approximation, proportional to the volume flow rate.

A simple theory of operation of turbine meters may be explained with reference to Figure 25.20, which shows a simplified velocity diagram at a blade of an ideal meter. From Figure 25.20 it can be seen that

$$\tan \beta = \frac{r\omega}{v}$$

where β is rotor blade angle, r is the radius of the rotor, ω is the angular velocity of the rotor, and v is the mean fluid velocity. Since

$$v = \frac{q}{A}$$

where q is the volumetric flow rate and A is the cross-sectional area of the

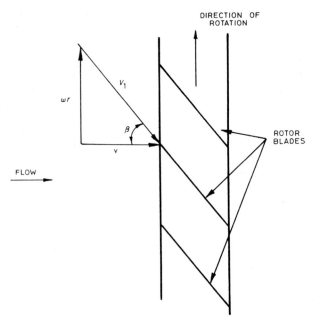

Figure 25.20   Simplified velocity diagram for ideal turbine meter

flow

$$q = \frac{rA\omega}{\tan \beta} = K\omega$$

where K is a constant.

To obtain the above result the following main assumptions have been made:

(a) No retarding forces are exerted on the rotor.
(b) Unform axial velocity distribution exists in the annular passage between the hub and the rotor casing.
(c) The helical rotor blades have no thickness.
(d) The total force on the blades is concentrated on an average radius.

A more complete analysis may be found in Jepson and Bean (1969).

Turbine meters for use with gas and liquid differ, especially in rotor design. A gas meter can pass a much greater volume of fluid in a given time than a liquid one. To maintain the rotation at approximately the same rate for maximum flow in each case and avoid overspeeding the bearing, gas turbine meters have blades set at a much smaller angle to the meter axis.

A pulse output is produced by sensing the passage of each blade past one or more sensors fixed in the meter body. An early sensor consisted of a magnet and a coil, and as a blade passed through the magnetic field it generated an emf in the coil which was subsequently detected. The disadvantage of this system is that a small retarding force acts on the blade. A variety of alternative methods have since been developed which reduce the drag to almost insignificant levels and thereby increase the useful working range of the meter at the bottom end.

The normal range of output devices is available, that is, *instantaneous flow rate* readings from a frequency counter, *totalized flow* from an integrator, and a range of *analog indicators* compatible with control equipment.

Off-the-shelf turbine meters are available in sizes ranging from 25 to 201 mm and may be obtained in smaller and larger sizes. The cost increases rapidly with increasing size for the larger meters. Flow range may be obtained from manufacturers' literature but typical maximum flow velocities are $10 \, \mathrm{m \, s^{-1}}$ for liquids and $50 \, \mathrm{m \, s^{-1}}$ for gases. Most types are available with a range of screwed or flanged ends. Line pressure is not a problem with those meter bodies available which can withstand several hundred atmospheres. Temperature capability can extend from the cryogenic region to 600 °C or more.

A correctly chosen turbine meter should have a linearity within ±0.5 per cent over a flow range of from 15:1 to in excess of 100:1 depending on fluid and meter type. Repeatability is particularly good with a typical value better than ±0.1 per cent. Calibrated accuracy should be within ±0.25 per cent. Response time is typically of the order of milliseconds.

The turbine meter has now become accepted as a standard flowmeter and is used for fiscal metering purposes. Amongst its advantages are its good calibrated accuracy, its linearity, and its repeatability. The flow range is larger than many comparable devices and when used with care its calibration is retained for long periods of service (Scott, 1971).

25.4.2 Electromagnetic

The electromagnetic flowmeter has been an established device since the early 1960's. While it is suitable only for liquids, given an initial calibration and used with care, especially as regards zero drift, it is useful both as a flow control and monitoring device and as a flowmeter *per se* with claimed accuracies of up to ±0.2 per cent. Much useful information on the calibration and long-term behaviour of electromagnetic flowmeters can be found in Scott (1975).

Operation of the device is based on the familiar theory of the generation of an emf in a wire moving in a magnetic field. Figure 25.21 shows a copper wire cutting the flux of a permanent magnet. The wire is moving in a direction perpendicular to its length and perpendicular to the magnetic field with velocity v, and the result is a voltage, V, generated between its ends of value Blv, where l is the wire length and B the magnetic flux density.

Figure 25.22 shows a diagram with the essential features of an electromagnetic flowmeter. The fluid flows in a circular tube. Electrodes make contact with the fluid to obtain the potential difference across the fluid and a

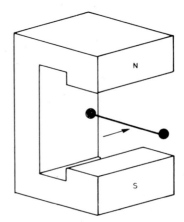

Figure   25.21 Wire   moving
through magnetic field

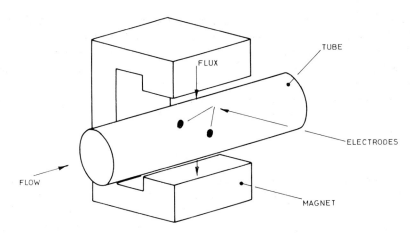

Figure 25.22   Simple electromagnetic flowmeter

magnetic supplies the magnetic flux. In this diagram we may imagine filaments of liquid spanning the tube from one electrode to the other and moving through the magnetic field so that a voltage is generated between their ends as was the case with the copper wire. Referring now to Figure 25.23, we have depicted the case where more than one wire is moving through the magnetic field. Wire P is in a region of strong magnetic field B and has a velocity v. Wire Q is in a region where the field is about the same size but it only has a velocity of $v/2$. Wire R, while moving with velocity v, is in a region of weak magnetic field, say, $B/4$. Thus the magnetic induction in each case will be different:

$$\text{In P} \quad V_P = Blv$$
$$\text{In Q} \quad V_C = Blv/2 \qquad\qquad (25.46)$$
$$\text{In R} \quad V_R = Blv/4$$

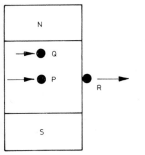

Figure 25.23   Three wires (Q, P, R) moving through magnetic field

Now if the ends of the wires are connected currents will flow since the potentials differ, and the result will be that V_P will be reduced by the ohmic loss. The application of this to Figure 25.22 results in a complicated picture of voltages and circulating currents. However, the actual operating equation of the flowmeter may be represented quite simply for a range of conditions by

$$\text{voltage between the electrodes} = BDv_m \qquad (25.47)$$

where D is the pipe diameter and v_m is the mean fluid velocity. This is the basic equation for the flowmeter and it is valid providing the magnetic field is uniform and the velocity profile is axisymmetric. In practice of course neither of these conditions is perfectly met.

Shercliff (1962) suggested a means of predicting the effect of distorted profiles using what he called a *weight function* (Figure 25.24). The velocity at each point in the cross-section of the flowmeter is multiplied by the weight function. If, as an extreme case, all of the flow passed by one of the electrodes and the rest of the flow was stationary then all the flow would be multiplied by 2.0 or more and the apparent signal would be twice as great as for the same flow with axisymmetric profile.

In practice magnetic fields are of finite length and this leads to a non-uniformity of field called by Shercliff *end-shorting*. A value S, the

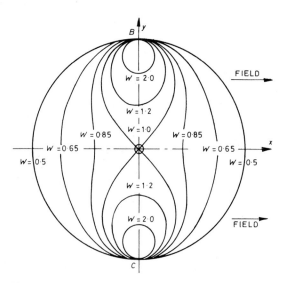

Figure 25.24   Shercliff's weight function for a uniform field point electrode flowmeter

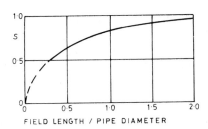

Figure 25.25 Sensitivity variation
with meter length

flowmeter sensitivity, can be defined by

$$S = \frac{\text{voltage generated}}{BDv_{\mathrm{m}}} \qquad (25.48)$$

and variation of S for short fields is shown in Figure 25.25. Design advances have minimized these effects.

Figure 25.26 shows the main components of a typical flowmeter with point electrodes. It is a combination of primary and secondary elements. The primary element consists of the metering tube (with insulating liner), flanged ends, coils to produce an AC magnetic field, and electrodes. In addition, some means to produce a reference signal proportional to the magnetic field is usually provided, such as the current transformer shown in

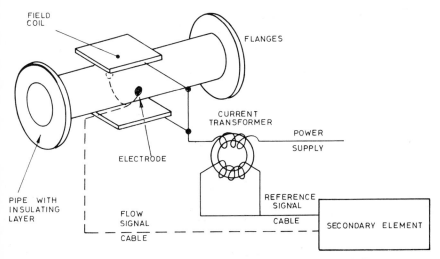

Figure 25.26   Components of typical commercial electromagnetic flowmeter

Figure 25.26. The metering tube will usually be non-magnetic to allow field penetration. It is normally available in a range of diameters from about 3 mm to about 1830 mm. The inside surface is insulated to prevent signal shorting. The electrodes have to make contact with the fluid and, therefore, break through the tube lining. The secondary element is sometimes known as the *converter*; it amplifies and processes the flow signal, eliminates spurious electromotive forces, and compensates for variations in the electricity supply.

Some of the main advantages have been noted already. When the flowmeter is calibrated and used with care, an accuracy of better than one per cent can be obtained over a typical flow range of $0.5-10 \text{ m s}^{-1}$. The size range available is very wide, though the larger instruments are expensive. Having no moving parts the device is less prone to breakdown than some flowmeters. The main disadvantage is in the difficulty of maintaining calibration and zero setting. Useful comments on this are given in Scott (1975).

25.4.3　Ultrasonic

One major advantage of ultrasonic systems is that the flow is not disturbed as there need be no obstruction within the pipe. A further advantage of the more sophisticated arrangements is that the actual flow velocity profile is accounted for directly with minimal assumptions about its form.

Proposals for the use of ultrasound in flow measurement have been put forward during at least the last 50 years (Jespersen, 1973), but only in the last decade or so has the technology become available to allow the development of practical industrial meters. Commercial instruments are still at an early stage of development and it is probable that with time the present wide range of types will narrow as experience is gained in their use. Discussion here will be limited to the two most widely developed types.

Most types of ultrasonic flowmeter depend upon the fact that the velocity of a sound wave in a moving fluid is modified. If the wave has a component in the *direction of flow* then the magnitude of the *velocity is increased* and the direction of the wave is altered towards the flow direction. The *reverse* happens if the wave is travelling *against* the flow. Within limits the frequency of the ultrasound itself is not important so long as good transmission is obtained.

A very simple type of flowmeter is illustrated in Figure 25.27. A pair of ultrasonic transducers are separated by a distance D. If the acoustic pulses are transmitted through the fluid in either direction, the transit time t for uniform fluid velocity is given by

$$t = \frac{D}{C \pm V_p} \approx \frac{D/C}{1 \pm V_p/C} \text{ for } C \gg V_p \qquad (25.49)$$

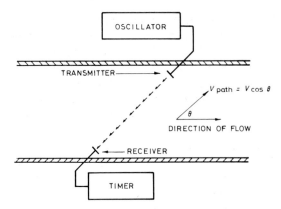

Figure 25.27 Simple ultrasonic flow measuring device

where C is the velocity of sound in the still fluid, and V_p is the velocity component of the fluid in the direction of the acoustic path. The negative and positive signs of the right-hand term refer to the downstream and upstream transit times respectively. With no flow, the datum $t_0 = D/C$ and the difference $\Delta t = t_0 - t = DV_p/C^2$. Δt could, therefore, be a measure of the fluid velocity.

The method is now only of historical interest as t_0 and t must be measured simultaneously, and it cannot be assumed that t_0 is constant, since C may be subject to small variations due, for example, to temperature variations. Furthermore, the fluid velocity can only be measured along the acoustic path. Thus, to obtain the volume flow rate, the effect of the velocity profile of the fluid has to be considered.

Velocity profile effects

In the preceding simplified analysis it was assumed that the velocity component of the fluid was constant along the acoustic path. In practice this is never so and the expression for transit times must be replaced by a line integral.

Thus

$$t = \int_{\text{path}} \frac{dl}{C \pm V_1} = \frac{D}{C}\left(1 \pm \frac{1}{C}\right)\int_{\text{path}} \frac{V_1}{D}\,dl \qquad (25.50)$$

where V_1 is the component of velocity distribution along the acoustic path of length l.

The measured velocity is, therefore, equivalent to the mean fluid velocity component along the acoustic path, and hence the true volume flow rate

1108 HANDBOOK OF MEASUREMENT SCIENCE

cannot be obtained simply from the measured velocity multiplied by the cross-sectional area of the pipe. Equation (25.49) is, however, still valid if V_p is considered as the mean velocity component along the measurement path. For fully developed flow the velocity profiles within the pipe are stable in time and symmetrical about the pipe axis. Thus, for a given profile, the average velocity V_a is related to the average velocity across a diameter V_d by a constant. Thus $V_d = KV_a$, and to measure the volume flow rate Q it is merely necessary to employ a single acoustic path and implement the relationship $Q = KAV_d$, where A is the cross-sectional area of the pipe.

The relationship between the correction coefficient, K, and the Reynolds number, Re, for fully developed turbulent flow in smooth circular ducts has been widely discussed. For example, Kritz (1955) has proposed the expression

$$K = 1 + 0.19Re^{-0.1} \tag{25.51}$$

For laminar flow where $Re \leqslant 2 \times 10^3$ with a parabolic velocity distribution the correction factor is 4/3.

Frequency difference method

The frequency difference method is also known as the *sing-around* method; it is illustrated in Figure 25.28. Originally it was developed at the National Bureau of Standards (Greenspan and Tschiegg, 1957) for measuring sound velocity to one part in 5000. Two pairs of transducers are arranged in an X configuration. A short pulse is emitted from one transducer to the other where it is amplified and triggers off another pulse in the same direction,

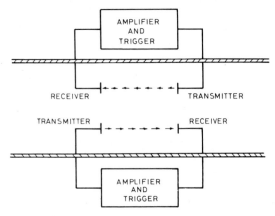

Figure 25.28 Frequency difference *sing-around* system

generating a sequence of pulses in a closed loop. The second pair of transducers in the other leg of the X forms a similar loop whose pulses travel in the opposite direction.

From equation (25.49), the time interval t_1 between pulses in the downstream direction is $D/(C+V_p)$ and the pulse repetition frequency f_1 becomes $(C+V_p)/D$. Likewise the frequency f_2 of the other loop, transmitting in the upstream direction, becomes $(C-V_p)/D$. The frequency difference between the two loops will therefore be $2V_p/D$ which, being independent of C, is a desirable result. Only when practical values are substituted in this expression are the limitations of the technique obvious. A reasonable value for the transducer separation D would be 0.25 m. If the flow velocity V_p is $0.5 \, \mathrm{m \, s^{-1}}$ the difference in frequency will only be 1 Hz, which is not easily measured accurately without sacrificing response time. Furthermore, if there is the slightest coupling between the two loops, one loop will tend to lock to the frequency of the other.

Leading edge (LE) ultrasonic flowmeter

Another principal type of meter is the *leading edge* ultrasonic flowmeter (Fisher and Spink, 1971). As its name suggests, the leading edge of the pulse transmitted through the flowing medium activates a time-measuring circuit.

Equation (25.50) shows that the measured transit time depends on the mean fluid velocity as well as the speed of sound. To overcome this disadvantage the upstream and downstream transit times, t_u and t_d respectively, are measured simultaneously and the algorithm $f = (t_u - t_d)/(t_d \times t_u)$ is computed; the mean fluid velocity along the acoustic path can be determined from $V_p = f/2D$.

As this is a purely digital method, more than one path can be readily incorporated into the system to cater for distorted flow with only a slight increase in cost. Chordal paths are commonly used (Figure 25.29). The spacing of the paths has been chosen according to the Gaussian integration formulae (Scarborough, 1966) so that the sum of the measured mean velocities, multiplied by a constant weighting factor, equals the volume flow rate.

The advantage of this system is that a variable correction coefficient, dependent on the Reynolds number, is not required to maintain reasonable accuracy for flow rates ranging from laminar flow to those approaching the speed of sound in the measured media. Furthermore, it can be used in conditions where the flow profile is considerably distorted. Calibration appears to be unnecessary if electrical delay times and acoustic path lengths can be measured independently with reasonable accuracy. The only necessary precaution is to ensure that a single acoustic path does not encompass a localized velocity aberration.

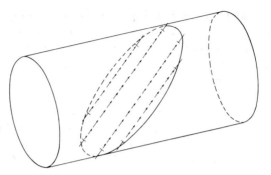

Figure 25.29 Transducer arrangement and measurement plane for typical LE ultrasonic flowmeter showing four chordal acoustic paths

Apart from velocity profile effects, the performance of ultrasonic flowmeters of this type will be mainly determined by the *resolution* and *accuracy* of
the *timing circuit*. With a reduction in the distance between the transmitter
and receiver or of the flow rates, timing errors will become correspondingly
more significant. Independent tests (ARL, 1971) have shown this is indeed
the case. On the other hand these tests have also shown that accuracies in
the order of 0.1 per cent (of the true flow rate) are possible even for
asymmetrical velocity profiles if the ultrasonic transit times are two orders of
magnitude greater than that of resolution of the time interval unit.

Features of ultrasonic flowmeters

The general features of no obstruction and good potential accuracy have
been noted. The technique is well suited to large-diameter ducts and has the
advantage that cost is independent of increasing size. Path lengths of several
hundred feet have been used. Difficulties occur with smaller pipe sizes from,
say, 150 mm downwards.

  Most liquids can be handled provided that the transducers are given a
window of a suitable material. Good transmission depends closely on
matching this material to the fluid refractive properties and the problem can
be acute in organic fluids such as oil. The presence of contaminants and
temperature fluctuations can affect signal quality even in those systems
designed to minimize these problems. A major limitation at present is the
measurement of gas flow where signal strength would be expected to be
much less. Meters are available which claim to measure gas flow satisfactorily, but such claims have not yet been fully substantiated. Pressure and
temperature ranges are wide. In normal systems where the removable

transducers have windows flush with the pipe wall sealing is the weak point, though pressures of 400 atmospheres have been attained. With careful choice of the window temperatures from cryogenic to several hundred degrees centigrade may be dealt with. The velocity range covered is generally that found in liqud flows.

A development which could prove valuable is that of *clamp-on* systems using existing pipework. These require very careful alignment and acoustic matching with the pipe wall is a severe problem in all but ideal cases. Manufacturers often claim accuracy within one per cent, but these claims also remain to be substantiated.

25.4.4 Vortex Shedding

The development of the *vortex-shedding* flowmeter has been a rapid one from the first commercial instruments in the early 1970s to a position at the end of that decade where it is challenging seriously for a major share of the flowmeter market.

Vortex shedding may occur when a body, especially one which is non-streamlined or *bluff*, is placed in a stream of fluid. Where the flow round the body separates from the surface circular *eddies* begin to form and under certain circumstances a pressure feedback mechanism synchronizes these eddies and produces a regular shedding from alternate sides of the body. This regular patttern of vortices is known as the *Von Karman vortex street* (Figure 25.30).

The first detailed experimental study into the generation of vortices behind a bluff body was carried out by Strouhal (1878). He showed that the frequency, f, of a vibrating wire of diameter d, placed in a cross-stream of air with a velocity past the wire of V, was equal to $V/6d$.

Later, Rayleigh introduced the non-dimensional relationship $S = fd/V$, where S is known as the *Strouhal number*. If S is constant, then for a chosen diameter, d, the frequency, f, would be linearly related to velocity, v, which of course is an ideal relationship for a flowmeter. Experimental work has

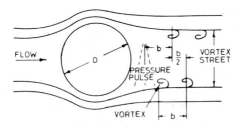

Figure 25.30   Vortex shedding

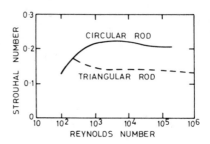

Figure   25.31   Strouhal   number
against Reynolds number

shown that over the range of Reynolds numbers 30 000 to 100 000, S is
constant for most types of bluff bodies, but the value of S is dependent on
the shape of the body as shown in Figure 25.31 where the Strouhal number
has been plotted against Reynolds number for a circular rod and a triangular
rod. The latter is preferable as a flow-metering device since the Strouhal
number is constant over a wider range of Reynolds number.

When the fluid stream encounters the body, it must divide to pass around
the barrier (Figure 25.30). Because of viscous adhesion, the *boundary layer*
(fluid moving along the surface of the obstruction) moves slower than the
outer layers. After the obstruction has been passed, both the boundary layer
and the fast moving layer must separate from the obstruction before they
can recombine. At very low flow rates the recombination occurs without
much turbulence because the velocity difference between the two layers is
small. As the flow rate increases, however, the boundary layer tends to cling
longer to the obstruction than the outer layer, resulting in the formation of
local separation in the fluid behind the obstruction.

As the flow increases further the vortices become relatively stable and
long-lasting so that they line up directly behind the bluff body and *alternate*
in sequence from one side to another. This is because a strong pressure
pulse accompanies the formation of a vortex which in turn temporarily
inhibits the formation of a vortex from the opposite edge. As soon as the
newly formed vortex has moved downstream, the pressure at this edge
suddenly falls and a vortex is produced there.

The result is the *street* of vortices alternating from side to side and having
equal spacing. The *frequency* of oscillation is determined by the *dimensions*
of the object, the *rate of flow*, and the Strouhal number according to
Rayleigh's equation.

Although the effect of vortex shedding has been known for many years
there were technical problems to be overcome before it could be embodied
into a practical flowmeter. There were two main problems: the first was to
find a method of obtaining regular vortex shedding which had a linear

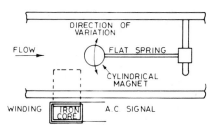

Figure 25.32 Early type of vortex-
shedding flowmeter

frequency relationship with the fluid velocity, and the other was to develop a technique of detecting the rate of vortex shedding under noisy hydraulic conditions.

The earliest flow instruments based on this principle used flexible obstructions whose vibration rate was an index of the flow rate. Figure 25.32 shows a vortex-shedding device with an external pick-up coil for converting the pressure-induced vibrations of a cylindrical magnet into electrical pulses with a frequency proportional to the flow rate. The main disadvantage with this type of arrangement is that the motion of the bluff body tends to alter the nature of the vortex shedding, and as we have already seen (Figure 25.31) a circular cross-section is not the ideal shape for a bluff body.

The vortex meter shown in Figure 25.33 has a triangular bluff body, and a heated pair of thermistors placed in the paths of the vortices are cooled at a faster rate when they encounter the swirling fluid. Consequently the thermistors alternately change resistance to create a train of electrical pulses with a frequency linearly proportional to the flow rate (Figure 25.31) and since a rigid obstruction is used in this sytem relatively stable vortices are generated. A variant of the same theme is to replace the thermistors with a transverse ultrasonic beam (Figure 25.34). The vortices interrupt the beam and produce an oscillatory voltage at the receiver. The main advantages

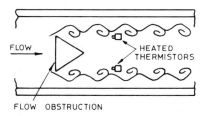

Figure 25.33 Vortex meter with
triangular bluff body and thermistor
detectors

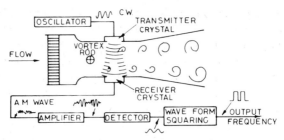

Figure 25.34   Ultrasonic vortex meter

claimed for this type of meter (Miller *et al*, 1977) are good linearity (1 per cent) coupled with a wide operating range (1:30). In addition a high degree of sensitivity is claimed for the detector, permitting the use of a very slim vortex-shedding body with consequent low head loss.

In another design (Burgess, 1977) shown in Figure 25.35 the vortices are initially formed by a fixed, specially shaped upstream obstruction. A second obstruction, attached by a cantilever beam to the first, vibrates in response to vortices passing it. Some of the vortex energy generated by the first obstruction is absorbed by the second with the net result that the stability of the vortex street is enhanced. The vibration of the second obstruction is detected by a strain gauge embedded in the first obstruction and the cantilever. A signal-conditioning system converts the strain gauge output into an analog signal which is proportional to the flow. The claimed uncertainty for this flowmeter is within 1 per cent of full range, with a repeatability of ±0.15 per cent and a rangeability of 15 to 1.

The range of forms of bluff body and sensor types is now wide. Some manufacturers offer a choice of sensor types for a given body. Instruments are available off-the-shelf in sizes from 25 mm to at least 200 mm and smaller and larger ones are obtainable. For sizes above 0.5 m insertion devices can be obtained to measure the velocity at the pipe centre line or some other selected point. Temperatures from cryogenic to 400 °C and a

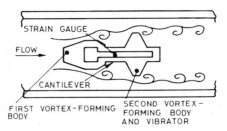

Figure 25.35   Cantilever vortex meter

wide pressure range can be handled. Some manufacturers claim that a single instrument will meter liquids and gases equally well, but current experience does not always support this claim. Flow ranges of up to 100:1 are also claimed, but in most cases something like 15:1 would be more realistic, for the lower end of the range is limited to a Reynolds number of around 2000 and the upper end is also generally limited, for instance, by cavitation in water or turbulence levels in gas flow. Linearity may be within ±1 per cent over the working range and this feature is being improved as bluff body forms are developed. Most, if not all, meters currently need calibration to attain optimum accuracy levels, which can be as good as ±0.15 with ±0.1 repeatability for the best meters.

The potential advantages of vortex meters are numerous. Good accuracy, linearity, and repeatability and a wide working range are coupled with low relative cost. Few or no moving parts tends to lead to good reliability with little maintenance. The theoretical non-dependence on fluid makes them potential transfer standards.

The disadvantages stem largely from a failure in the present generation of meters to meet all of these claims in an unequivocal fashion. Meter factor is found to vary with fluid in many cases and vibration is a problem which may never be solved fully since the vortex shedding itself is an oscillatory phenomenon. However, there is little doubt that during the 1980s the vortex meter will attain a pre-eminent position in the flow-metering field.

25.5 POINT VELOCITY METHODS

This section deals with four methods of *point velocity measurement* incorporating *hot-wire anemometers*, *vane-type* anemometers, and *currentmeters*, *pitot tubes* and *laser velocimeters*. The first three are well established methods with standards relating to their use, whereas the laser technique is recent.

Each method requires the measurement of a *point* velocity or speed and if an accurate flow measurement is required *several measurements* across the duct are made and then integrated. Further information on the selection of measurement locations and on integration techniques is given in Ower and Pankhurst (1977).

25.5.1 Hot-wire and Hot-film Anemometers

These devices depend upon the relation between the rate of loss of heat from a heated body and the speed of flow of a fluid in which it is immersed. Although termed point velocity methods, measurements are made over a *finite length or area* in the flow. The probes tend to be delicate and require considerable skill in their use. They are thus largely confined to research work often involving turbulence studies for which the hot wire was the only suitable instrument before the advent of the laser anemometer. However, a number of more robust versions of the instrument are available for mean speed measurements. Liquids and gases may be measured, though the delicate hot-wire probes tend to break very quickly in liquids. A useful review of calibration and use is given in Ower and Pankhurst (1977).

The fundamental work on hot-wire anemometry was done by L. V. King in the early years of the twentieth century (King, 1914). King found that the rate of heat loss from a cylinder maintained at a greater temperature than the surrounding fluid was

$$H = k\theta + (2\pi k C_v \rho \, dv)^{1/2}\theta \qquad (25.52)$$

where H is the rate of heat loss per unit length of cylinder, d is the cylinder diameter, θ is the temperature difference, k is the thermal conductivity of the fluid, C_v is the specific heat capacity at constant volume of the fluid, ρ is the fluid density, and v is the fluid velocity. At velocities below about $100 \, \text{m s}^{-1}$ for air this relationship alters, and King and others (Collis and Williams, 1959) have given improved formulae. The basic relationship of equation (25.52) has also been found to be improved by replacing the index $\frac{1}{2}$ by the exponent 0.45 (Bradshaw, 1971).

A simple hot-wire arrangement is shown in Figure 25.36. A typical diameter of wire is 0.02 mm and the minimum length to diameter ratio is usually 200:1. The wire may be made of platinum in the form known as Wollaston wire or of tungsten. Normal operating temperatures are 900 and 250 °C respectively.

The device may be operated either in *constant-current mode* or in *constant-temperature mode*, which in practice means keeping the resistance constant. The latter method has tended to be preferred because it is simpler

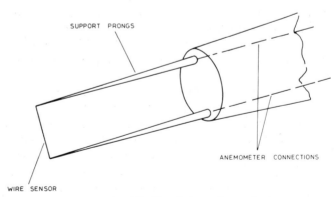

Figure 25.36   Simple hot-wire probe

to do and allows a simpler form of calibration (Ower and Pankhurst, 1977). There are various forms of the more robust shielded hot-wire systems but these will not be described here. In one example (Simmons, 1949) the hot wire is enclosed in a protective tube and formed into a thermocouple to compensate for the much reduced sensitivity. Another way to overcome the fragility of the unshielded hot wire is illustrated by the hot-film anemometer. Here an electrically conducting thin film is deposited on a sturdy insulating support (platinum on glass is a typical arrangement). The device is unsuitable for turbulence measurements but is adequate for mean flows and can be used in impure liquids and even slurries.

The simple unshielded hot wire is a difficult instrument to use and its main application was in turbulence research where it has now been superseded to a considerable extent by the laser Doppler velocimeter. However, the various more robust forms can be useful in some cases where, with care, an accuracy of better than ±1 per cent on velocity can be obtained, but they are mainly used as low-cost lower-accuracy velocity indicators. An appropriate choice of instrument will cover gas velocities from 0.1 to 500 m s^{-1} and a temperature range of up to 1000 °C. Hot films may measure liquid velocities from 0.01 to 25 m s^{-1}.

25.5.2 Vane Anemometers and Currentmeters

These meters perform essentially the same tasks in gas and liquid respectively. Vane anemometers are mainly used for measuring air speeds in large ducts or ventilating shafts, while propeller-type currentmeters are widely used for large-scale water flow measurements.

These instruments operate in a similar fashion to the turbine meter. The vane anemometer generally has an eight-bladed rotor of the form shown in Figure 25.37 enclosed in a close-fitting cylindrical sleeve. In early versions the motion of the spindle was transmitted to a pointer by mechanical means, but in modern versions there is generally some form of electromagnetic pick-up as in turbine meters.

A typical currentmeter is illustrated in Figure 25.38. The propellor is mounted on an axle linked, in turn, to a support member containing the pick-up device.

Vane anemometers can be obtained in sizes ranging from 0.02 to 0.4 m diameter to measure air speeds in the range 0.1–100 m s^{-1}. A typical speed range for a single instrument is 20:1 generally read on three or four analog scales. Accuracies of better than ±2 per cent may be obtained after calibration, but bearing wear can affect the calibration and the instruments are generally used as flow indicators rather than accurate devices. They are normally used in air flows at conditions close to ambient. Details of their use can be obtained in Ower and Pankhurst (1977).

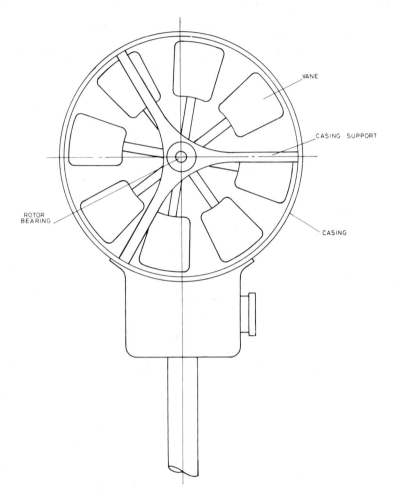

Figure 25.37   Vane anemometer

Currentmeters are obtainable in sizes from 0.02 to 0.125 m and are often mounted in banks in large conduits or open channels. A typical range is 0.2–5 m s^{-1} with an accuracy of ±2 per cent. A guide to the use of currentmeters is given in the standard test codes (IEC, 1963).

25.5.3 Pitot Tubes

The simple pitot tube and the pitot-static tube are extensively used for the measurement of gas speed and to a lesser extent the measurement of a point *velocity* in liquids. Much research has been carried out into their design so

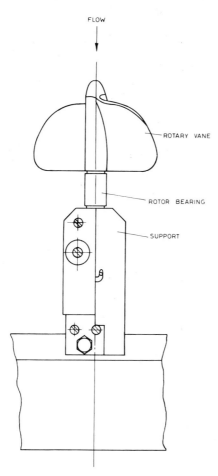

FLOW

ROTARY VANE

ROTOR BEARING

SUPPORT

Figure 25.38   Currentmeter

that a carefully made tube can provide an accuracy within ±1 per cent without calibration.

The pitot tube exemplifies the application of the familiar Bernoulli's theorem. This states that

$$\tfrac{1}{2}\rho v^2 + p = C \tag{25.53}$$

where ρ si the fluid density, v is the fluid velocity at a point, p is the static pressure at the point, and C is a constant, with the simplifying assumptions that the flow is incompressible and that no significant change in altitude is undergone. When the fluid is brought completely to rest at a point this is known as a *stagnation point,* and the pressure there is the *stagnation pressure*, p_0. From equation (25.53)

$$p_0 = p + \tfrac{1}{2}\rho v^2 \qquad\qquad (25.54)$$

and p_0 is equal to the total pressure at the point. Thus if the flow is brought completely to rest by an upstream-facing pitot tube linked to a manometer, an estimate can be made of the velocity provided that the static pressure p can also be measured.

A pitot tube can be any hollow tubing with a small orifice facing into the flow so that a stagnation point is formed and its pressure can be measured (see Figure 25.39). In practice the construction generally follows similar lines to modern pitot-static tubes with an ellipsoidal or hemispherical nose. The corresponding measurement of static pressure is more difficult as there must be no component of flow into the tapping hole nor must the local velocity pattern be altered by its presence. The static pressure is most commonly measured at the wall, but care must be taken with hole size and position and pipe roughness. Details of the use of pitot tubes are given in BSI (1973).

A convenient means of measuring static pressure is to incorporate measurement orifices in the body of the pitot tube itself. Early versions suffered from Reynolds number dependency. Owen and Pankhurst, (1977) give an interesting account of these problems and of the development of the now standard NPL *modified ellipsoidal* pitot-static tube whose form is shown in Figure 25.40. When the total head and static pressure leads of a pitot-static

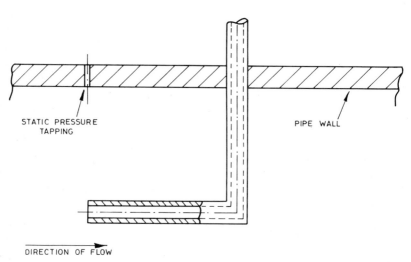

Figure 25.39   Simple total-head pitot tube

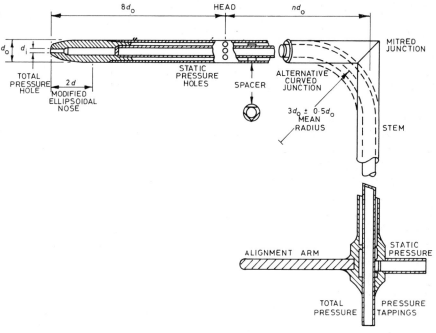

Figure 25.40   NPL standard *modified ellipsoidal* pitot-static tube

tube are connected across a manometer to give a differential pressure Δp, then

$$\Delta p = K\tfrac{1}{2}\rho v^2 \tag{25.55}$$

where K is a calibration factor. For an ideal tube K will be identically 1 (unity) and for the modified ellipsoidal NPL tube it is within ± 0.5 per cent of 1.000 over a wide Reynolds number range. As the tube behaviour varies slightly with scale, minimum and maximum diameters of tube are specified in the standard (BSI, 1973).

Pitot tubes (taken here to include pitot-static tubes) are inexpensive and available off-the-shelf in a wide range of sizes. The standard NPL form, used with care, will give accuracies of better than ± 1 per cent. Probably the main disadvantage is the difficulty of measuring low air speeds. Below 5 m s^{-1} the calibration factor K increases rapidly, and while correction factors may be applied the measurement of low pressures presents an additional problem. Above 60 m s^{-1}, in air, compressibility becomes a factor though corrections are straightforward. In liquids perhaps the main problem is breakage of the probe. In both fluids response time is a problem with smaller tubes because of the small-diameter internal ducts. Extremes of temperature and pressure and hostile fluids may be measured with the obvious limitations.

25.5.4 Laser Velocimeter

The *laser velocimeter* or LDV was first proposed in 1964 (Yeh and Cummins, 1964) and its development received considerable impetus from the NASA space programme where it was hoped to use the device to examine the complex three-dimensional flow at the jet of a Saturn V rocket at take-off. During the 1960s optical and electronic developments tended to remain in the guidelines of the NASA work and the first commercial instruments were of limited applicability because of this. After 1970 new optical and electronic techniques were evolved which combined to give a more practical instrument. However the LDV still requires an operator with a considerable amount of skill and has a high capital cost so that its sales have been confined almost exclusively to research departments in universities and large laboratories. In these respects it is comparable to the hot-wire anemometer. The advent of microprocessor technology may lead to a third generation of LDVs which are easier to use and relatively less expensive, but the cost of a system is likely to remain larger by a factor of ten than the cost of the common full flowmeters such as turbine and vortex.

There are two alternative explanations of LDV operation which appear very dissimilar but are in fact equivalent. Unless very special precautions have been taken, all fluids contain small impurity particles, such as specks of dust, algae, or air bubbles. The laser Doppler velocimeter makes use of contaminants about 1 μm or less in size which are, therefore, small enough to follow closely the flow pattern. The light scattered by these particles undergoes a slight frequency shift according to their velocity. This is detected, and is a measure of the flow velocity.

The velocity of light, c, is given by the product of wavelength and frequency

$$c = \lambda \nu \qquad (25.56)$$

where λ is the wavelength and ν the frequency. Consider a particle moving with velocity v. The number of wavefronts of light incident upon it per second is ν_p, given by

$$(c - v \cdot k_0) = \lambda_0 \nu_p \qquad (25.57)$$

where k_0 is a unit vector in the direction of the incident wave and $v \cdot k_0$ represents the scalar product of this and the particle velocity v ($v \cdot k_0 = |v| \cos \theta$ where θ is the angle between v and k_0). The quantity ν_p is also the number of wavefronts scattered per second by the particle. However, since the particle is moving, the apparent distance between scattered wavefronts (and hence the apparent frequency) to a stationary observer will be the frequency incident on the moving particle, modified by a factor $[(c - v \cdot k_s)/c]$. Thus the observer will see a *scattered frequency* v_s given by

$$v_s = v_p \frac{c}{c - v \cdot k_s} = \frac{(c - v \cdot k_0)}{\lambda_0} \frac{c}{(c - v \cdot k_s)} \tag{25.58}$$

Thus the Doppler frequency shift Δv is given by

$$\Delta v = v_s - v_0 = \frac{c}{\lambda_0} \left(\frac{c - v \cdot k_0}{c - v \cdot k_s} - 1 \right)$$

$$= \frac{v \cdot (k_s - k_0)}{\lambda_0 \left(1 - \frac{v \cdot k_s}{c} \right)} \tag{25.59}$$

If the particle velocity v is small compared with c,

$$\Delta v = \frac{v \cdot (k_s - k_0)}{\lambda_0} \tag{25.60}$$

Consider now a moving particle scattering light at the intersection of two beams, as shown in Figure 25.41. An observer viewing along the direction k_s will detect a frequency shift $\Delta v_1 = [(v/\lambda_0)(k_s - k_{0,1})]$ in the light scattered from beam 1, and $\Delta v_2 = [(v/\lambda_0)(k_s - k_{0,2})]$ in the light scattered from beam 2. The difference frequency is

$$\Delta v_2 - \Delta v_1 = \frac{v}{\lambda_0} \cdot (k_{0,1} - k_{0,2}) \tag{25.61}$$

which is independent of k_s. Thus the two-beam system will give a component of this frequency when viewed from any direction. In the symmetrical system shown in Figure 25.42 the difference frequency will be

$$\Delta v = (\Delta v_2 - \Delta v_1) = 2 \frac{v \sin \frac{1}{2}\theta}{\lambda_0} \tag{25.62}$$

This will be superimposed on a background spectrum of frequencies which

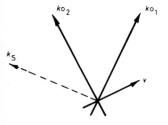

Figure 25.41   Incident and scattered beams – general case

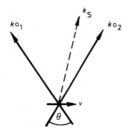

Figure 25.42   Incident and scattered beams – symmetrical case

depends on the properties of the detector and the scattering properties of the particles, and on the range of angles over which the scattered light is collected.

When a beam of light passes obliquely from one medium to another of different refractive index, its direction is altered and its velocity and wavelength also change. Thus the Doppler frequency is different, for example, if the beams cross in water.

If θ and θ' are the angles of intersection of two beams in air and a fluid of refractive index n respectively, then $n = \sin \frac{1}{2}\theta / \sin \frac{1}{2}\theta'$. If two beams intersect in a fluid of refractive index n, but the angle of intersection is measured in air:

$$\Delta \nu = \frac{2\nu (\sin \frac{1}{2}\theta / n)}{\lambda_0 / n} = \frac{2\nu \sin \frac{1}{2}\theta}{\lambda_0} \tag{25.63}$$

If θ' is measured in the fluid

$$\Delta \nu = \frac{2\nu \sin \frac{1}{2}\theta'}{\lambda_0 / n}$$

$$= \frac{2n\nu \sin \frac{1}{2}\theta'}{\lambda_0} \tag{25.64}$$

However, in the optical system described here it is *not necessary* to know either the explicit value of the *angle of intersection* or the *refractive index* of the fluid concerned. The Doppler frequency depends both on the *wavelength of light* used and on the *angle of intersection* of the beams. Fortunately, the change of angle with refractive index exactly compensates for the change of wavelength with refractive index, so that, if P is the separation of two parallel beams and l is the focal length of a lens bringing them to intersection, the Doppler frequency is given by

$$\Delta \nu = \frac{\nu P}{\lambda_0 \left(l^2 + \dfrac{P^2}{4} \right)^{1/2}} \tag{25.65}$$

The same formulae for the Doppler frequency may be derived by an alternative approach based on the assumption of the existence of fringes at the intersection of the beams. Where the path lengths travelled by the two beams are equal, or differ by a whole number of wavelengths, the intensities of the beams will add, while the beams combine destructively where the path lengths travelled by the two beams differ by an odd number of half wavelengths. Thus the intersection region will contain a series of light and dark bands or fringes.

If the three coordinate system (x, y), (x', y'), (x'', y'') shown in Figure 25.43 is considered, then the path difference at the point $P(x, y)$ is $y' - x''$. In a medium of refractive index n, this optical path difference will be $n(y' - x'')$. Using the transformation

$$y' = x \cos \tfrac{1}{2}\theta + y \sin \tfrac{1}{2}\theta$$
$$x'' = x \cos \tfrac{1}{2}\theta - y \sin \tfrac{1}{2}\theta$$

the path difference is thus $2y \sin \tfrac{1}{2}\theta$. There will be a bright fringe wherever this value is a multiple of λ', where $\lambda' = \lambda_0/n$. Hence the fringe spacing b is given by

$$b = \frac{\lambda_0}{2n \sin \tfrac{1}{2}\theta} \tag{25.66}$$

Thus a stationary set of fringes with spacing b can be imagined to exist in the fluid. If a small particle passes through this region, there will be a flash of light at each fringe, and, since the fringes form a fixed grid, the rate of flashing will be directly proportional to the particle velocity. The interval

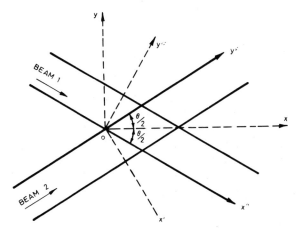

Figure 25.43 Coordinates at region of beam intersection

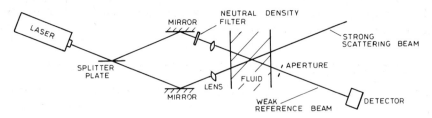

Figure 25.44   Goldstein–Kreid system

between flashes will be $t = (b/v)$ and so the frequency will be $\Delta\nu = (2vn \sin \frac{1}{2}\theta)/\lambda_0$ as before.

Two principal types of optical system have evolved. One, known as the *reference beam* system, is based on the early NASA work. It uses a strong scattering beam and a weak reference beam impinging directly on the photodetector, as shown in Figure 25.44 (Goldstein and Kreid, 1967). Alignment is critical and it is susceptible to vibration. It is mainly used now for applications where there is a heavy concentration of particles in the flow. The other type, known as the *fringe* or *dual-scatter* method, employs beams of equal strength. A typical arrangement is shown in Figure 25.45. (Blake and Jespersen, 1972). It is simple to align, stable, and preferable to the reference beam system in most situations. Both methods may be extended to two- and three-dimensional measurements, to turbulence studies, and to backscattering work.

The electronics system required by an LDV is complex and will not be discussed in detail here. The earlier concept of a *frequency tracker* is preferable in some cases, but for most applications the current tendency is to use *timer-counter*-based systems.

The LDV has many good features. It gives an absolute measure of velocity, independent of fluid, at an effective point; it is truly linear over a range of typically $10^5 : 1$; accuracy levels of ± 0.1 per cent may be obtained without calibration; and velocity readings are almost instantaneous so that it can be used for turbulence studies. As well as high cost, it has other

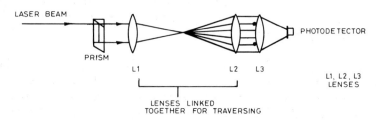

Figure 25.45   NEL optical system

disadvantages in requiring a pipe with windows and a relatively clear gas or liquid. Also, a skilled operator is required.

25.6 FLOW MEASUREMENT REFERENCE STANDARDS

Before concluding this chapter it is important to review the various standard test facilities which are used to evaluate the performances of flowmeters. Clearly it is only possible to make a very brief mention of these facilities and attention here has been restricted to the main *gravimetric* and *volumetric* methods used in primary standard test facilities available for liquid and gas measurements. Further information on the main types of standard reference test rigs for flow measurement may be found in Brain (1978) and NEL (1976) where primary and secondary calibration methods are considered. A primary standard method is one in which reference flow is determined by measurements of the basic dimensions of the SI units mass, length, temperature, and time. A secondary standard method is one in which reference flow is determined using a flowmeter which has been calibrated by a primary method. Many flow calibration facilities exist throughout the world, examples being the NEL laboratory mentioned above; plus those of the Sira Institute, England, the National Bureau of Standards (NBS) USA, and the Kent–BBC facility in Australia.

25.6.1 Gravimetric Systems

Static weighing system for liquid flows

For liquid flows the gravimetric system may be arranged as shown in Figure 25.46. At the beginning of a test, liquid from the constant-head tank flows through the meter on test at a rate controlled by valve x. From this control valve the liquid flows to the diverter system where it is directed into the sump tank and then pumped back into the constant-head tank to complete the loop. When a steady flow is obtained in the loop the diverter system is operated and the flow is diverted into a previously weighed weighing tank for a measured period of time before being diverted back into the sump. During this diversion period readings are recorded at the flowmeter on test, and at the end of a test these readings can be compared with the reference flow rate which is obtained by dividing the mass of liquid collected by the diversion time.

Clearly, with this type of system it is most important that flow should remain steady during the diversion period. Operation of the diverter should be uniform in either direction and preferably swift in relation to the time of diversions since, if the time of diverter operation is not taken into account, it can cause errors particularly when the diversion time is small.

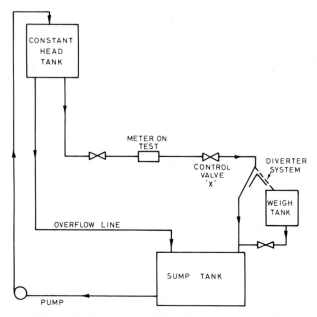

Figure 25.46   Layout of liquid gravimetric facility

Flow rates of up to approximately $1.5 \, \text{m}^3 \, \text{s}^{-1}$, at the pressures of up to 5 bar, can be measured on this type of gravimetric rig. Reference flow rates can generally be measured to within an estimated uncertainty of better than ±0.2 per cent of flow rate.

Static weighing system for gas flows

A simple gravimetric gas flow test rig is outlined in Figure 25.47. In this system a steady flow of gas is passed through the test pipeline in which the meter to be calibrated is installed. By diverting the gas flow from the meter on test to a gas-collecting vessel over a measured period of time, and weighing the collection vessel before diversion and the collection vessel plus enclosed gas after diversion, the flow of gas may determined.

Due to compressibility effects, problems encountered in maintaining a closely controlled flow rate are considerably greater in gravimetric gas flow systems than in liquid flow gravimetric systems. In gas flow systems it is vital that the critical flow nozzle section should be designed such that nozzle performance is not affected by changes in downstream pressure throughout a test run. Also, since gases are very much less dense than liquids, further difficulties are encountered in accurately weighing the diverted mass and even very small leaks in the system can prove to be serious sources of error.

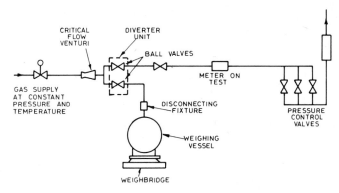

Figure 25.47 Gravimetric system for gases

Collins and Selby (1965; 1966) have described a system which can be used to measure dry air flows, at pressures and temperatures close to ambient, in the range 10^{-6} to $2.5 \times 10^{-1} \, \text{kg s}^{-1}$, and at the National Engineering Laboratory (NEL) in the United Kingdom a gravimetric system (Brain and Reid, 1980) has been commissioned to measure flow rates of up to $5 \, \text{kg s}^{-1}$ at pressures up to 50 bar. In these systems the mean reference flows can be determined to within ± 0.2 per cent of flow rate.

25.6.2 Volumetric Systems

Meter prover for liquid flows

Meter provers experience wide usage for measuring the volumetric flow rates of oils. While both *piston*-type and *ball*-type provers are used, only *ball* provers are considered here. A *ball*-type meter prover may be set up as shown in Figure 25.48 and operates as follows. Liquid is passed through the loop in the direction shown until a stable flow is obtained. The sphere, which is usually an inflatable rubber ball which fits closely inside the pipe, is then inserted into the flow by means of a specially designed valve. When the sphere passes the first detector a pulse is generated which switches on an electronic timing device and when it has swept through the calibrated volume the second detector pulse switches the timer off. The sphere is then diverted back towards the valve where it is collected ready for the next insertion.

It is clear that by knowing the time of traverse of the sphere as well as the value of the calibrated volume the volumetric flow rate may be obtained and compared with the reading indicated by the test meter during the calibration run.

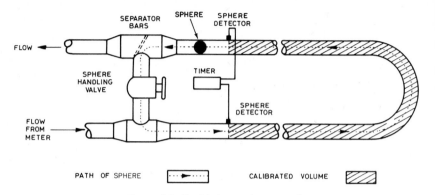

Figure 25.48   Volumetric prover loop

In meter provers it is important that the valve switching operation should be carried out with deliberate and uniform speed and sufficient length of pipe should be allowed between the valve and the calibrated section so that the switching operation is completed some time before the sphere enters the calibrated section.

Errors can be caused by leakage past the sphere, valve leakage, air entrainment, and incorrect compensation for pressure and temperature effects on the prover volume. When this method is used, liquid flow rates of up to approximately $1 \text{ m}^3 \text{ s}^{-1}$ can be measured to better than ± 0.1 per cent of flow rate.

Bell prover system for gases

In this method of gas flow measurement a cylinder closed at one end is inverted over a water bath as shown in Figure 25.49. The cylinder is counterbalanced by weights on a line over a near frictionless pulley such that a predetermined pressure is created in the trapped volume. A pipe passing through the water communicates with the trapped volume and, as the cylinder is lowered, leads the displaced gas to the meter on test. By timing the fall of the cylinder, and knowing the volume–length relationship for the cylinder, the volume flow of gas through the meter may be determined and compared with the meter reading.

In order to minimize expansion or contraction of the gas, the water, gas, and air temperatures should not differ by more than $1 \,^{\circ}\text{C}$. Errors can also arise from leaks, incorrect compensation for change in buoyancy on the counterbalance pulley, and the fact that the gas is not fully saturated. At present, for flows up to some $10^{-2} \text{ St m}^3 \text{ s}^{-1}$, this method, which is usually used to meter air flows at conditions close to ambient, can be used to

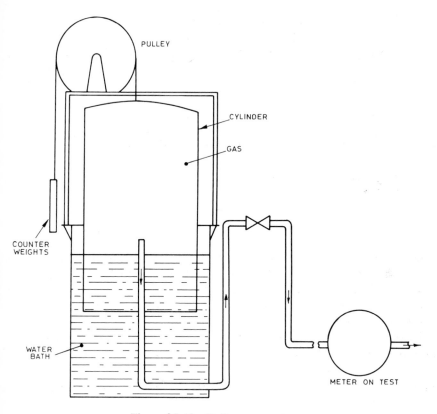

Figure 25.49   Bell prover system

measure flows to within ±0.25 per cent if strict precautions are taken to minimize the errors mentioned above.

25.6.3   Points to be Noted When Calibrating Flowmeters

The calibration method to be used, as well as the care with which the meter calibration is to be undertaken, depends largely on the accuracy of calibration required. Devices used for quantity measurement, which are often rate meters in which the output is integrated with time, usually need to be calibrated with greater accuracy than devices used for rate measurement alone. The main reason for this is that quantity meters are frequently used when *buying* or *selling* fluids. For rate meters repeatability or linearity of output can, in certain cases, be more important than absolute accuracy, although in the majority of cases absolute accuracy is still the most important performance requirement.

The conditions of use of the flowmeter must also be noted, and if the meter is calibrated in a test facility in a laboratory care must be taken to ensure that the laboratory calibration remains valid when the meter is used in the field. For instance, variations in the velocity profile upstream of the meter, due to pipework installation effects, can have a marked effect on the performance of most meters and serious errors would result if, say, an orifice plate, calibrated with 100 diameters of straight pipe upstream of it in a laboratory test, was installed in an industrial flow line just downstream of a bend. To avoid possible installation effects laboratory calibration should, therefore, be carried out with the flowmeters installed in their immediate pipework configurations.

In most flow laboratory calibrations air or water is used as the test fluid, and if the meter on test is to be used to measure a different fluid then, obviously, the effects on the meter of using this different fluid should be considered. Reynolds number similarity holds for a large range of flowmeters, but this similarity does not hold for several meters such as variable area meters and certain thermal flowmeters. To obtain the best accuracy, these devices should be calibrated in the fluids in which they are to be used.

The pressure and temperature at which the flowmeter is to operate are other important factors which must be taken into account, since clearly variation in pressure and temperature can significantly affect fluid density, particularly if the fluid is a gas, as well as the dimensions of, say, the throat of a pressure difference device or the clearance volume of a positive displacement meter. It should also be noted that unless the density is measured directly the composition of the fluid must also be considered. For gases in many cases humidity can prove to be a problem since significant errors can result if it is ignored; while for liquids, air entrainment can seriously affect flowmeter readings.

ACKNOWLEDGEMENTS

The figures in this chapter have been taken from NEL Reports, Lectures, and Papers, and are reproduced by permission of The Director, NEL.

REFERENCES

ARL (1971). 'Evaluation of the Westinghouse leading edge flow measurement system', *Alden Research Laboratories USA Test Report* no. ARL-M71-83, Worcester Polytechnic Institute, Holden, Mass.

Arnberg, B. T. (1962). 'Review of critical flowmeters for gas flow mesurement', *J. Bas. Engng*, **84**, 477-57.

ASME (1971). 'Fluid meters, their theory and application', The American Society of Mechanical Engineers, New York.

Benedict, R. P. (1969). *Fundamentals of Temperature, Pressure and Flow Measurements*, Wiley, New York.

Benson, J. M. (1971). 'Survey of thermal devices for measuring flow', in *Symp. on Flow – its Measurement and Control in Science and Industry*, 10–14 May, Paper 2-1-213, Instrument Society of America (ISA), Pittsburgh Pa.

Blake, K. A. and Jespersen, K. I. (1972). 'The NEL laser velocimeter' *NEL Report* no. 510, National Engineering Laboratory, East Kilbride, Glasgow.

Bradshaw, P. (1971). *An Introduction to Turbulence and its Measurement*, Pergamon, Oxford.

Brain, T. J. S. (1969). 'Mass flow measurement methods', *Metron*, **1**(1), 1–6.

Brain, T. J. S. (1978). 'Reference standards for gas flow measurement', *Measurement and Control*, **11**, 283–8.

Brain, T. J. S. and MacDonald, L. M. (1977). 'Methods of flow measurement in closed conduit systems', *Technical Data on Fuel*, pp. 112–26, Scottish Academic Press, Edinburgh.

Brain, T. J. S. and Reid, J. (1978). 'Primary calibrations of critical flow venturis in high-pressure gas', *Flow Measurement of Fluids*, pp. 55–64, North-Holland, Amsterdam.

Brain, T. J. S. and Reid, J. (1980). 'Primary calibrations of critical flow venturi nozzles in high-pressure gas', *NEL Report* no. 666, National Engineering Laboratory, East Kilbride, Glasgow.

Brand, D. and Ginsel, L. A. (1951). 'The mass flowmeter – a method for pulsating flow', *Instruments*, **24**, 331–5.

BSI (1964), BS 1042: Part 1: 1964. 'Methods for the measurement of fluid flow in pipes', *Part 1: Orifice Plates, Nozzles and Venturi Tubes*, British Standards Institution, London.

BSI (1973), BS 1042: Part 2A: 1973. 'Methods for the measurement of fluid flow in pipes', *Part 2: Pitot Tubes. Class A accuracy*. British Standards Institution, London.

Burgess, T. H. (1977). 'Flow measurement using vortex principles', *Proc. Symp. Application of Flow Measuring Techniques, Flow-Con '77*, Brighton, England, April, pp. 321–39, Institute of Measurement and Control, London.

Collins, W. T. and Selby, T. W. (1965). 'A gravimetric gas flow standard – Part 1: Design and construction', *Report* no. K-1632, May, Union Carbide Corporation, Oak Ridge Gaseous Diffusion Plant, Tenn.

Collins, W. T. and Selby, T. W. (1966). 'A gravimetric gas flow standard – Part II: Performance and evaluation', *Report* no. K-1632, April, Union Carbide Corporation, Oak Ridge Gaseous Diffusion Plant, Tenn.

Collis, D. C. and Williams, M. J. (1959). 'Two-dimensional convection from heated wires at low Reynolds numbers', *J. Fluid Mech.*, **6**, 357.

Decker, M. M. (1960). 'The gyroscopic mass flowmeter', *Engineers' Digest*, **1**, 92–3.

Dowden, R. R. (1972). *Fluid Flow Measurement – a Bibliography*, British Hydromechanics Research Association, Cranfield, Beds.

Fisher, S. G. and Spink, P. G. (1971). 'Ultrasonics as standard for volumetric flow measurement', in *AERE/NEL Symp. on Modern Developments in Flow Measurement*, Paper no. 13, September, Harwell.

Goldstein, R. J. and Kreid, D. K. (1967). 'Measurement of laminar flow development in a square duct using a laser Doppler flowmeter', *J. Appl. Mech.*, **34**, 813–18.

Greenspan, M. and Tschiegg, C. E. (1957). 'Sing-around' ultrasonic velocimeter for liquids', *Rev. Sci. Instrum.*, **28**, 897.

IEC (1963). 'International code for the field acceptance test of hydraulic turbines', *Publication* 41, Bureau Central de la Commission Electro-technique Internationale, IEC, Geneva.

IMEKO (1980). *A Working List of Books Published on Measurement Science and Technology in the Physical Sciences*, ed. Sydenham, P. H., Department of Applied Physics, Delft Technical University, Netherlands.

ISO (1979). ISI 5167. 'Measurements of fluid flow by means of orifice plates, nozzles and venturi tubes inserted in circular cross-section conduits running full', International Organisation for Standardisation, ISO, Geneva.

Jepson, P. and Bean, P. G. (1969). 'Effect of upstream velocity profiles on turbine flowmeter registration', *J. Mech. Engng Sci.*, **11**, 503–10.

Jespersen, K. I. (1973). 'A review of the use of ultrasonics in flow measurement', *NEL Report* no. 552, National Engineering Laboratory, East Kilbride, Glasgow.

Johnson, R. C. (1965). 'Real-gas effects in critical flow nozzles and tabulated thermodynamic properties', *NASA* TN 0-2565, US National Aeronautics and Space Adminstration, Washington, DC.

Katys, G. P. (1964). *Continuous Measurement of Unsteady Flow*, Pergamon, Oxford.

King, L. V. (1914). 'On the convection of heat from small cylinders in a stream of fluid', *Phil. Trans. R. Soc.* A, **214**, 373.

Kritz, J. (1955). 'An ultrasonic flowmeter for liquids', *Proc. Instrum. Soc. Am.*, **10**, no. 11, 1–2, Paper no. 55-16-3.

Li, Y. T. and Lee, S. Y. (1953). 'A fast response true mass-rate flowmeter', *Trans. Am. Soc. Mech. Engrs*, **75**, 835–61.

Linford, A. (1961). *Flow Measurement and Meters*, Spon, London.

Miller, R. W., De Carlo, J. P. and Cullen, J. T. (1977). 'A vortex flowmeter calibration results and experiences', in *Proc. Symp. Application of Flow Measuring Techniques, Flow-Con '77, Brighton, England*, April, pp. 341–372, Institute of Measurement and Control, London.

NEL (1976). 'Flow measurement facilities at NEL on 1 January 1976', *NEL Report* no. 608, National Engineering Laboratory, East Kilbride, Glasgow.

Newcombe, J. (1969). 'Secondary instruments for flowmeters', in *Seminar on Gas Measurement*, Paper 1, 36, Institution of UK Gas Engineers, University of Salford.

Norton, H. N. (1969). *Handbook of Transducers for Electronic Measuring Systems*, Prentice-Hall, Englewood Cliffs, NJ.

Orlando, V. A. and Jennings, F. B. (1954). 'The momentum principle measures true mass rate', *Trans. Am. Soc. Mech. Engrs*, **76**, 961–5.

Ower, E. and Pankhurst, R. C. (1977). *The Measurement of Air Flow*, Pergamon, Oxford.

Scanes, P. (1962). 'The mass flowmeter', in *Symp. on Flow Measurement in Closed Conduits*, National Engineering Laboratory, 1960, Paper G-1, HMSO, London.

Scarborough, J. B. (1966). *Numerical Mathematical Analysis*, p. 152, Johns Hopkins University Press, Baltimore, Md.

Scott, R. W. W. (1971). 'Turbine flowmeters', in *Purchasing Directory for Process Industries* 1971, Morgan Grampian Books, West Wickham, Kent.

Scott, R. W. W. (1975). 'A practical assessment of the performance of electromagnetic flowmeters', *Conf. on Fluid Flow Measurements in the mid* 1970s, 8–10 April 1975, National Engineering Laboratory. East Kilbride, Glasgow.

Shercliff, J. A. (1962). *The Theory of Electromagnetic Flow Measurement*, Cambridge University Press, Cambridge.

Simmons, L. F. G. (1949). 'A shielded hot-wire anemometer for low speeds', *J. Sci. Instrum.*, **26**, 407.

SIRA (1978). 'Fluid mechanical measurements', *Information Review*, **IR12**, Sira Institute, Chislehurst.

Stratford, B. S. (1964). 'The calculation of the discharge coefficient of profiled choked nozzles and the optimum profile for absolute air flow measurement', *J. R. Aeronaut. Soc.*, **68**, 237–45.

Strouhal, V. (1978). 'Uber eine besondere Art der Tonaregung', *Ann. Phys. Chem.*, **5**, 216–51.

Sydenham, P. H. (1980). *Transducers in Measurement and Control*, Adam Hilger, Bristol and ISA, Research Triangle Park.

Tullis, J. P. and Smith, J. (1979). 'Coriolis mass flowmeter', Paper presented to *NEL Reynolds Building Silver Jubilee Conference, November* 1979, National Engineering Laboratory, East Kilbride, Glasgow.

VDI (1969). DIN 1952. 'Flow measurement with standardised nozzles, orifice plates and venturi nozzles', Verein Deutsch. Ing. (VDI), Flow Measurement Committee, May.

Walker, R. K. (1966). 'A historical review and discussion on the design features of positive displacement gasmeters', *Instrum. Control Syst.*, **39**, Part 1, 141.

Yeh, Y. and Cummins, H. Z. (1964). 'Localised fluid flow measurement with a He-Ne laser spectrometer', *Appl. Phys. Lett.*, **4**, 176–8.

Handbook of Measurement Science, Volume 2
Edited by P. H. Sydenham
© 1983 John Wiley & Sons Ltd

Chapter

26 P. H. SYDENHAM

Transducer Practice: Thermal

Editorial introduction

Temperature, and related variables, commonly need to be measured either as the prime interest of or as an influence variable acting on a process.

As with all measurements temperature measurement also requires a good understanding of the nature of the temperature measurand so that sensors are designed to provide adequate static and dynamic response, and measurement precision and accuracy.

The fundamental units and scales associated with temperature are distinctly different in physical manifestation to those of the previous two chapters' subjects, i.e. length and flow.

Here too, sensor design invariably can rest upon well defined mathematical modelling of the physical principle involved. Proper appreciation of that theory enables efficient design of instruments and correct interpretation of results.

This chapter concentrates largely, but not solely, upon electrical output methods, introducing, in the short space available, the most commonly used methods.

26.1 NATURE OF THERMAL MEASURANDS

26.1.1 Historical Understanding

Unlike the obvious nature of the length measurand thermal quantities posed severe philosophical problems to those interested in studying and making use of these entities. There can be little doubt that early peoples could appreciate the physiological manifestations of heat flow and temperature difference. Despite much interest, especially from the 16th century AD onward, it was not until the late 19th century that the nature of heat was finally resolved as having the attributes of a form of energy.

Early experimenters were able to devise instruments that measured the temperature variable (called thermometers); many early ones—the thermo-scopes—really measured some combination of barometric pressure and temperature.

Thus in the area of measuring thermal attributes man had created units, scales, and standards well before he had an adequate appreciation of the nature of the measurand. (This is not an uncommon situation!)

Once the nature of heat was understood great advances in the practices of its measurement were possible. It is also of importance to note that such understanding occurred concurrently with the emergence of electrical sens-ing methods. A paper by E. Mach 'Die Principien der Wärmelehre' pub-lished in 1896 gives insight into the state of thinking, even at that late period (Ellis, 1968).

Much has been written on the historical aspects of thermal parameter measurement; for a guide to that literature see Sydenham (1979).

Although satisfactory definition of such parameters as temperature is now possible, the definition is not explicable in the same simple and clearly evident way that we experience with length measurement.

In essence temperature is understood, at the conceptual level, in one of two ways. The first is in terms of thermodynamics, a microscopic concept; the second with gas-molecule physics, a microscopic viewpoint (Woolsey, 1974). These bases are expanded in Section 26.2.1.

26.1.2   Temperature and Related Measurands

Good measurement practice rests on proper understanding of the thermal entities and thermal properties of materials. In relation to the following discussion and to Section 26.2.1, Table 26.1 lists often met measurands and their units.

Temperature

This is the measure of the state of *hotness*, or *coldness*, of a body relative to some standard state. Some authors define it as a *degree of heat*, or *level of heat*. Combination of two quantities of a fluid at the same temperature (without heat loss) results in the same temperature, not the sum of their values. In this way the temperature measurand is an *intensive* variable. Temperature should not be confused with *heat*.

Heat

The *Oxford Dictionary* defines this as the *quality of being hot* thereby avoiding defining what heat is in a physical, objective, manner.

Heat is more usefully defined as the quantity of heat energy that exists

Table 26.1 Thermal measurands and their SI units

Concept group	Measurand	SI unit	Name (if any)	Other permissible unit (if any)
Temperature	thermodynamic temperature	K	kelvin	
	Celsius temperature	—	—	°C (degree Celsius)
	temperature interval	K	—	°C (optional, provided not mixed with K)
Heat	quantity of heat	J	joule	
	heat flow rate	W	watt	
	heat flux density	$W\,m^{-2}$		
Performance parameters of system	linear, or volume thermal expansion coefficient	dimensionless number, K^{-1}		
	many other thermal coefficients exist, e.g. of electrical resistance, of elasticity, etc.			
	thermal conductivity	$W\,m^{-1}\,K^{-1}$		
	thermal conductance (heat transfer coefficient)			
	on area basis	$W\,m^{-2}\,K^{-1}$		
	on defined system basis	$W\,K^{-1}$		$K\,W^{-1}$ also used
	Heat capacity			
	total basis	$J\,K^{-1}$		
	volume basis	$J\,m^{-3}\,K^{-1}$		

within a given physical boundary. Thus the two quantities of fluid mentioned above, when combined, will hold the sum of the heat energies, even though the temperature has not changed. Often the single term *heat* is used to describe the quantity of energy involved.

Two quantities of fluid at different temperatures will again, when combined without losses occurring, provide a heat energy quantity equal to the sum of each but in this case the temperature will change to a proportionate value between the two temperatures. These simple examples suffice to exemplify that temperature and heat content are quite different measurands. Furthermore, *heat* should not be confused with *hot*, a temperature attribute. In popular language, however, they are often used synonymously—the heat of the day often implying the hottest time (but in reality describing the time when the heat energy content of the environment is at a maximum).

Heat flow

Thermal energy flows from the hotter to the colder state where two different temperature regions occur. The transfer of heat energy is loosely termed the *heat flow*. *Heat flow rate* is the amount of heat energy flowing in a unit time.

Heat flux density

The amount of heat flowing through a defined area is the *intensity of heat flow rate* or *heat flux density*.

The above terms relate to the energy source, heat. The following terms relate to material performance parameters that are dependent upon thermal state, or that resist heat flow or store heat.

Thermal expansion

The extent to which a piece of material changes length with temperature is called the linear expansion coefficient. This is expressed as a relative length change.

A volume change thermal expansion coefficient is sometimes also of value in quantifying a system change due to temperature change.

Thermal influence on systems is probably the most-met, unwanted, system-perturbing parameter. As well as length change, temperature state can also effect mechanical elasticity, magnetic properties, electrical resistivity, thermoelectric generation magnitude, and many more system parameters (Sydenham, 1980a, Chap. 2). In many thermosensors this can be a severe source of error, for example a traditional mercury-in-glass thermometer suffers some degree of temporary zero-point reduction after a large cyclic temperature excursion because the glass requires considerable time to recover its pre-transient shape.

Thermal conductivity

This describes the relative ability of a substance to conduct heat. The reciprocal, less used, statement is *thermal resistivity*.

Heat transfer coefficient

A material's heat transfer capability through a unit area is termed the *thermal conductance*. The *heat transfer coefficient* quantifies this in a normalized manner.

Heat capacity

The quantity of heat that a system can store is its *heat capacity*. It can be expressed in a total or on a per unit volume basis.

Specific heat capacity

This gives a measure of the ability of a material to store heat energy in a given mass, allowance being made for its temperature.

These examples will illustrate the subtle differences in the nature of thermal measurands. In many cases what might appear to be a single thermal measurement problem may require measurement of other thermal quantities.

26.1.3 Heat Energy Transfer

Application of hardware to measure thermal parameters correctly in both the static and dynamic states requires good grounding in the mechanisms of heat transfer. Unless adequate design precautions are taken measurements will be inaccurate due to the sensor disturbing the system state or because what is actually sensed is not a true representation, particularly during transient conditions, of what the sensor is supposed to map. These commonly met sensing difficulties are particularly noticeable in thermal measurements.

Heat is transferred from a hot body to a cooler one by conduction, convection, and radiation mechanisms. Adequately simple descriptions now follow that will prepare the reader for application of thermal sensors.

Conduction

This mechanism, which relates to substances that cannot move relatively at a macroscopic level, is best understood on the basis of the *kinetic energy principle* which states that particles impart energy to those of lower energy state. Energy exchange continues in an attempt to balance temperatures until *thermal equilibrium* is established. Steady heat flow (the case of a sensor body acting to shunt heat continuously from a source), by conduction q_d perpendicular to the surfaces of two infinite, parallel, sink and source surfaces, follows the law

$$q_d = kA \frac{T_1 - T_2}{x}$$

where k is the thermal conductivity of the medium between the surfaces, A the area of transfer, T_1 the source surface temperature, T_2 the sink surface temperature, and x the distance between source and sink surfaces.

For cases where two surfaces at temperatures T_1, T_2 are brought into contact (a probe stabilizing to a changed temperature, for instance) the heat flow is not steady. The material's thermal conductivity and the temperature difference existing decide the instantaneous heat flow rate, the rate of temperature equalization reducing exponentially with time. In theory two bodies starting at different temperatures, that are coming into equilibrium, will never reach thermal equilibrium. In practice 3 to 5 time constants of this first-order dynamic system can be regarded as the time for adequately close equality to be reached—see Chapter 17, Coughanowr and Koppel (1965), and Jakob and Hawkins (1957) for expanded theoretical considerations.

Thermal conductivity can be regarded as analogous to electrical conductivity and temperature difference to electric potential (voltage). Thus thermal conductivity problems can be modelled and mentally viewed using simple series electrical circuit analogies. The steady voltage source case

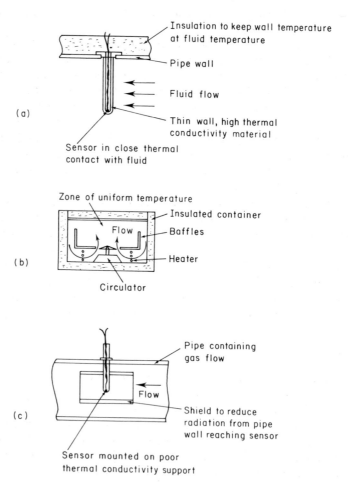

Figure 26.1   Examples of good thermometer mounting practice: (a) Design of a thermometer well to provide best transient response where thermal conductivity dictates design. (b) Convective sensing errors can sometimes be reduced by using forced circulation to ensure uniform temperature distribution within a fluid system. (c) Shields applied to reduce radiation background errors

represents steady heat flow. The changing heat source situation requires a time variant voltage source analogy. Furthermore, heat modelling often requires a distributed, not a point source of energy. Approximations can often be made to provide adequate understanding of how a temperature probe will respond.

A fact often overlooked is that heat transfer across a boundary between two different materials (even an impurity film between two pure identical materials) can provide thermal resistance that is often far greater than that of the materials themselves.

As an example consider air flowing on the outside of a steel thermometer tube at standard ambient conditions. For this case consider the tube to have 100 arbitrary units of thermal conductivity between the inner surface and the outer. The air–steel outside interface will have some 20 units and the inside steel interface to the internally contained thermometer probe itself only 0.02 units. (Values vary widely depending on actual application.)

A first-order response design (the simplest dynamic mode) of the transient response would regard, in this case, the probe's thermal resistance as the significant dissipative source that should be connected with the thermal storage capacity of the whole probe.

Good thermometer probe dynamic design thus requires careful interfacing to reduce thermal resistance and mass. Figure 26.1a portrays the elements of good probe design where thermal conductivity parameters are dominant.

An inadequately fast-response probe will introduce significant unwanted amplitude and phase-shift errors and even oscillatory output. See Chapter 17, Coughanowr and Koppel (1965) or Doebelin (1975) for explanation of the magnitude of such errors for a first-order response thermometer bead responding to sinusoidal temperature excitation.

Convection

Probe design often also requires, or is predominantly concerned with, convective heat transfer. It occurs in substances that can flow. Heating of part of the substance reduces its density, which then allows it to be displaced by colder substance. Circulation occurs when the hotter substance moves into colder regions to rise back into hotter ones, thereby constantly representing changing source–sink surface heat transfer conditions by conduction between molecules. This is called *free convection*. It is a most complex process involving fluid dynamics mathematics. When the fluid is purposefully made to pass the heat exchange surface it is called *forced convection*.

The useful, if simplistic relationship,

$$q_v = hA(T_1 - T_2)$$

connects steady heat flow q_v between the fluid and the transfer plate to the *film coefficient of heat transfer h*, the area of surface A in contact with the moving medium, and the temperature difference between the surface at T_1 and the bulk of the moving medium at T_2, the surface assumed here to be hotter than the moving medium.

To complicate the thermal probe designer's task unexpected conditions can occur.

When the vertical thermal gradient reduces sufficiently convection ceases and a cold layer can then exist on top of, not below, the hot layer. Tall thin columns of air exhibit this phenomenon most markedly.

At very high relative speeds (beyond Mach 2) between a surface and the fluid, surface heating effects produce a *thermal barrier:* the boundary layer becomes an *energy separation* not a transfer interface.

The electrical analogy described for conduction also applies for convection, the two descriptive equations being of the same form (Jakob and Hawkins, 1957).

When attempting to maintain unchanging temperature distribution within, say, an instrument enclosure, convective currents are controlled by reducing the cell size of all available air spaces. This can be done using baffles, or by filling space with cellular material.

Figure 26.1b is an illustrative example of minimization of convective errors by using forced circulation in a temperature controlled water bath.

Radiation

By now, it must be clear that making correct temperature measurements is fraught with interface problems. However, the third mechanism of heat transfer, *radiation*, introduces still more variables that might need consideration.

Unlike conduction and convection which require material contact for energy transfer, radiation can occur across a vacuum. It transfers energy by virtue of the electromagnetic radiation mechanism.

The basic descriptive equation giving the rate of heat transfer q_r between an emitter and some point external to it is given by,

$$q_r = \varepsilon k A T^4$$

where ε is the surface emissivity factor, k the Stefan–Boltzmann constant, A the radiating area, and T the surface temperature (in kelvin).

Thus heat transfer is not linearly related to temperature of the surface: it also depends greatly on the *emissivity* (or *absorptivity* for the reversed situation of energy absorption by a plate) of the surface.

Emissivity is, except for metallic surfaces, generally independent of the wavelength and temperature of the radiation. In such cases the surface is called a *gray* radiator.

The perfect emitter/absorber is called a *black* radiator. For this the emissivity is 1.

Often emissivity is a function of the temperature. It is lack of adequate knowledge of this factor that presents problems for accurate assessment of heat flow by radiation and high precision measurements based on this principle (called *pyrometry*)—see Section 26.3.5.

The following list gives an appreciation of the wide variance of emissivity and unexpected values, 'blackness' not being as the eye views it.

Material	Typical ε	Material	Typical ε
Oxidized aluminium	0.11	Steel	0.80–0.95
Copper (not oxidized)	0.20–0.30	Brick (rough red)	
Brass	0.60	Porcelain (glazed)	0.91–0.94
Copper (oxidized)	0.60	Paints	
Iron	0.65–0.80	Lampblack	
Carbon	0.80	Ice	0.95–0.99
Paper	0.90	Water	
		Black body	1.00

In each case the value will vary with temperature and the uncertainty of the values can vary by as much as 30%. For a more complete listings see, for example, Plumb (1972) and the several works on infrared engineering and pyrometry mentioned in Sydenham (1980b).

The space between the emitter and a sensor will not interact with the energy transfer if it is a vacuum. Total radiation received at a point in such cases is decided by the polar radiation properties of the radiator and any flux gathering system used with the detector.

The existence of molecules in the intervening space, such as water vapour, carbon dioxide, fumes in general, protective instrument windows, lenses and the like, can greatly modify the received energy levels.

Thus a thermometer sensor will also 'see' its surroundings and its mountings as thermal radiation adds energy transfer to the sensor in an unwanted manner. Often (see Figure 26.1c) a probe is shielded with an enclosure to ensure that any such energy pick-up is minimized. In general the aim is to ensure that all surrounding surfaces that can radiate onto the probe have constant emissivity near unity and that they are near to the temperature of the ambient of the probe. This may require shields in which the inner surface is thermally well insulated from the outer to prevent the inner surface from re-radiating at the outer's temperature. Over long time periods steady background radiation will give rise to such re-radiation at some level even in well insulated systems.

In practice, then, the placement position, method of mounting, surface treatments, materials used, and actual application are all factors of thermometry system design that can be as exacting in need for careful consideration as are the sensors. Too often the sensors alone are considered, in which case gross errors of measurement can arise. Like many measurements it is not easy to verify the measurement data, the best that can be done being to ensure that the above theoretical factors have been considered and allowed for if their potential error budget warrants it.

In consideration of heat balance of sensor systems, as will be seen later, it is also important to give consideration to the generation of heat by the sensor. Many electrical sensing methods pass current through the sensor to interrogate its latent information. This produces losses that will cause the sensor to be different to the surrounding: this is known as the *self-heating* effect. Where such heat sources cannot be tolerated methods are used that bring the sensor into thermal balance in order to make a measurement.

26.1.4   General Literature on Thermal Measurement

A number of sources of general account have already been listed (Doebelin, 1975; Plumb, 1972; Woolsey, 1974).

Further general texts on instrumentation containing chapters on thermal parameter measurement are Benedict (1969), Jones (1974), Mansfield (1973), Norton (1969), O'Higgins (1966), and Sydenham (1980c).

Specific works on temperature in general are Baker *et al.* (1961), Barber (1973), Hall (1969), and Merrill (1963).

Further texts will be introduced within subsections on specific methods.

26.2   UNITS, SCALES, AND STANDARDS OF TEMPERATURE

26.2.1   Units and Scales

Since the 16th century many units and scales of temperature have been introduced. One thermometer made in the 17th century had over a dozen scales to accommodate a wide range of units.

In principle a fundamental unit of temperature could be to tie the magnitude of a temperature scale to the energy of a chosen molecule, this conforming with the microscopic concept of temperature. For this $\frac{1}{2}mv^2 = \frac{3}{2}kT$ giving T in terms of the molecule's velocity v, mass m, where k is Boltzmann's constant. This has not proven practicable.

On the other hand it is possible to make use of the macroscopic property of matter to define scales and units in terms of thermodynamic theory. In this case the ideal gas law

$$PV = RT$$

where P is pressure, V volume, R the gas constant, and T temperature, yields a linear relationship between temperature and pressure (or volume) that can be used to define a temperature scale in practice. No ideal gas exists but a few come close enough; for others corrections can be made. Constant volume thermometer forms are usually used.

The unit applied to this is the base SI thermodynamic unit, the kelvin; the scale is called the absolute or Kelvin scale.

The above expression yields a means to define a scale in terms of thermodynamic theory but the numerical values of points upon the gas scale need to be defined in more practical terms.

For reasons of convenience the ice and steam points of water historically have been used to form an interval that is divided into 100 steps (others exist) called degrees (written °).

Careful experiment has established that the zero of the Kelvin scale (a theoretical abstraction because absolute zero can only be approached, not reached) relates to the SI scale of degree Celsius units as

$$t \,°C = T\,K - 273.15$$

Many other scales were proposed—body temperature down to brine, steam to ice with $100°$ at the ice point and $0°$ at the steam point. The more familiar Fahrenheit system has $32\,°F$ at the ice point and $212\,°F$ at the steam point.

In countries adhering to metrication only K, °C units are now acceptable.

The range of need for temperature measurement magnitude (in the absolute sense) spans from thousandths of a degree kelvin, at 1 K, up to the temperatures of plasmas such as are found in man-made systems or in nature (millions of degrees kelvin).

The magnitude of discrimination sought from sensing systems depends upon the magnitude of the temperature. At the ambient level millikelvin temperature difference measurement is common in science, the microkelvin level being at the limit of detection systems.

26.2.2 Standards

In principle some form of apparatus that produces absolute zero can be used to set the zero of a gas thermometer system, the latter then producing the scale.

The gas thermometer, however, cannot yield the reproducibility possible from other more practical forms of thermometer.

Over the years the International Practical Temperature Scale (IPTS) has been generated and improved. The current form, IPTS 1968, is shown in Table 26.2. Each of these points of definition requires an apparatus to produce the necessary reference point. Such equipments are maintained in the various national physical standards laboratories. They are used to calibrate the more workable thermometers to be described in Section 26.3.

Table 26.2 Development of the International Practical Temperature Scale (IPTS)[a]

	1927 ITS-27 (°C)	1948 ITS-48 (°C)	1948 IPTS-68 (°C)	1968 IPTS-68 (°C)	1968 IPTS-68 (K)
tp hydrogen				−259.34	13.61
bp hydrogen, 25/76 atm				−256.108	17.042
bp hydrogen				−252.87	20.28
bp neon				−246.048	27.102
tp oxygen				−218.789	54.361
bp oxygen	−182.97	−182.970	−182.97	−182.962	50.188
fp water	0.000	0			
tp water			+0.01	+0.01	273.16
bp water	100.00	100	100	100	373.15
fp zinc				419.58	692.73
bp sulphur	444.60	444.600	444.6		
fp silver	960.5	960.8	960.8	961.93	1235.08
fp gold	1063	1063.0	1063	1064.43	1337.58
fp tin		231.9	231.91	231.9681	505.1181
fp lead		327.3	327.3	327.502	606.652
fp zinc		419.5	419.505		
bp sulphur				444.674	717.824
fp antimony		630.5	630.5	630.74	903.89
fp aluminium		660.1	660.1	660.37	933.52

[a] tp, triple point; bp, boiling point; fp, freezing point.

Standards scales and units are discussed in Barber (1973), Jones (1974), and Plumb (1972). Historical development is covered in depth in Klein (1974).

26.3 PRINCIPLES OF PRACTICAL THERMOMETERS

26.3.1 Expansion and Vapour Pressure Devices

Expansion principle

Most, but not all, substances increase in volume with increase in temperature.

The linear, rather than volume, expansion of solid materials is that property often employed to form thermometers. These thermometers make use of a change in length, occurring in response to an imposed temperature difference of ΔT, from the original length l_0 to the new length l according to

$$l = l_0(1 + \alpha \, \Delta T)$$

where α is the *coefficient of linear thermal expansion.*

Values of α, for the various materials, vary from small negative values through a wide range of positive values (-0.3 to $+50$ parts per million per kelvin)—see Jones (1974) and Sydenham (1980a).

In application the length change can be sensed directly by viewing a driven pointer, or indirectly by means of a linear displacement sensor. Linear and rotary forms have been devised (see Figure 26.2). Bearing in mind that all materials exhibit some degree of length change with temperature it can be seen that a second mechanical member (required to act as the

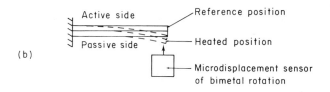

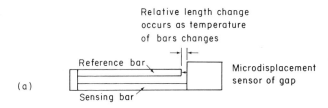

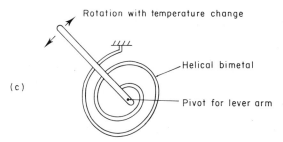

Figure 26.2 Examples of solid material expansion thermo-sensing elements: (a) basic differential linear expansion rod system; (b) bimetal strip, small deflection; and (c) bimetal strip, large rotation

reference length) also changes length. The net effect will be some combined value of α.

Often the reference length is made of the nickel–iron alloy Invar which has an expansion coefficient close to zero.

Design and application of this principle might need to recognize that α is not necessarily a constant: it can vary with temperature level, with age of the material, and with the material's past history of stress. This class of thermometer is also somewhat generally slow to respond due to the thermal mass of the sensing element. The integration effect may, however, be advantageous in providing smoothing.

Expansion thermometers can also be built that make use of expansion of liquids and gases. In this case the fluid nature of each allows *volume* or *cubical* expansion to be harnessed, with its greater sensitivity to temperature change.

Expansion to a new volume V from an initial volume V_0 for an imposed temperature difference of ΔT follows

$$V = V_0(1 + \beta \, \Delta T)$$

where β is the *coefficient of thermal cubical expansion*. The design of liquid and gas filled expansion thermometers must allow for the temperature induced volume change of the containing system which forms the reference volume. However, this is not as critical a system parameter as its reference counterpart in linear expansion thermometers because thermal volume expansion is proportional to α^3 making container volume changes less significant. Additionally higher values of β occur for non-solids, this being especially so for gases.

Design must be carefully carried through to allow for temperature induced dimensional changes to communicating capillary tubes.

Vapour pressure principle

The pressure of saturated vapour increases as temperature of the fluid and vapour rise. Given a containing system, of fixed volume, the internal pressure will be an indication of temperature. The pressure changes can be used to drive a Bourdon tube as in Figure 26.3 or a pressure sensor.

Various liquids are employed in this way to span a wide range of temperatures. These can variously cover temperature intervals from 50 to 150 °C. More details on the above forms of thermometer are available in Jones (1974) and O'Higgins (1966).

26.3.2   Thermocouples

In 1821 Seebeck discovered that when two dissimilar metals are joined together in a loop, as shown in Figure 26.4a, creation of a temperature

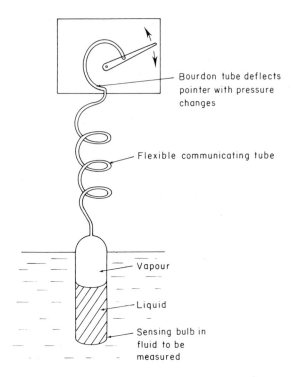

Bourdon tube deflects
pointer with pressure
changes

Flexible communicating tube

Vapour

Liquid

Sensing bulb in
fluid to be
measured

Figure 26.3 Principle of vapour pressure thermometer

difference $\Delta T = T_2 - T_1$ between the two junctions causes an electric current
to flow. Peltier, in 1834, observed that the reverse also applies: passing a
current around such a circuit will cool one junction and heat the other.

The Seebeck voltage E is that generated and used by the *thermocouple*
range of temperature sensors. It can be related to the temperature difference
between the two junctions $(T_2 - T_1 = \Delta T)$ according to

$$E = a\, \Delta T + b(\Delta T)^2 + c(\Delta T)^3$$

where a, b, and c are constants appropriate to the two materials used to
form the couple.

Given controlled purity metals and standardized couple combinations it is
possible to provide tables of the voltage E for a couple system in which one
junction (called the *reference* junction) is held at $0\,°C$. Most texts on
instrumentation contain examples of these tables. Many combinations of
metals are in use. Figure 26.4b provides a summary of the kinds and their
ranges of use.

Junctions are formed by twisting the wires in contact, by welding, or by
deposition of one material onto another.

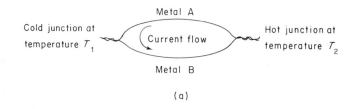

(a)

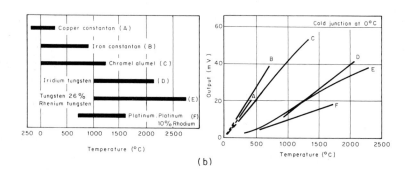

(b)

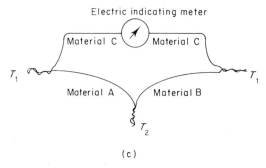

(c)

Figure 26.4   Thermocouple temperature sensing: (a) basic circuit of an
iron–copper thermocouple system; (b) ranges and outputs of common
couples; and (c) basic reading circuit using d.c. meter movement

Maintaining the reference junction at 0 °C is usually impracticable. Special
compensating electronic circuitry, that senses the ambient temperature, is
used to generate the required loop voltage to give a 0 °C simulation for a
reference junction.

Reflection on the above principle will show that to sense the current in
such a circuit requires insertion of connections and circuits that, in themselves, may constitute unwanted thermocouples.

Although the principle is straightforward correct design requires considerable detailed practical knowledge when high precision and low uncertainty

levels are sought and where the thermocouple is placed at some distance from the indicating unit.

Advantages of thermocouples are their low cost, extremely wide temperature range (different materials can progressively span from absolute zero to near 3000 °C) and, if need be, very small size. Also possible is a high speed of response where protective enclosures can be made with small thermal mass.

In many applications simple indicating circuitry can be used. The simplest measuring system is as shown in Figure 26.4c.

Note that the connecting leads (usually of copper) give rise to the need for an additional junction to overcome the dissimilar nature of the leads with the other metals. More accurate results (to nanovolt discrimination) can be obtained by using a detector that does not require current flow and hence does not produce Peltier error effects. In such cases a null-balancing potentiometer is often used (see Chapter 20). Good introductions to thermocouple theory and practice are available in Baker *et al.* (1961), Barber (1973), Benedict (1969), Hall (1969), Jones (1974), O'Higgins (1966), and Plumb (1972). Specific attention to thermocouples is given in Kinzie (1973).

Thermocouples are used extensively in industrial process sensing and the user, because of industry standardization, can generally install proprietry couple-material, transmitters, compensators, and controller circuits without needing any specialized skills or depth of understanding.

Where greater sensitivity is needed it is possible to form several couples into a series circuit holding all reference junctions at ambient (or 0 °C) and all sensing junctions at the temperature to be sensed. This assembly is called a *thermopile*. With such methods it is possible to discriminate microkelvin temperature changes over short periods of time.

Good practice dictates that thermocouple systems (although apparently inherently calibrated) should be calibrated by an independent traceable sequence. The various IPTS datum temperature points are the ultimate reference points, but in the laboratory or factory use is made of a standardized thermometer to transfer a calibration using a suitable constant temperature enclosure to hold both standard and field thermometer at the same temperature.

26.3.3 Metallic Electrical Resistance Temperature Sensors

Although the principle was realized in 1821 by Sir Humphrey Davy, it was not until 1871 that application was first made of the change of electrical resistance of an electrical conductor as a means to measure temperature. In the period following to 1887, work by Sir William Siemens, Callendar, and Griffiths laid the foundations of resistance thermometry based on the use of metallic resistance materials. It has proven to be a most reliable method, is relatively simple to use, has a good degree of linearity over a wide range and if properly designed and maintained can provide assured long term stability.

The resistance R_T of a sensing element, for the commonly used metals (copper, nickel, and platinum) is related (from 0 °C upward) to temperature T (in °C) by the following expression

$$R_T = R_0(1 + \alpha T + \beta T^2)$$

where R_0 is a reference point resistance, usually taken at 0 °C, and α, β are curve fitting constants. (A slightly different formulation applies for temperatures less than 0 °C.)

As the required precision of determination is increased it becomes necessary to make use of an increasing number of the terms of the series. For example two constants often suffice for copper sensors but three are usually needed for the platinum and nickel forms. Choice of the number of terms also depends upon the temperature range required of the sensor.

The values α, β etc. are usually specified as part of the calibration certificate of commercial resistance thermometers.

Even with this curve fitting procedure the range will not necessarily agree well enough with the independently defined IPTS fixed points. Correction values are calculated using equations that have been developed to correct measured values to the desired IPTS values.

Resistance thermometers, also called RTD's, are often specified by their *fundamental interval*, this being the change in resistance that occurs between 0° and 100 °C ($R_{100} - R_0$). The Callendar equation uses this concept to express the resistance–temperature relationship:

$$T = \frac{R_T - R_0}{R_{100} - R_0} 100 + \delta(T - 100)T$$

where T is the temperature being sensed, R_T the resistance of the sensor at that temperature, R_0 the resistance at 0 °C, R_{100} the resistance at 100 °C, and δ a characteristic constant for the winding (for example, this is, for platinum 1.5).

The first term is known as the *platinum temperature*. The difference between the platinum temperature and the degrees Celsius temperature is, therefore, given by

$$T - T_{Pt} = \delta(T - 100)T$$

Tables of this difference are available that enable degrees Celsius to be obtained easily from those indicated by the sensor metal used.

The ideal metal for such probes is yet to be devised. Design aims to select a material having a high temperature coefficient of electrical resistance, high melting point, great purity, high resistance to corrosion, and ease of fabrication. Figure 26.5 summarizes the working ranges, melting points, and temperature coefficient of resistance of the three commonly used metals.

Construction of the sensor aims to produce a sensing element that retains stable electrical and mechanical properties. As depicted in Figure 26.6 sensing elements are formed either by winding coils, that are suitably mounted or, a more recent innovation, by thick or thin film deposition on a ceramic or other suitable base. A protective sheath is generally used which is made of metal, quartz glass or ceramic materials. Such encapsulation usually degrades the dynamic response of the probe. Time constants (systems assumed to be of first-order kind) vary from 0.1 s to 10 s or more. Special film and wire probes have been built with millisecond time constants.

Measurements with platinum resistance thermometry can be made traceable to the IPTS points to an accuracy of the order of 0.01 °C. If absolute calibration is not required, as for example when measuring small temperature variations, sophisticated sensing elements and associated circuitry can yield stability, with similar discrimination, to within 0.00002 °C per day.

The sensing element is usually interrogated by placing it in some form of modified Wheatstone bridge. Resistance values are usually reasonably low (order of 20–500 Ω).

The finite and significant values of the connecting lead resistances also vary with temperature, and, therefore, are a possible source of error. With special precautions bridges can be used to detect 1 $\mu\Omega$ changes in 100 Ω sensors. Typical basic bridge circuits are shown in Figure 26.7. The most basic *two-lead* configuration (Figure 26.7a) suffers from such lead errors.

Addition of an extra connecting lead to one end of the sensor (Figure 26.7b), the so-called *three-lead* method, provides compensation that deteriorates as the bridge moves away from balance. This largely overcomes lead errors for many bridges operate into null balance detectors.

The Mueller bridge (Figure 26.7c) provides for reversal of the circuit connections thereby indicating errors as the difference between the two readings. It cannot, however, yield continuous read-out and only yields the differential action for adequately steady temperatures.

A *four-wire* system (Figure 26.7d) provides compensation for the leads connecting the sensor to an unbalanced bridge.

These d.c. configurations properly only apply to measure sensors in bridges that are operating as true zero-order systems, that is, having no a.c. signals present. This will hold for d.c. excitation and steady-state bridge currents but not for a.c. excitation or during transients or rapidly fluctuating thermal measurements.

In many cases a.c. excitation is used to improve the signal-to-noise ratio and to increase the sensitivity. Phase sensitive detection (see Chapter 11) is generally used. Here the reactive elements of the circuit—sensor inductance

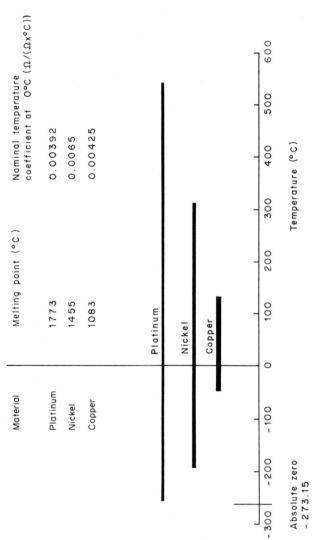

Figure 26.5   Data of commonly used resistance thermometer materials

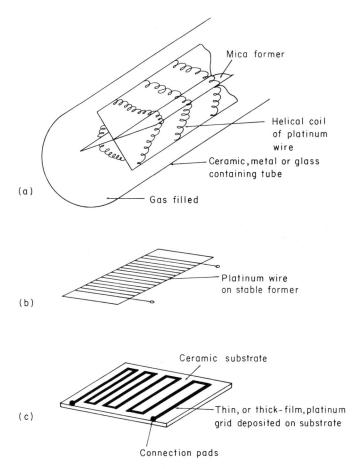

Figure 26.6 Some forms of metallic resistance thermometer sensors: (a) coiled platinum on mica former; (b) platinum wound on flat former; and (c) metal deposited on ceramic substrates

and capacitances (see Chapter 20 for discussion of a model of a resistance, lead capacitances, bridge capacitance, and so on)—are potential sources of error. Forms of bridge other than those shown are therefore employed. The transformer ratio bridge is commonly used.

At balance a bridge circuit drives no current into its detector. However, in that condition currents are flowing in each of the four resistances. These currents will give rise to *self-heating* of the sensor and, to a lesser effect, to the resistance values in the bridge.

Thus, as is explained in some detail in Section 26.3.4, where a thermistor

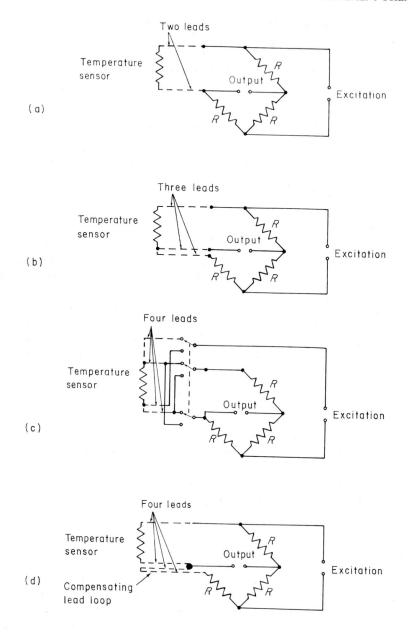

Figure 26.7 Typical bridge circuits used to interrogate resistance temperature sensors: (a) standard basic Wheatstone bridge—two-wire connection; (b) three-wire connection; (c) Meüller reversing connection bridge; and (d) four-wire connection

thermometer design is outlined, the bridge excitation must be kept adequately low to obtain an *offset* temperature that is tolerable

This is a major reason why metallic resistance temperature sensing systems can only be used to yield relatively low bridge sensitivities. The easy availability of electronic gain overcomes this potential source of limitation.

26.3.4 Semiconductor Sensors

Semiconducting materials and devices can also be used to sense temperature. There are basically two forms of semiconductor sensor: the thermistor and the semiconducting diode junction.

Thermistor systems

Bulk semiconducting material is here used to form a resistance element. As temperature rises the resistance of a thermistor resistance element reduces; they have a *negative temperature coefficient* (NTC). (Positive coefficient (PTC) thermistors also exist but their use is not discussed here.)

The rate of change of resistance with temperature is typically more than 10 times greater than that of a metallic temperature sensor but the characteristic is highly non-linear and less stable especially where large temperature excursions are experienced. The relationship between resistance and temperature is related exponentially to absolute zero. Expressed in terms of degrees Celsius the relationship is

$$R_t = R_{20} \exp (B/T - B/293)$$

where R_t is the thermistor resistance, at temperature T in degrees Celsius, R_{20} the resistance at 20 °C. B is the *characteristic temperature* (K) for the actual device. Unlike the case for metallic resistance characterization this formulation is only used in an approximate sense, for example, B and R_{20} values quoted for a particular device are usually toleranced ±20% or greater.

Thermistor bridge design

The use of a resistance element in a bridge detection circuit requires determination of the allowable excitation voltage magnitude (to limit, self-heating, offset temperatures), choice of the optimum resistance value to use in the bridge circuit, and assessment that the element is not dissipating excessive power beyond its design limit. Design also requires calculation of the sensitivity of the sensor in the bridge.

Now follows a condensed version of the design of a thermistor bridge as published by Bell and Hulley (1966).

The incremental rate of change of resistance with temperature of a thermistor is given by

$$\frac{dR_t}{dT} = -R_t \frac{B}{T^2}$$

This can be seen to depend upon the temperature of the thermistor.

Sensitivity to temperature of a thermistor placed in a standard Wheatstone bridge (such as that shown in Figure 26.8a) having an excitation level of V volts, fixed resistances of R, and output voltage V_o is given by

$$\frac{dV_o}{dT} = V \frac{R}{(R + R_t)^2} \frac{dR_t}{dT}$$

The value of R that makes the differential of this expression equate to zero

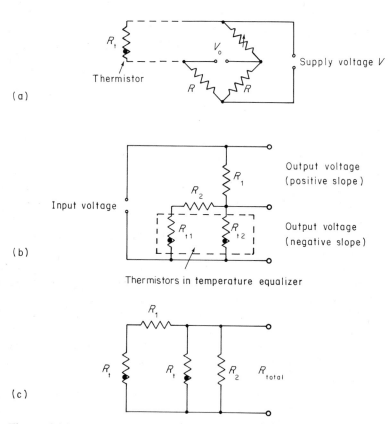

Figure 26.8 Circuits for thermistor use: (a) basic Wheatstone bridge circuit; (b) extended linearity of output voltage with temperature circuit; and (c) extended linearity of resistance with temperature circuit

is the optimal value of bridge resistance. The value changes with the temperature at which the bridge is balanced and thus cannot be uniquely optimized for widely varying temperature situations. Bell and Hulley (1966) discuss this problem.

It can be seen that increasing V proportionately increases the bridge sensitivity. However, this also increases the self-heating of the thermistor at a faster rate.

Power dissipated P (in watts) in the thermistor is given by

$$P = \frac{V^2}{(R + R_t)^2}$$

This should not exceed the maximum dissipation P_{max} usually quoted for the device.

Published data for the particular thermistor encapsulation usually provide the *thermal resistance* θ of the element as degrees Celsius temperature rise per watt dissipated, or the reverse, *dissipation constant* k, the milliwatts required to raise the element one degree Celsius. Using the former method the temperature rise, in degrees Celsius, of the element above its ambient is given by

$$T_{offset} = \frac{V^2}{(R + R_t)^2} R_t \theta$$

where θ is the thermal resistance in degree Celsius/watt. Note that the offset is proportional to the square of the bridge excitation whereas sensitivity has a linear characteristic.

The above formulations enable a design to be generated; a certain amount of iteration is needed to arrive at a suitable thermistor nominal resistance value. For each geometrical form manufacturers usually offer a range of R_{20} values covering from 200 Ω to 500 kΩ. An example calculation is provided in Bell and Hulley (1966).

General opinion is that thermistors are not suited to thermometry where low discrimination and high reproducibility are needed. Experience has shown that the drift to less than millikelvin per day rates is achievable provided the thermistor is being used in a fixed temperature situation (for example, to maintain fixed point temperature control). An extensive study of stability has been published (Wood *et al.*, 1978). The non-linear characteristic can be overcome by computational compensation or by linearization circuitry such as those shown in Figures 26.8b, c.

Thermistors are not regarded as suitable replacements for metallic resistance sensors except where low precision is required. Their key advantages are ability to be made in a variety of sensor shapes including some small enough to be embedded in hypodermic needles. Their nominally high resistance values also reduce the effect of change of resistance of connecting leads but in cases where adequate signal-to-noise ratio is hard to achieve design requires low R_{20} values to be used to reduce the electronic thermal resistance noise of the thermistor.

Semiconductor junction sensors

One of the shortcomings in normal use of solid-state semiconductor junction devices is their dependency on operating temperature. The voltage drop provided across a forward biased p–n junction is temperature dependent and, therefore, offers the potential to be used as a temperature sensor.

Since around 1960 many relevant papers have been published and several devices marketed making use of this principle. The principle is now commonly used within monolithic integrated electronic circuits as a means to sense temperature for circuit compensation. In the main, systems are based upon use of a transistor rather than a diode device.

The particular features that make junction semiconductor temperature detectors attractive are their small size, ability to be integrated into monolithic circuitry, and also their potentially excellent linearity over the temperature range from nominally −50 to 120 °C. Furthermore it can be shown (as is discussed below) that temperature measurements can be tied to fundamental physical constants without need to know the practical features of individual transistors—the so-called *calculable* thermometer which, in principle, needs no traceable calibration to temperature reference points.

Following the account by Verster (1968) it is there shown that the temperature coefficient, the change of the base–emitter voltage V_{be} (refer to Figure 26.9a) of a forward-biased junction transistor to change in temperature T (K) is given by

$$\frac{\partial V_{be}}{\partial T} = \frac{k}{q}(\ln I_c - \ln K - r - r \ln T)$$

where k is Boltzmann's constant, q the electron change (both fundamental invariant physical constants); I_c the collector current, K a factor related to geometrical aspects of the device, and r a fixed value dependent upon the diffusion state of the device and material used. (K and r are independent of temperature but are fixed values varying between individual devices.)

Assuming that I_c is held constant with additional circuitry, study of the equations shows that, in this form, use of V_{be} as an indicator of temperature has several shortcomings. For this case, the above equation becomes

$$\frac{\partial V_{be}}{\partial T} = A \ln T$$

where A is a constant peculiar to the individual device. From this it can be

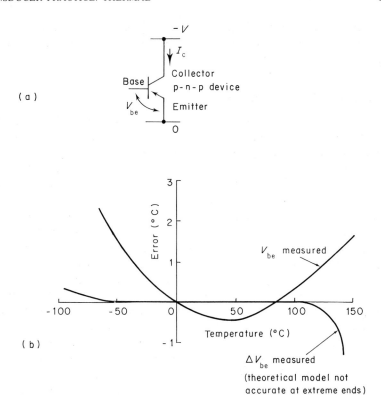

Figure 26.9   Use of a transistor as a temperature sensor: (a) basic transistor circuit; (b) calibration curves for V_{be} and ΔV_{be} methods

seen that the characteristic is not theoretically linear and that the V_{be} output will differ for a given temperature from device to device (by as much as ± 100 mV). Figure 26.9b shows this expression as plotted by Verster for a particular transistor.

Means to reduce the deficiencies of this principle have been devised that enable mass-reproducible elements having the same calibration and sensitivity. Here we now consider two of these approaches: use of switched I_c levels (after Verster, 1968), and a self-adjusting bridge method that corrects for V_{be} variations (Ruehle, 1975).

Instead of supplying a constant collector current I_c, Verster's method switches I_c between two not equal levels, I_{c_1}, I_{c_2}. For each current level there exists a corresponding V_{be_1}, V_{be_2} level. Thus, using the expression for V_{be}

(see Verster, 1968),

$$\Delta V_{be} = V_{be_1} - V_{be_2} = \frac{kT}{q} \ln \left(\frac{I_{c_1}}{I_{c_2}} \right)$$

differentiation of ΔV_{be} with respect to temperature T (K) yields

$$\frac{\partial (\Delta V_{be})}{\partial T} = \frac{k}{q} \ln \left(\frac{I_{c_1}}{I_{c_2}} \right)$$

The constants k and q are fundamental, invariant values. If the ratio I_{c_1}/I_{c_2} is held constant (this is achievable to high precision) then the slope of the $\Delta V_{be}-T$ characteristic is constant to a high order.

Thus, means are provided to produce thermometer sensing elements that are interchangeable regardless of type and batch; that have a calibration that is calculable from physical constants; and a current ratio (set by electrical standards) that is highly linear.

Sensitivity of the ΔV_{be} method is partly sacrificed compared with the V_{be} alternative. Verster quotes suitable I_{c_1}, I_{c_2} currents as 100 μA and 0.1 μA for which $\partial(\Delta V_{be})/\partial T = 600$ μV C^{-1} (regardless of device). Using the steady current V_{be} method the sensitivity would be in the vicinity of 2.4 mV C^{-1} (dependent upon device). Figure 26.9b includes the calibration curve for the switched current method showing the greatly reduced error of linearity.

This principle was adopted by National Semiconductor (devices LX5600 and LX5700) but in a slightly different form. They used two separate matched transistors operating at different constant collector currents taking the difference between the two V_{be} values to generate ΔV_{be}. In principle it would be inferior to switching the same device. In both cases the heating effect of the sensing transistor(s) and circuitry produce offset temperature errors.

A second approach to make the V_{be} method practical is that reported by Ruehle (1975). By use of feedback circuitry the method reported by Ruehle (1975) holds the base current constant which, in turn, holds the collector current constant. Incorporation of *linearity* and *sensitivity* trimming potentiometers enables the system to be set up after manufacture to provide a scale for which 0 °C corresponds with 0 volt output with a 5.00 volt output at 100 °C and a high degree of linearity.

To reduce self-heating offset errors the temperature sensor is mounted remote from the circuitry. The account provides a comparison table between various thermometry methods.

Thus in this alternative, high gain and satisfactory operation are obtained at the expense of requiring set-up calibration when first made and presumably during its life.

Microelectronic forms of sensors, data processing, and data highway distributed control schemes are allowing great changes to be introduced into temperature measurement and control systems. Appreciation of these changes is provided in Bailey (1981).

26.3.5   Radiation Pyrometry

As explained in Section 26.1.3, in the discussion of heat energy transfer by the radiation mechanism, all substances radiate electromagnetic energy. The wavelength of the radiation and its intensity are related to the temperature of the radiation source.

Temperature sensing systems built to use this method are generally now known as *pyrometers* or, less commonly, *radiometers.*

Instruments either respond to the *total energy* received (all wavelengths are utilized) or to specific narrow bands of radiation (such as the optical spectrum).

Pyrometers are traditionally regarded as being useful for temperatures beyond around 1000 °C but improvements to long-wavelength detectors have made the method suited to temperatures as low as normal ambient (20 °C). Those used for sensing the lower temperatures are often assembled into image forming arrangements rather than single point temperature sensors: these are called *thermal imaging* systems.

High temperature pyrometry and thermal imaging are both based on use of the black-body radiation laws. In these methods the temperature determination must always be adjusted for the emissive behaviour of the surface.

Introductory descriptions are available in Benedict (1969), Jones (1974), O'Higgins (1966), Plumb (1972), and Sydenham (1980c). Hudson (1969) and Vanzetti (1972) deal specifically with the low temperature infrared methods.

Basic elements

The basic elements that might be used in a pyrometer are shown in Figure 26.10. Not all are used in all designs. A radiation gathering system (a *refracting* or *reflecting* radiation concentrator) focuses the radiation received from a given area of the subject onto a, smaller area, appropriate detector. The gathering system provides gain in temperature resolution but in doing so may trade off spatial resolution. A *window* is often used to maintain the cleanliness required by the internal 'optical' system.

Systems often make use of phase sensitive detection (Chapter 11) to improve the signal-to-noise ratio (Blair and Sydenham, 1975). In this case the radiation is modulated (usually by a mechanical chopping disc) prior to detection.

The detector is followed by electronic circuits that amplify the signal,

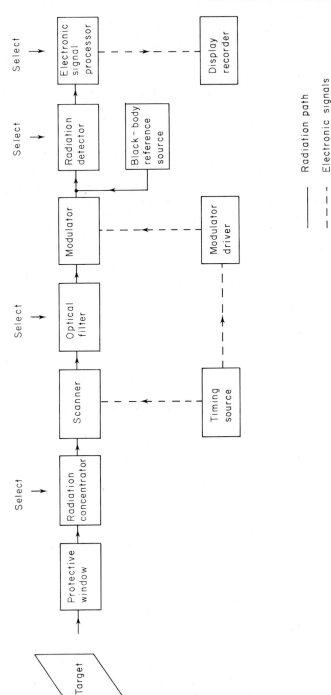

Figure 26.10   Basic elements that might be used in a pyrometer

perhaps filtering the frequency spectrum and implementing the phase sensitive detection process. Display might be in numerical or analog meter format. Signal output, in some instruments is available for closed loop control or automatic recording of values.

Thermal imaging systems generally make use of a single detector that is *scanned* across the target area. Synchronization is used to turn the time-sequential signal into a time-varying, two-dimensional, picture format—the *thermal image.*

Depending upon design the radiation path often contains wavelength filters. These may be either broad or narrow band transmission elements or the filtering may be obtained by using dispersion elements (prism or grating) to isolate the required wavelength at a given spatial position for transmission to the detector. (Chapter 23 introduces optical elements.)

The emissivity of the target's surface will often require objective and perhaps automatic determination of the gray level of a signal. This can be done using a standard *black-body* radiation source which is periodically sensed by the detector: it provides a fixed reference level for automatic signal adjustment.

Total radiation instruments

In these the detector is chosen to absorb and respond to all wavelengths of energy presented to its surface. Detectors are usually blackened to absorb maximum radiation.

Transduction usually makes use of the heating effect on a thermopile to

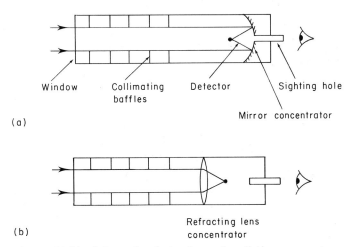

(a)

(b)

Figure 26.11   Schematic of simple total radiation pyrometers:
(a) reflecting mirror system; (b) refracting lens system

generate an output voltage or to change the resistance of the detector (thermocouple and metallic resistance thermometer principles, respectively).

Typical total radiation pyrometer systems are shown in Figure 26.11. In order to obtain consistent results a parallel radiation beam input is needed. A series of holes in discs is often used as a simple means to obtain a closely parallel beam of radiation from the target: lens collimators, unless made of expensive materials, will act as selective wavelength filters.

Mirrors are generally used as they are able to transmit a wide range of wavelengths at comparatively low cost. A sighting hole is provided to allow the operator to decide which part of the object of interest is being measured. Filters may be provided in the eyepiece to aid sighting of very hot targets.

Calibration of the system against standard heat sources is required periodically: black body 'furnaces', packaged as a laboratory heat source, are available commercially. For details see Hudson (1969).

Narrow band optical pyrometers

The most common form of optical pyrometer is that in which the brightness of a standardized tungsten lamp filament is adjusted to be equal to that of the target: the so-called *disappearing filament* pyrometer. Figure 26.12 shows the basic elements involved. In use the operator adjusts the rheostat until the filament just disappears. Temperature is then read from the meter measuring the circuit current which is convertible to temperature.

These are used for normal measurement of temperatures in the region from 600 °C to 3000 °C. Beyond 1400 °C an absorbing screen is placed between the target and the filament to attenuate the energy to within the range that can be produced by continuous heating of a filament. Manually balanced pyrometers cannot be used for automatic recording or continuous control functions. Autobalancing systems are available for those purposes.

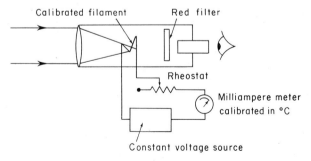

Figure 26.12   Principle of disappearing filament optical pyrometer

Emissivity can be estimated using radiation at two different wavelengths that are emitted from the object (Benedict, 1969; Sydenham, 1980c; and Bailey, 1981).

Thermal imaging

Thermal contour pictures of a surface (of continental size or features of a microelectronic circuit) can be produced using a *thermal imaging* system. Figure 26.13 shows the basic elements of an imaging system in which one axis is provided by movement of the imaging system relative to its target. (Fixed units require a second scan axis.)

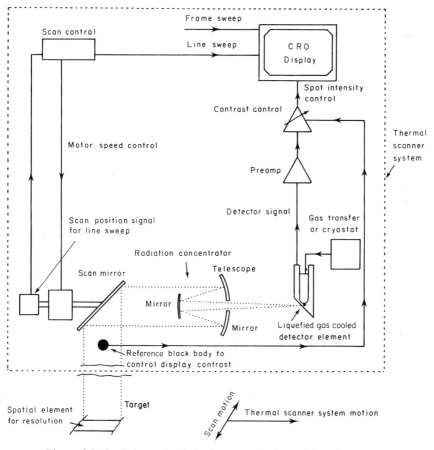

Figure 26.13   Schematic of single-scan axis thermal imaging system

A single, highly sensitive (and usually expensive) detector is sequentially shown points on the target using an oscillating (or spinning) mirror that systematically varies the optical path. The detector usually remains stationary.

Infrared systems are needed because thermal scanners generally map ambient level temperatures for which the majority of the energy radiated will be in the 1–14 μm wavelength region.

The size of target seen is set by the optical system parameters: the final mirror, detector calibration, and picture composition controls remaining basically the same for all systems.

Temperature discrimination is generally in the region 0.1–0.01 °C but it is possible to detect 0.001 °C differences if enough target area is available to provide an adequate energy level to the detector.

26.3.6   Miscellaneous Methods

Methods discussed in the above sections are those most generally used. Additional to those are numerous other alternatives that may be more suited to specific applications. Some examples follow.

Acoustic thermometry

The velocity of sound usually decreases with temperature rise in solids and liquids and increases with temperature rise in gases.

Given adequate knowledge of the substance composition it is possible to relate temperature to acoustic velocity. The method is useful from cryogenic temperatures near absolute zero temperature up to gas plasma temperatures of 15,000 °C.

Two methods of applying the principle are in use. In the first, the sound wave is directed in the substance itself. The alternative method launches the sound wave in a secondary substance that is in temperature equilibrium with the substance of interest. Electrical signals are usually coupled into the medium using piezoelectric transduction crystals.

Errors arise in this method from inadequate knowledge of the temperature–velocity relationship for the material, timing phase error, and from refracted sound wave paths which can lengthen the distance travelled beyond the point-to-point line symplistically assumed.

Fluidic sensors

A mechanical cavity resonates at a characteristic frequency dependent upon the temperature of the gas in the cavity. For cavities used it varies as the square root of the Kelvin temperature. Typical output frequencies lie in the range 5–40 kHz.

Gas, of which the temperature is to be measured, is fed through the cavity, the resonance frequency being monitored.

Johnson electrical noise sensors

Random noise voltage V generated by an electrical existence varies as

$$V^2 = 4kTR\,\Delta f$$

where k is Boltzmann's constant, T the temperature in kelvin, R the resistance in ohms and Δf the frequency bandwidth of the signal measured.

Thus noise voltage is related to temperature. Allowance is also needed for the temperature dependency of resistance with temperature. Signal levels are in the microvolt region. This method is used, with different sensing resistors, for temperatures ranging from near absolute zero to 1500 °C.

Other, less usual, methods are outlined in Mansfield (1973) and Norton (1969).

REFERENCES

Bailey, S. J. (1981). 'Temperature measurement 1981: sensors expand loop influence', *Control Engineering*, **28**, No. 2, 67–71.

Baker, H. D., Ryder, E. A., and Baker, N. H. (1961). *Temperature Measurement in Engineering*, Wiley, New York.

Barber, C. R. (1973). *Temperature Measurement at the National Physical Laboratory*, HMSO, London.

Bell, E. C. and Hulley, L. N. (1966). 'Precision temperature control', *Proc. IEE*, **113**, No. 10, 1671–7.

Benedict, R. P. (1969). *Fundamentals of Temperature, Pressure and Flow Measurement*, Wiley, New York.

Blair, D. P. and Sydenham, P. H. (1975). 'Phase sensitive detection as a means to recover signals buried in noise', *J. Phys. E.: Sci. Instrum.*, **8**, 621–7.

Coughanowr, D. R. and Koppel, L. B. (1965). *Process Systems: Analysis and Control*, McGraw-Hill, New York.

Doebelin, E. O. (1975). *Measurement Systems: Application and Design*, McGraw-Hill, New York.

Ellis, B. (1968). *Basic Concepts of Measurement*, Cambridge University Press, Cambridge.

Hall, J. A. (1969). *The Measurement of Temperature*, Chapman and Hall, London.

Hudson, R. D. (1969). *Infrared System Engineering*, Wiley, New York.

Jakob, M. and Hawkins, G. A. (1957). *Elements of Heat Transfer*, Wiley, New York.

Jones, E. B. (1974). *Instrument Technology*, Vol. 1, *Measurement of Pressure, Flow and Temperature*, Newnes–Butterworths, Sevenoaks, UK.

Kinzie, P. A. (1973). *Thermocouple Temperature Measurement*, Wiley, Chichester.

Klein, H. A. (1974). *The World of Measurements*, Simon and Schuster, New York.

Mansfield, P. H. (1973). *Electrical Transducers for Industrial Measurement*, Butterworths, Sevenoaks, UK.

Merrill, H. J. (1963). *Light and Heat Sensing*, Pergamon, London.

Norton, H. N. (1969). *Handbook of Transducers for Electronic Measuring Systems*, Prentice Hall, Englewood Cliffs, NJ.

O'Higgins, P. J. (1966). *Basic Instrumentation: Industrial Measurement*, McGraw-Hill, New York.
Plumb. (1972). *Temperature, its Measurement and Control in Science and Industry*, Instrument Society of America, Pittsburgh.
Ruehle, R. A. (1975). 'Solid-state temperature sensor outperforms previous transducers', *Electronics*, **48**, No. 6, 127–30.
Sydenham, P. H. (1979). *Measuring Instruments; Tools of Knowledge and Control*, Peter Peregrinus, London.
Sydenham, P. H. (1980a). '*Mechanical design of instruments*', *Measurement and Control*, **13**, series beginning October.
Sydenham, P. H. (Ed.) (1980b). 'A working list of books published on measurement science and technology', International Measurement Confederation IMEKO (Applied Physics Dept., Technische Hogeschool Delft, Delft).
Sydenham, P. H. (1980c). *Transducers in Measurement and Control*, Adam Hilger, Bristol; ISA, Research Triangle.
Vanzetti, R. (1972). *Practical Applications of IR Technique*, Wiley, New York.
Verster, T. C. (1968). 'p–n junction as an ultra-linear calculable thermometer', *Electron. Lett.*, **4**, No. 9, 175–6.
Wood, S. D., Mangum, B. W., Filliben, J. J., and Tillet, S. B. (1978). 'An investigation of the stability of thermistors', *J. Res. NBS*, **83**, No. 3, 247–63.
Woolsey, G. (1974). 'Temperature measurement', in *Introduction to Stathmology*, P. H. Sydenham (Ed.), University of New England, Armidale, New South Wales.

Handbook of Measurement Science, Volume 2
Edited by P. H. Sydenham
© 1983 John Wiley & Sons Ltd

Chapter

27 R. S. WATTS

Transducer Practice: Chemical Analysis

Editorial introduction

Analytical instrumentation, used to determine the chemical properties of materials, relies heavily upon optical techniques combined with electronic signal control and processing, and fine mechanics to structure the optical components in relation to the source, detectors, and materials under test.

This class of measurement system, although based upon the same physical and mathematical processes as other classes, incorporates a wide range of arrangements in its own characteristic way. It, therefore, provides another facet of measurement system design which is unlike those presented in Chapters 24, 25, and 26. Of particular relevance to chemical measurement, for signal energy levels are usually quite low, is the material presented in Chapter 11.

This class of measurement system contains much information about numerous variants. The field is sufficiently broad in its types of transducers and its application to make a general comprehensive account virtually impossible in one reasonable sized text. Here the author provides only an introduction to the methods. As a guide, texts tend to concentrate on specific areas within the class, for example on spectroscopy, on chromatography, on air or water purity, on process control instruments, and on principles of instrumental analysis. A list of some 30 texts appears in IMEKO (1980).

27.1 INTRODUCTION

Chemical measurement presents problems of classification and definition. We shall restrict this chapter to those transducers which can distinguish and/or quantify different chemical species. This is an empirical definition, and excludes, for instance, the chemical applications of density measurement, but this is adequately covered elsewhere.

Chemists usually classify their techniques partly by field of application, and partly by principle. The classification given here is restricted to the principle of the transducer; this will sometimes lead to problems. One could, for example, entertain a lengthy debate on the principle of mass spectrometry.

Since this chapter concentrates on transducers, rather than entire measurement systems, the discussion of peripheral devices is brief, and only given where the operation of the transducer is necessarily integrated with other equipment.

Space permits only a condensed summary of commonly available chemical transducers. Where possible, references are given to industrial measurement applications of chemical transducers, although these are not extensive compared with laboratory applications. A systems approach to chemical measurement will greatly assist in taking laboratory methods into industrial processes (Malmstadt, 1980; Verdin, 1980).

27.2 OPTICAL TRANSDUCERS

27.2.1 Use of Incident Radiation

The interaction of electromagnetic radiation with a molecule is the basis of many measurements in chemistry.

A molecule, in a state i, interacts with a photon of frequency ν as shown in Figure 27.1. At some time later, a photon of frequency ν' may be re-emitted, leaving the molecule in state f, which may be different from i. The measurement process usually takes place on the re-emitted photon, and provides several pieces of information, for example, *frequency, intensity*

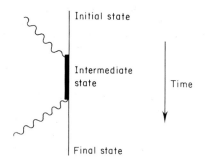

Figure 27.1   The interaction of a photon of frequency ν with the initial state i of a molecule. The molecule is left in a final state f and the re-emitted photon has frequency ν'

(number), *state of polarization.* Knowledge of the incident photon can then be used to infer the *identity* or *concentration* of the molecule.

Now given are some specific examples of optical transducers, classified according to the measurement made on the re-emitted photon.

27.2.2 Intensity Measurement

This is by far the most common type of optical transducer. The intensity measurement itself is made by a photomultiplier (Figure 27.2), or for larger numbers of photons, a solid state detector (for example, photodiode, photovoltaic cell). In the infrared region, the low energy of the photon necessitates careful choice of detector. Several types, together with their characteristics, are listed in Table 27.1.

In qualitative analysis, energy dispersion (*monochromation*) of the incident photon is usually carried out prior to intensity measurement. An exception to this is in X-ray fluorescence spectrometry (see below). Some examples of optical systems are given below.

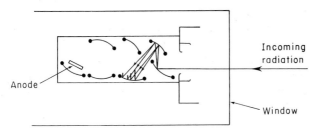

Figure 27.2 The photomultiplier: Incident radiation strikes an electron emissive surface; secondary electrons collide with successive electrodes in a many-from-one cascade process

Table 27.1 Infrared detectors

Name	Type	Detection sensitivity $(cm\,Hz^{1/2}\,W^{-1})$	Response time (ms)
Bolometer	ambient thermal	10^8	1.0
Golay	ambient thermal	1.5×10^9	15
Thermocouple	ambient thermal	1.5×10^9	15
Pyroelectric	ambient thermal	10^8	2×10^{-3}
Photoconductor	cooled semiconductor	10^9	10^{-1}–10^{-6}

27.2.3   UV–Visible Absorption

Fundamental considerations (see Hargis and Howell, 1980), show that the intensity of incident photons I_0 can be related to the transmitted intensity I by the Beer–Lambert law:

$$\log (I_0/I) = abc$$

where

   a = molar absorptivity (a constant for a given species under given condi-
      tions) ($\text{g}^{-1}\,\text{cm}^{-2}$)
   b = cell path length (cm)
   c = concentration (g cm^{-3})

This relation is commonly used by chemists in quantitative analysis: solutions of known concentration are used to plot a calibration curve, from which unknown solution concentrations may be read (Yotsuyanagi, 1979).

   There are many variations in instrument design (Lue, 1979), depending on the purpose of the instrument, for example

(a)   Quantitative analysis, with infrequent wavelength changes and low spectral bandpass requirements (Figure 27.3);
(b)   quantitative analysis, with frequent wavelength changes and stringent spectral bandpass requirements (Figure 27.4);
(c)   qualitative analysis and structural studies, with wavelength scanning and stringent bandpass requirements (Figure 27.5).

Many other configurations are possible; some are mentioned below. Table 27.2 lists some major instrument components and their characteristics.

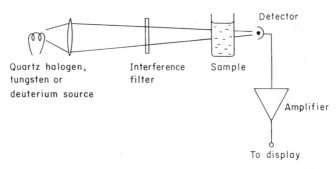

Figure 27.3   A simple colorimeter, or filter photometer, commonly used in repetitive quantitative analysis in the laboratory or process stream

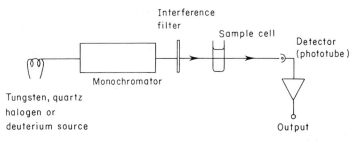

Figure 27.4   A single-beam spectrophotometer of a design used commonly in routine quantitative analysis

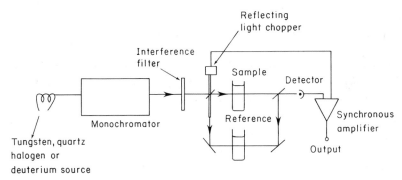

Figure 27.5   A double-beam UV–visible spectrophotometer of a design used for obtaining absorption spectra for qualitative analysis, structural work, and multicomponent quantitative analysis

Table 27.2   Major instrument components in UV-visible spectrometers

Item	Description
Light source	Visible: tungsten or quartz halogen lamp
	UV: hydrogen, deuterium, or (more expensively) xenon lamp
	Usually requires stabilized power supply
Monochromator	Almost always reflection grating—quality of grating determines resolution. Optical design can reduce stray light
Filter(s)	Can be absorption (colour) or interference; used to remove unwanted orders of diffraction due to grating
Modulator	May be rotating sector disc (chopper) or modulation achieved by switching the light source power supply
Detector	Solid state in simple instruments, phototube in intermediate, and photomultiplier in most sensitive instruments

27.2.4 Atomic Absorption

This special case of UV–visible absorption spectrometry is of particular importance because of its wide use in trace metal analysis (Koirtyohann, 1980), and because of the transducer design (Figure 27.6).

The light source is an atomic emission lamp (usually a hollow cathode lamp) whose cathode is of the same metal as the analyte. Its emission spectrum is, therefore, coincident on the absorption spectrum of the vaporized element in the sample. Since atomic spectra are sharp, line spectra, this provides an inherently high degree of freedom from interelement spectral interference. Thus the monochromator need only possess resolution sufficient to remove neighbouring lines (Figure 27.7).

The source is modulated, and the detector amplifier is synchronous with this modulation. This removes much of the ambient light interference from the amplified signal, although photomultiplier saturation must be avoided. The sample is vaporized either by flame, inductively coupled plasma, or in an inductively heated graphite furnace. The heating serves the purpose of removing solvent and organic sample matrix, but it also renders the analyte into the unionized atomic state which then absorbs the source radiation.

Atomic absorption is primarily a quantitative tool, although the freedom from interelement interference can allow it to be used as a qualitative tool as well. The Beer–Lambert law applies strictly to atomic absorption in a flame, although self-absorption and the vagaries of the sample vaporization process mean that a linear plot of absorbance versus concentration is rarely obtained. A calibration curve is usually plotted from measurements on solutions of known concentrations.

Some common analytes, limits of detection, and relative precision are listed in Table 27.3. The most important difficulty with the method is interference due to the sample matrix, which can take many forms (for instance, molecular band spectra, failure of the analyte to dissociate, matrix suppression).

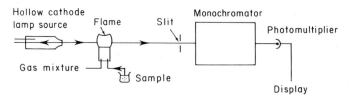

Figure 27.6   A simple design of atomic absorption spectrometer. Solutions are aspirated, nebulized, and swept into the flame where neutral analyte atoms are formed. These absorb the characteristic radiation of the source by an amount proportional to the original solution concentration

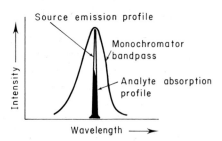

Figure 27.7 Spectral profiles of atomic spectra and the low resolution monochromator typically used in atomic absorption spectrometry

Table 27.3 Common analytes and limits of detection in atomic absorption spectrometry (AAS). Precision is typically 1–4% relative, depending on sample matrix and instrumentation.

Element	Wavelength (nm)	Flame type	Typical limit of detection ($\mu g\,cm^{-3}$)
Aluminium	309.3	N_2O/C_2H_2	0.2
Arsenic	193.7	air/C_2H_2	0.4
Barium	553.5	N_2O/C_2H_2	0.3
Cadmium	228.8	air/C_2H_2	0.002
Calcium	422.7	air/C_2H_2	0.005
Chromium	357.9	air/C_2H_2	0.008
Cobalt	240.7	air/C_2H_2	0.02
Copper	324.8	air/C_2H_2	0.002
Gold	242.8	air/C_2H_2	0.02
Iron	248.3	air/C_2H_2	0.02
Lead	217.0	air/C_2H_2	0.02
Magnesium	285.2	air/C_2H_2	0.0005
Manganese	279.5	air/C_2H_2	0.002
Molybdenum	313.3	N_2O/C_2H_2	0.1
Nickel	232.0	air/C_2H_2	0.02
Potassium	766.5	Ar/H_2	0.004
Rhodium	343.5	air/C_2H_2	0.01
Selenium	196.0	air/C_2H_2	0.2
Silicon	251.6	N_2O/C_2H_2	0.5
Silver	328.1	air/C_2H_2	0.006
Sodium	589.0	air/C_2H_2	0.001
Strontium	460.7	N_2O/C_2H_2	0.006
Tin	224.0	N_2O/C_2H_2	0.2
Titanium	264.3	N_2O/C_2H_2	0.4
Zinc	213.9	air/C_2H_2	0.001

N_2O = nitrous oxide, Ar = argon, C_2H_2 = acetylene, H_2 = hydrogen.

27.2.5   Derivative Spectra

One disadvantage of instruments designed to relate transmittance $(T = I/I_0)$ to concentration using the Beer–Lambert law is the logarithmic dependence between the two. This restricts the sensitivity of simple absorption methods compared with, say fluorescence, where the relationship is approximately linear.

This problem can alternately be expressed as the limited usable range of the change in concentration with transmittance (or *specific response*)

$$\frac{dc/c}{dT} = \frac{0.4343}{T \log T}$$

In practice, measurements above about 1.2 or below about 0.1 absorbance units will be difficult.

An alternative is to record $\partial^2 T/\partial\lambda^2$, the second derivative of transmittance with respect to wavelength. The relationship between the normalized second derivative of transmittance and concentration is now approximately linear:

$$-\frac{1}{T}\frac{\partial^2 T}{\partial\lambda^2} \simeq bc\frac{\partial^2 a}{\partial\lambda^2}$$

The problem of recording second derivatives is one of avoiding noise amplification. For gaseous pollutant studies, this has been done optically (Figure 27.8) (Kroon, 1978). Many modern recording spectrometers, however, now perform numerical differentiation on the digitized absorption spectrum. This is useful not only for extending the limit of detection in certain cases, but also for removing baseline drift due to broad, interfering absorption.

The effect of taking the second derivative with respect to wavelength is to emphasize those regions of the spectrum which have a rapid change in

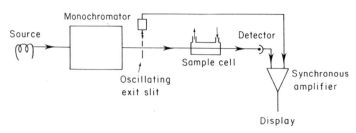

Figure 27.8   A second derivative spectrometer with optical differentiation. The wavelength of the source is modulated at frequency ω, and the detector signal at frequency 2ω is synchronously amplified. This is proportional to the second derivative of transmittance with respect to wavelength

transmittance with wavelength, that is, sharp peaks. This increases the ease with which species which have interfering absorption spectra may be resolved.

27.2.6 Parallel Spectrometers

The conventional UV–visible recording spectrometer, using a continuum light source, monochromator, and scanning mechanism is still widely used. However, the availability of photodiode array detectors with the requisite sensitivity has now made possible *parallel spectrometers*, in which a large region of the spectrum is recorded simultaneously by allowing the dispersed radiation to fall on a high resolution detector array after it has passed through the sample (Figure 27.9).

This design has many advantages:

(a) Accumulation of entire spectra in short time intervals, thus facilitating, for example, the measurement of spectra on chromatograph eluent.

(b) Reduction of instrument noise and complexity through the greater use of solid-state components.

(c) Absence of a monochromator and slits, with resultant increase in light throughput and relative simplicity of optical design.

Advances in the technology of photodiode arrays have also meant better signal-to-noise ratios in light measurement, and a reduction in *blooming* (the effect where one light sensitive element scatters light into a neighbouring element, thus decreasing the effective resolution of the spectral information). Conventional scanning instruments measure with an absorbance precision of typically no better than 10^{-3} units. Careful design in a parallel

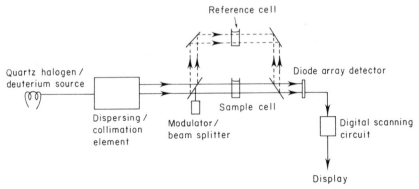

Figure 27.9 A parallel UV–visible spectrometer. The dispersed radiation is not resolved by slits into small wavelength regions, but is allowed to fall on a photodiode array. This allows simultaneous measurement of intensity at all wavelengths

spectrometer can reduce this to 10^{-4} units. This has important implications for multicomponent analysis of compounds with overlapping spectra. The least squares analysis of such spectra can be extended to species with almost-coincident spectral profiles, and to mixtures of 6–8 components.

Such advances must restore UV–visible spectrometry to its former pre-eminence as a general method of trace inorganic analysis.

27.2.7 Infrared Absorption

Until relatively recently, infrared spectrometry was used mainly in qualitative and structural chemistry. Several instrumental developments have changed this situation radically. They are:

(a) Introduction of low cost, large bandpass, single-beam instruments intended for quantitative analysis of gases and liquids (Figure 27.10) (Frant, 1980; Pochon, 1980).
(b) Development of differential reflectance instruments for solids analysis (Figure 27.11).
(c) Introduction of better infrared detectors and sources (notably, for the latter, the use of diode lasers).

Instrumental design for the infrared has always been plagued by the low energy of the infrared photon. This limits the sensitivity of semiconductor detectors, which rely on the incident photon promoting an electron into the conduction band. Efforts to use thermal detectors meet with another problem: their slow response limits source modulation frequencies to regions where ambient infrared contributes significantly to instrument noise. Nevertheless, the conventional double-beam scanning design has found wide

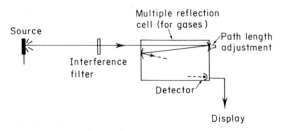

Figure 27.10 A simple single-beam infrared spectrometer suitable for quantitative analysis. The filter may be fixed or variable; in the latter case, a crude spectrum can be obtained, but the chief use is in quantitative work, where the high energy throughput gives a low limit of detection

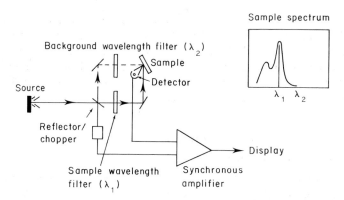

Figure 27.11 A differential reflectance infrared spectrometer suitable for rapid quantitative analysis in solid samples. The diffuse scattering caused by the uneven solid sample requires that the absorbance at the analyte wavelength be corrected by a simultaneous baseline absorbance (see inset)

application in chemistry. Table 27.4 lists some of the common components of this design and their characteristics.

Recent developments of the instrument, concomitant with digital recording and control, include on-line spectral library search, and multicomponent mixture analysis and derivative spectra facilities exactly analogous to those discussed in Section 27.2.5 in connection with UV–visible absorption.

The use of tunable diode lasers, whilst bringing higher routine resolution to infrared spectrometry, also brings instrumental complexities, such as the need for cryogenic tuning and multiple sources. Table 27.5 lists some of the more common diode lasers, their wavelength range and typical power output.

Table 27.4 Major components of scanning infrared spectrometers

Item	Description
Source	Electrically heated element, for example, Globar, Nernst glower
Monochromator	Reflective grating
Modulator	Rotating sector disc (low rotation speeds used because of large time constant of some detectors). Switching of source possible only with lasers
Sample cell	Usually sodium chloride (fragile, water soluble); can use silica for some spectra, also calcium fluoride. Sample may be pressed into potassium bromide disc
Detector	See Table 27.1

Table 27.5   Common diode lasers[a]

Type	Available range (cm^{-1})	Typical tuning range (cm^{-1})	Continuous power (mW)
$Pb_{1-x}CD_xS$	5000–2600	50	0.25
$PbS_{1-x}Se_x$	2600–1200	200	0.50
$Pb_{1-x}Sn_xTe$	1500–300	50	0.10
$Pb_{1-x}Sn_xSe$	1200–300	50	0.10

[a] Changing the doping of a given semiconductor type alters the range which the diode will cover. For a specific diode, tuning is achieved by varying the temperature within a cryogenic range.

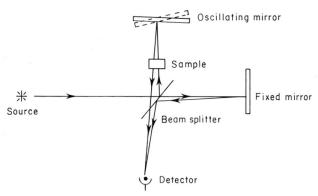

Figure 27.12   A Fourier transform infrared spectrometer. The sample spectrum, as the Fourier transform, is obtained in a time of the order of a mirror oscillation. Resolution is very high (of the order of 0.01 cm^{-1})

Fourier transform infrared spectrometers (Figure 27.12) offer resolution higher than conventional spectrometers. The interferometric principle also gives high optical throughput (*Fellgett's advantage*), and a very short scan time, which makes them suitable for chromatograph eluent monitoring.

27.3   OPTICAL EMISSION

Here are discussed three main types of optical emission:

(a)   Rayleigh scattering (elastic photon scattering);
(b)   Raman scattering;
(c)   fluorescence.

The photon scattering processes involved are illustrated in Figure 27.13.

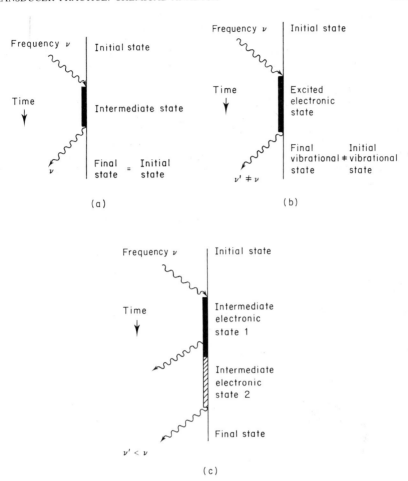

Figure 27.13   The photon scattering processes involved in: (a) Rayleigh
scattering; (b) Raman scattering; and (c) fluorescence

27.3.1   Rayleigh Scattering

This process is commonly referred to simply as light scattering or
nephelometry, and is used, for example, to measure suspended solids concentration in liquids and gases. The basis of this measurement is that the
number of scattered photons can be related to the particle size distribution.
The nature of the relationship depends on the size of the scattering centre
(for example, molecule or particle) and the model used to represent the
process.

Measurements of scattering intensity and polarization give information
about average size and size distribution. For this purpose, a nephelometer

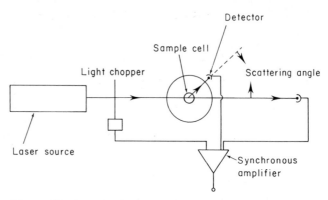

Figure 27.14   A nephelometer. Scattered intensity is measured as a function of scattering angle

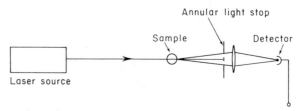

Figure 27.15   A small-angle scattering nephelometer
(or turbidimeter). Limit of detection is good, and the
scattered intensity is largely independent of particle size
within the 1–10 μm range

(Figure 27.14) is used. For simple measurements of suspended solids
number concentration, a fixed-angle nephelometer can be used. However, it
is also possible to measure scattered intensity indirectly, as the loss in
transmittance, provided absorption can be neglected. For this purpose, it is
best to measure the transmittance over a small solid angle (Figure 27.15).
This avoids undue dependence of the observed scattering on particle size.

27.3.2   Raman Scattering

A photon which undergoes a change of vibrational state on scattering
emerges at a frequency shifted from the incident photon frequency by an
amount comparable with a vibrational (or infrared) quantum. The Raman
spectrum from this process may yield three types of information: (a)
structural; (b) qualitative; and (c) quantitative.

For all three purposes, the method offers advantages of ease of sample

Table 27.6 Lasers commonly used in Raman spectrometry[a]

Laser	Common wavelengths (nm)	Typical power (W) (per line)
Argon ion	457.9, 488.0*, 574.5*, 528.7	0.1–10.0
Helium–neon	632.8	0.001–0.1
Helium–cadmium	441.6*	0.01–0.05
	325.0	0.002–0.005
Krypton ion	647.1*	0.2–0.8
	530.9	0.1–0.4
	568.2	0.1–0.2
	676.4	0.1–0.2

[a] It is also possible to use tunable dye lasers as sources; these can be tuned to an absorption band to obtain *resonance* Raman spectra. An asterisk denotes the principal line(s).

handling over infrared spectrometry—glass cells can be used, since the incident and scattered photons are usually in the visible, and aqueous and solid samples, including samples of biological origin, may be handled easily.

All instruments now use laser light sources, Ar^+, He–Ne, and Kr^+ being common (Table 27.6). Instrument design, as for UV–visible absorption, may be either scanning (Figure 27.16) or parallel (Figure 27.17). The requirements on detector sensitivity are stringent, but the higher information throughput of the parallel instrument allows the longer integration times necessary for photodiode arrays.

The monochromator on the scanning instrument needs to have high resolution and low stray light. Careful optical design is required, with double or even triple monochromators being used.

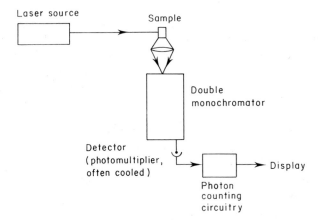

Figure 27.16 A scanning Raman spectrometer

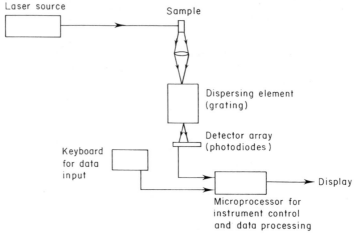

Figure 27.17   A parallel Raman spectrometer

Source intensity modulation and synchronous amplification (Chapter 11) is the usual signal processing design used in the less expensive instruments. For better limits of band detection, photon counting (Chapter 11) is used: a single incident photon strikes the detector and results in a current pulse at the detector preamplifier. Discrimination circuitry allows the rejection of pulses below a present threshold. The pulse is converted to a *clean* TTL pulse which increments a counter N_1. A background count N_2 is accumulated when the source modulator switches the incident photons off. After a suitable counting time (which may even be quantum-statistically limited), a corrected reading $(N_1 - N_2)$ can be recorded. The main controls over system sensitivity are laser power, detector thermal noise (which may be reduced by cooling), and optical losses. The latter can be expected to be greater in the monochromator instruments.

Quantitative analysis usually requires the use of an internal standard because of laser power fluctuations and the difficulty of reproducible beam alignment. Inorganic ions are particularly well quantified or identified using Raman spectrometry, and there are few other transducers suitable for them (Landon, 1979). Raman spectrometry also finds use in general organic qualitative and quantitative analysis, being complementary to nuclear magnetic resonance, mass spectrometry, and infrared absorption.

27.3.3   Fluorescence

We restrict our definition of fluorescence to those processes of relative rapid timescale ($\sim 10^{-7}$–10^{-9} s) resulting from the radiative de-excitation of electronically excited molecules or atoms.

Fluorescence intensity I_f is related to incident intensity I_0 by

$$I_f = \phi_f I_0 (1 - 10^{-abc})$$

where ϕ_f is the fluorescence quantum efficiency and a, b, c are the Beer–Lambert law parameters of absorptivity, path length, and concentration (see Section 27.2.3). For small concentration and small absorption, we may write

$$I_f \simeq 2.303 \phi_f I_0 abc$$

and the intensity–concentration plot is almost linear. It is evident that increasing the incident intensity I_0 decreases the limit of detection in a manner not possible in optical absorption: fluorescence is typically used at the trace level (10^{-4}–10^{-9} g—and lower), with a high selectivity because of the fact that fluorescence is much less common than absorption.

Fluorescence gives a two-dimensional surface spectrum (intensity as a function of both excitation and fluorescent radiation wavelengths), (Figure 27.18). The information content of this spectrum is large, and appears not to have yet been fully exploited as a molecular fingerprint.

Simple instrument design can be intended for recording excitation spectra or emission spectra (Figure 27.19). Interference, or even absorption filters, are used to remove unwanted wavelengths in the dispersion element not associated with recording the spectrum.

More complex designs (for molecular solution spectra) use two monochromators to record the surface spectrum such as in Figure 27.20. Light collection efficiency is important; collection lenses and fibre optics are used.

The choice of source is difficult. Simpler instruments may use deuterium lamps (UV excitation is the most fruitful), but for better sensitivity, a xenon

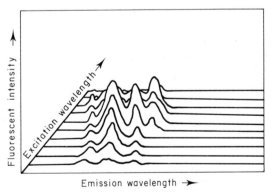

Figure 27.18 A fluorescence spectrum. The fluorescent intensity is plotted as a function of excitation wavelength and emission wavelength

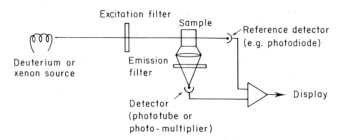

Figure 27.19 A simple filter fluorimeter. The reference detector allows compensation for fluctuation in the source intensity. A modulator is also used in some designs

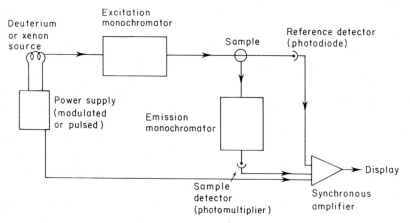

Figure 27.20   A double monochromator fluorimeter

Table 27.7   Species that may be quantified by molecular fluorescence spectrometry

Analyte	Fluorescent moiety	Typical limit of detection
Metals (e.g. Al, Be, Ca, Cd, Cu, Mg, Rh, Sn, Zn, Zr)	Complexing agent (e.g. 8-hydroxyquinoline for Al, Be; flavanol for Zr, Sn; benzoin for Zn, Si)	$\mu g\,cm^{-3}$ to $ng\,cm^{-3}$
Aromatic hydrocarbons	Aromatic ring, especially fused rings	$ng\,cm^{-3}$
Certain food additives (e.g. quinine, riboflavin)	Aromatic ring	$ng\,cm^{-3}$

lamp is preferable. Some recent commercial instruments use pulsed xenon lamps with boxcar integration (Chapter 11) to improve the signal-to-noise ratio and reduce ozone production.

Boxcar integration (like synchronous amplification) may be viewed as a special case of cross-correlation: a fluorescent photon pulse is gated into a counter at a time determined precisely by the source pulse and the lifetime of the fluorescent state. This greatly reduces integration of spurious light from background, incident photons not eliminated by the monochromator, and phosphorescent processes.

Some common analytes and their characteristics are listed in Table 27.7.

27.3.4   X-ray Fluorescence Spectrometry

X-ray fluorescence is no different in principle from fluorescence in optical spectra and so is treated here. The differences in practice are, however, considerable.

A scanning monochromator X-ray fluorescence spectrometer is illustrated in Figure 27.21. X-rays from a broad-band X-ray tube are collimated and directed at the sample. The fluorescent X-rays are passed through a monochromating crystal and fall on a detector (see Table 27.8) (Muggleton, 1978). The sample spectrum of fluorescent intensity against wavelength serves as a qualitative analysis tool. The fluorescent intensity is proportional to concentration; this is the basis of quantitative analysis. Most spectrometers can also be operated in absorption mode, in an exactly analogous manner to optical absorption. For practical reasons, these measurements are generally less useful than fluorescence measurements.

A more advanced, parallel design of spectrometer for routine quantitative analysis is shown in Figure 27.22. The dispersed fluorescent radiation is allowed to fall on a set of detectors placed so that any one detector receives

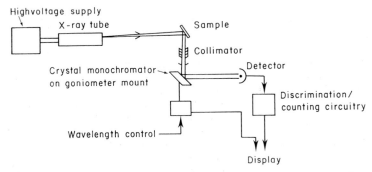

Figure 27.21   A diagram showing the principal components of a scanning monochromator X-ray fluorescence spectrometer

Table 27.8   Detectors used in X-ray fluorescence spectrometry

Type	Description
Scintillation detectors	Thallium iodide doped crystal of sodium iodide; fluorescence detected by photomultiplier; suitable for energy dispersive spectrometer
Proportional counter	A gas ionization detector: the X-ray photon ionizes a gas flowing through a chamber with a central electrode at a potential with respect to walls. Gas ions carry a current which is proportional to the incident photon energy (so suitable for energy dispersive spectrometer)
Semiconductor detectors	Incident photons induce current flow across a biased junction suitable for energy dispersive spectrometer

only fluorescence characteristic of a specific element. The monochromating crystal is usually fixed, and therefore so is the set of elements for which the instrument may analyse. Such designs are usually highly automated and suitable for repetitive multi-element analysis in, for example, metal or mineral samples (Whitelaw, 1980). Up to 30 elements may be quantitated simultaneously (Table 27.9).

A major difficulty with X-ray fluorescence is the matrix effect: each element present in the (usually solid) sample may absorb (attenuate), or fluoresce in addition to (enhance), the analyte fluorescence.

This problem is sometimes lessened by diluting the matrix with a non-fluorescent medium such as sucrose, carbon black, or lithium metaborate. In

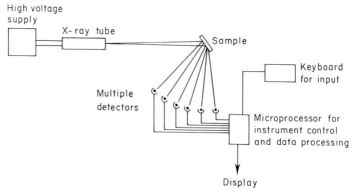

Figure 27.22   A parallel (or simultaneous) X-ray fluorescence spectrometer

Table 27.9   X-ray fluorescence analytes

Generally applied to all elements with atomic number greater than that of magnesium, but has been extended to boron.

Precision is matrix and element dependent, but the following examples are a guide:

Element	Precision (% absolute)
Au	0.04
Ag	0.02
V	0.06
U	0.006
Fe	0.3–1.0
Cu	0.1–0.5
Ca	0.4
S	0.05
Cr	0.2

routine analyses, however, it is more usual to treat the problem as a least squares, multicomponent spectral analysis problem.

A common procedure is to treat the mass fraction C_i of element i as being represented as

$$C_i = (AI_i + B)(1 + \alpha I_j + \cdots)$$

where the first term in parenthesis is the fluorescence (and background) due to the analyte, and the second term is a model of the attenuation or enhancement due to all other elements in the matrix. By making measurements of the fluorescence at different wavelengths, one arrives at a multiple

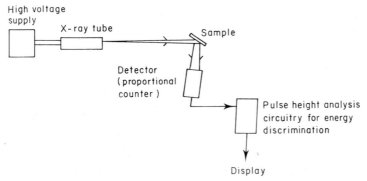

Figure 27.23   An energy dispersive X-ray fluorescence spectrometer

regression problem very similar to that discussed in connection with UV–visible spectra.

An alternative design of instrument, known as an energy dispersive X-ray fluorescence spectrometer, is shown in Figure 27.23. The spectrum is obtained by pulse height discrimination (see Chapter 11) on the fluorescent photon, rather than by a crystal monochromator. The resolution is less but this type is more suitable, through its simplicity, for portable use or process stream monitoring. Some instrument configurations are suitable for borehole analysis. These simple instruments often use radioisotopes as X-ray sources, with ^{241}Am being a common choice.

27.4 MASS SPECTROMETRY

Mass spectrometry is the process whereby a molecule in the vapour state is broken into charged fragments which are then separated according to their mass/charge ratio, and detected according to their number.

A mass spectrometer is therefore a complex set of modules for fragmenting, resolving, and detecting. It can nevertheless be viewed, as are optical spectrometers, as a transducer for converting a unique chemical structure into a highly individual fingerprint of response versus mass/charge ratio of its daughter fragments.

Figure 27.24 illustrates the structure of a mass spectrometer. Table 27.10 lists some common types of fragmentation systems. It may be added that the task of bringing the molecule into the vapour state prior to fragmentation is usually accomplished by using a gas chromatograph as sampling and/or separating device (Watkins, 1977). Samples may also be introduced by direct injection.

The fragments are accelerated by a series of electrostatic slits to pass into the mass–charge separator. The most common types of separator are listed

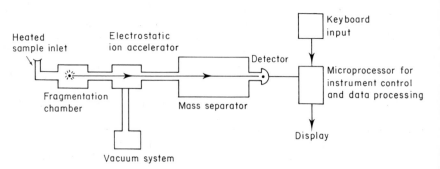

Figure 27.24   The principal components of a mass spectrometer

Table 27.10 Fragmentation systems used in mass spectrometry

Type	Description
Electron impact	Molecules bombarded with 70 eV (usually) electrons; fragmentation is violent, and molecular ion may not be observed
Chemical ionization	An ionized reactant gas (e.g. methane) forms reactive intermediates with itself; these react with sample to produce a simple spectrum. Quasi-molecular ion (at molecular weight plus one) often formed
Spark source ionization	RF spark (e.g. 800 kHz, 100 kV) applied to high melting point solids; equal sensitivity for all elements; high resolution spectrometer required because of wide energy spread of ions; useful for minor elements in semiconductors
Laser microprobe	Focused pulsed laser (e.g. 10^8 W cm^{-2} on 50 μm^2) used to vaporize sample; electron impact used to fragment; time-of-flight spectrometer used; can use small samples and plot element distributions

in Table 27.11. Resolution here is defined as

$$R = m/\Delta m$$

where Δm is the mass difference between peaks at m and $(m + \Delta m)$ when the peaks overlap at 10% of their mean height (this is referred to as 10% valley definition). Detection is accomplished using one of the devices listed in Table 27.12.

A mass spectrum can be used for: (a) qualitative analysis; (b) quantitative analysis; and (c) structural work. There are two common ways in which

Table 27.11 Mass separators used in mass spectrometry

Type	Resolution (see text)	Comment
Magnetic deflection:		
(a) single focusing	200–600	Available with permanent magnet
(b) double focusing	to 35,000	
Quadrupole	400–1200	Rapid, linear scan; simple relative to magnetic types
Time-of-flight	400–600	50 μs scan time; suitable fast kinetic studies

Table 27.12   Detectors used in mass spectrometers

Type	Description
Faraday cup	Ions impinge on insulated cup and absorbs an electron—results in ion current, e.g. 10^{-12}–10^{-15} A
Electron multiplier	Beryllium–copper dynode chain used to amplify primary emitted electron caused by impact of fragment ions; e.g. 10^{-18} A limiting current
Channeltron	Secondary electron multiplier in the form of fine tubes coated with internal layers of secondary emissive material—typical limiting current 10^{-19} A

qualitative analysis can be performed:

(a)   Using the known abundances of isotopes, together with rules derived from the valence (binding power) of atoms, to fix the numbers of atoms of a given type in a molecule whose spectrum is given. This method depends on accurate mass/charge ratio measurement, and high resolution.

(b)   Using the observed mass spectrum as an empirical fingerprint of the molecule, which may then be compared with a library of spectra using a statistical matching criterion. Such is the individuality of the information about a molecule contained in a mass spectrum that this method is demanding of instrument reproducibility rather than resolution (Self, 1978).

Quantitative analysis is achieved, as with many other spectral methods, by comparison of peak heights or areas per unit for standards to those of unknowns. The inherent high resolution of mass spectra does allow multicomponent analysis, particularly in gas mixtures, rather more easily than in common forms of optical spectroscopy.

Structural work is of limited interest in analysis and will not be discussed further here.

A useful process monitor based on mass spectrometric design is the quadrupole gas analyser. This is a simple quadrupole mass spectrometer designed for quantifying light molecules (up to say, molecular mass 100). Samples are pumped directly from the process stream through a restrictor into the fragmentation chamber; the mass separator is merely required to separate whole number masses. A resolution of 150 is often adequate. Quantification is obtained by monitoring the parent ion current (that is, the detector current at molecular mass of the compound).

27.5 GAS AND LIQUID CHROMATOGRAPHY

Formally, chromatography is a separation technique which makes use of spectroscopic and other transducers as detectors of separated chemical species. It would, however, be remiss not to discuss one of the most widely used chemical analysis techniques here. For this purpose, consider the chromatographic system as a chemical transducer which operates on a mixture of compounds and produces quantitative data on the components. Using scanning detectors (for example, fluorescence, UV–visible absorption, mass spectrometry), it is also possible to produce qualitative data.

The basis of chromatographic separation is the differential migration rates of chemical species as they are swept over an active medium by an inactive carrier. The emergent molecules are separated in time from others with a different affinity for the active medium, and subsequently show a detector peak which is clearly differentiated from the baseline signal.

The correspondence between this general description of chromatography and the chemical terms used in gas and liquid chromatography is given in Table 27.13.

A diagram of a gas chromatograph is given in Figure 27.25 (Cram *et al.*, 1980). Mixtures are injected into the injection port and vaporized, if necessary, before being swept by the carrier gas on to the column. The active medium is usually a liquid coated on to a solid support (often finely divided silica). The column may be either packed with this, coated solid, or be of fine bore and have the active medium coated on its internal walls (capillary column). Some characteristics of common packings are given in Table 27.14.

Resolution R in chromatography is defined as

$$R = 2\Delta/(w_A + w_B)$$

where Δ is the separation of two neighbouring peaks, and w_A and w_B are their respective Gaussian widths. R is an indicator of the ability of a column

Table 27.13   Terms used in gas chromatography

Descriptive term	Chromatography term
Active medium	Stationary phase (usually a liquid, coated on a solid support)
Inactive carrier	Carrier gas (usually helium, nitrogen, or argon/methane)
Migration rate	Expressed as retention volume (the volume of carrier gas passed through the column from injection to emergence of the species in question)

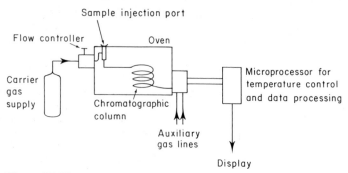

Figure 27.25　Diagram of a gas chromatograph. Dual columns are often used so that detector signal drift can be compensated for when the temperature of the column is varied during a chromatogram

to resolve certain standard mixtures—the greater its value for a given column and type of mixture, the greater the column's ability to separate two compounds of similar affinity for the active medium.

Since separation with finite resolution does not constitute an unequivocal identification of a compound, it is desirable to use a detector with discriminatory powers. Common gas chromatography detectors are listed in Table 27.15.

Quantitative analysis by gas chromatography is based on integration of detector response over a peak. This area is a function of the mass fraction of the compound in the mixture injected, but is insufficiently reproducible for precise quantitative analysis without the use of an internal standard in the mixture. The respective areas A_x, A_s, and mass fractions m_x, m_s of unknown and standard are related by

$$k = m_x/m_s = A_x/A_s$$

Table 27.14　Column packings (stationary phases) used in gas chromatography

Increasing polarity	dimethyl silicone (e.g. OV-101, SE30)
	50% phenyl methyl silicone (e.g. OV-17)
	polyethylene glycol (e.g. Carbowax)
	diethyl glycol succinate
	3-cyanopropyl silicone
	trifluoropropyl methyl silicone
Special phases	porous polymer beads (e.g. Porapak, Tenax)
	triethyl hexyl phosphate

Table 27.15   Detectors used in gas chromatography

Type	Limit of detection $(g\,s^{-1})$	Linear range	Description
Flame ionization	10^{-10}	10^6	Column eluent, including sample, is burnt in a hydrogen/air flame The resulting ions flow across a potential of c. 300 V and the current is amplified. Universal detector, except for a few species (e.g. H_2O, H_2S, N_2, CO_2)
Thermal conductivity	10^{-8}	10^4	Column eluent passed over heated platinum filament; decreased conductivity of a sample molecule increases filament temperature and resistance; universal detector
Electron capture	10^{-13}	10^2	Column eluent ionized by radioactive source; carrier gas ions yield slow electrons which are captured by sample molecules; this gives a drop in the ion current between detector electrodes; highly selective for, e.g. halogen and sulphur compounds
Flame photoionization	10^{-13}	10^3	Similar to flame ionization, but a photomultiplier is used to detect luminescence resulting from combustion of sample molecules. Highly selective for, e.g. phosphorus and sulphur compounds

k is determined from a mixture contining known quantities of analyte and standard.

Liquid chromatography is in many ways analogous to gas chromatography. The sample is transported in a solvent stream across the active medium (Figure 27.26). Greatly increased usage of this technique has resulted from the development of bonded stationary phases (see Table 27.16 for typical columns). In these, the stationary phase molecules are chemical bound to the silica support, and the choice of polarity of these molecules allows the separation of a wide range of organic and inorganic species.

Detection is usually by UV absorption (see Table 27.17). As in gas chromatography, the time of emergence of a compound is usually insufficient for definite identification; a UV or fluorescence spectrum is a useful adjunct to qualitative analysis.

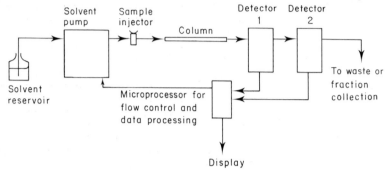

Figure 27.26   Diagram of a high performance liquid chromatograph. Two
solvent streams are sometimes used to vary solvent composition during a
chromatogram. This is called gradient elution, and it improves resolution

Quantitative analysis is the same as for gas chromatography, except that
an internal standard is often unnecessary due to the greater reproducibility
of the separation.

The technique has grown rapidly in recent years, and is applied to a wide
range of products from foods, pharmaceutical products, and metabolites, to
pesticide residues, polymers, plant growth hormones, and molecules of
biological interest such as peptides and proteins. The range of applicability is
similar to, yet wider than that of gas chromatography, as it embraces
involatile substances. Liquid chromatography often involves only dissolution
and filtration of the sample whereas gas chromatography may require

Table 27.16   Typical columns used in high performance liquid chromatography

Type	Use
C_{18} hydrocarbon bound to silica support.	General purpose analytical and preparative column; used for probably 70% of all hplc work, but preferred for separating non-polar species
Phenyl group, and derivatives, bound to silica support	General analytical and preparative column for species of intermediate polarity
Cyano, or amine group bound to silica support	Special purpose analytical column for separating polar species, e.g. sugars
Size separation; cross-linked polymer packing which selectively impedes the elution of species according to molecular volume	The two basic types are suitable for proteins (aqueous size separation column) and synthetic polymers (solvent size separation column). Used in sequence with above columns for very high resolution

Table 27.17   Detectors used in high performance liquid chromatography

Type	Description
UV–visible absorption	A special-purpose double-beam spectrometer, with small sample volume, high sensitivity and low resolution. Usually covers wavelength range 190–700 nm, where many molecules, but especially those with conjugated double bonds, absorb strongly. Requires use of non-absorbing solvent (e.g. hexane, methanol, water) can detect typically 10^{-6}–10^{-8} g
Fluorescence	A simple filter fluorescence spectrometer, with both excitation and emission filters changeable to give selectivity and enhanced sensitivity. Useful for fluorescent analytes occurring in a complex, absorbing matrix. Can detect typically 10^{-6}–10^{-9} g
Refractive index	A differential refractometer with small volume flow cells; usually based on critical angle or beam deviation principle. Poor limit of detection (e.g. 10^{-4}–10^{-6} g), but universal in detectability. Difficult to use where solvent composition is changed during chromatogram
Electrochemical	Based on polarography (q.v.); low limit of detection (e.g. 10^{-6}–10^{-8} g) for selective analytes (see Table 27.20)
Electrical conductivity	Used for conducting analytes (e.g. inorganic ions) in conjunction with modern ion separation columns; can detect 10^{-6}–10^{-8} g

extensive sample clean-up and the preparation of volatile or detectable derivatives.

Process gas chromatographs are available for multicomponent analysis of liquid or gas process streams, and process liquid chromatographs are just emerging commercially. Their use should spread widely in the next few years and enable far more comprehensive chemical monitoring than has hitherto been possible.

27.6   ELECTROCHEMICAL TRANSDUCERS

Here are considered the two most common types of electrochemical transducer, namely, potentiometry (ideally, a zero current measurement) and voltammetry (a small current measurement).

The first type is where the potential of a galvanic cell is measured and correlated with the activity (or, approximately, concentration) of the species of interest. This measurement is based on the Nernst equation

$$E = E^s - \frac{0.059 \text{ V}}{n} \log \left(\frac{[A]^a[B]^b}{[C]^c[D]^d} \right)$$

for an electrochemical reaction

$$aA + bB + ne \rightarrow cC + dD$$

where e denotes an electron and the square brackets denote activity of the species (approximately, concentrations). E is the measured potential, and E^s is the standard potential, measured at defined conditions and activities.

Since a single chemical species is generally of interest, it is necessary to provide a comparison, or reference, electrode against which the electrode process of interest may be measured. Some common reference electrodes are given in Table 27.18.

Direct potentiometry is the most rapidly developing technique in electroanalysis (Kalvoda, 1978). The potential of a selective electrode relative to a standard is measured at very low, or zero, current. The selectivity of the electrode is usually conferred by a transport process through a membrane. The most common example is the glass electrode, which measures pH:

$$\text{pH} = -\log a_{H^+}$$

where a_{H^+} is the activity (approximately, concentration) of the hydrated hydrogen ion in solution. The glass electrode has a thin-walled membrane of carefully doped glass and an internal electrode (often silver/silver chloride). The membrane is immersed in the test solution, together with a reference

Table 27.18   Electroanalytical reference electrodes

Type	Description
Calomel	Two concentric glass tubes, the inner one containing mercury and a paste of mercury, mercury (I) chloride, and potassium chloride. The outer tube contains a solution of potassium chloride connected to the sample solution via a glass frit. The half-cell reaction is: $$Hg_2Cl_2 + 2e \rightarrow 2Cl^- + 2Hg$$
Silver/silver chloride	A silver wire coated with silver chloride and immersed in a solution of potassium chloride saturated with silver chloride. Contact with the sample solution is via a porous glass frit. The half-cell reaction is: $$AgCl + e \rightarrow Cl + Ag$$

Table 27.19 Ion selective electrodes—examples

Type	Analyte	Membrane	Limit of detection (mol dm^{-3})	Reference
Glass	H_3O^+ (pH)	doped alkali metal-silicate glass	10^{-11}	—
Solid-state (crystalline and non-crystalline)	F^-	LaF_3 crystal	10^{-6}	—
	Cl^-, Br^-, I^-	Ag/AgX ($X = Cl^-, Br^-, I^-$)	10^{-6}	(Papeschi et al., 1978)
	SO_4^{2-}	$Pb/PbSO_4$	5×10^{-4}	(Gorina and Ryvkina, 1978)
	Ag^+	Ag/AgS	10^{-7}	—
	Cu^{2+}	Cu_xSe	10^{-6}	—
	Ca^{2+}	CaF_2	10^{-6}	(Ghosh et al., 1978)
Liquid	K^+	valinomycin	10^{-5}	—
	NO_3^-	4,4'-diphenyl 2,2'-bipyridine nickel	10^{-6}	—

electrode and the potential between the two is measured. The very high impedance ($\sim 100\,M\Omega$) of the circuit has provided measurement difficulties in the past, although the advent of field-effect transistor amplifiers has simplified matters.

The principle of other ion-selective electrodes (Table 27.19) is the same, although the membranes depend upon diverse transport processes for their selectivity. There are now hundreds of different ion-selective electrodes, although few have proven reliable in routine laboratory or factory measurements. Perhaps the best is the fluoride electrode (Figure 27.27), which has a solid-state membrane with a higher selectivity than most. Nevertheless,

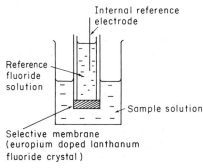

Internal reference electrode

Reference fluoride solution

Sample solution

Selective membrane (europium doped lanthanum fluoride crystal)

Figure 27.27 The fluoride electrode. Selectivity is gained because the rigid crystal lattice favours transport of an ion with the correct radius/charge ratio

some minutes can be necessary for a stable reading and the response is logarithmic with concentration.

Ion-selective electrodes have an advantage of simplicity, in that concentrations can be read directly from test solutions. They should, therefore, be useful in unattended process stream measurements, but much more development is required before they will possess the desired stability, ruggedness, and freedom from interference.

Voltammetric techniques are dominated by polarography and its variants. This is the measurement of the diffusion-limited current between a dropping mercury electrode and a reference. The dropping mercury electrode (DME) is simply a continuously renewed (every 3–5 seconds) drop of mercury which forms under gravity at the open end of a fine glass capillary. The electrode is, therefore, not easily contaminated, and depleted solution is carried away with the falling drop. The DME, therefore, has the virtue of reproducibility. Also, mercury has a large hydrogen overpotential, which means that a wide range of cathode potentials is available without the danger of solution electrolysis.

The theoretical and practical basis of the polarograph is the Ilkovic equation:

$$i_d = knD^{1/2}Cm^{2/3}t^{1/6}$$

where

i_d = (measured) maximum diffusion-limited current
k = constant
n = number of electrons transferred
D = diffusion coefficient of electroactive species
C = concentration of electroactive species
m = mass flow rate of mercury
t = drop time

The current flow in a practical measurement is limited by the rate of

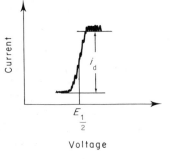

Figure 27.28 A typical d.c. polarogram. The short-period oscillations correspond with the growth and decay of the mercury drop. The voltage $E_{1/2}$ at half current step height is known as the half-wave potential, and is used to select different analytes. The current step height i_d is the diffusion-limited current used in the Ilkovic equation for quantitative analysis

Table 27.20   Limits of detection in polarography[a]

Mode	Typical limit of detection ($mol\, dm^{-3}$)
d.c. sweep	$10^{-4}-10^{-6}$
a.c. modulated sweep	$10^{-4}-10^{-8}$
Anodic stripping	$10^{-6}-10^{-10}$

[a] The limit of detection (LOD) is highly dependent on analyte and experimental conditions.

diffusion of the analyte molecule and its concentration. For given solutions and electrode conditions, a plot of i_d versus C will be a calibration graph from which unknown concentrations may be read.

The measurement can be made simply by sweeping the electrode potential across the values at which an analyte may be reduced, and the result is a d.c. polarogram (Figure 27.28). An improved signal-to-noise ratio can be obtained by modulating the sweep voltage and using phase-locked amplification of the measured current. This is a.c. polarography, and results in an improved limit of detection (see Table 27.20).

A further refinement is anodic stripping voltammetry, where the analyte is first reduced at a fixed electrode (often a mercury film on a platinum or graphite surface) and then oxidized by a rapidly sweeping reversed potential. The advantage is that the limit of detection is improved by effectively integrating the measured current over a larger proportion (perhaps all) of the analyte in the sample. A disadvantage is that only species which undergo reversible electrode processes may be quantitated.

Voltammetric methods are widely used and are relatively inexpensive (Johnson, 1980). They are generally not robust enough for unattended process stream measurement; selectivity is often conferred by altering the buffer in which the measurement is made—a difficult procedure to automate.

A summary of limits of detection for the various polarographic methods is given in Table 27.20.

27.7   STANDARDS

In chemical analysis, standards are required for qualitative and quantitative analysis. In qualitative analysis, only the purity and identity of the substance are important. These are usually assured by modern fine chemical suppliers for simple molecules, although some doubts as to the optical purity (that is, presence of unwanted optical isomers) of samples can occur. These are best resolved by measuring the optical rotation or circular dichroism spectrum of the sample. Indeed, it will often only be in such measurements that optical purity will have significance. Most other chemical transducers are insensitive to optical isomerism.

In quantitative analysis there are often real problems and expense in obtaining analysed samples whose precision and accuracy of analysis are assured. For many common acids and bases, ampoules of carefully standardized solutions are commercially available. Mixtures of known composition are much harder to obtain and the analysis can often be established only by careful collaborative work, involving different laboratories and methods. Such samples, called standard reference materials (SRM), can sometimes be obtained commercially (for example, British Chemical Standards) or from the appropriate national standards body (such as, National Bureau of Standards (NBS), Washington; Physikalisch-Technische Bundesanstalt (PTB), Braunschweig). This applies particularly to standard ores, cements, or alloys which are essential for the proper standardization of methods such as X-ray fluorescence spectrometry. A range of such standards will often be necessary.

In organic analysis, the problems are much more diverse and it is true to say that rigorous standardization of organic analysis is less well developed than in inorganic analysis. There is a tendency for industries requiring organic analyses to have relied on empirical measures, rather than the concentration of a specific substance. For this and other reasons, the commercial demands for precision and accuracy in quantitative organic analysis are usually not as rigorous as those in inorganic analysis.

International bodies such as the Organisation Internationale de Métrologie Légale (OIML) and Codex Alimentarius have laid down specific analytical methods and reagent purities for both organic and inorganic analysis. National standards bodies have also laid down many standard analytical procedures for local contractual and legal purposes. These are usually classified according to the product of interest rather than the method and often rely on the traditional empirical measures referred to above.

There is a great need to establish rigorous standard methods which are largely independent of sample preparation. Recent developments in instrumentation have now made possible this desideratum for a number of important analytes and we must hope that their number increases.

REFERENCES

Cram, S. P., Risby, T. H., Field, L. R., and Yu, W. L. (1980). 'Gas chromatography', *Anal. Chem.*, **52**, 324R–60R.

Frant, M. S. (1980). 'Process analytical instruments', *Anal. Chem.*, **52**, 1252A–3A.

Ghosh, M., Dhaneshwar, M. R., Dhaneshwar, R. G., and Ghosh, B. (1978). 'Response to calcium ions of the calcium fluoride crystal membrane electrode', *Analyst (London)*, **103**, 768.

Gorina, M. Y. and Ryvkina, L. E. (1978). 'Ion selective electrode for determining sulfate ion', *Zh. Anal. Kim.*, **33,** 2269.

Hargis, L. G. and Howell, J. A. (1980). 'Ultraviolet and light absorption spectrometry', *Anal. Chem.*, **52,** 306R–23R.

IMEKO (1980). 'A working list of books published on measurement science and technology in the physical sciences', P. H. Sydenham (Ed.), Applied Physics Department, Delft Technological University, Delft.

Johnson, D. C. (1980). 'Analytical electrochemistry: theory and instrumentation of dynamic techniques', *Anal. Chem.*, **52,** 131R–8R.

Kalvoda, R. (1978). 'Modern polarographic and voltammetric techniques', *Euroanal.*, **3,** 197–208.

Koirtyohann, S. R. (1980). 'Current status and future needs in atomic absorption instrumentation', *Anal. Chem.*, **52,** 736A–8A.

Kroon, D. J. (1978). 'The analysis of ambient air', *J. Phys. E: Sci. Instrum.*, **11,** 497.

Landon, D. O. (1979). 'The development of instrumentation for microparticle analysis by Raman spectroscopy', *Proc. 14th A. Conf. of The Microbeam Anal. Soc.*, pp. 185–90.

Lue, J. T. (1979). 'Wavelength modulated spectrometer implemented with photodiodes', *J. Phys. E: Sci. Instrum.*, **12,** 833.

Malmstadt, H. V. (1980). 'Analytical instruments for the 1980s', *Analyst (London)*, **105,** 1018–31.

Muggleton, A. H. F. (1978). 'Semiconductor x-ray spectrometers', *Report* ANU-P-704 (*Chem. Abs.* 92: 103625m).

Papeschi, G., Bordi, S., and Carla, M. (1978). 'Solid state chloride, bromide, or iodide selective electrodes', *J. Electrochem. Soc.*, **125,** 1807.

Pochon, M. (1980). 'Practical experience with the use of IR instruments as on-stream analyzers', *Chimia*, **34,** 385–96.

Self, R. (1978). 'Developments in mass spectrometry', *Prog. Flavour Res.*, 135–43.

Verdin, A. (1980). 'Process analyzers', *Chem. Eng. (London)*, **362,** 683–6, 690.

Watkins, P. (1977). 'Recent advances in instrumentation for analytical mass spectrometry', *Proc. Symp. Med. Biol. Appl. Mass Spectrom.*, pp. 85–92.

Whitelaw, C. (1980). 'Influence of computer development on the design and application of analytical x-ray instrumentation', *Can. Res.*, **13,** 48–52.

Yotsuyanagi, T. (1979). 'Absorption spectrophotometry: instruments and methods', *Bunseki*, **9,** 592–7.

Handbook of Measurement Science, Volume 2
Edited by P. H. Sydenham
© 1983 John Wiley & Sons Ltd

Chapter

28 F. G. PEUSCHER

Design and Manufacture of Measurement Systems

Editorial introduction

This chapter begins by reviewing the physical design process through which a measurement system is created to suit a particular measurement need.

Most measurement systems involve extensive use of electronic componentry. The material presented here deals with designing systems which contain numerous component parts, discussing how to realize chosen yield systems that will perform to chosen levels of reliability and serviceability.

Also outlined are factors concerned with sales, service, and product promotion of instruments and their specification.

Further relevant readings concerned with the very broad subject of design are available in Chapter 2 (systems in general), Chapter 3 (standards and specifications) and Chapter 14 (closed-loop systems) of Volume 1. All other chapters of this volume are, as is this one, concerned with practical realization of measurement systems.

28.1 INTRODUCTION

Like all technical equipment, measurement systems are intended to perform certain defined functions. In most cases the performance of a technical system S (Figure 28.1) results from interaction between the factors:

Technical equipment (a in Figure 28.1)

Human operators (b in Figure 28.1)

Instructions and procedures (c in Figure 28.1)

The technical equipment can perform as required, under the given circumstances of the system S, if:

(a)　Equipment has been procured with a specification that is suitable for the purpose under consideration and if the equipment does perform as specified at switching-on for the first time (t_0 in Figure 28.2) and will continue to do so during the use period ($t_0 - t_x$ in Figure 28.2).

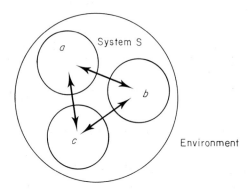

Figure 28.1   Simplified system representation

(b)   The system is operated in the right manner.

(c)   The operating instructions and procedures are workable.

Imperfect performance may occur at t_0, because of a *defect* (equipment does not perform as specified at switching-on the first time) and in the *use period* $t_0 - t_x$ because of a *failure* (termination of performance to specification). In this chapter influences on the system performance caused by imperfections of the human operator, and of the instructions and procedures are not discussed.

Technical equipment can be obtained by buying it or by making it in-house. In both cases it has to be *manufactured*; this can only be done properly using a working plan, called a *design*. The design should be based on a target-specification that is related to the foreseeable function of the equipment in the future-system (Figure 28.3).

In this chapter we will discuss how to design and manufacture measuring equipment in such a manner that, at t_0, it has a high probability of conforming with the specification (*high conformity*), and that it has a high probability to do so during use (*high reliability*).

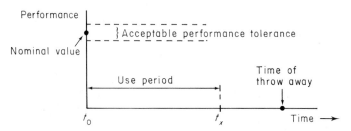

Figure 28.2   System performance versus time

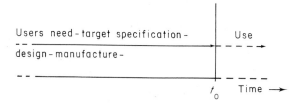

Figure 28.3   System manufacturing aspects

The total lifecycle of technical equipment is illustrated in Figure 28.4 (a combination of Figures 28.2 and 28.3). For more information see Chestnut (1966, 1967).

Except for the do-it-yourself world, two totally different *domains* of thinking and acting have been attached together here: the domains of the *users* and the *suppliers* of equipment. These domains interact in the Figure 28.4:

(a) selling–buying region—a commercial activity;
(b) marketing–specifying region—a technical-commercial activity.

A third *domain of control* regulates the use of technical equipment. This is becoming very important. It sets the requirements to the users (Figure 28.4), and to the specifications.

Regulatory mechanisms control the use and construction of technical equipment, in many cases they are introduced because of the social implications of the expanding use of and human dependence on equipment. One should remember that every system has wanted and unwanted inputs and

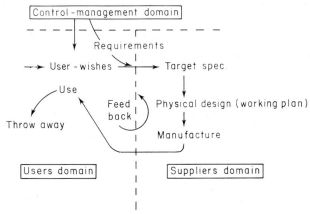

Figure 28.4   Process of making and using equipment

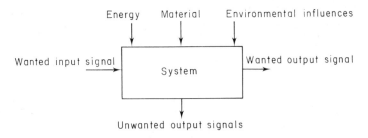

Figure 28.5   Inputs and outputs to system

outputs; Figure 28.5 depicts this. Solving a problem with technical aids often begets other problems.

It is not difficult to find examples of technical systems that do perform in society in a good and responsible manner; nor is it difficult to find examples where they do not. In the latter case the necessary multidisciplinary cooperation between the three domains did not function adequately. In such cases not only the technical domain is to blame.

The interaction of the three domains with the *time domain* has not been illustrated in Figure 28.4, but it is relevant. The time when the user formulates his needs is quite different from that when the user receives his equipment.

As there are many different basic measurement tools, based on different principles of operation, modern measurement equipment can be seen as an application of a variety of different technologies, many of them being in a continuously evolutionary or revolutionary development situation. Electronic technology (in transduction, signal processing and display systems) is one of the most used disciplines in measuring equipment. In many cases, due to the promising properties of newly marketed circuits, the ratio of new to tried and proven portions of electronic equipment can be relatively high. As a consequence high conformity and high reliability are not obtained without special design effort.

A methodical design and manufacturing approach, using modern (computerized) tools and methods (which includes reliability engineering), is a must in electronic system realization. The various documents, such as MIL (1974), written for military systems purposes, are invaluable sources of relevant information.

The emphasis in this chapter will, because of their high proportion of use, be put on the design and manufacture of *electronic* measuring systems.

28.1.1   From User's Need to Marketed Product

This section is concerned with general aspects of the design process. The process starting with a user's need and ending with instructions about how to manufacture a system (Figure 28.4) is called the *design process*. It is not a straightforward, single-pass, process but is an iterative procedure generally consisting of the following:

(a) *Establishing* the requirements of the system, resulting in a target specification.

(b) *Physical design process*, mostly divided into:
  (1) an *early-study phase*, giving an idea about the technical feasibility of performance, reliability, and cost;
  (2) a *consequence-study phase*, on investments, price, manpower, profit;
  (3) a *final-design phase*, development and engineering.

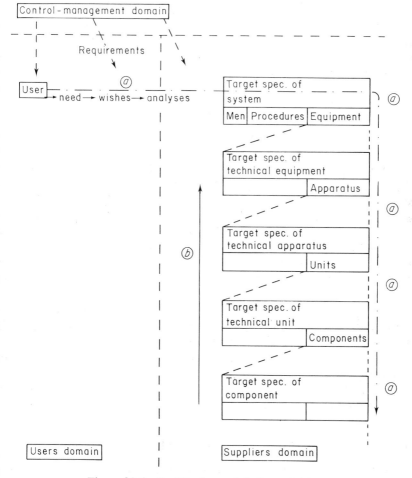

Figure 28.6 Partitioning and design activities

The supplier's domain of Figure 28.4 has been elaborated in more detail in Figure 28.6. The activity '*a*' takes place on every system level (see arrow in Figure 28.6) starting with establishing the system requirements and ending with establishing the component requirements.

Arrow '*a*' illustrates the partitioning of the system (see Section 28.1.3). The physical design activity '*b*' takes place in the opposite direction (arrow '*b*' in Figure 28.6). For more detail see Baker *et al.*, (1970, pp. 1–17) and Baker *et al.*, (1972, pp. 395–440).

Some remarks and explanations are appropriate on these activities:

(a) Complete system requirements in most cases are rarely known at the outset. Usually a rough draft of the requirements is formulated in terms of the basic system parameters *performance, reliability*, and *cost*.

Initial performance requirements are broadly defined, specific detail being added and modified as the development proceeds, possibly in agreement with the capabilities of current technology. It is very important to formulate the requirements punctually in terms that are known to both user and designer. International normalization of definitions and terms can be of great help. A careful up-to-date statement of the requirements should be available at all times to avoid designing a system that meets the first target specification but does not meet the ultimate user's need. It also provides a means for measuring progress.

(b) A general architecture of electronic measuring systems (Figure 28.7) shows the several kinds of electronic functions which may be involved— amplification, modulation, memory, logic, and software.

The equipment realizing the architecture can be made up of a large number of materials, processes, structures, circuit configurations, interconnections, software, and other aspects. A system optimized in every aspect generally is not achievable: however, careful selections can be made to

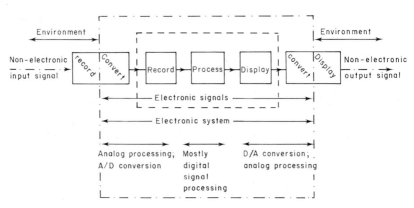

Figure 28.7   Generalized architecture of electronic measuring systems

obtain partial optimization. Factors of importance in connection with such compromises are:

(a) performance tolerances under standard conditions of use;
(b) environmental conditions under which successful (specified) operation should be guaranteed;
(c) compatability with other equipment;
(d) human ability to design, manufacture, test, operate, and maintain the equipment;
(e) flexibility for change;
(f) patents position.

Reliability requirements depend upon the consequence of failures and the cost required to achieve and maintain a specified level of reliability and performance. Sometimes availability is of importance for repairable systems.

The cost involved in a design project should be carefully estimated to establish an appropriate budget and time schedule. The cost price of the system should be estimated at the outset and be regularly calculated and compared with the target during the design phase. This helps avoid unpleasant surprises at the end of the final design phase.

The organization of the design process is illustrated in more detail in Figure 28.8.

The design activity, in most cases, is done by a team of specialists: the design team. The design team includes electronic designers, mechanical designers, quality and reliability and cost-price specialists, and others as needed to form a core. The team should cooperate very closely with, or be extended by, technical–commercial staff members, manufacturing specialists, and other skills that are appropriate.

The work needed to design electronic equipment that is to be manufactured in medium or large production series not only includes the electrical and mechanical construction, choice of realization technology, computer simulation, and laboratory testing but also includes calculation of cost-price,

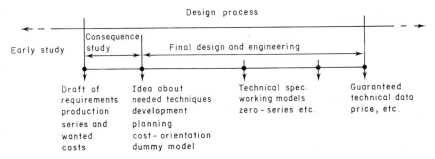

Figure 28.8   Organization of the design process

production of prototypes and zero-series, development of factory test equipment and its use, plus generation of operating and service instructions. The designer's cost per hour has to be calculated. These costs are related to all costs in and out of the factory that deal with the design activity and the total amount of directly worked hours.

The design budget includes the total designer's costs (designer's cost per hour multiplied by the totally planned design hours) plus other costs related to the electrical and mechanical design, the test programs, computer use, materials for prototypes and the zero-series, advice of specialists, engineering services, mechanical and electronic factory tools, (foreseeable) redesign that might be needed to update the design, financial costs, and more.

Design expenditures during the design phase should be carefully compared with the budget and the time schedule.

The design investment can be recovered by sales. An example of the percentage composition of the price of an electronic equipment system manufactured in medium size numbers is as follows. Total sales price is allotted 100%.

Transport and delivery inspection	5%	
Insurance and guarantee service	15%	
Sales costs	25%	
Factory gross profit	15%	
Total	60%	(variations of $\pm 20\%$ are possible)
Costs of electronic components	15%	
Costs of mechanical parts	$7\frac{1}{2}\%$	
Factory salaries	$7\frac{1}{2}\%$	
Packing and handling	$2\frac{1}{2}\%$	
Design budget	$7\frac{1}{2}\%$	
Total	40%	(variations of $\pm 20\%$ are possible)

28.1.2 The Physical Design Process

The process to design, manufacture, and apply electronic equipment can be characterized by (see Figure 28.4):

(a) the design of the target specification;
(b) the physical design;
(c) manufacturing;
(d) application.

Physical design is a matter of teamwork; it has to deal with many disciplines. It draws upon the following groups of knowledge.

Materials technology. A designer should understand structures and the behaviour of complex modern material systems.

Device technology. To use modern circuits effectively requires a fundamental understanding of structure, fabrication, capabilities, and limitations of such devices under all relevant conditions of use.

Contacts and connections. The ever-increasing demands of reliability and small size are continuing to produce new techniques. The continually increasing production rates of contacts and the increasing rate of problems arising from new ranges of application significantly influence the quality of electronic equipment.

Circuit theory. Device modelling to computer simulation and analysis of performance, reliability, thermal design aspects, and optimization of tolerance allocation can be very helpful tools to increase the effectiveness of the design process.

System design, mechanical engineering, human factors, ergonomics, styling, economics, manufacturing process, and market position are also important factors of physical design.

Reliability, maintainability, serviceability. Equipment reliability and durability results directly from design choices and the factory's quality management.

Typical objectives aim to minimize deterioration due to environment and stress, reaching proper ambient conditions within the equipment so as to allow it to operate within design limits, selection of appropriate devices of required (and known) quality level, all of these sought in order to avoid catastrophic and drift failures. Applying proper system structures and redundancy optimizes the probability of the equipment to perform as specified and guaranteed.

Application of proper manufacturing techniques, using skilled and experienced factory workers who are well quality-minded helps provide high production yields and good reliability.

28.1.3 The Aim of Physical Design

Apart from economic and social aspects, the aim of physical design is to produce a working plan detailing how to make a product in a given factory that has a fair chance of meeting the specified performance and cost, and that has a fair chance of maintaining its specified performance in the use period (Figure 28.9) bearing in mind that the equipment quality often is related to the required use-performance and possibly foreseeable misuse!

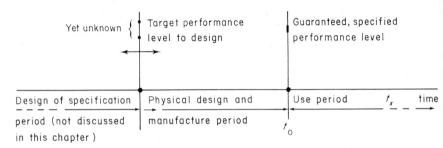

Figure 28.9   Physical design as a time function

During the physical design and manufacturing-period

Activities to be done in this period are as follows:

(1) Establishing the equipment requirements related to performance, reliability and costs, and trade-offs or compromises between them all being based on the target specification (Figure 28.10).

(2) Partitioning, that is the separation of functions (and reliability) and their allocation to various units or modules to satisfy the design requirements; making decisions about which portion of a circuit should be included in a certain module, production technology, hardware or software (Figure 28.11).

Every partitioning decision has to be made as a result of minimizing costs constraints to performance and reliability requirements and in the context of the organization and the permissible tolerances.

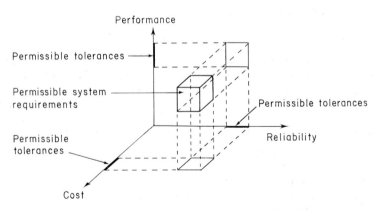

Figure 28.10   Permissible system requirement and basic parameter tolerances

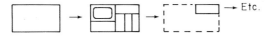

Figure 28.11 Partitioning

(3) Circuit development and application of known (or normalized) circuits or devices; analyses of performance and reliability, permissible tolerances and contacts with component suppliers in the early development phase.

(4) Mechanical development.

(5) Organizational activities, engineering, design of factory tools and methods, incoming goods inspection, mounting, soldering, measuring, trimming, service; selling activities and exhibition-models.

(6) Delivery of equipment (including user and service manuals) to the user.

Generally speaking n different actions are to be performed in this period up to t_0 (Figure 28.9). Every action should give a successful result but it may give an unwanted result measured at t_0 (defined as a defect)—more information is available in Brewer (1972).

The mean probability of an unwanted result may be h defects per action; h should have a very low value. When performing n design and manufacturing actions one can expect a mean of $\mu_n = nh$ defective results. Applying Poisson statistics, the probability of x defects by a mean of μ_n, is given by

$$f(x, \mu_n) = \frac{\exp(-\mu_n)(\mu_n)^x}{x!}$$

In an ideal situation $x = 0$; so the probability of obtaining zero defects in a mean expectable number of μ_n is given by

$$f(0, \mu_n) = \exp(-\mu_n) = \exp(-nh)$$

This is the probability of obtaining the required level of guaranteed specified performance at time t_0 (Figure 28.9), the so-called conformity. Conformity can be maximized by minimizing n and minimizing h.

The number of design and manufacturing actions n, can be minimized by

(a) not developing a new product, but producing (and selling) an older one;
(b) only partially redesigning the equipment;
(c) applying normalized circuits;
(d) applying hybrid or silicon integrated circuits;
(e) efficient organization;
(f) good teamwork.

The number of possible defects per action h, can be minimized by using:

(a) good process control, good analysing and measuring tools and proce-
     dures in both the design and the manufacturing periods;
(b) working experience and know-how about materials, material handling
     and devices;
(c) well educated and skilled, designers and manufacturers;
(d) automation;
(e) application of normalized, known circuits;
(f) application of hybrid or silicon integrated circuits;
(g) efficient organization and quality-minded management.

During the period of use

Apart from unforeseen misuse, equipment may fail in the use period $t_0 - t_x$
(Figure 28.9) by catastrophic or drift failures. These failures can be initiated
by components not being able to withstand the applied stress caused by
temperature, moisture, vibration, voltage or current.

Observing technical equipment in use can characterize this as a mean
probability of trouble of λ failures per hour (λ should be a very low value).
In performing t hours of use one can expect a mean of $\mu_t = \lambda t$ failures to
occur. Following Poisson statistics, the probability of zero failures occurring
in a mean expectable number of μ_t is given by

$$f(0, \mu_t) = \exp(-\mu_t) = \exp(-\lambda t)$$

This stands for the probability of no failures occurring in t hours; it defines
the equipment reliability, $R(t)$. Reliability $R(t)$ can be maximized by:

(1) Minimizing t by not, or seldom, using the equipment. As a rule the
     equipment has been built to be used so this is not a fair proposition.
(2) Minimizing λ. In general the mean number of equipment failures per
     hour, the equipment failure rate λ, is related to the following factors:
     (a) The complexity of the equipment construction; (b) The influence of
     an elementary imperfection on the overall equipment performance; and
     (c) The environmental conditions.
     Minimizing the failure rate λ can, therefore, be achieved by simplifying
     the equipment construction, taking care not to effect its performance.
     This can be done by minimizing the total number of processes needed
     for production, the total number of different elements, the total amount
     of material, the number of interconnections, the area of chip surface
     per element or per subsystem, the number of different production
     techniques, and by using well tried, known, and well controllable
     techniques.

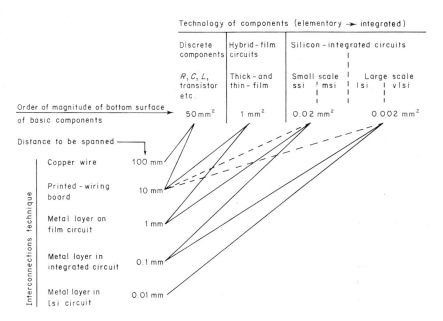

Figure 28.12   Component and interconnection techniques

A general solution is to apply a hybrid or silicon integrated circuit on a much larger scale.

For illustrations of modern discrete components, hybrid thick and thin film circuits and silicon integrated small, medium large, and very large-scale circuits (VLSI) consult the many journals on electronics, electronics design, electronic components, micro-miniaturization and sensors.

In Figure 28.12 the relationship is illustrated between these modern components and their (mostly used) connection techniques.

Illustrations are available in journals on packaging, printed wireboard techniques, interconnection techniques, and manufacturing.

Reliability is also improved by minimizing the total number of components. This can be done by the application of hybrid or silicon integrated circuits: by choosing devices made by well known and qualified suppliers, by well known and tried techniques using automated production processes having a high yield plus good process control and inspection: by maintaining close contact with component suppliers and users and by engendering a good quality minded organization.

Another possibility is the introduction of redundant structures at the system level (such as spare parts and stand-by redundancy) or at the substrate level. Development of very clever and clean circuits, their performance being dependent on well understood principles, can also help as can excellent design, use of well characterized components and adoption of well chosen reliability and stress targets.

28.2 DESIGN FOR PRODUCTION, RELIABILITY, AND SERVICEABILITY

28.2.1 Introduction

Technical products should be designed in such a way that they are able to perform a specified function without giving trouble: the user seeks a product on which he can rely—it should be reliable.

In general we can define the reliability of any technical product or item as its ability to perform a required function reliably. A designer, however, has to specify the products more accurately than is possible by use of the word ability; a working definition of reliability is:

The *reliability* of an item is the characteristic of that item, expressed by the probability that it will perform a specified function under stated conditions for a stated period of time.

For more information about reliability definitions see IEC (1969).

Introduction into the technical specification of the characteristic reliability adds a new dimension to the specification: the dimension of time. The well known characteristics of an item such as colour, weight, output voltage, and temperature mostly can be measured at the instantaneous moment; observed values can be compared with the specified ones. The observed value of the ability to perform as required, for a stated period of time, on the contrary, can only be discussed after that time period has elapsed. The reliability in the specification is an expectation of the extent to which the user of the item can, or will, trust or rely on this expectation. It has to do with, among other things, the following:

(a) Construction of the item: Is the design concept a sound one, can it withstand normal use without too much degradation or wear?

(b) Components: Are they being used in a proper way so as to avoid the occurrence of too many failures?

(c) Know-how: The care, the motivation, and the know-how-not of the designer and all other people dealing with design, production, service, and use of the product.

Many activities in the lifetime cycle of a product, for example formulating and analysing a target specification based on needs and requirements, design development and engineering, production, are creative ones. They all represent parts of the conception–realization stage of a technical product.

They are activities that can be analysed as time functions: writing (that is,

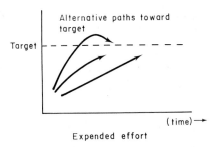

Figure 28.13 Performance growth curves

designing) a target specification, starting with a more or less unclear defined need (at t_0), and ending with a target specification (at t_e) that is understandable to designers as well as to users, without interpretation difficulties, is one of these time functions.

In practice the time function is not a straightforward curve. In Figure 28.13 some practical possibilities have been illustrated, showing that one should take sufficient time to work out the design.

Forms of learning curves have been published in connection with reliability (Bird and Herd, 1976; Duane, 1964; Green, 1976; Summerlin, 1976).

These curves are called reliability–growth curves, but in fact they are design– or development–growth curves as they are illustrating how, during the design stage, the designer is getting closer and closer to the target.

Suppose one is designing a system and at time t (Figure 28.14) it has reached a certain level of performance and reliability, p_t. There is a significant, but unknown, number of imperfections left in the system under design $(\varepsilon_r(t))$ at this moment. In the timespan $(t, t+dt)$ are discovered a number of imperfections of $d\varepsilon_r(t)$—by means of, say, testing and simulation. These imperfections are eliminated by appropriately changing the design.

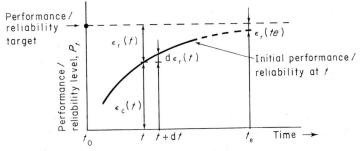

Figure 28.14 Elimination of imperfections

It can be understood that the discovery rate of the imperfections, seen as a decay, $d\varepsilon_r(t)/dt$, has a fair chance of being proportional to the remaining undiscovered defects, $\varepsilon_r(t)$, at time t which is given by

$$-\frac{d\varepsilon_r(t)}{dt} \propto A\varepsilon_r(t)$$

Thus

$$-\frac{d\varepsilon_r(t)}{\varepsilon_r(t)} \propto A\ dt$$

or rewritten,

$$\ln \varepsilon_r(t) \propto -At$$

giving

$$\varepsilon_r(t) \propto Be^{-At}$$

where A and B are constants (see Figure 28.15).

Such an optimization process occurs in many lifetime stages of a technical product: in producing a target specification, in designing a hardware product or in debugging a software program: in having ideas, working out these ideas, realizing them, and comparing the results with the target, making corrective actions and the like. The designer can get very close to the target, but it will take time and care.

When the target can be described very clearly and if the system under design can be tested on all specified characteristics, then $\lim_{t \to \infty} \varepsilon_r(t) = 0$, or very close to 0.

For very complicated items such as large-scale integrated circuits (LSI's), complicated hybrids, and software programs, a 100% testing procedure during the design and manufacture period is virtually impossible to perform. There will remain some imperfections.

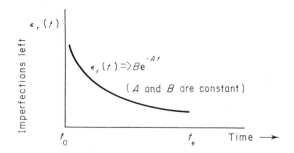

Figure 28.15   Optimization and debugging

These remaining imperfections, $\varepsilon_r(t_e)$, can manifest themselves as bugs, defects or early failures, to use common jargon, during the first operating period.

Reliability is a difficult characteristic to measure and to test. We may expect a certain number of imperfections (from a reliability point of view) to remain in the finished design, manifesting themselves during the first operating time. Reliability is based on the probability of future behaviour of items, about which exact knowledge of their individual behaviour is not known. We can, therefore, also expect a spread in the reliability properties of technical products.

Experience with systems in use, practical behaviour of components, possibilities of service, overhaul and repair, influence of system structure on reliability, mathematical, computer, and other design tools, can be used to help the designer.

28.2.2 Learning from Established Systems

A user of a system, as a rule, is interested in it performing reliably. Therefore, let us first consider from a designer's point of view a repairable system in the user's period. In general a system will consist of technical subsystems, hardware, and software parts, operating procedures and people dealing with the system. Here we only deal with the hardware part of it.

Such a system, in principle, has only one target condition: the condition S_u (short for 'system up'), when it is performing as required. There may be an alternative condition S_d (short for 'system down'), where the system is unable to perform because of needed repair after the occurrence of a failure or overhaul (see Figure 28.16).

Experience with the use of man-made products demonstrates that this is the realistic approach. Among all probabilities there is one certainty: failures will occur. A failure exists when the ability of the system to perform its required function is terminated for one reason or another.

Analysis of the occurrence of failures in systems shows that in systems composed of many and different components, each having different failure

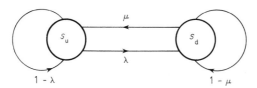

Figure 28.16 System up and down conditions

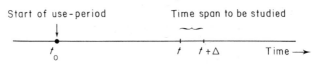

Figure 28.17   Use period

mechanisms, there appears to be a mean number of, say, q failures occurring per N hours of use. One can expect a mean probability of $q/N = \lambda$ failures occurring per hour (compare with Section 28.2.3).

In analysing a repairable system let us assume a service organization repairs the failures with a mean probability of μ repairs an hour. The user is interested in the probability of the system being in an up-condition. This probability is called the system availability, $A(t)$. In conjunction with Figure 28.17 it can be seen that

$$
\underbrace{P_u(t+\Delta t)}_{\substack{\text{Probability} \\ \text{of being in} \\ S_u \text{ at } (t+\Delta t)}} = \underbrace{P_u(t)}_{\substack{\text{Probability} \\ \text{of being in} \\ S_u \text{ at } t}} \quad \underbrace{-\lambda \Delta t P_u(t)}_{\substack{\text{Probability} \\ \text{of failures} \\ \text{occurring} \\ \text{in } (t,\, t+\Delta t)}} \quad \underbrace{+\mu \Delta t P_d(t)}_{\substack{\text{Probability} \\ \text{of repair in} \\ (t,\, t+\Delta t)}}
$$

where the overbraces indicate: Probability of being in S_u ($=a$ failure sensitive situation) over $-\lambda \Delta t P_u(t)$; Probability of being in S_d ($=a$ repair needed situation) over $+\mu \Delta t P_d(t)$.

also

$$
P_d(t+\Delta t) = P_d(t) + \lambda \Delta t P_u(t) - \mu \Delta t P_d(t)
$$

giving

$$
\frac{P_u(t+\Delta t) - P_u(t)}{\Delta t} = -\lambda P_u(t) + \mu P_d(t)
$$

and

$$
\frac{P_d(t+\Delta t) - P_d(t)}{\Delta t} = \lambda P_u(t) - \mu P_d(t)
$$

which, in the limit,

$$
\lim_{\Delta t \to 0} \Rightarrow \frac{dP_u(t)}{dt} = -\lambda P_u(t) + \mu P_d(t)
$$

also

$$
\frac{dP_d(t)}{dt} = \lambda P_u(t) - \mu P_d(t)
$$

A solution can be

$$P_u(t) = C_1 + C_2 e^{-(\lambda+\mu)t}$$

Suppose $P_u(t = t_0) = p$ (for $p \leqslant 1$). Then

$$P_u(t) = \frac{\mu}{\mu+\lambda} + \frac{\lambda}{\mu+\lambda} \, p e^{-(\lambda+\mu)t} = A(t)$$

$A(t)$ is plotted in Figure 28.18. Ignoring trouble in the initial period $t_0 - t_x$, the long-term availability is given by

$$A(t)_{ss} = \frac{\mu}{\mu+\lambda}$$

However, it can be seen that there exist two most disturbing facts:

(a) the quantity p in $P_u(t = t_0) = p$;
(b) the time $t_0 - t_x$, where worst availability can occur.

In the literature it is generally supposed that $P_u(t) = 1$ at time t_0, that is, the probability that the system will be in a working condition at the time $t_0 = 100\%$.

This is, however, not a natural fact at all. The designer and the producer should not rely on p being 1 for it will usually become excessively costly to try and approach $p = 1$.

For example, in a system consisting of 1000 components all of them must be in a working condition for the system to be able to perform as required.

Suppose there exists a probability of $q = 0.1\%$ that one of the 1000 components has a defect (at t_0). A defect is a lack of the ability to perform when called upon at time $t = t_0$. What is the probability of the system having no defects at t_0? Taking into account that a defect should be a rare event, Poisson statistics will be applicable. Let $f(x, N)$ be the probability of

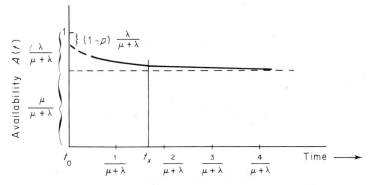

Figure 28.18   Availability versus time (theoretical)

occurrence of x defects with a mean expectation of N defects, thus

$$f(x, N) = \frac{e^{-N}(N)^x}{x!}$$

In our case x should equal 0 and $N = 0.1 \times 0.01 \times 1000 = 1$, resulting in $f(x, N) = f(0, 1) = e^{-1} = 0.37$ or 37%. We can, therefore, expect that 37% of the systems built to these figures are able to operate at t_0 and that 63% of these systems have no ability to perform at t_0. They are defective; in this case $p \neq 1$, but $p = 0.37$.

Would it be feasible to achieve $p = 0.99$? This would require $e^{-N} = 0.99$, resulting in $N = 0.01$ and $q = 0.0001\%$. The probability needed of a defect in a component can be seen to be very low. The designer must pay attention to this and have tools to attack the problem.

The time $t_0 - t_x$, in most literature $A(t)$, is supposed to reach the steady-state value following the initial exponential slope (Figure 28.18). In practice this is not true in many cases unless special effort is expended. In the start-up period of the use time $(t_0 - t_x)$ service people have little experience and initially the repair time $1/\mu_i$ can be very much greater than $1/\mu$. Moreover, it is general experience with systems that in the start-up period more failures will occur than in the steady-state time, so $\lambda_i \gg \lambda$. With $\mu_i = b\mu$ and $\lambda_i = a\lambda$, a more realistic $A(t)$ curve results, as shown in Figure 28.19.

Possible actions that can be employed by the designer to avoid this early-use effect, or at least to minimize it, are as follows:

(a) Use easily repairable systems by incorporating unit construction, easy fault-tracing procedures, easily removable units. For example, introduction of hybrids or integrated circuits can greatly help ease of repair.

(b) Provide good instruction to the service organization, using components and spare parts already known to them.

(c) Minimize λ for the components used.

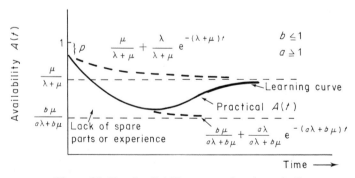

Figure 28.19   Availability versus time (practical)

Minimizing λ means carefully analysing the quantity of λ. What system quality influences λ? How can it be measured? How can it be minimized? How can one compare the system under design with the target? Is λ a meaningful system parameter?

28.2.3 Hazard Rate and Failure Rate

The termination of the ability to perform is called a failure. There are catastrophic failures (performance totally broken down) and marginal failures (system still performing, but out of tolerance), drift failures, degradation failures, wear-out failures, misuse failures, early failures, and more, all of which have quite different failure mechanisms, origins, and definitions.

In Section 28.2.2 it was defined that there is a probability of λ failures occurring each hour in an operating system. A more general approach (a more correct one) is to assume a system having failure-free times, t_s, under the condition of which the termination of each failure-free time is due to different failures which are mutually independent. For this t_s is a stochastic variable $\underline{t_s}$ having a continuous failure distribution function $F(t)$ and a probability density function $f(t) = dF(t)/dt$ (Figures 28.20 and 28.21). $F(t)$ is a cumulative distribution function of a non-negative random variable, for which

$$F(t) = 0 \quad \text{for } t < 0$$

$$0 < F(t) < F(t_t) \quad \text{for } 0 < t < t_t$$

$$F(t) \rightarrow 1 \quad \text{as } t \rightarrow \infty$$

The system reliability $R(t)$, or the probability of being in condition S_u by time t, can be represented as

$$R(t) = 1 - F(t)$$

For practical purposes (for, say, operation, maintenance figure), we are often concerned with the rate at which failures occur.

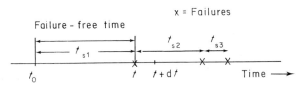

Figure 28.20   Failure free times

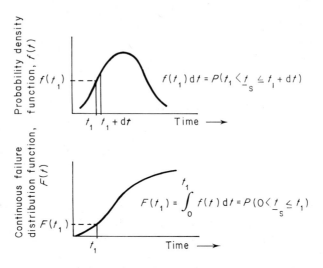

Figure 28.21 Failure distribution and failure density
curves

The probability of failure in an interval between t, $t+dt$ is given by

$$F(t, t+dt) = P(t < \underline{t}_s \leq t + dt/\underline{t}_s > t)$$

$$= \frac{P(t < \underline{t}_s \leq t + dt)}{P(\underline{t}_s > t)} = \frac{f(t)\,dt}{1 - F(t)}$$

$$= \frac{f(t)\,dt}{R(t)}$$

calling $f(t)/R(t) = h(t)$, the hazard rate, we have finally

$$F(t, t+dt) = h(t)\,dt$$

The mathematical product $h(t)\,dt$ is the conditional probability of failure in the interval between t, $t+dt$ for items that were performing at time t; the hazard rate $h(t)$ is a rate of decrease of $R(t)$. The parameter $h(t)$ can be evaluated from

$$R(t) = \exp\left(-\int_0^t h(t)\,dt\right)$$

The mean time to failure (MTTF) is the mean value of $\underline{t}_s$ which is given by

$$\text{MTTF} = \int_0^\infty t f(t)\,dt$$

If the system is operated until it fails, then repaired and operated again until

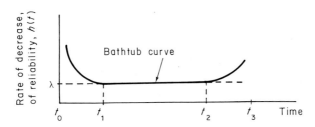

Figure 28.22   Hazard rate bathtub curve

it fails, with alternating periods of repair and operation thereafter, it is conceivable that the same $F(t)$ applies independently to all periods of operation. If so, the MTTF may also be considered as the mean time between failures (MTBF).

Experience with many systems usually shows that $h(t)$ has a character similar to that shown in an idealized form in Figure 28.22, the so-called bathtub curve. In the timespan $t_1 - t_2$, $h(t)$ can be considered as being nearly independent of time. The number of failures occurring each hour is constant and the mean value of $h(t)$ in this timespan is λ, the failure rate. This is the same λ discussed in Section 28.2.2.

In the interval (t_1, t_2) the expression for the reliability is given by

$$R(t) = e^{-\lambda t}$$

for which the MTTF and the MTBF approach $1/\lambda$.

In summary:

(a)   the constant failure rate concept only holds for interval $t_1 - t_2$;
(b)   the $h(t)$ concept also holds for subsystems and components;
(c)   interval $t_0 - t_1$ is called the early failure period;
(d)   interval $t_2 - t_\infty$ is called the wearout time;
(e)   interval $t_1 - t_2$ is called the random failure period.

Design approach

According to the definition of $R(t)$, a system should perform its function (it should perform and should not have completely broken down) as specified. It should not be out of tolerance under all relevant conditions during a stated period of time. Misuse is excluded unless it was a design parameter.

In Figure 28.23 failures of catastrophic and drift form are represented. The probability of not having broken down can be connected with catastrophic failures $P_c(t)$. The probability of not being out of tolerance can be connected with drift or marginal failures $P_m(t)$. Evaluation mathematically

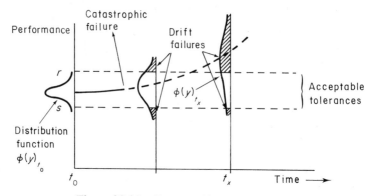

Figure 28.23   Catastrophic and drift failures

results in

$$R(t) = P_c(t) P_m(t)$$

Remembering that

$$P_m(t) = \int_r^s \phi(y)\,\mathrm{d}y$$

the definition of a marginal failure depends on the chosen limits r and s and on the sensitivity of the system structure to drift. The drift of system characteristics can mostly be minimized by a proper design.

It is of no use, however, designing a system with an high $P_m(t)$ when it is too weak to withstand normal use without breaking down due to a catastrophic failure.

A wise design approach, therefore, is to design a system that is able to withstand normal use with a low probability of catastrophic failures, and high $P_c(t)$. This leads to criteria concerning the basic system concept and the basic quality level of components. Ensure that the system will be as insensitive as is required to drift of components.

28.2.4   Influence of System Structure

A technical system can be characterized by its architecture. For instance a medical intensive-care system has to pick up, transport, and display information concerning patients.

It can be described by its outfit. For example for patient number 1, the patient system P_1, will consist of electrodes and amplifiers for measuring electrical heart actions (e) as well as the mechanical heart actions (p). There are cables C_e to transport electrical information and cables C_p to transport

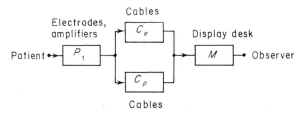

Figure 28.24   Schematic of a patient monitoring system

mechanical information. There is also the main display desk M with alarm units, the power supply, and recorders (Figure 28.24). Technical realization of the system consists of components and other subsystems constrained to work together by means of hard wired copper conductive wiring, by thin or thick film conductors or the like, IC's, conductive elements, mechanical switches, electronic switches triggered by software, and so on.

What is the influence of the system structure (see Figure 28.24) on the reliability expressed as a hazard or a failure rate? Both rates represent a conditional probability of failure. What is the probability of no failure and how can the system reliability be described in terms of subsystem characteristics?

A handy tool for doing this is the so-called failure tree analysis (FTA), which leads to a Boolean expression for failure. The system of Figure 28.24 leads to the failure tree of Figure 28.25.

Before starting to construct a failure tree one should carefully define what constitutes a failure.

In the above example no failure can be said to occur if the electrical (e) and mechanical (p) patient information is available on the desk and also if electrical or mechanical information is available on the desk. Failure occurs if there is no information available on the desk.

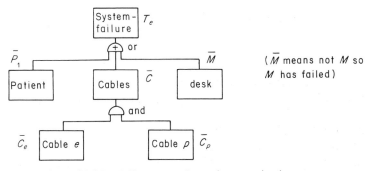

Figure 28.25   Failure tree of a patient monitoring system

Be mindful that the top event T_e represents a fault condition of the system (S_d). There is a failure ($= T_e$) if P_1 has failed or C_e and C_p have both failed, or M has failed, that is:

$$T_e = \bar{P}_1 + \bar{C}_e\bar{C}_p + \bar{M}$$

This results in:

$$T_e = \bar{P}_1 + \overline{C_e + C_p} + \bar{M}$$

From which

$$\bar{T}_e = \bar{P}_1 + \overline{C_e + C_p} + \bar{M}$$

leading to

$$\bar{T}_e = P_1(C_e + C_p)M$$

As T_e stands for the condition of failure (S_d), $\bar{T}_e$ gives the condition of required performance (S_u). The system reliability is the probability of $\bar{T}_e$ with respect to time:

$$R(t) = P(\bar{T}_e, t)$$

Reliability is:

$$R(t) = P\{(P_1(C_e + C_p)M), t\}$$

In the case of a simple system, as in Figure 28.24, it is simple to identify the components of this equation for $R(t)$. When dealing with much⁺ more complicated systems failure tree analysis is said to be a handy tool (Barlow et al., 1975).

Series and parallel configurations in general

A system, composed of a number of components that will fail as soon as any single component fails, is said to have a series configuration from a reliability point of view; for example, the units P_1 and M in Figure 28.24.

A system, composed of a number of components, that will fail only if all components fail is said to have a parallel configuration. Often the word redundancy is used in this context, for example, the units C_e and C_p in Figure 28.24.

The expression for $R(t)$ related to Figure 28.24 involves a chain of series and parallel factors. Most systems can be divided into series and parallel factors, sometimes into combinations. With this approach we can assess the system reliability from the individual component reliabilities with the help of series and parallel calculating rules. Now follows derivation of calculating rules for series and parallel configurations and means to assess the reliability of components expressed in failure rates.

Series configuration: probability of performing (without time consideration)

The system of Figure 28.26 consists of n series components of x_i. All units x_i are supposed to have two conditions: S_u (working) and S_d (failed). The probability of x_i being in S_u is $P_u(x_i)$. With all units independent, the probability of the system being in S_u is the intersection of $P_u(x_i)$ (Figure 28.27):

$$P_u(\text{system}) = \bigcap_{i=1}^{n} P_u(x_i) = \prod_{i=1}^{n} P_u(x_i)$$

This intersection should not be empty. When dealing with series structures be sure that the working environmental conditions of components overlap. Optimizing $P_u(\text{system})$ means choosing components having a high probability of performing under the same environmental conditions.

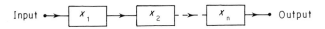

Figure 28.26   Series system for overall reliability calculation

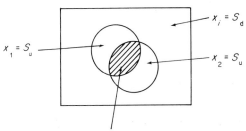

Intersection of x_1 and x_2 : $x_1 \cap x_2 \Rightarrow P_u(x_1) P(x_2)$

Figure 28.27   Series system reliability Venn diagram

Series configuration: reliability (a time function)

The probability of a unit x_i being in S_u for a given period of time $(0, t)$ is given by:

$$P_u(x_i, t) = R_{x_i}(t) = \exp\left(-\int_0^t h_i(t)\, dt\right)$$

For the series configuration of n units x_i the probability of good performance for a given period of time is given ny:

$$P_u(\text{system}, t) = R_s(t) = \prod_{i=1}^{n} \exp\left(-\int_0^t h_i(t)\,dt\right) = \exp\left(-\sum_{i=1}^{n} \int_0^t h_i(t)\,dt\right)$$

When the hazard rate is constant and equal to λ_i this results in:

$$R_s(t) = \exp\left(-t\sum_{i=1}^{n} \lambda_i\right)$$

In Section 28.2.3 it was shown that system reliability

$$R_s(t) = \exp\left(-\int_0^t h(t)\,dt\right)$$

From the last two equations can be derived

$$\int_0^t h(t)\,dt = t\sum_{i=1}^{n} \lambda_i$$

This means that $h(t)$ is not a time function and that the system failure rate equals

$$\lambda_s = \sum_{i=1}^{n} \lambda_i$$

with $\lambda_i =$ the failure rate of series component x_i. The number n influences the probability of being in S_u. This is shown in Figure 28.28.

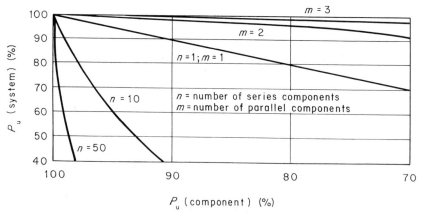

Figure 28.28   Influence of n series and m parallel units on system reliability

Parallel configuration: probability of performing (without time consideration)

Here, for Figure 28.29,

$$P_u(\text{system}) = \bigcup_{i=1}^{m} P_u(y_i) = 1 - \bigcap_{i=1}^{m} \bar{P}_u(y_i) \quad \text{(Figure 28.30)}$$

When optimizing the reliability of parallel systems ensure that the parallel components do perform as required under different environmental conditions and timespans.

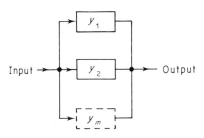

Figure 28.29 Parallel system for reliability calculation

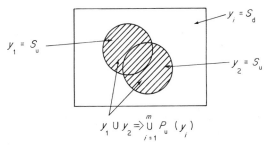

Figure 28.30 Parallel system reliability Venn diagram

Parallel configuration: reliability (a time function)

$$P_u(\text{system}, t) = R_s(t) = 1 - \prod_{i=1}^{m} [1 - R_{yi}(t)]$$

so

$$\exp\left(-\int_0^t h(t)\, dt\right) = 1 - \prod_{i=1}^{m} [1 - \exp(-\lambda_i t)]$$

The system failure rate is not constant in this case. The influence of the number m of parallel components is shown in Figure 28.28.

With the help of such equations it is possible to predict the reliabilities of systems and express them in terms of failure rates of components. To do this it is necessary to have a thorough knowledge of the failure rate of components (see also Section 28.2.5).

The literature contains many tables, lists, and curves representing failure rates of components under various use conditions. Refer to IEC (1969), MIL (1974), ESRO (1975), and suppliers' catalogue sheets on components.

Failure rates of components are obtained as a function of environmental stress (for instance, ambient temperature), internal stress (for instance, dissipated power), and other parameters of interest. Very little, however, is said about the time interval during which these values are applicable (in Figure 28.22, $t_1 - t_2$).

Providing a reliability prediction procedure is reasonably simple. The main problem is providing adequate component-part-identifying information on stress and other influence parameters that will enable the designer to select the appropriate failure rate.

Complete reliability prediction using stress analysis is within the capabilities of computer systems. The bottleneck in using the programs is obtaining and storing all of the component information that is needed.

28.2.5 Hazard Rates and Failure Rates of Components

Measurements and experience with the use of large quantities of all kinds of components resulted in the typical $h(t)$ curves of Figure 28.31; also indicated are $F(t)$ and $R(t)$.

How can we comprehend this different behaviour of components with respect to time and component type? For an electronic designer it is very fortunate that many electronic components show a constant failure rate over a very long usable period. Thus for electronic equipment, with the help of this constant (and published) failure rate, we can design systems that are strong enough to withstand normal use without catastrophic random failures. (This is a main factor in the practical possibility of developing vastly complex systems in the electronic regime.)

Using components in technical equipment means subjecting them to all kinds of stress, for instance voltage, current, humidity, temperature, and mechanical stress, both constant and cyclic. Stresses can manifest themselves in chemical reactions, corrosion, electrolysis, whisker building, and many unexpected ways. Such unwanted processes can influence the ability of a system to perform reliably.

Most modern components are made by a mass production process, that is they are liable to tolerance spread. Not all components made in one

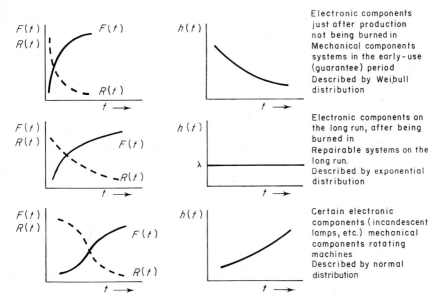

Figure 28.31   Curves of $h(t)$, $R(t)$, and $F(t)$ of components

production run will have the same quality level. Just following production many defective components (not performing at t_0 due to such things as open circuits, leakage, bad bonding, and scratches) are observed. Further it is found that components that do perform a while, but that are not strong enough to withstand normal (specified) uses, fail after a short period of use. Fortunately it is possible to produce components that do meet the specification if used properly. Designers must take into account the distribution of individual component parameters.

We can divide these parameters into those dealing with strength and those with performance (see Figures 28.32 and 28.33). The rated stress B should be less than the strength S and the required performance G should include the component performance F.

Stress is a very complicated concept. It is generally not a constant parameter so due to stress variations a small number of components fail a short time after being put into service at t_0 (section ef in Figure 28.32). The initial $S(y)$ distribution changes somewhat from $S(y)_{t_0}$ to $S(y)_{t_x}$.

Stress also initiates a deterioration process over time which decreases the strength of the components. It can be arranged that the probability of no failures $R(t)$ (in the period when deterioration is proportional to time) leads to a failure rate which is independent of time and only a function of stress and strength (Shooman, 1968; Umedo, 1978). The occurrence of random

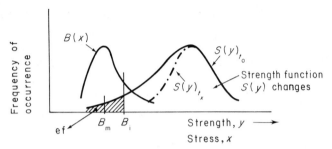

Figure 28.32   Probability density of stress and strength for a
typical component

failures in this period can be understood as the failure of components that
have become too weak under peak stress conditions. They are random
failures occurring under random stress conditions. This model holds in the
time interval $t_1 - t_2$ (Figure 28.22). When the components have reached a
certain degree of weakness the chemical and other deterioration processes
may cause an increase of degradation of strength in time. Then we see an
increasing hazard rate ($t > t_2$, Figure 28.22).

In the same time the performance F may change (Figure 28.33) and drift
failures can occur provided we did not eliminate their effect by feedback or
other well known techniques.

As is illustrated in Figure 28.23, and discussed in Section 28.2.3, apart
from catastrophic failures, drift failures, originated by drift of electrical
parameters of components, can have considerable influence on system
reliability.

Circuit tolerance analyses for variations in component characteristics are
preferably checked by calculation based on worst-case conditions (considering the effect of simultaneous combinations of components characteristics at
the worst-direction limits), or based on Monte Carlo or other probabilistic

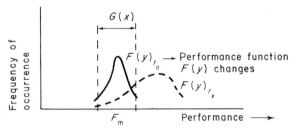

Figure 28.33   Probability density of required and typical performance of components

approaches. A number of computer programs are available for these kinds of analysis on various system characteristics.

With the help of such programs circuits can be designed in such a way as to avoid drift failures to a chosen confidence level of acceptance. Annoying failures such as intermittants, failures that only occur at some special sequence of pulses in digital systems or at some hot-spot temperatures are not discussed in this chapter. For information refer to Bird and Herd (1976).

Some interesting questions for a designer are as follows:

(a) How are defective components eliminated at t_0 (Figure 28.22)?
(b) How are weak components (showing failures in $t_0 - t_1$) eliminated?
(c) How are components prevented from appearing in the situation corresponding with $t_2 - t_\infty$?
(d) How accurate are the values of λ in the tables?

Consider these in turn.

(a) Eliminating defective components at t_0

The answer is to choose good and well tried components from a well known supplier, one having a good quality management organization. Also choose the right assurance quality system (AQL) (Beckmann et al., 1977; Pomplum, 1977; Wetherill, 1977).

(b) Eliminating weak components showing failures in interval $t_0 - t_1$

Eliminating components that are not strong enough is a process that will proceed without any purposeful action being needed during the first months of operation. This is not a very useful method (costs of guarantee, trouble during the running-in period, for instance). Using the components for a short time at a stress of say B_i (Figure 28.32) would not help much as it will take time for components to break down. There is, however, a possibility to accelerate this process and have it ended (or nearly ended) at the moment of delivery to the user.

In many cases the processes initiating the deterioration of components are of a chemical nature. As chemical processes are dependent of temperature, so are the deterioration processes.

The Arrhenius Law states that:

$$r = A e^{-E/kT}$$

where the time rate of chemical processes depends on A, a constant, E is the activation energy of the degradation process in electronvolts (eV), k is the Boltzmann constant and T the absolute temperature in kelvin (K). When such a chemical process initiates degradation of a component, r can be seen as degradation rate for temperature activated degradation.

In the case of an exponential distribution of (change) failures in time, then, as rule, the reliability (which equals the probability of survival) of a component can be characterized by its (constant) failure rate λ.

In the case of a time-linear deterioration process it can be understood that the probability of failure must be proportional to r and that the MTTF is proportional to $1/r$. Hence $1/MTTF_i (= \lambda_i)$ is proportional to $A \exp(-E/kT_i)$. This equation opens up a possibility of accelerating the process of degradation of potentially weak components and eliminating them (that is before being soldered onto the printed circuit board) by a burning-in procedure operated at a higher burning-in temperature, T_a.

Measuring the MTTF at two temperatures, T_n (normal) and T_a (accelerated) yields for the deterioration-acceleration factor τ the following:

$$\tau = \frac{MTTF_n}{MTTF_a} = \frac{\lambda_a}{\lambda_n} = \exp\left[-\frac{E}{k}\left(\frac{1}{T_a} - \frac{1}{T_n}\right)\right]$$

Other possibilities to obtain and to interpret acceleration can be described by the concept of complete acceleration and the Eyring model (see (Shion, 1978; Lycoudes, 1978). In this way we are able to eliminate weak products in a very short time. It is important to be mindful that many deterioration processes having quite different activation energies may be involved.

Burning-in is a tool for getting rid of weak components and for discovering weak points in the system. It can be seen as a part of a more extensive screening process consisting of a number of appropriate tests aimed at revealing hidden potential failures and defects.

All of these tests should be coupled with known or expected failure and defect mechanisms. Therefore, it is very important to study and to analyse all components and subsystems used and to obtain a good working relationship with the supplier in order to get the much needed information and trust.

(c)   Avoiding coming into the timespan $t_2 - t_\infty$

In general this problem can be tackled by introducing a suitable organization of inspection, data collection and corrective or preventive maintenance (see Baker *et al.*, 1972).

(d)   Accuracy of the published values of λ

For a supplier assessment of component failure rate data is a question of working (sometimes for many years) on design, qualification testing, reliability monitoring, failure analysis, and product processing. Over the lifetime of

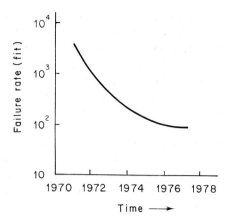

Figure 28.34   Improvement of failure
rate of a typical component versus
time. 1 fit = 10^{-9} failures/hour

a product one can obtain reliability improvement. Figure 28.34 shows a
trend in reliability improvement for a semiconductor of a special type.

The values of λ published are as accurate as they can be at the publishing
date. A designer should always use the latest publication for reasons
mentioned above.

In the early production stage of a component it is not possible to measure
and to assess failure rates with high confidence levels. Accelerated testing
can help obtain an idea of λ for a rather unknown new component (see
Manabe, 1978).

Component failures are mostly connected to wanted and specified qualities. The definition of failure in such a case has to do with the termination
of the wanted component function, either completely or beyond a stated
limit.

Sometimes it is possible to define failure by observing secondary or
unwanted qualities such as noise and third harmonic signal. Noise measurement (sometimes noise injection) and third harmonic indices can give
information about oncoming degradation and lack of strength, even before
desired qualities show effects of deterioration.

28.2.6   General Comment on Ways to Improve Reliability

The more defects and potential failures one can detect in an early (and not
complicated) system stage the lower the costs will be.

So apart from excellent quality inspection of incoming goods, good AQL
contracts when applicable, and well designed screening, including burning-in

procedures, it is important to control the custom in-house components, bondings, and soldered connections. Considerable knowledge exists on high-reliability connections.

Avoid constructing too large a unit with too many complicated components, one that is difficult to test, in which it is nearly impossible to define a failure, has high fault-finding costs, and other factors based on unsufficiently maturated technologies. It is worth analysing the advantages of hybrid LSI and VLSI circuits in these contexts.

Another aspect of burning-in is the ageing effect on components as a certain stabilization of component properties can be reached. Proper use of components can help better the reliability. Derating is a helpful method in this context. Hybrids and IC's can increasingly be expected to be, and to become, items that can help to keep in balance electronic systems of the future.

28.3 SALES, SERVICE, AND PROMOTION

The increased use of highly integrated circuits and subsystems and the use of complicated computerized measurement systems inevitably leads to the requirement of very close cooperation between the user domain, the management domain, and the design and manufacturing domain (Figure 28.4). Two interfacing disciplines between them are the sales and the service disciplines. They have a central role in operating measurement systems satisfactorily.

Satisfactory operation of technical products depends upon using products with:

(a) specified performance meeting the requirements;
(b) good appeal, terms, and time of delivery;
(c) acceptable costs;
(d) ability to continue to perform as specified and that are easily serviced for moderate costs.

The sales discipline covers the factors (a)–(c), service covers factor (d).

In a, perhaps, somewhat idealized way one can say that the service discipline helps the users to continue to use the system and that the sales discipline helps the potential user to decide what system has to be bought.

The purchasing decision is not an easy one to make. It should be based on an economic analysis assessing the total cost of system acquisition and ownership, taking into consideration the operational and performance requirements under all relevant conditions of such factors of use as product appeal, reliability, and the like.

The total cost of ownership includes operating, maintaining, and servicing costs, sometimes agreed by a service contract.

The sales discipline should specialize in understanding the user's problem and helping a buyer decide how to solve it, more than to study what can be sold.

Understanding the user's problems requires a basic technical understanding of the user's discipline and a mutually identical interpretation of concepts of both disciplines. Normalization of definitions can help very much as is shown in Section 28.4.

Making an analysis of life-cycle costing of a technical system requires a very systematic approach. All costs of the real life situation should be counted, for all reasonable alternative solutions. In order to distinguish between advantages and disadvantages it is necessary to employ weighting factors, and a system for continuing the results of each measure, in order to optimize the result. These optimization studies are an extension of the life-cycle costing studies (see Kiang, 1978).

The sales discipline can help evaluate the user's need and decide what system should be used.

This approach is not a cure-all approach as there are other factors of influence such as economics, climatical, and political environment, technological developments and market regulations.

The literature contains easily understandable curves illustrating the influence of reliability on costs, such as Figure 28.35. The combined cost curve C_1 for the manufacture and repair of defects in the factory points to an optimization point at O_1. The cost in this picture is factory price of acquisition or at another scale, selling price. The cost curve for manufacturing and repair of defects C_1, combined with customer repair and service costs to eliminate failures during the use period, C_2, depicts another optimum point at O_2. The cost in this picture is total life-cycle cost.

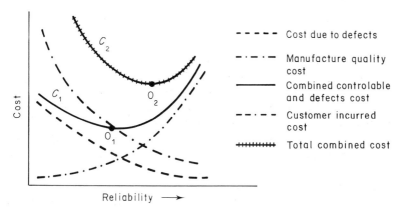

Figure 28.35 Survey of quality costs (theoretical)

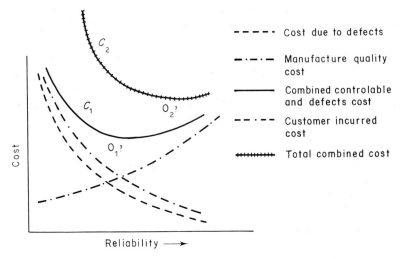

Figure 28.36   Survey of quality costs (practical)

The optimum O_2 only exists in the case where there is no other competitive supplier on the market. In general there are, and one has to take into account such factors as customer dissatisfaction costs and loss of market deliveries. On the other hand the rising curve of manufacturing cost with reliability and the falling curves of repair cost are very dubious. In many cases the so-called points of optimum costs, O_1 and O_2, cannot be denoted at all (Figure 28.36).

In view of the rising cost of labour, activities such as repair of defects and repair of failures should be avoided as much as possible. This factor points in the direction of providing high reliability, which can be provided by modern integrated electronics, in hybrid as well as solid state totally monolithic form.

28.4   SPECIFICATIONS AND NORMALIZATION

In the total process of formulating a user's need, writing and developing a target specification, designing, manufacturing, selling and buying measurement equipment and using it, many different disciplines are involved. It is of utmost importance to harmonize the information flow through these disciplines. This can be done by *speaking the same technical language*, for instance by using generally accepted and understandable terms and definitions. Of great help are the normalization publications, such as IEC 278: Documentation to be supplied with measuring apparatus; IEC 348: Safety requirements for electronic measuring apparatus; IEC 359: Expression of the functional performance of electronic measuring equipment. They are a guide for manufacturers and users and contain requirements about how to express performance quantities using uniformly agreed terms and values.

In general the technical specification of equipment is based on reference conditions, the conditions under which the apparatus meets the requirements, concerning intrinsic errors. Reference conditions can be defined in terms of:

(a) ambient temperature;
(b) input voltage, frequency, waveform, source impedance;
(c) output voltage, frequency, impedance;
(d) position of the apparatus;
(e) ventilation;
(f) relative humidity of air;
(g) external magnetic field;
(h) vibrations, dust, and other external influences.

Other concepts of importance are: the nominal conditions of use, limit conditions of operation, and limit conditions of storage and transport.

A number of tests have been specified to verify whether an apparatus meets its specification.

The instruction manual is a technical document containing information for the correct application, operation, maintenance, and repair of the apparatus and for understanding its operation; it also includes guarantee obligations.

Correct application, for instance, requires a correct statement of the apparatus specifications. These specifications are to be given in accordance with the recommendations.

Correct application also requires a correct statement of

(a) principle of operation;
(b) preliminary instructions (unpacking and repacking);
(c) installation of the apparatus;
(d) preparing the apparatus for operation;
(e) operating instructions;
(f) making measurements;
(g) calibration;
(h) mechanical construction;
(i) detailed description of the circuitry (for skilled users).

A great many very skilled specialists from the user, as well as from the manufacturing and the control management domains, have spent and are spending much time and effort in this normalization work.

ACKNOWLEDGEMENT

My very sincere thanks to my friend J. L. Leistra for reading and commenting on this chapter.

REFERENCES

Baker, D., Fleckenstein, W. O., Koehler, D. C., Roden, C. E., and Sabia, R. (1970). *Physical Design of Electronic Systems*, Vol. I, *Design Technology*, Prentice Hall, Englewood Cliffs, NJ.

Baker, D., Fleckenstein, W. D., Koehler, D. C., Roden, C. E., and Sabia, R. (1972). *Physical Design of Electronic Systems*, Vol. IV, *Design Process*, Prentice Hall, Englewood Cliffs, NJ.

Barlow, R. E., Fussell, J. B., and Singpurwalla, N. D. (1975). *Reliability and Fault Tree Analysis*, Society for Industrial and Applied Mathematics, Philadelphia.

Beckmann, M., Reichelt, C., and Spenhoff, E. (1977). 'Stichprobenpläne für Attribute—Auswahl unter Berücksichtigung der Kosten', *Qualität und Zuverlässigkeit*, **22**, No. 12, 274–9.

Bird, G. T. and Herd, G. R. (1976), 'Avionics reliability control during development', and 'Experiences in flight avionics malfunctions', *Agard Lecture Series No. 81*, NATO, Neuilly sur Seine, France.

Brewer, R. (1972). 'Reliability terms and definitions', *Micro Electronics and Reliability*, **II**, 435.

Chestnut, H. (1966). *System Engineering Tools*, Wiley, New York.

Chestnut, H. (1967). *System Engineering Methods*, Wiley, New York.

Duane, J. T. (1964). 'Learning curve approach to reliability monitoring', *IEEE Trans. Aerospace Electron. Syst.*, **AES-2**, 563–6.

ESRO (1975). 'QRA-14 specification, failure rates', European Space Research Organisation, 1975, Product Assurance Division ESTEC, Noordwijk, The Netherlands.

Green, J. E. (1976). 'Reliability growth modelling for avionics', *Agard Lecture Series No. 81*, NATO, Neuilly sur Seine, France.

IEC (1969). 'IEC 271: Preliminary list of basic terms and definitions for the reliability of electronic equipment', International Electrotechnical Commission, Geneva.

Kiang, T. D. (1978). 'Life cycle costing—the key to future survival', *ICQC 78 Tokyo*, TS. C6-04, Union of Japanese Scientists and Engineers, Tokyo.

Lycoudes, N. (1978). 'The reliability of plastic microcircuits in moist environments', *Solid St. Tech.*, **21**, No., 10, 53.

Manabe, N., Niikawa, T., Kajiyama, M., Takaide, A., and Goto, T., (1978). 'Decreasing failure rate observed in IC's', *ICQC78, Tokyo*, TS. D2-08, Union of Japanese Scientists and Engineers, Tokyo.

MIL (1974). *Military Standardization Handbook, Reliability Prediction of Electronic Equipment*, MIL-HDBK-217B (20-9-1974), Dept. of Defense, USA.

Pomplun, W. (1977). 'Grundgedanken zur Auswahl von AQL- und LQ-Werten in Prüfunterlagen für Fertigung und Einkauf', *Qualität und Zuverlässigkeit*, **22**, No. 12, 280–2.

Shioni, H. (1978). 'On lifetime acceleration and environmental application factor for electronic components', *ICQC 78 Tokyo*, TS. D2-09, Union of Japanese Scientists and Engineers, Tokyo.

Shooman, M. L. (1968). *Probabilistic Reliability: an Engineering Approach*, McGraw-Hill, New York.

Summerlin, W. T. (1976). 'Illusory reliability growth', *Agard Lecture Series No. 81*, NATO, Neuilly sur Seine, France.
Umeda, N. K. (1978). 'The control of equipment to ensure the reliability of electronic equipment', *ICQC 78 Tokyo*, TS. D2-15, Union of Japanese Scientists and Engineers, Tokyo.
Wetherill, G. B. (1977). *Sampling Inspection and Quality Control*, Chapman and Hall, London; John Wiley, New York.

Handbook of Measurement Science, Volume 2
Edited by P. H. Sydenham
© 1983 John Wiley & Sons Ltd

Chapter

29 E. BONOLLO

Quality Control and Inspection of Products

Editorial introduction

Controlling the quality of measuring instruments is an important aspect of measurement systems practice. Using measurements made with instruments to control the quality of other kinds of product or service is also relevant to the theme of this Handbook. This chapter continues our discussion of manufacturing aspects of measurement systems following on from the material of Chapter 28.

This chapter introduces the reader to developments that have occurred in recent years in quality control philosophy, including moves to standardize procedures through the national standards of specification mechanism.

The author then, using an area of industry where such developments are possibly the most utilized—gauge room metrology—shows how products can be inspected to provide data for quality control schemes.

Measuring instruments within themselves, and within their uses, are most diverse and include manufacture in numerous regimes. The gauging metrology emphasis provided here will form a basis on which the reader can base his or her own needs of quality control and inspection.

29.1 INTRODUCTION

During the 1960 and 1970 decades many social and industrial communities showed an increasing awareness of problems relating to the quality of goods and services and of the need for more quality control. From a social point of view, this awareness has been manifested by the formation of consumer groups, associated press coverage and relevant actions from governments. It is apparent that lay consumers are becoming better educated about, and conscious of, their quality rights and options for redress in cases of product quality failures. The educational institutions have also moved in a comparable, but mainly industrially oriented, direction by introducing higher level courses in the *quality technology* field of knowledge. It can be expected that these courses will have a significant beneficial impact on quality control activities and functions in industry and commerce.

In an industral sense, considerable quality awareness has been demonstrated by the efforts of such bodies as the American Society for Quality Control (ASQC), the European Organization for Quality Control (EOQC), and the International Organization for Standardization (ISO). As an example familiar to the author, the Australian Organization for Quality Control (AOQC) has continually stressed, through training courses, workshops, and seminars, the important contribution which quality technology can make to company profitability and survival. These efforts have been supported by the Standards Association of Australia (SAA) which has published a number of standards (SAA, 1971, 1972, 1973, 1975, 1977) on quality control terminology, methods, and systems, following the expressed need for such documents on quality technology from various industrial firms and educational establishments.

Many technically well advised managers and engineers, having realized the costs and benefits of quality technology, are planning, organizing, and budgeting for quality control activities as a necessary part of company operations. Some estimate the expected *product recall* and *product liability* suit cost in view of the known *make or break* company implications of these costs. In addition, it is widely accepted, in the context of manufacturing industry, that quality technology has improved the quality of products and services whilst providing gains in productivity, profitability, and other areas affecting company survival. However, as witnessed by the findings in a published survey (AOQC–DOP, 1977) there is a consensus of opinion which believes that quality technology is underutilized and its relative merits not properly understood, particularly in many small and medium sized manufacturing firms. Further work which considers the implications of this technology appears warranted, especially in terms of the quality control and inspection of products.

In this chapter, those aspects of quality technology related to the inspection of products will be considered and discussed with emphasis on the engineering measurements made of, or with, those products. Before doing this, however, the first part of this work will review some quality technology developments up to the publication of recent national standards and research work so as to provide a framework for better appreciating the important place of quality technology in an organization and the related contributions of quality control and inspection. The material of the chapter is relevant to both measurement system hardware design and to its use in performing measurements.

29.2 REVIEW OF DEVELOPMENTS IN QUALITY TECHNOLOGY

29.2.1 Introductory Remarks

It is appropriate to take a brief overview of some historical and background information to quality technology. This will be followed by a more detailed review of selected *quality* concepts, typical *quality control activities*, and current developments in *quality control systems*.

29.2.2 Background Information

Many of the concepts inherent in modern quality technology have been known for centuries as witnessed by the hieroglyphics of the ancient Egyptians (Thoday, 1973). These graphics show a clear acquaintance with the principles of conceptual design, the practicalities of manufacturing, and the quality control aspects of measurement, particularly in terms of assuring that architectural modules conformed to the design specifications. Similar principles and controls were used by the Romans who, for instance (Thoday, 1973), standardized their military chariot wheel tracks to 1.435 m (which, incidently, coincides with the British ralways imperial measure gauge of 4′ 8½″). Over the years, these concepts developed steadily through the various craft and guild organizations where product quality was often the responsibility of individual craftsmen who were close to and highly influenced by the customer's requirements.

With the advent of the principles of *interchangeability*, the industrial revolution and the *quantity production* of goods, quality control activities were often associated with the now traditional and well known *inspection tasks*. However, with the increasing product throughput and complexity of manufacturing industries, it was realized that inspection functions were *time consuming* and, hence, *expensive* in themselves and only concerned, primarily, with the *detection* and *removal* of quality *defective* products, often after the manufacturing processes had taken their course. Therefore, they tended to be inadequate in terms of appraising the quality capabilities of processes, preventing defectives from occurring and providing the necessary assurance (or protection) that the customer's quality requirements for the finished product had been met. As these were (and still are) crucial quality control problems affecting company survival it is not surprising to find that they became the subject of serious research.

Over the period 1930–1980 various research workers, such as Shewhart (1931), Tippet (1950), Dodge and Romig (1959), and Duncan (1965) investigated the above problems. They found that some of the ideas of the early mathematicians on *probability theory* could be developed and combined with *inspection tasks* in practice, leading to economies in inspection costs as well as to improvements in product quality. This relatively new area of knowledge is commonly known as *statistical quality control*. The associated techniques are widely applied and current to process and product control situations despite recognized shortcomings.

More recently, many investigators, such as Feigenbaum (1961), Juran (1964, 1974), and Hayes and Romig (1977), realized that a broader view of quality control was required by management if quality control techniques and activities were to be more effective in practice. They surveyed individual developments in fields like *systems design, product design, manufacturing, inspection, statistical quality control, purchasing, marketing, and management* and suggested a *total quality control* management philosophy which could be applied to many organizational levels and activities in a company's operations. This philosophy includes an underlying notion concerning a company's position as the supplier link in a customer–supplier chain, namely that the quality of a supplier's products (or services) is highly dependent upon the supplier's proper control over purchasing, design, manufacturing and inspection activities together with all the necessary steps required to ensure that the supplier's products conform to the customer's quality requirements. These vital concepts and developments will now be considered in more detail.

29.2.3   Selected Quality Concepts

From an industrial point of view, the *control* part of *quality control* is a relatively easy notion to understand since it suggests a classical feedback cycle. These cycles involve (Feigenbaum, 1961; Juran, 1964, 1974; Hayes and Romig, 1977) a *planning phase* (including the setting of standards), an *implementation phase* (producing certain results), a *comparison phase* (where the results achieved are compared to the standards), and a *feedback* of this comparative information to the planning and implementation phases so that *corrective action* may be taken. As such, feedback cycles may be associated with different activities at various levels in an organization.

The *quality* part of *quality control* is, however, worth reconsidering in terms of its definition and interpretation. The term *quality* has been defined (SAA, 1971) as the *fitness for purpose* of a product, device or service. This definition may be expanded to better appreciate the many aspects of quality control in its widest sense as well as the more specific meanings. The word *purpose* implies the existence of a set of primary functions (Miles, 1961; ASTME, 1967; Jones, 1970), or the work that a product (or service) is primarily designed to do, for which there is a stated or potential need. For instance, with a flashlight the primary function of the globe is to provide light while a lens, if used, can provide the secondary function of protecting the globe. In general, it has been suggested (Miles, 1961; ASTME, 1967; Jones, 1970) that *use functions* coincide with the designer's interpretation of the properties of a product required to accomplish a use, work or service, whereas *esteem functions* are related to the features or attractiveness of a product which promote an independent wish to acquire it. An underlying idea is that a product is a means to an end (not the end itself) of providing given sets of functions under certain operating conditions and for a given period of time.

The word *fitness* implies the existence of related criteria or yardsticks, namely:

(a)   The short and long term adequacy of the design of a product, with respect to providing for given use and esteem functions. This may include comparison with existing designs for the same functions, field testing, and so on.

(b)   The fidelity with which the product, at the point of acceptance, conforms to the design specifications for that product. For a unit of product the dichotomy is *conformance* or *non-conformance*, although, in some cases, non-conformance is subject to interpretation with respect to *fitness for purpose*. Also, there are further criteria which have taken on an increasing importance in recent years; that is,

(c)   The fidelity with which the product in actual operation in the market place conforms to predetermined policies on warranty, maintenance and product service.

Associated with the above three points are the basic requirements that the product must be designed, manufactured, and marketed in ways which are both practical and economically worthwhile. An assessment of the product with respect to criteria (a) is often referred to as *quality of design* (including reliability, particularly if long term performance is involved). A measure with respect to (b) is known as *quality of conformance* and of criteria (c) as *quality of* (*sales and*) *service*. This interpretation, while identifying aspects of product quality which need to be quantified, also suggests that the quality of a service can be approached in a similar way to the quality of a product.

The three sets of criteria relating to the qualities of design, conformance and service highlight the problems which a manufacturer must adequately solve in order to meet customer requirements while remaining economically viable. Failure in satisfying any one set of criteria can obviously result in disastrous consequences for a manufacturing organization. This has been recognized in the *total quality control* concept or management philosophy which has been defined (Feigenbaum, 1961) as

> a management system for integrating the quality-development, quality-maintenance and quality-improvement efforts of the various groups in an organization so as to enable production and service at the most economical levels which allow for full customer satisfaction.

More specifically, total quality control aims to achieve the three qualities mentioned by means of the relevant body of knowledge (quality technology) continually developed over the last forty years. In keeping with Feigenbaum's ideas quality technology should be used to formulate quality policies and objectives, to analyse and plan product quality, and to design and implement those quality control systems which will yield full customer satisfaction with respect to the qualities realized.

Apart from quality planning and system design, the scope of quality technology, which also reflects its multidisciplinary nature, includes aspects on design and manufacturing communication, human factors, and economic analysis. It also includes quality control methods, reliability techniques, measurement science, inspection methods, industrial experimentation, and applied quality control. Since much of this knowledge has been classified into techniques, quality technology tends to be easier to understand than the broad concepts embodied in total quality control.

Further insights into the nature and importance of quality technology may be obtained by reviewing some typical quality control activities and functions, including inspection of products.

29.2.4 Typical Quality Control Activities

It is a well known proposition (Drucker, 1964; Juran and Gryna, 1970) that with any company, or industrial facility, objectives need to be set in every area where performance and results directly and significantly affect the survival of the company—otherwise, the company cannot hope to establish and maintain an economically viable position. These *survival areas* (Drucker, 1964) include market standing, innovation, productivity, profitability, employee development, and social responsibility. Furthermore, within any company, there exists a fundamentally important set of on-going activities which interact with and affect these survival areas. For example, there are the activities of marketing and sales, personnel, product design, process planning, production, quality control, inspection, production control, financial control and so on. The term *on-going* is used here to stress the fact that each activity (sphere of action) or group of activities exist, in a formal or informal way, in a changing environment whilst providing functions (work or tasks) which are essential and continual in nature.

To proceed further, quality control activities may be considered in terms of providing essential sets of functions which, in turn, are characterized by sets of performance elements. In this context, *process control* and *product control* are functions; *process averages, quality costs* and *average out-going qualities* are typical performance elements. Process control functions are mainly concerned with minimizing variations in the quality level (often expressed as a fraction or percent defective) which occur during the production process. Process control may employ such statistical quality control techniques as natural, modified, fraction defective and cusum control charts (Feigenbaum, 1961; Duncan, 1965; Knowler *et al.*, 1969; Grant and Leavensworth, 1972; Juran, 1977; Hayes and Romig, 1977).

In contrast, product control is concerned with maintaining an acceptable quality level at the input, output, and sometimes intermediate terminals of groups of production processes where the product is often assembled in lots or batches. Product control lends itself to the application of statistical sampling techniques such as single, double, multiple and sequential sampling plans (Feigenbaum, 1961; Duncan, 1965; SAA, 1972, 1973; Juran, 1974). There are apparently no firm rules for the application of these techniques and this can create confusion especially with small manufacturing companies engaged in low volume production. In many cases, it has led to the misconception that the above techniques represent total quality control and reluctance on the part of small business to formalize quality control functions.

In general, some quality control functions will be pre-occupied with quality of *design*, some with quality of *conformance* and some with quality of *service*. A number will be essentially technical while others will be concerned with the planning, supervision and assessment of other functions (that is, managerial). In practice, quality control functions may be so extensive and specialized as to require grouping as separate activities. This may be particularly the case with *metrology, non-destructive testing, inspection*, and *quality assurance* (sections, departments or laboratories).

However, the descriptions of these activities and related functions can vary, leading to difficulties in visualizing their properties and their intended place in a company's organizational structures. For example, AS1057 (SAA, 1971) has defined 'quality assurance' as 'the provision of evidence or proof that the requirements for quality have been met', which implies an associated place in various levels of organization.

By comparision, 'inspection' has been defined (SAA, 1971) as 'the activity of measuring, testing or otherwise examining products and services for determining conformance to the stated requirements', which may also imply a lower organizational level than quality assurance. As reported in the literature (Fetter, 1967), quality assurance and inspection may exist as independent activities although they require similar information (it is also realized that inspection is usually an integral part of production). Fetter's version of quality assurance is interesting in that this term is supposed to mean 'the total set of operations and procedures included within the production system whose goals are conformance of product output to design specifications'. As a final example, the versions provided by (Hartz, 1974) are worth noting since he has proposed that quality control activities can be classified into sets of *planning* and *assurance* functions. Planning functions are reportedly those concerned with decisions about plans and specifications and the implementation of these plans and specifications. In contrast, assurance functions have been referred (Hartz, 1974) to inspection (or measurement) and the preparation of the groundwork for taking corrective action (by analysis and evaluation of conformance to plans and specifications).

Allowing for the versions of activity description which exist, it is usually possible for companies to arrive at a consensus of opinion regarding the description and properties of quality control activities in the relevant industry, paying due regard to preferred terminology, such as found in SAA (1971). What appears to be much more difficult to determine is the form of hierarachy required in the logical grouping, relevance, and organization of activities (functions) into a quality control framework (or system) suitable for particular companies, bearing in mind the nature of a company's business. In this apparently difficult problem area, there may be no unique solutions for the characteristics of the quality control systems adopted, while the solutions adopted may not be optimal. To emphasize these aspects, try to visualize a given manufacturing company which can implement any one of a number of different quality control systems to achieve the same quality objectives. Although outside the scope of this work, similar problems may occur in determining quality control systems which are appropriate to such organizations as a bank, hospital, and large retail store.

Nevertheless, from a quality control and inspection viewpoint it is important to consider quality technology with respect to the design and implementation of quality control systems, particularly as this is consistent with recent trends to formally align quality control activities with such systems. In addition, the structure of and rationale behind these systems ought to be considered especially as these have not been widely debated in the research literature, even if various standard documents, detailing system requirements, have been published (SAA, 1975, 1977).

29.2.5   Quality Control Systems

A quality control system with prescribed requirements is a relatively new concept in many countries although considered earlier by various workers (Feigenbaum, 1961; Hall, 1962; Juran, 1964; Hartz, 1974; Juran, 1974). Hence, a brief review of developments is in order before considering in more detail the current approach at the national standards level such as that in Australia (SAA, 1975, 1977), where system requirements have been standardized. This will be followed by a review of a research approach.

Feigenbaum (1961) has defined a quality control system as 'the network of adminstrative and technical procedures required to produce and deliver a product of specified quality standards'. Another version, given in a more recent research paper (Hartz, 1974), is 'a quality control system is the total amount of technical and adminstrative activities that contribute to the task of quality control'. Strangely enough, recourse to recent standards documents (SAA, 1975, 1977) has not detected a direct definition of the term, although detailed system requirements have been specified.

Combining the above thoughts with previous comments regarding the hierarchy and logical grouping of activities and functions, it is evident that a quality control system can provide the procedures, organizational structure and resources upon which company quality control activities are based. In essence, a system creates a place for quality technology in an organization. The need for standardizing the requirements of quality control systems appears to have developed with the refinement of the total quality control philosophy and the noted multidisciplinary aspects and interactions existing in complex production. In particular, as an example, the Australian Standards AS1821–23 on 'supplier's quality control systems' were prepared (SAA, 1975) as a 'result of the need expressed by many organizations for national standards which can be used both by Australian industry and government instrumentalities as a condition of contract when purchasing goods or services of an assured quality'. Furthermore, it has been proposed (Custance, 1977) that general reasons in favour of such systems include the adoption and widespread use of the similar MIL-Q-9858 A (USDD, 1963) document in the United States, the adoption in the United Kingdom of defence standards DEF STAN 05–21, 24, and 29 (UKMD, 1973) and, finally, 'the increasing difficulty an Australian manufacturer will have in supplying sophisticated equipment to overseas markets unless he can demonstrate a quality capability similar to AS1821–23'. Hence, without questioning the validity of the last quotation, sufficient grounds exist to consider the standards approach particularly as this may become an integral part of quality technology in world-wide practice.

The standards approach to quality control systems

In reviewing the standards approach, the Australian trends will be taken as indicative of the requirements applying to quality control and inspection in any particular country. According to the author's readings, the current (at time of writing) Australian Standards (SAA, 1975), published in the form of one, twenty-two page document on three quality control systems, refer to the following objectives:

(a) To provide statements of requirements which may be nominated by individual customers as a condition of contract when seeking to buy goods for which they have need for assurance of quality before acceptance.

(b) To act as recommendations to suppliers whose products are distributed for general sale as a guide to identifying the requirements of effective and economical management systems to control and assure the quality of their products.

It is not surprising to note that these are laudable objectives consistent with the basic principle (SAA, 1975) that '(Australian) standards are prepared only after a full enquiry has shown that the project is endorsed as a desirable one and worth the effort involved'.

The three standards above have been identified (SAA, 1975) as AS1821 (supplier's quality control system—level 1), AS1822 (supplier's quality control system—level 2) and AS1823 (supplier's quality control system—level 3). In AS1821, system requirements have been specified which would 'normally be applied to contracts where the customer considers quality control to be essential in all phases' of the life cycle of a product (in keeping with the total quality control philosophy and the application of quality technology to all relevant levels). In this case, it is evident that a supplier (or vendor) is expected to have a comprehensive quality control organization and resources, including responsibility for design and development.

Standard AS8122—level 2 has specified system requirements where the quality control activities or functions are mainly directed towards the manufacture, assembly, and testing of supplies and services. Though similar to AS1821, it is comparatively less complex since the requirements of activities related to design are excluded. This complexity is significantly reduced in the third standard AS1823 where the listed system requirements apply only to the case where conformance to contract requirements can be established by a supplier's inspection operations, subject to control and surveillance by the customer's authorized representative. The relevance of the reported requirements with respect to the three systems is summarized in Table 29.1.

The requirements listed in Table 29.1 indicate the basic differences between the three systems. It is also realized that each requirement has been provided as a section or clause heading. Hence, it is important to note that the standards publication contains additional information on the respective requirements. For example, the 'Work Instructions' heading (in AS1821) includes the statement that 'the supplier shall develop and maintain clear and complete documented instructions that prescribe the performance of work affecting quality which would be adversely affected by the lack of such

Table 29.1 System and associated requirements summary; as interpreted from AS1821–23: 1975 (4)[a]

Requirements (as given under section headings)	Systems		
	AS1821— level 1	AS1822— level 2	AS1823— level 3
1 General	A	A	NL
2 Organization	A	A	NL
3 Planning	A	A	NL
4 Work instructions	A	A	NL
5 Records	A	A	A
6 Corrective action	A	A	NL
7 Design control	A	NL	NL
8 Documentation and documentation change control	A	A	NL
9 Inspection equipment	A	A	A
10 Control of supplier-procured supplies and services	A	A	NL
11 Manufacturing control	A	A	NL
12 Control of non-conforming supplies	A	A	A
13 Indication of inspection status	A	A	NL
14 Statistical quality control	A	A	NL
15 Handling, storage, delivery and use	A	A	NL
16 Verification of conformance	A	A	A
17 Facilities and assistance to be provided by the supplier	A	A	A
18 Reponsibility for inspection	NL	NL	A

[a] Notes: (1) 'A' denotes that the requirement applies to the relevant system.
(2) 'NL' denotes that the requirement has not been listed with respect to the particular system.
(3) With respect to levels 1 and 2, requirement 18 appears to be catered for under item 6, 'corrective action'.

instructions'. Considerably more qualifying information can be found in the draft standard guide to AS1821–23, 'suppliers quality control systems, (SAA, 1977), which contains an extensive series of questions aimed at assessing the extent to which a supplier's system conforms to AS1821–23 (Custance, 1977).

A notable feature of the standards approach (SAA, 1975; 1977) is the attempt to associate the stated *requirements* with the *elements* of the system to be designed, implemented and managed by a supplier. However, the decision-making and organizational levels of these elements do not appear to have been explained qualitatively with respect to the three systems. In addition, it is not clear if an equal weighting should be allocated to the Table

29.1 requirements in establishing their relative importance to the effectiveness of a system (for instance, should corrective action and design control share an equal weighting?). Notwithstanding these apparent ambiguities, the basic requirements inherent in the total quality control concept are clearly evident from Table 29.1, namely: organization and planning for quality together with control over purchasing, design, manufacturing, inspection, distribution, and service activities are critical consideratons. These standards have made a meritorious attempt to describe and explain the essential features of three quality control systems of differing complexity whilst allowing companies the flexibility to decide upon how such systems are to be implemented.

The fact that national standards on quality control systems have been published implies the basic need for regulating quality procedures involved in the transfer of goods and services between any two parties in the supplier–customer chain. As such, it must affect all parties in this chain of companies depending upon the prevailing attitudes to quality. It also implies that, while such quality standards are worthwhile in the national interest, previously existing company procedures were perhaps inadequate and not amenable to industrial self-regulation. Hence the need for formulating a uniform industrial code of *quality* practices. Most importantly, the standards demonstrate the high degree of development of quality technology and its strong interrelationship with the survival areas of a company.

A most interesting aspect of the standards approach is the micro or synthetic tactics adopted in identifying (by a questioning technique) the desirable features or properties (activities, functions, and more) of a quality control system. This approach is reminiscent of the source quality assurance (or vendor quality facilities evaluation) views often reported in the literature (Miller, 1962; BSI, 1975; OIML, 1978) where company quality capabilities are assessed in detail. From the micro level, it should be possible to *build up* a system appropriate to a particular company, although few papers dealing directly with the quality systems design problem have been detected in the research journals. With respect to the latter, the research approach reported by Hartz (1974) is worth reviewing briefly, particularly as it may serve to amplify the standards viewpoint on quality technology.

A research approach to quality control systems

Hartz (1974) has proposed that a quality control system may be described in terms of a 'decision structure', a 'flow of information' and the 'administrative and technical activities'. It has also beeen suggested (Hartz, 1974) that, down to a certain level of details, this description may be made almost independently of the specific form of company organization. According to Hartz (1974), the system quality control activities and functions may be classified, with respect to the nature and degree of uncertainty of the associated decisions, into strategic, tactical, and operative levels of control. He has also suggested that the most important quality control activities may be grouped into various main elements of a quality control system based on these three decision levels. These elements include the determination of inspection specifications and inspection planning.

Detailed information has been provided (Hartz, 1974) on how the main elements in the system could be subdivided into groups of activities (and associated functions) at the various levels including data on the flow of information and materials. Although an in-depth review is outside the scope of this work, some further comments appear warrranted on the proposed levels of control, having noted that these are not immediately evident in the standards approach (SAA, 1975).

The reported strategic level of control concerns (Hartz, 1974) decisions on such matters as objectives, restrictions, and resource development and allocation. Typical examples quoted include deciding upon the quality policies and objectives, the quality needs of the market place, the quality levels of products and services and decisions concerning the design, development and administration of the quality control system. (It appears that quality technology at this level involves ill-defined problem areas and decisions characterised by a high degree of uncertainty.)

The strategic level has been claimed to provide guiding information to the tactical control level where decisions are relatively more detailed occurring more frequently. Typical examples listed include the working out of programmed sets of instructions, rules and procedures for process and product control techniques together with the allocation of inspection stations and resources (quality technology at this level involves reasonably defined problem areas and less uncertainty in decisions).

Finally, Hartz has suggested that the third (operative) level of control is concerned with working out decisions in detail from the tactical level. Decisions at the operative level tend to be repetitive and occur with a high frequency, for instance, analysis of measurements and maintenance of records and equipment (here, quality technology involves well-defined problem areas and decisions characterized by a high degree of certainty).

An appealing feature of this research approach is the apparent macro or analytical tactics employed to identify the elements common to many quality control systems and the associated hierarchy (control levels), before attempting progressively more detailed descriptions of activities and functions. When combined with the systems information noted in the standards and guides, such as SAA (1975, 1977), a clearer appreciation may be obtained of the technical and administrative skills required for using quality technology in the systems context. In closing this review some further overall comments appear useful.

29.2.6 Comments

This review has shown that quality technology has grown into a comprehensive and multidisciplinary field of knowledge upon which modern quality control techniques, activities, and systems are based in practice. Significantly, this includes activities concerned with the measurement and inspection of products. In keeping with the total quality control philosophy, quality technology has been developed for a specific purpose. It is essentially designed to assist an organization to achieve quality objectives assuming that these have been incorporated in the company policies. It is evident that there are few areas in a company which quality technology does not affect and, ideally, it should permeate many levels of a company's operations. It also provides strong evidence supporting the belief that quality control and inspection of goods and services is necessary. There is little doubt that quality technology has developed to a point that it is a vital part of management. This is witnessed by the quality control systems documents reviewed, which also reflect the degree of sophistication involved.

One of the significant benefits of quality technology is that it provides a systematic means of identifying, formulating and solving quality control problems. This may, in addition, create side benefits or an atmosphere relevant to examining other activities, that is, the careful approach to quality problems could generate more attention to such facets as financial and production control problems. That quality technology is relevant to company survival has been amply shown in this review. Just how effective it is in practice, however, will vary according to a number of considerations. Basically, quality technology demands a management policy presupposing that quality is considered to be important to the company. This review has also highlighted the strong interconnection which exists between quality technology and the inspection of products. The inspection of products from the point of view of engineering measurements will now be considered.

29.3 MEASUREMENT AND INSPECTION OF PRODUCTS

29.3.1 General Comment

It is well known that the science of measurement (metrology) covers a very broad field of knowledge (Miller, 1962; Barry, 1964; Puttock, 1965; BSI, 1975; Doebelin, 1975; OIML, 1978) which considers all kinds of measurements. The term 'engineering (dimensional) metrology' (SAA, 1980) is more specific since it refers to that field which caters, primarily, for the measurement of parameters concerned with the length dimension, that is, length, angle, and geometric relationships. This field may include the examination of engineers limit gauges, jigs, fixtures, cutting tools and components together with gears, splines, and serrations. Also included is the measurement of surface topography, the examination of engineers measuring instruments and the calibration (all of the operations required to determine the values of the errors) of length and angle standards.

It is the engineering metrology activity and related functions which form the basis, in terms of providing traceability (SAA, 1980) to the national and international standards of measurement, for the inspection of numerous products during manufacture. Although closely interrelated, the metrology and inspection activities tend to be organized separately (as noted earlier) in terms of contributing to the quality control of products and consumer goods (SAA, 1980; Bonollo, 1978). However, these activities are still very extensive in scope. Therefore, only a selected combined treatment is presented below beginning with a brief review of some typical areas where inspection and measurement functions are vitally important. The emphasis of this account is on dimensional measurement parameters: this is the prime area of industrial activity. Although restricted in this way, the philosophies involved are relevant to the other measurements made.

29.3.2 Review of Measurement Areas

With respect to inspection functions, these quality control relevant areas (Bonollo, 1977) may include:

(a) The inspection (measurement) of the specified dimensions, or quality characteristics (SAA, 1971), of inward and outward goods, products and materials to various deterministic or statistical sampling plans (SAA, 1972; Dodge and Romig, 1959). Sampling plans could be incorporated in contractual agreements drawn up between supplier and consumer manufacturing organizations.

(b) The measurement of product quality characteristics at the input and output terminals of a group of production processes using sampling plans similar to those mentioned above. Here the requirements of high volume manufacturing may necessitate automation of inspection and measuring operations in contrast to the classical, manually controlled approach.

(c) In-process measurement of product quality characteristics with respect to quality control charts of the attributes or variables type (Shewhart, 1931; Duncan, 1965).

(d) Employment of process variability or capability studies (SAA, 1971) for process selection, design (tolerance) investigations or control chart procedures. This may be the function of inspection or quality control staff depending upon the practices of particular companies.

In terms of metrology there are a number of problem areas which need to be mentioned, namely:

(e) Measurements associated with product design, development and prototype testing, and the implementation phase of engineering design. These may include the critical examination of product dimensions and parameters to design specifications at the tryout or initiation stage of product manufacture, as well as in-service tests.

(f) Inspection and quality control service and support functions. For example, initial and periodic calibration and control of inspection gauges, comparators, test sets, and other measuring apparatus as used in various areas of a manufacturing facility.

(g) Calibration and control of working standards and measuring equipment to requirements prescribed by various national bodies. In Australia, the relevant body is the National Association of Testing Authorities (NATA, 1975, 1979). Other examples are discussed in Chapters 3 and 31.

(h) Design and development of measuring systems and methods, assessment of uncertainty of measurement, and laboratory management to nationally approved standards of competency.

(i) Measurement services offered to other organizations not connected with a particular metrology laboratory. These services may include some of the items mentioned above but will normally be defined by the terms of registration (or similar policy) for a particular laboratory. For example, the classes (NATA, 1979) of tests may include, examination of engineer's limit gauges and the examination of jigs, fixtures, cutting tools, and components. However, the examination of engineer's measuring tools and instruments, testing of machine tools, assessment of surface topography, examination of working standards of length and angle, and examination of precision measuring instruments may not be included in a particular laboratory's terms of reference. Hence, these measurements would require the services of a higher echelon laboratory.

(j) Selection and training of signatories to accept technical responsibility for the measurements performed within the terms of registration. It is of interest to note that some signatories may be invited to join panels of 'Assessors in Metrology', preoccupied with national laboratory registration programs (Bonollo, 1977).

The above review serves to outline the scope and importance of measurement and inspection functions and, therefore, related measurement problems. Additional information can be gained by considering the characteristics of a conventional measuring system since such a system forms an essential part of many measuring processes.

29.3.3   A Conventional Measuring System

Measurement processes may be considered (Bonollo, 1977) in terms of four main parts: namely the measuring system, the measuring methods, the environment, and, of course, the human factors. Application of the process, with respect to a given system, will enable the size or shape of a workpiece or other measurands of importance to be described in terms of suitable units. More specifically, a measurement process is characterized by the series of system related operations required to execute a measurement and, hence, determine the value of a physical quantity or measurand (Bonollo, 1975; SAA, 1980). In general, a measuring system can be considered (Bonollo, 1975) as the assembly of physical elements necessary for achieving the objectives of a measurement with the application of a measuring process in a given environment. For conventional engineering measurements for instance, these systems may be analysed with respect to five basic elements, as shown in Figure 29.1, namely:

(a) The workpiece (system) incorporating the measurand or quantity to be measured.
(b) The measurement standards physically embodying the units or concepts by means of which the quantity is described.
(c) The sensors (or feelers) under the direct influence of the physical quantity and the standards.
(d) The indicating device to display or transmit to the human senses the response or position of the sensors.
(e) The reference frame or components required to interrelate the above elements of the system.

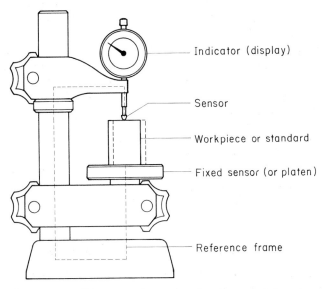

Figure 29.1 Elements of a conventional measuring system
(consistent, in this case, with an engineer's comparator)

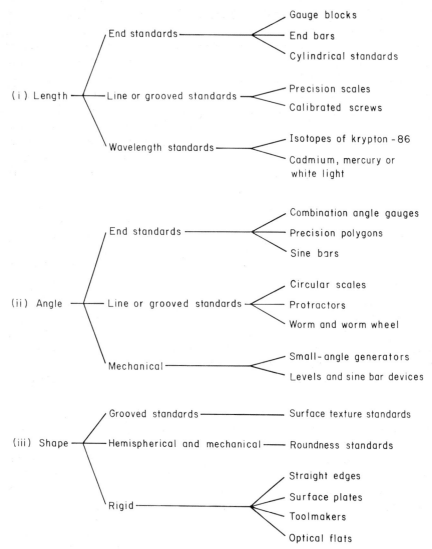

Figure 29.2   Simple classification of commonly used mechanical dimensional
measurement standards (NATA, 1979)

In brief, an unlimited variety of workpieces are encountered in quality
control and measurement situations and this makes classification difficult.
However, a classification on the basis of shape, or geometrical characteristics, is known (Bonollo, 1977) to provide useful starting points for mechanical objects. A similar approach can be used for the classification of commonly used standards of length and angle, although the variety is relatively

restricted, as partly illustrated in Figure 29.2. Needless to say, standards of length are fundamental to measurement since all measurements are really comparisons. It is interesting to note that only a few centuries ago the inch was defined as equal to 'three barley corns, dry and round, laid end to end' (Edward I, 1272–1307). It is only since the 18th century that any real progress has been made in the field of calibration. As is well known, the international standard of length is the metre, which is defined (but see p. 1045–6 for redefinition) in terms of wavelengths, subject to certain conditions, of the orange-red radiation from krypton-86. This was a very significant advance in the field of fine measurement since, by means of interferometry, reproducible standards of length are now available world-wide.

Various types of sensors (or position indicators) are used in measuring systems. Mechanical, optical, and pneumatic type sensors are commonly used but there are also other types. It is a well known rule that the geometry of standards used should be similar to that of the workpieces. Other considerations include thermal effects and the interpenetration of the sensors with the standards and workpieces. It has been noted (Doebelin, 1975) that a typical sensor contains a transducer, amplifier, and display element. This means that the method of amplification, leading to various values of magnification, can be used to classify different types of position indicators. The make up of measurement systems is dealt with in more detail in Chapter 15. Standards are covered in depth in Chapter 3.

There is a very extensive range of measuring systems commercially available. In some cases, the workpiece and standards are treated independently so that the measuring system is synonomous with the measuring instrument, particularly if the former is simple, as shown in Figure 29.1. Thus, it follows that various systems may be employed to measure a measurand, say, associated with the size of a given workpiece, while different measuring methods (Bonollo, 1977) can be applied to a given system. Although not new, these ideas can form a useful basis for a study of measuring methods and the related measured size equations.

29.3.4 Measuring Methods and Related Equations

Some interesting features of measuring methods may be appreciated by applying a simplified measurement process to the system depicted in Figure 29.1 (which will be recognized as essentially an engineer's comparator type of instrument). Assuming steady-state conditions, the length of the workpiece, L_w, can be expressed as a function of the calibrated (known) length of the standard, L_s, the respective readings of the indication on workpiece and standard, R_w and R_s, and the respective sets of systematic errors $\{E_w^*\}$ and $\{E_s^*\}$. For the workpiece, the systematic errors relevant to R_w may be considered further in terms of the easily determined or identifiable (well-defined) elements, for instance:

$$\{E_w^*\} = \{E_{0w}, E_{tw}, E_{dw}, E_{fw}, \ldots, E_{iw}\} \tag{29.1}$$

where

E_{0w} = the zero or datum error

E_{tw} = error due to thermal effects (that is, as reflected by the influence quantity of temperature deviating from its reference value of 20 °C)

E_{dw} = error due to elastic deformation or the interpenetration of sensors and workpiece

E_{fw} = error in the transfer function of the measuring system or instrument, also known as response error in SAA (1980)

E_{iw} = the ith or last systematic error in the set which can be determined by estimate, calculation or experiment for the purposes of the measurement

A similar set of systematic errors may be associated with R_s, that is,

$$\{E_s^*\} = \{E_{0s}, E_{ts}, E_{ds}, E_{fs}, \ldots, E_{js}\} \tag{29.2}$$

where the elements in this set correspond, but are not necessarily equal, to those in $\{E_w^*\}$ above.

As suggested in equations (29.1) and (29.2), the composition of $\{E_w^*\}$ and $\{E_s^*\}$ poses interesting questions as to the proper form of the relationship which interrelates the elements in these sets. However, in order to consider the measurement process noted above, it may be assumed (as is often done in practice) that a simple algebraic relationship applies so that:

$$\{E_w^*\} = \{E_{0w}\} + \{E_w\} \tag{29.3}$$

and

$$\{E_s^*\} = \{E_{0s}\} + \{E_s\} \tag{29.4}$$

where the systematic errors $\{E_w\}$ and $\{E_s\}$ represent the sum of the remaining elements in $\{E_w^*\}$ and $\{E_s^*\}$ respectively.

Since the value of L_s is known and R_s and R_w are system indicated or displayed values, the process can be modelled initially on the basis of two equations. Firstly, with the standard in the system,

$$\{E_s^*\} = R_s - L_s = \{E_{0s}\} + \{E_s\} \tag{29.5}$$

or, if $\{E_s\}$ is known from previous tests, then

$$\{E_{0s}\} = R_s - (L_s + \{E_s\}) \tag{29.6}$$

where $\{E_{0s}\}$ is a parameter.

Secondly, with the workpiece in the system, it follows that,

$$L_w = R_w - [\{E_{0w}\} + \{E_w\}] \tag{29.7}$$

Adopting the frequently applied assumption that the independent zero errors, $\{E_{0s}\}$ and $\{E_{0w}\}$, are equal, equations (29.6) and (29.7) can be combined to give,

$$L_w = L_s + (R_w - R_s) + (\{E_s\} - \{E_w\}) \tag{29.8}$$

where the terms may be single values or arithmetic averages in practice. It is important to observe that this is the measured size equation which models the objective of the measurement (L_w, the length of the workpiece). It is also worth noting that equation (29.8) reflects the length measuring method used (in this case, the widely applied direct comparision method (SAA, 1980)). Furthermore, equation (29.8) is often simplified in practical cases where the shape, length, and material composition of the workpiece and standard are approximately the same. Under these conditions, it is assumed that $E_{tw} = E_{ts}$, $E_{dw} = E_{ds}$, and $E_{fw} = E_{fs}$ so that, neglecting other elements, $\{E_s\} = \{E_w\}$. Hence, equation (29.8) reduces to:

$$L_w = L_s + (R_w - R_s) \tag{29.9}$$

This procedure is consistent with the null method of measurement (SAA, 1980) provided L_s is selected so that, for all practical purposes, $R_w = R_s$. Further simplification follows with direct methods of length measurement (SAA, 1980), for example, when using vernier calipers, micrometers or systems, such as that depicted in Figure 29.1 which have been set to zero or a datum position. Here, equation (29.7) is applied directly assuming that E_{0w} and E_w are contained within reasonable limits, or do not exceed the values listed in the relevant standard specifications for the instruments involved.

The above equations are typical of the methods associated with measuring instruments and systems used in quality control. They indicate the importance of regular calibration to ensure that the systematic errors are kept within predetermined limits, otherwise the accuracy of measurement of the product quality of conformance to design specifications may be seriously affected. They may also be used to demonstrate the often little known effects, as shown below, of influence quantities such as temperature.

It was implied earlier, as suggested by the description of E_{ts} and E_{tw}, that the remaining systematic error elements were independent of thermal effects with the latter catered for separately. However, to better consider these effects, a more direct approach may be required, in practice, with E_{tw} and E_{ts} not included in $\{E_w^*\}$ and $\{E_s^*\}$ respectively. For discussion purposes, and neglecting other thermal effects, let

t_0 = standard measurement reference temperature (20 °C)
t_w = temperature of the workpiece
t_s = temperature of the standard
C_w = coefficient of linear expansion of the workpiece material
C_s = coefficient of linear expansion of the standard material.

Then, equation (29.7) may be written as

$$(L_w)^{t_w} = (R_w)^{t_w} - (\{E_{0w}\} + \{E_w\})^{t_w} \tag{29.10}$$

where $(L_w)^{t_w}$ denotes the length of the workpiece at temperature t_w and the other temperature-dependent terms are similarly described. It also follows that, at the standard temperature,

$$(L_w)^{t_0} = (L_w)^{t_w} + (L_w)^{t_w}(t_0 - t_w)C_w,$$
$$= (L_w)^{t_w}[1 + (t_0 - t_w)C_w] \tag{29.11}$$

Equating for $(L_w)^{t_w}$ in equations (29.10) and (29.11) gives

$$\frac{(L_w)^{t_0}}{1 + (t_0 - t_w)C_w} = (R_w)^{t_w} - \{E_{0w}\}^{t_w} - \{E_w\}^{t_w} \tag{29.12}$$

Similarly, it can be shown that

$$\frac{(L_s)^{t_0}}{1 + (t_0 - t_s)C_s} = (R_s)^{t_s} - \{E_{0s}\}^{t_s} - \{E_s\}^{t_s} \tag{29.13}$$

Assuming that $\{E_{0w}\}^{t_w} = \{E_{0s}\}^{t_s}$ will permit equations (29.12) and (29.13) to be combined to give,

$$(L_w)^{t_0} = (L_s)^{t_0}\left(\frac{1 + (t_0 - t_w)C_w}{1 + (t_0 - t_s)C_s}\right) + [(R_w)^{t_w} - (R_s)^{t_s}][1 + (t_0 - t_w)C_w]$$
$$+ (\{E_s\}^{t_s} - \{E_w\}^{t_w})[1 + (t_0 - t_w)C_w] \tag{29.14}$$

where $(L_w)^{t_0}$ is the length of the workpiece at the reference temperature t_0. Although subject to the noted assumptions, equation (29.14) is a more general case of equation (29.8). As might be expected, substituting the condition $t_s = t_w = t_0$ will yield an identical result. Among the various other possibilities, the condition $t_s = t_w = t$ ($\neq t_0$) with $C_s = C_w = C$ and $\{E_s\}^{t_s} = \{E_w\}^{t_w}$ is very interesting since it corresponds to a common practical belief that, for a workpiece and standard made from the same material, the difference in length is not important provided both articles are kept at the same temperature. Making the appropriate substitutions in equation (29.14) will give

$$(L_w)^{t_0} = (L_s)^{t_0} + (R_w)^t - (R_s)^t + [(R_w)^t - (R_s)^t][(t_0 - t)C] \tag{29.15}$$

which clearly shows that the difference in length, as reflected by the term $[(R_w)^t - (R_s)^t]$, as well as the deviation from the reference temperature could well be important depending upon the relevant numerical values. Hence, if a minimum uncertainty of measurement is required for the measurand, the effects of influence quantities should be carefully considered.

All of the measuring methods listed above may vary with respect to the number of elements contained in the size equations depending upon the refinement required and the variations in procedure used. For example, the length measurement process may employ two standards, L_{s_1} and L_{s_2}, where $L_{s_1} < L_w < L_{s_2}$. Another interesting feature is that, for a given method, the number of these elements will increase as the geometrical complexity of the workpiece increases. A classical example is the measurement of simple effective (pitch) diameter of external screw threads where the p-value of the thread measuring cylinders employed, together with the related systematic errors, need to be incorporated. Although it may appear elementary at first sight, these facts are often overlooked when considering the accuracy of determination, or uncertainty, of engineering measurements.

29.3.5 Uncertainty of Measurement

In contrast to theoretical expectations, an absolute measurement is, of course, impossible to achieve in practice. A statement such as 'this shaft is exactly 20.00 mm in diameter' is well known to be not absolutely true or correct. The exception occurs, for instance, when the statement is a discrete count such as, 'twenty pistons' or 'five water pumps' (see also Barry, 1964). While a correct value may exist, there is no absolute confidence that the value obtained is the truth. On the other hand there is no assurance that it is incorrect. It could be the true value but there is always, to a varying degree, a measure of uncertainty involved since, whenever measurements are made, systematic and random errors are generated. It is this measure of uncertainty which is of critical importance when measurements are made in a factory or metrology laboratory. Whether the value obtained for a measurement is the absolutely correct value does not really matter as long as the accuracy of the measurement can be estimated within reasonable limits of confidence which are consistent with the accuracy objectives. Then the measurements will be meaningful. In this respect the terms *accuracy of determination* or *uncertainty of measurement* are widely used to highlight an important measurement area. The degree of accuracy with which measurements can be made will vary with a great many factors including those related to all of the elements of the measuring process. Consequently, only a brief discussion of well-known approaches is given hereunder.

While some industrial measurement laboratories (Bonollo, 1977) tend to avoid the question 'To what accuracy?' and work to arbitrarily selected or assigned values, there are a number of deterministic and statistical methods available which enable the uncertainty of measurement to be estimated. All of these involve a reasonably close study of errors and error contributing sources typical of measuring processes. A useful clue is provided by the nature of the size equations mentioned previously. Thus the accuracy equation relevant to equation (29.8) may be expressed in deterministic form as,

$$\Delta L_{\mathrm{w}} \approx \frac{\partial L_{\mathrm{w}}}{\partial L_{\mathrm{s}}}\,\Delta L_{\mathrm{s}} + \frac{\partial L_{\mathrm{w}}}{\partial R_{\mathrm{w}}}\,\Delta R_{\mathrm{w}} + \frac{\partial L_{\mathrm{w}}}{\partial R\mathrm{s}}\,\Delta R_{\mathrm{s}} + \frac{\partial L_{\mathrm{w}}}{\partial \{E_{\mathrm{s}}\}}\,\Delta \{E_{\mathrm{s}}\} + \frac{\partial L_{\mathrm{w}}}{\partial \{E_{\mathrm{w}}\}}\,\Delta \{E_{\mathrm{w}}\}$$

(29.16)

where the Δ's are the estimated limits (positive and negative) of the respective random errors and ∂ denotes the respective partial derivatives of the independent terms in equation (29.8). Since the partial derivatives can also be positive or negative, the limiting values of ΔL_{w} (positive and negative) can be more readily obtained by taking the modulus or absolute value of each of the relevant terms, thus

$$|\Delta L_{\mathrm{w}}| \approx \left|\frac{\partial L_{\mathrm{w}}}{\partial L_{\mathrm{s}}}\right|\,|\Delta L_{\mathrm{s}}| + \left|\frac{\partial L_{\mathrm{w}}}{\partial R_{\mathrm{w}}}\right|\,|\Delta R_{\mathrm{w}}| + \left|\frac{\partial L_{\mathrm{w}}}{\partial R_{\mathrm{s}}}\right|\,|\Delta R_{\mathrm{s}}|$$

$$+ \left|\frac{\partial L_{\mathrm{w}}}{\partial \{E_{\mathrm{s}}\}}\right|\,|\Delta \{E_{\mathrm{s}}\}| + \left|\frac{\partial L_{\mathrm{w}}}{\partial \{E_{\mathrm{w}}\}}\right|\,|\Delta \{E_{\mathrm{w}}\}|$$

(29.17)

Equation (29.17) will be recognized as being consistent with the rule for combining errors based on the calculus. Although only approximate, this approach may be widely used in practice for estimating the uncertainty of measurements and for comparing different measuring methods having the same objective. For this case, the rule is also very simple to apply since the partial derivatives are all equal to unity. Therefore, in terms of equation (29.8) the accuracy equation (29.17) reduces to a simple form analogous to the tolerance equations found in the dimensional analysis of engineering designs. Further simplification can follow if the random error components of the systematic errors (that is, $\Delta \{E_{\mathrm{s}}\}$ and $\Delta \{E_{\mathrm{w}}\}$) are ignored on the basis that systematic errors are constants by definition, or that random variations would be inherent and included in the values ΔR_{s} and ΔR_{w} respectively.

An alternative estimate of uncertainty of measurement, gaining favour nowadays (Bonollo, 1977; SAA, 1980), can be obtained by applying a statistical method, based on the law of addition of variances, to extend equation (29.17). Neglecting systematic error terms, the standard deviation S_{lw} (related to L_{w}) may be obtained from

$$S_{\mathrm{lw}} \approx \left[\left(\frac{\partial L_{\mathrm{w}}}{\partial L_{\mathrm{s}}}\right)^2 S_{\mathrm{ls}}^2 + \left(\frac{\partial L_{\mathrm{w}}}{\partial R_{\mathrm{w}}}\right)^2 S_{\mathrm{rw}}^2 + \left(\frac{\partial L_{\mathrm{w}}}{\partial R_{\mathrm{s}}}\right)^2 S_{\mathrm{rs}}^2\right]^{1/2}$$

(29.18)

where the partial derivatives were noted above and S_{ls}^2, S_{rw}^2, and S_{rs}^2 are the variances associated with L_{s}, R_{w}, and R_{s} respectively. It is then often customary to state the arithmetic average value for L_{w} together with an appropriate number of standard deviations to provide an estimate of the uncertainty at a stated confidence level ($2S_{\mathrm{lw}}$ is often associated with a 95% confidence level). An apparent difficulty with this procedure is the problem of obtaining reliable estimates for the various standard deviations so that they can be properly pooled according to their respective degrees of freedom (in keeping with statistical theory (Hald, 1952; Duncan, 1965)). This would then permit S_{lw} to be determined to a known number of degrees of freedom and, hence, make the estimate of uncertainty far more meaningful. Despite this difficulty, the statistical approach shown by equation (29.18) is often considered to provide a more realistic estimate of uncertainty of measurement compared to the noted deterministic model. Both methods, however, are dependent on proper recording and filing of empirical laboratory data so that the noted variances can be found.

Following up on the previous remarks about workpiece geometry, it could be expected that, as the geometry increases in complexity, the number of terms in the accuracy equation would also increase. Thus for a given

Figure 29.3 Gauges of the attributes type (for components of relatively simple geometry): (a) Fixed, full-form 'go' and 'no go' gauges; (b) full-form ring gauge; (c) adjustable, single-dimension limit gauges; (d) single-dimension caliper gauges with set limits. (Photograph courtesy of G. Branson (Design Department) and R. Fraser (Metrology Laboratory), RMIT)

measuring process, system and method the uncertainty of measurement could be expected to increase as the workpiece geometry became more complex with the resultant deterioration in accuracy of determination.

29.3.6 Some Measuring Systems and Hardware Considerations

In addition to selection problems, it is realized that the extensive range of available hardware may constitute a considerable capital outlay for a given quality control and measurement activity. In addition, many contemporary developments have occurred in terms of electronic, interferometric and laser-based manual and automatically controlled measuring systems. Consequently, the following discussion is restricted to noting a few examples of the more traditional types of quality control equipment.

(a) *Gauges of the attributes type.* As suggested by Figures 29.3 and 29.4, many examples of full-form and single-dimension limit gauges may be found in the literature. These gauges are often used in production to simulate the limiting dimensional assembly conditions of workpieces. They are widely

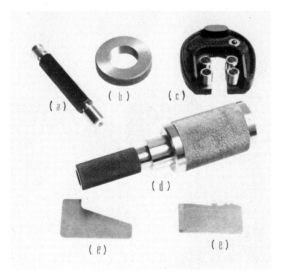

Figure 29.4 Gauges of the attributes type (for components of relatively complex geometry): (a) Screw plug gauge ('go' full-form and 'no-go' effective diameter); (b) full-form screw ring gauge; (c) adjustable screw caliper/snap gauge; (d) taper plug and ring gauges (shown assembled); (e) profile gauges. (Photograph courtesy of G. Branson (Design Department) and R. Fraser (Metrology Laboratory), RMIT)

(a)   (b)

Figure 29.5   Indicating gauges of the variables type: (a) for an external dimension; (b) for an internal dimension (shown with setting piece in position). (Photograph courtesy of G. Branson (Design Department) and R. Fraser (Metrology Laboratory), RMIT)

used with sampling schemes and control charts of the attributes type (Duncan, 1965; SAA, 1972) where the dichotomy is conformance or non-conformance (go or no-go) to the design specifications.

(b) *Indicating gauges of the variables type.* Figure 29.5 shows two examples of indicator fixture gauges as used with control charts of the variables type for single dimensions, whereas Figure 29.6 illustrates a similar but more flexible arrangement suitable for multiple inspection situations. Both systems incorporate mechanical methods of amplification although electronic hardware is often substituted in production.

(c) *Process capability systems.* An example of this type of system is illustrated in Figure 29.7. It may be used for both process capability studies and control chart applications. For ease of operation, special scales are usually fitted to the indicating device so that the deviation from the standard setting can be read off in convenient units. The data plot is representative of the output which can be obtained when microprocessors or programmable desk-top calculators are used to process the results of the measurements.

Figure 29.6   Indicating gauges of the variables type (arranged as a multidimensional measuring system). (Photograph courtesy of G. Branson (Design Department) and R. Fraser (Metrology Laboratory), RMIT)

As suggested in the review of measurement areas, all of the above
examples serve to reinforce the basic relationships which exist between
measurement and quality control functions in industry.

29.4   CONCLUDING REMARKS

In the first part of this work, a review of selected concepts and developments
in quality control has been presented, including an examination of important moves concerning the standardization of quality control systems. Some
relevant benefits to an organization were noted together with the strong
interconnection which exists between quality technology and the inspection
of products. This review was found useful in paving the way for a study, in
the second part of the chapter, of the inspection of products from the
viewpoint of engineering measurements. Various aspects of inspection and
measurement have been discussed. It was shown that measuring processes
and their comparative nature could be appreciated by considering a conventional length measuring system and its related elements. This led to a

(a)

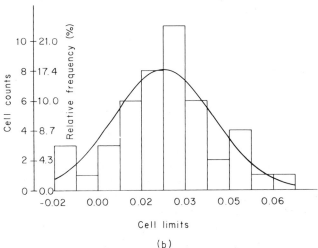

(b)

Figure 29.7 Process capability/control chart measuring system (for a number of measurands on a single component): (a) apparatus; (b) typical data plot. (Photograph courtesy of G. Branson (Design Department) and R. Fraser (Metrology Laboratory), RMIT)

discussion of measuring methods and the associated size equations whilst noting the possible effects of influence quantities. The importance of these equations in estimating the uncertainty of measurement has been demonstrated and the large variety of avilable hardware has been acknowledged. Although the above considerations have been presented qualitatively, it has been proposed that quality control and inspection of products are key issues which are fundamentally associated with product design and manufacture in industry.

REFERENCES

AOQC-DOP (1977). *Survey of Quality Control in the Manufacturing Industry*, Australian Organization for Quality Control/Commonwealth Department of Productivity, Canberra.

ASTME (1967). *Value Engineering in Manufacturing*, F. M. Wilson (Ed.), Prentice-Hall, Engelwood Cliffs, NJ.

Barry, B. A. (1964). *Engineering Measurements*, Wiley, New York.

Bonollo, E. (1977). *Measurement and Quality Control, Institution of Engineers, Australia, Victoria Division, Quality control symposium, Melbourne, May*.

Bonollo, E. (1978). 'The contribution of quality technology to company survival', *Proc. 6th Natl. Conf. Inst. Indust. Engrs. Australia, Hobart*.

BSI (1975). *British Standard, BS5233: 1975. Glossary of Terms Used in Metrology*, British Standards Institution, London.

Custance, H. M. (1977). 'Examination of quality control in an explosives factory and the method of introducing suppliers quality control in accord with AS1821–22', *Quality Technology Report MP502/77*, Department of Mechanical and Production Engr, Royal Melbourne Institute of Technology. (See also Commonwealth Government, Department of Supply, EFM Work Report No. 1977/7).

Doebelin, E. O. (1975). *Measurement Systems: Application and Design*, McGraw-Hill, New York.

Dodge, H. F. and Romig, H. G. (1959). *Sampling Inspection Tables*, 2nd edn, Wiley, New York.

Drucker, P. F. (1964). *Managing for Results*, Pan Paperback.

Duncan, A. J. (1965). *Quality Control and Industral Statistics*, 3rd edn, Irwin, Illinois.

Feigenbaum, A. V. (1961). *Total Quality Control*, McGraw-Hill, New York.

Fetter, R. B. (1967). *The Quality Control System*, Irwin, Illinois.

Grant, E. I. and Leavensworth, R. S. (1972). *Statistical Quality Control*, 4th edn, McGraw-Hill, New York.

Hald, A. (1952). *Statistical Theory with Engineering Applications*, Wiley, New York.

Hall, A. D. (1962). *A Metholology for Systems Engineering*, Van Nostrand, New Jersey.

Hartz, O. (1974). 'Quality control systems—elements of a theory', *Proc. 2nd Int. Conf. on the Development of Production Systems Denmark*, 1973, Taylor & Francis, London.

Hayes, G. E. and Romig, H. G. (1977). *Modern Quality Control*, Benziger, Bruce and Glencoe, California.

Jones, J. E. (1970). *Design Methods*, Wiley, New York.

Juran, J. M. (1964). *Managerial Breakthrough*, McGraw-Hill, New York.

Juran, J. M. (1974). *Quality Control Handbook*, 3rd edn, McGraw-Hill, New York.

Juran, J. M. and Gryna, F. M. (1970). *Quality Planning and Analysis*, McGraw-Hill, New York.

Knowler, L. A., *et al.* (1969). *Quality Control by Statistical Methods*, McGraw-Hill, New York.

Miles, L. D. (1961). *Techniques of Value Analysis and Engineering*, McGraw-Hill, New York.

Miller, L. (1962). *Engineering Dimensional Metrology*, Edward Arnold, London.

NATA (1975). *NATA and Your Laboratory*, National Association of Testing Authorities, Sydney.

NATA (1979). *Metrology*, National Association of Testing Authorities, Sydney.

OIML (1978). *Vocabulary of Legal Metrology, Fundamental Terms*, International Organisation for Legal Metrology, Geneva.

Puttock, M. J. (1965). *Notes on Engineering Metrology*, William Brooks, Sydney.

SAA (1971). *Australian Standard, AS1057–1971. Terms Used in Quality Control*, Standards Association of Australia, Sydney.

SAA (1972). *Australian Standard, AS1199–1972. Sampling Procedures and Tables for Inspection by Attributes*, Standards Association of Australia, Sydney.

SAA (1973). *Australian Standard, AS1399–1973. Guide to AS1199 Sampling Procedures and Tables for Inspection by Attributes*, Standards Association of Australia, Sydney.

SAA (1975). *Australian Standard, AS1821–1823—1975. Suppliers Quality Control Systems*, Standards Association of Australia, Sydney.

SAA (1977). *Draft Australian Standard, DR76103, 1976. Guide to AS1721–1823, Suppliers Quality Control Systems*, Standards Association of Australia, Sydney, published as AS2000–1977. (See also AQAP-1, *NATO Quality Control System Requirements for Industry*, 1970.)

SAA (1980). *Draft Australian Standard, DR79051, 1979, Glossary of Terms Used in Metrology, Part 1—General Terms and Definitions*, Standards Association of Australia, Sydney. (Published as AS1514, part 1, 1980).

Shewhart, W. A. (1931). *Economic Control of Quality of Manufactured Product*, Van Nostrand, New Jersey.

Thoday, W. R. (1973). 'Motivation for quality: The significance of quality terminology', *The Quality Engineer (UK)*, **37**, 11.

Tippett, L. H. C. (1950). *Technological Application of Statistics*, Wiley, New York.

UKMD (1973). DEF STAN 05–21, 1973, *Quality Control System Requirements for Industry. DEF Stan 05–24, 1973, Inspection System Requirements for Industry. DEF STAN 05–29, Basic Inspection Requirements for Industry*, United Kingdom Ministry of Defence, London.

USDD (1963). *Quality Programme Requirements, MIL-Q-9858A*, United States Department of Defense, Washington.

Handbook of Measurement Science, Volume 2
Edited by P. H. Sydenham
© 1983 John Wiley & Sons Ltd

Chapter

30 J. W. HOBSON

Management of Existing Measurement Systems

Editorial introduction

In recent decades measurement systems have grown from a handful of sensors and control loops of an *intensive* arrangement to *extensive* systems—an 800 control loop process plant is not uncommon—that cannot be managed by solving problems on an *ad hoc*, as they occur, basis.

As yet there are no commonly agreed hard and fast rules or procedures for managing an existing extensive measurement system. Perhaps there never will be, for like all management facets it does not lend itself entirely to mathematical and physical modelling.

It is desirable, therefore, in order to assist understanding of this measurement system aspect, to provide a discussion of the needs and possible procedures, as opposed to the means, leaving it to individuals to design and implement their own strategies to meet their own circumstances.

It cannot be overstated that a large system's integrity rests heavily upon the quality of its measurement system—a point often overlooked by general management.

This chapter, compiled by a person responsible at a senior level, for such matters in a nation's government supply factories, provides a summary of his extensive experience for others to use as appropriate.

30.1 INTRODUCTION

30.1.1 Rationale

Dependent upon the way in which, 'existing measurement systems', are defined, discussion of their management could require a textbook in its own right. *Homo sapiens* normally starts 'measuring' activities before birth and generally continues them in some form until death. This measurement resource is rightly left to its possessor for its management, and is therefore beyond the scope of this text. Nonetheless those who bear the responsibility for management of measuring systems would do well to keep this truth in the forefront of their thinking.

The proper management of existing measurement systems (meaning those physical appurtenances used to convert certain variables to designated signals convenient to human activities and capabilities) is dependent entirely on the skill and judgement of the operators effecting the task, and also relies on their ability to continue carrying out such tasks to the utmost of their skill and ability. Attempts to 'calibrate' this resource are not possible because there is no known way of establishing a 'standard measurer' against which standard other operators (measurers) could be calibrated.

The most common way of achieving this is by means of an assessment of the measurer carried out by persons recognized as having appropriate capabilities in the particular field of measurement. (Refer to Chapter 31.) No amount of high quality equipment will produce 'good' measurements if the measurer is incompetent. Nor can a highly proficient operator produce good measurements from an inferior piece of equipment. Management therefore should have arrived at the point where an operator (or series of operators) who is proficient in his work is supported by the best equipment available within the financial limits imposed by the directors of the organization.

The relevance of the foregoing to the management of existing measurement systems is that it sets a scene not utopian but realizable, and if such conditions exist, that is a combination of 'good' measurers and 'good' equipment, then the management of a measuring system will have as its goals the continuance (and improvement where possible) of those conditions. Where these conditions do not exist then to the management tasks must be added the necessity to establish such conditions.

Plans to continue using such measuring systems will require development over a period, and should project as far as the dismantling of the particular process plant, the removal of the need to have such measurement as a data component, the end of the useful life of the various components of the measuring system, or the end of the useful life of the plant itself.

30.1.2 Necessity for Early Consideration

Measuring systems, all too often, are left out of the primary considerations for new plant, tending to be 'tacked on' or left as afterthoughts, for example, 'tack in' a pressure gauge or thermometer, rather than coordinate all of the details as the design work progresses.

The best place to start managing a measuring system for any plant is right at the beginning of the design for the new plant. The hardware and management control systems and relevant measuring equipment can then become an integral part of the overall design.

Factories and establishments must be governed by the principle that the goal of maximum production of acceptable items with minimum expenditure of resources is achieved. In any process measurements are essential at various stages of the process to ensure that the process is proceeding satisfactorily, and to provide information on deviations from the pre-determined requirements. Appropriate corrections can then be applied to minimize process deviations. The measuring and control systems themselves are subject to examination against the criteria that maximum production of acceptable items is obtained with a minimal expenditure of resources. Additionally, the measurement and control systems are subject to external influences insofar as the need to ensure either the uniformity of the same product from several sources or the interchangeability of different items engenders the requirements for calibration of the individual parts of each measuring and control system. All measurements are, therefore, referenced back (made *traceable*) to the appropriate national standard via a known series of checks.

In all of this activity there is implicit the necessity to maintain all systems in proper and effective working order, to ensure that breakdowns or failures are reduced to low levels, conditions optimized, and all necessary servicing, calibrations, and other back-up operations are effected in due time. The necessity to coordinate this work to achieve an ordered and effective programme covering all of these needs is therefore clear.

Throughout all of the actual working life of control systems and instruments there is a need for calibration to *ensure* that the measurements are what they purport to be, to give some certainty regarding such factors as interchangeability of manufactured parts, and to provide the base for quality control management. Hence it can be seen that the product quality control function is determined in large degree by the aptness of measuring instruments and the systems they support. Many prominent public figures are on record in this regard throughout recent history, attesting to the view that nations with good measuring operations within their industrial supports are usually in good trading positions.

30.1.3 Place in Overall System

The need for an early start on the process of managing an existing measurement system can be seen as starting with the planning of that system. To obtain the best performance from a system an adequate determination of the resources to be used is essential. Care and attention to fine detail at this stage leads to the lowest resource consumption consonant with the desired output. Also, the establishment of detailed plans to manage the system, from an early stage, provides a means of monitoring progress, and a means of taking effective measures to correct adverse trends which may appear. A basis is established which permits a ready assessment of all options in the event of changes forced by considerations not known or predictable in the initial considerations.

Accumulation of information on the various component parts of the system will also be of use in providing performance information on the instrument itself. There is a lack of information (and much misinterpreted information) regarding the real performance of instruments. This is particularly true in areas where the environment is particularly adverse, for example, in acid concentration plants there are problems not only of selective attacks on materials by the various liquids but the problems of acid vapours in the surrounding atmospheres. More will be said later in this chapter about the subject of *instrument evaluation*, particularly in the light of present trends towards the 'throw-away' instrument, with its promise of low cost and the danger of reduced product quality. Procedures of instrument evaluation are discussed in Chapter 31.

30.1.4   Value to Operations

Measuring equipment is readily accepted in the production plant provided that it is accepted and understood as an aid, is simple to use, and there is no overdue stress placed on its use. The workers at the plant will take great pride that, 'the needle's been kept right on the line', or 'its easier now we've got a gauge on it'. Yet management so often takes up the view presented earlier using a bare minimum of instrument aids rather than ensuring an adequate set of instruments, including recorders, exists at key points.

The decisions are often not based on a well reasoned cost–benefit analysis but are based purely on total cost, not being related to the cost of the plant or to the likely costs to be incurred in the event of damage to life or property, nor to the total cost change reflected in the product pricing. Also, where there will be, at some time, a requirement to have that instrument calibrated, the calibration is regarded as an unnecessary expense regarding it as a waste of money or other resources. Very often the need for an instrument to be calibrated is not found until the product becomes inferior or unsaleable. At this stage the cost recovery process can be well in excess of the cost of calibration of the instrument, without considering the damage to product prestige. The resultant losses can be quite substantial in the event that the inferior product may be involved with the safety of human life and recall and repair or replacement action must be carried out.

30.1.5   Some Factors and Influences on the Measuring System

In an endeavour to show in a concise form the factors which influence, and hence need to be considered in, the management of an existing measuring

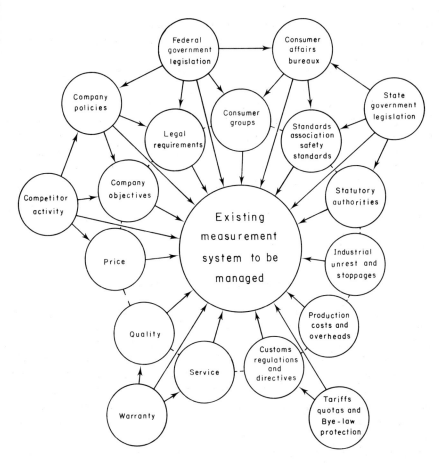

Figure 30.1 Web of some important factors and influences which require consideration in order to sustain optimum performance of an existing measurement system

system, the web of some important factors has been set out, as shown in Figure 30.1. There may be other factors which should be included. These will vary between the various nations but in general the coverage given should be effective in demonstrating the relationship of the various factors and allow the reader to set up a reasonable picture of the situation appertaining in his immediate concern of managing the measuring system.

Eleven principal factors would seem to govern the management of the measuring system, and these are set out in the diagram as the immediately orbiting satellites. However, the outer seven orbiting satellites, whilst not exerting control on the immediate situation to the same degree, are probably of equal importance, as they influence both the measurement system as well as some of the inner satellites.

Although, 'industrial unrest and stoppages' and, 'production costs and overheads' are shown as not having any governing factors or satellites, they could be related to 'company policies', to 'federal government' or to 'state government' legislation. However, most commonly, they are not entirely predictable in any given set of circumstances although, production costs and overheads can be influenced to a noticeable degree by industrial unrest and stoppages.

The inclusion of 'competitor activity' may not be relevant in some nations, but nonetheless it is an important factor as its absence means that the 'price' factor can have a different influence to that which it would exert in the situation as depicted, that is, where such competition truly exists. It can also be said that the 'warranty' factor is governed by the 'quality' and 'service' factors but these all go hand-in-hand.

30.2 REPLACEMENTS

30.2.1 Planning for Replacements

From time to time it will be necessary to replace instruments, sensors, and so on for a wide variety of reasons. These range from catastrophic failures, to sheer obsolescence or an inability to maintain the item economically. Where it is intended that the process controlled will have a long lifetime planning should include an appropriate number of replacements, determined at the design stage, to better evaluate the overall economics of the process plant. The rapid technological changes which presently are reducing the expected lifespans of many items, not just instruments, are causing many problems in this regard. Should an instrument manufacturer find that his business is no longer viable, and ceases manufacture, then processes employing that manufacturer's items are caught in a cleft stick. Either there will be unplanned buying of available instruments and spare parts, adding to storage and handling problems, or the process instrumentation will be changed to use instruments from another manufacturer with all of the problems attendant on making and controlling such a changeover.

The exercise becomes a rather sophisticated set of planning manoeuvres to provide the required level of control at minimum cost, and with adequate provision for possible replacements or an ample supply of spare parts. Reasons to plan for replacements in the early stages are, therefore, largely economic in terms of the instruments or process control components. This is further complicated by the possibility of technology changes forcing a re-examination of the entire plant to maintain or improve efficiency and relative costs.

30.2.2   Effects on the System or the Process

Where the effects of replacement planning are more obvious and direct in relation to the process, the overall effects on the system (meaning the whole of the operation) are not immediately apparent. In considering the effects of replacements on the system we must look at such problems as the deterioration of instruments if replacement stocks are purchased and stored. Appropriate storage methods for instruments and spare parts are discussed elsewhere. At this stage we need only to appreciate that such problems must be considered and resolved, with appropriate financial provisions being made. One consideration often ignored or neglected after a cursory oversight of the situation, is the regular calibration of instruments. There is a lack of appreciation that an instrument standing idle (that is, in store) can and does drift from its previously known calibration. A decision is therefore to be taken where instruments are held in store as 'changeover' spares. Should the instrument be kept in its calibrated state or be allowed to lie, a calibration being carried out just before it is put into service?

This brings us to consideration of the effects on the process itself of replacement planning. It is most likely that 'down-time' will be minimal when direct replacement instruments are held, and exchange action used where maintenance and/or calibration are required. Although the process must be interrupted in some way, it is usually the case that this shutdown is minimized where the faulty instrument is removed and a known, calibrated, replacement is installed in its stead. The notable exceptions to this will be those processes with a known prolonged shutdown procedure which must follow through without an interruption. In such case there is probably enough time to effect repairs and on-the-spot recalibration of the instrument, always provided that a competent, well-trained, instrument technician is employed to carry out all of these operations.

Reliance on extramural suppliers of the maintenance and calibration needs of process plant instrumentation is a mixed blessing to the manager. On one hand the employment of an instrument technician for a plant where he would not be fully employed is a waste of resources and can be depressing to the individual. On the other hand, the entire process plant operation is under the control of the manager, there will be greater uniformity of all operations, and a reduction of errors. The latter can be very costly when a faulty instrument calibration results in the loss of an entire production batch. All of these factors must be given due consideration and must not be written down as too expensive and a waste of monetary resources.

30.2.3   Evaluation of Measuring Equipment

With a few exceptions this problem does not yet appear to have been addressed fully (see Chapter 31). It is rare that information is readily available as to the performance of measuring equipment when, for example, the relative humidity reaches 95%. Some manufacturers issue statements that their particular instrument will perform reliably (that is within their specification figures) with environmental conditions not exceeding, say, 35 °C and 90% RH. Should the operational conditions exceed either of those figures individually or in combination then there is some guesswork as to the reliability of the measurements made at that time or thereafter. The relationships between the environmental influence parameters and the measuring equipment are usually not known.

Similar remarks apply to mechanical or durability requirements such as vibrations, bumps, and accidental mechanical shocks. Although many of these tests are described in military or defence specifications their introduction to civilian requirements is quite recent and not yet coordinated to meet all reasonable requirements. Most importantly, where such tests are applied to equipment or process plant, the required result is continued operation with little or no interruption. For measuring equipment that requirement extends to clear knowledge of the detailed performance of the instrument, before and after the test.

Also involved is the effect on the ability of the instrument to retain its calibration and performance in such more demanding circumstances. The reliability of the measurement is required to be known in all circumstances, to ensure that product quality and uniformity are known and sustained. The questions of environmental operating conditions and other factors of evaluation are of such importance in this field that a more detailed examination of this problem is made in Chapter 31.

The necessity for evaluation of measuring equipment can be met provided that the system is consciously 'locked-in' to a given set of instruments knowing that problems of supply will develop in later years. Tests to determine suitability for purpose can be carried out, but these also require a long-term commitment to the task. Evaluation then becomes a question also of selection of likely instruments from the multiplicity of items being regularly released on the market. This will be influenced by several factors including the reputation of the maker, preference for particular types of information presentation, reliability of previous products of the same makers, and word of mouth information from 'reliable' sources. Testing to determine suitability then is based on grounds which are not necessarily secure, as a very good instrument could be excluded by poor background factors, whilst a poor instrument will go into the test programmes because its background is good. Hence, the determination of suitability of an instrument is an essentially difficult and costly task which is, therefore, not carried out, or is ignored in favour of selection by preferential attributes which may or may not be soundly based.

30.2.4 Costs Evaluation

The evaluation of the costs related to replacement instruments will depend on the determinations and planning bases on which the whole project is carried out. A decision to hold replacement instruments incurs costs such as amortization, storage, checking, and maintenance; whereas the holding of stocks of spare parts will not, for example, incur checking and maintenance costs, but will reduce the storage costs. However, the heaviest single cost is the consequence of the undetected drift or failure of an instrument operating in the process control system. This highlights the great inherent value of an instrument which can, and often does, make thousands of measurements between its calibration intervals. The cost of lost or faulty production can be great. Properly maintained and calibrated instruments form one of the best insurances against such an eventuality.

Similarly, it can be seen that the importance attached to the functioning of the measuring device or its control should not be checked only against its cost but also against its role and the consequences of failure to carry out its role. If there are no data as to the performance and the reliability of a number of instruments being considered for a particular task, some evaluation is required, even if this is only an examination of competing items by an experienced technician. An initial cost saving at purchase time can result in large maintenance costs and possibly at the extreme situation, in loss of a reputation for high quality products. The consequential loss of sales, both of new instruments and spare parts, and the reflection on all of the manufacturer's products can be very severe, possibly leading to closure of the enterprise. As an incidental matter, it is of benefit to manufacturer and purchaser that these are the goals of the manufacturer, and that this can be demonstrated to the purchaser very clearly.

30.2.5 Value Judgements

The making of value judgements is a question of distinct importance where a complete measurement system is required; the whole being as good as the sum of the parts. Due consideration of the needs of the measurement system, the relevance of its separate parts, and the manner of functioning of the entire system, sets out the beginnings of the value judgement process.

However, the matter is considerably subjective. Regardless of relativities, there is always some degree of prejudice present in the considerations leading to a particular decision. Past experience tends to build in these prejudices so that the products of certain manufacturers will always be 'not good enough' whilst others will be, 'essential' or even 'to be obtained at all costs'.

This situation must be recognized and dealt with firmly. It may well be that an instrument or measuring item may be available to suit the purpose, and be of excellent quality. However, there is no reason to reject lower priced items if their performance is known to satisfy the task needs.

There is an attempt in many quarters to preserve a 'mystique' about measurements and control systems. In many cases an item is bought because it is 'the best'. This is very often a waste. The measuring expert has failed himself insofar as he has let his judgement be diverted from adhering to a fitness of purpose criterion by some imagined prestige to be gained or because of a belief that it must be adequate. Efficient use of resources is controlled by a number of factors, one of which is value judgement, that is the selection of the item or equipment, which, in the prevailing circumstances, affords the optimum combination of performance and reliability at lowest cost.

Although the manufacturers of the various measuring instruments produce items of good, and in many cases excellent quality, the pressures of maintaining commercially viable operations (as discussed in part in Sections 30.2.3 and 30.2.4) reinforces the situation where the user must study all circumstances, that is, make sound value judgements.

The place within the proposed operation of every item in a system has then to be scrutinized as to its value within the system, and then to its value as an intrinsic matter. Failure in either regard must lead to rejection and/or replacement of the item as a part of the system, or indeed in favour of another item. Unfortunately, although it is relatively easy to assess the cost of a measurement it is often most difficult to cost the benefit in order to make the cost–benefit analysis needed.

30.2.6   Rationalization

A brief scan of the commercial literature should start thought processes in regard to the improvement of an existing system, or to a new approach in resolution of a presently intractable problem. An entrepreneur may have entered the market with a device which will make these resolutions possible. However, this is a process which is categorized as normal commercial, and is occurring and recurring daily. The manager, in his bid to maximize his results, will study the relevant cost factors with regard to the item itself including installation and/or a changeover, integration within any existing system, effects on operations staff, and maintenance requirements. Engineering and/or scientific doubts being allayed, plans are then made to effect the necessary changes and obtain the resulting benefits. A neat, albeit oversimplified example, but essentially sound.

Rationalization is the term used here to deal with the activities controlling all of the processes set out in summary above. Beginning with a measuring system containing say eight or nine major elements, each of which contains a number of subelements, the maintenance of that system and the improvement of its overall performance should be regularly surveyed. Initially the whole system may have been supplied under tender action by one manufacturer. This would normally include all spares, handbooks for operation, maintenance, and calibration. Technical staff would familiarize themselves with the whole system.

Next comes the change illustrated earlier and for which all of the factors are not normally considered. To continue in orderly fashion the spares and handbooks for the new item(s) must be stored with the existing spares and handbooks, even though they are from another manufacturer. Repeating this process results eventually in a mass of technical data covering the, by now, greatly improved system. However, servicing requires the use of this data, and it must be readily understood, including all required cross-referencing, otherwise the servicing or maintenance activities occupy far too much time, and are far more prone to errors (resulting in more lost time).

The crux of the matter is shown to be the need to consider such subjects as training for service technicians, storage of additional information and spare parts from another manufacturer, a knowledge of or fully documented cross-referencing of, support information by managerial and technical staff, and a full understanding of the interrelationships in the revised system.

Much of this can be obviated by rationalization of the overall system. Although it requires second or third looks at the benefits of the proposed changes, some of the improvements will inevitably be penalties when all factors are considered. This is not to say, either, that a system once installed, should not be subject to change. Rather it is to say that the number of different makers products should be minimized within the overall system, whilst taking maximum benefit of performance improvement possibilities opened up by the introduction of new items in the market place.

So, the proposal to introduce an improvement must be studied from the viewpoint of rationalization. This means considering the effect of additional storage space, documentation, filing, staff training, and accounting and computer space requirements. These are all additional to any of the special requirements of the proposed new equipment.

30.2.7 Use of Automatic Data Processing

The manager of a measuring system very rapidly becomes aware that a large quantity of information in varying categories is being generated continually as a result of all of the operations in progress. Clearly, this information should be analysed from the various operational aspects of the process to aid in improving overall performance or provide cost reductions.

Generally there is little effective planning providing for the collation of all this information. Basically this occurs because the accumulating system records rapidly progress into a state of work overload, and in consequence the operators progress into a state of frustration. Operators complain regarding the quantity of documentation required for each item, both as an initial requirement, that is, the plant or stock record type information; and as the records on an item about which the detailed performance and any variation thereof must be known. There seems to be a general reaction against the sheer quantity of detailed information required about an item used for measuring, for example, its detailed calibration history, plus records of all maintenance and adjustments made. (This excludes such matters as zero-setting and chart changing.)

It would be desirable that this situation be exactly like that of the common steel rule. It is a highly precise measuring instrument which is very cheap, is easily carried around, requires very little checking and maintenance, can be severely abused, neglected, and ignored; yet can be picked up, possibly given a wipe-over, and will again give a precise measurement.

The manager will tend to look at large quantities of data, select the salient points, and form judgements based on them. Although these may be adequate judgements in a considerable proportion of cases, they will not provide the optimum obtainable results in all cases and hence will represent a failure to fully utilize the information resources available.

Changing from such manual methods to the use of automatic data processing (ADP) methods brings many problems in its train, not the least of these being the necessity to provide for adequate data storage capacity and possible expansion, secure verification procedures, and provide for the development of adequate programs to ensure that the information required by the manager is presented in an appropriate format or formats. It is also necessary to ensure that items entered into the ADP records are identically described if they are identical items.

This is an area of concern in constructing a set of records which will be adequate for the ADP functions. It is necessary either to use standard instrument nomenclature, either developed by in-house study or to adopt the standard nomenclature of another body. To this time there is no one coding or descriptive pattern system universally accepted for such purposes in the fields of measuring instruments or process control instruments.

To give some of the more obvious uses to which the ADP system can be put, such considerations as the development of routine maintenance systems, routine servicing systems, routine calibration systems, stock control and replenishment, and control of costs come readily to mind.

In any system where a decision is taken to introduce ADP it is essential that thorough study of the problem actually in existence be made. This should include determinations as to the questions the ADP system must continue to answer; the needs of the ADP system itself in the form of an adequate supply of correctly coded and verified information; the ability of the ADP system to expand or contract in line with the demands on the system; and, the ability to reprogram the ADP system where it is shown that the present system output is either unsuitable or must be changed to accommodate new or changed circumstances.

30.3   OPERATIONAL ENVIRONMENTS

30.3.1   Study of Maintenance Requirements

The operational environment within which the measurement system must operate is one of the factors with which the manager must contend. For a laboratory performing a reference or central laboratory function the environmental parameters most suited to the measuring or calibration function will be difficult to establish and then to sustain. Often, in the course of the operation of a laboratory, or even of an area in a plant where important process measurements are taken, the relevant parameters will not be adequately determined, and the penalties of the failure to observe and compensate for such lack must then be paid. These penalties will be either the necessity to repeat a calibration and verify the same, or the process batch of goods may be rejected on inspection, either becoming scrap or requiring rework, and in either case causing losses or delays.

The environmental requirements which must be maintained for the particular process, be they temperature, freedom from dust, absence of electromagnetic interference or any other relevant parameter, require attention in detail from the design stage onward. Every requirement of the environment has to be determined, including the methods of establishing the environmental conditions, and then controlling and measuring them so that user confidence is assured by the availability of recorded data of enough variables on the environment in which the process is carried out.

30.3.2   Need for Careful Observations

As mentioned in Section 30.3.1 there are penalties to be paid when accurate observations of the environmental parameters are not made and recorded. These penalties are modified as the necessary care and attention is given to the observations.

In most cases the space constituting the operating environment is subject to spatial and temporal variations caused by such extraneous means as change of ambient temperature and relative humidity, variations in electromagnetic field strengths and increased salt or dust concentrations in the ambient atmosphere. These, and other variables, all introduce their own particular influence changes for which the necessary compensations must be made, either by direct calculation of the effects (as in, say, change of temperature) or an estimation of the effect by the officer in charge (as in, say, unexpected additional dusts in the atmosphere).

The problems, if not consternation, caused when a product fails to pass inspection, or a test fails, when seemingly all precautions have been taken, are as nothing when compared to the painstaking task of systematically taking apart all of the work to determine with certainty the reason(s) for rejection. The examination of every test, and the environment in which the test was effected is a tedious affair, yet demands absolute concentration so that the root of the problem is in fact located and dealt with.

Well maintained and monitored duplicate log books become an essential in minimizing these problems. Such log books used by staff who are required to record all data resulting from, and occurring as, part of their work then become a useful key in unlocking the type of puzzle mentioned earlier. It is of great value if the Test Engineer, or Officer-in Charge, conducts a review of the recorded results with the Testing Officer, preferably on the same day. In this way salient points can be analysed, the relevant value of minor occurrences established in relation to the remainder of the work, and suggestions as to possible changes in equipment and procedures (or even staff where appropriate) discussed and the value determined in relation to the overall operational situation.

30.3.3   Protective Measures

The beginnings of what I call protective measures have already been partly described in Section 30.3.2. First and foremost is the requirement that the staff who carry out measuring and/or calibration activities must be trained into the habits of continuing to search, to question, to be particular to the point of pedantry, to examine in minute detail, and report in similar fashion. They should be quite prepared to log every item of information even though it may not seem entirely relevant at the time. Also they should expect of their superiors the third degree treatment from time to time, be prepared to answer in such situations from sheer logic developed from their work and to cooperate in furtherance of the development of the relevant techniques to assure, at least for the future, a basis for repeatability and reproducibility of a measurement, test, or calibration.

Such staff members then can be expected to be particularly observant of conditions and changes of conditions while test or measuring work is being carried out, and to act to either eliminate or minimize the problem condition(s). Also, they would report changes in their equipment where deterioration became evident and initiate action for maintenance work at an early stage so that very little work is required to restore the equipment to top-line condition.

Requirements in regard to calibration and servicing (which are dealt with in greater depth later in this chapter) will be set down in systematic fashion so that the overall condition of all measuring equipment is known or can be interpolated at the time of use, that is, a very close overall check is maintained to preserve intact the traceability of each measuring instrument. (The concept of traceability is relatively new and subject to several interpretations, but the basis of the concepts used herein are set out in Section 30.5.7 of this chapter. It is also discussed in Chapters 2 and 31.)

These measures are additional to all of the basic needs or requirements of an operating laboratory or measuring centre, such as safety of personnel in situations of electrical, chemical, or other environmental hazards; the protection of equipment from fire, electrical or mechanical hazards; the necessity to observe safety precautions as appropriate in the form of safety glasses; ear muffs; safety clothing and shoes; and in even more specialized measurement areas, observing extreme cleanliness, or following in exact detail the requirements of a measurement procedure.

30.3.4 Selection of Enclosures

Again the choice of an enclosure in which to house a measurement system is a simple task which when examined (and that not at great depth) takes on revised perspectives and reveals difficulties which, if ignored, will either plague the system with continual minor problems, or require a relatively expensive modification or repair job.

Generally the first task is to assess all of the relevant environmental features within which the proposed measuring system will operate, and then determine which of these are of an essential nature. In almost every situation it is preferable that the temperature be kept constant within certain limits. Other variables then will be desirable, that is, to be attained if reasonably possible. Others again will be there for consideration in the event of special tests or a special item maybe requiring unique consideration in its testing or checking.

For most laboratories performing a central or calibrating type function there are several features which are common to all enclosures used for the testing or calibration of systems. These general features include an air-conditioned environment, that is, the temperature and relative humidity are controlled accurately and there is mitigation of dust and salt problems because of the presence normally of an air-lock type entrance *cum* exit. Additional requirements may be the reduction of acoustic noise levels, radio frequency (r.f.) noise levels, and other electromagnetic interference (e.m.i.) levels; each of these requiring special and sometimes conflicting treatments.

Preferably also vibrations, both man-made and seismic, should be at least studied and mitigated where necessary to improve the quality and value of the required measurements.

Thus far the enclosures considered have been room size, rather than of the smaller variety in which all of the factors mentioned above are much more readily controlled. The penalty paid for using small enclosures is that a much more limited measuring system can be set up, or that a much smaller test can be conducted.

Selection of an enclosure for a test or measurement is often on a haphazard basis, or a selection is made as a matter of convenience. As a general rule there should be ample room for free atmospheric circulation within the selected enclosure, and extraneous influences should be reduced to levels consistent with the maintenance of steady conditions for test and measurement purposes.

30.3.5 Selection of Materials

This field demands, and has, specialist practitioners in its own right. This is because of the tremendous breadth of the materials spectrum and the variety of processes and interactions which can occur in the many and various industrial and commercial situations with which life is concerned. The uses of natural materials such as the various timbers, fibres, minerals (clays, rock) are nowadays inextricable from the uses of artificial materials such as the various plastics and elastomers.

The primary requirement is that the best results are achieved with the minimum use of resources. Therefore a good knowledge of materials and their relevant properties has either to be known or sought. For example, the chamber used to measure resistance to salt corrosion is effectively itself being salt-spray tested every time. Therefore, it has virtually to be a super-resistant item so that it will survive, and additionally will not introduce its own variations into the results by producing its own corrosion products, which will in turn interact with corrosion products of the item being tested. The need for great care in the selection and manufacture of the necessary item is obvious.

Similarly the selection of materials for measuring equipment may be governed by other considerations such as the need to be aware of galvanic corrosion in certain situations where unknown or unsuspected electrical potentials may exist, or may develop if there are changes in other circumstances such as the introduction of moisture in a previously dry situation.

This line of discussion leads to consideration of the various methods of fastening, joining or sealing which may be required. Similarly the various surface treatments and finishes which may be applied or utilized are covered in numerous texts and journals.

30.3.6 Corrosion Checks

Apart from the fact that corrosion of any sort on any equipment creates the impression of a lack of due care and attention, any signs of corrosion on measuring equipment is sufficient to destroy the confidence of the customer in the measuring or calibrating authority. They are justly entitled to assume a lack of proper care for such vital instruments and should insist on a full and clear demonstration that the equipment is sound, is calibrated, is properly functional, and hence may be the subject of a reasonable degree of confidence.

The essential point is that a measuring or calibration centre or laboratory must be entirely above suspicion. Good house-keeping is absolutely essential with every instrument having a place and every instrument in its place. This not only inspires customer confidence, but inculcates good working habits and provides a sound training for the young technician. He learns the right and proper way first time, not later when an inbuilt bad habit may take considerable effort to correct.

Technicians and tradesmen involved will also be about the business of checking regularly to ensure that there is no corrosion to be seen particularly at important places like the enclosures and terminals of standard cells (or other batteries for that matter). Also there should be a regular program of removing the battery packs from battery powered equipment and checking the state of charge as well as ensuring that there is no corrosion on the battery or in the battery compartment. Constant temperature, oil and water baths are other places where corrosion checks and hence regular cleaning should be carried out. Not so well known are the corrosive effects of high purity water (condensate is an example), electrostatic charges (dust gathering), and stray fields (r.f., e.m.i.). These are all readily dealt with and will be a part of the good house-keeping mentioned earlier.

30.4 TRAINING

30.4.1 Necessity for Training

With measurement and calibration, being the very foundation stones upon which quality control and our whole technology in general are built up and improved, it is essential to everybody that those who elect to engage in what can be at times our most tedious and boring tasks, and at other times our most exciting tasks, should receive a most detailed and thorough training, both academically and on the job. Some idea of the relative value to be assigned is given by the parallel illustration in which a machinist may make an error whilst measuring a part which he is making, rendering the single finished item useless. The measurement technician may make a similar sort of error in calibrating an instrument but if this error is undetected then thousands of other parts may be useless because of that one error on the part of the measurement technician. Due recognition of the needs, and of the responsibilities borne, will help to bring forward some necessary improvements in the quality of measurement practitioners. This applies to the management functions with even greater force, as the manager must be aware of all of these problem areas and be prepared to move to meet and solve the technical and human problems which are still largely unexplored.

Preferably, the managers will have graduated from, or trained through, the processes of measurement fundamental work in the laboratory or measurement centre. They will have worked with instrument fitters and technicians and so have developed a good appreciation of the human and working problems at these levels. However, in the case of the professionals there should not be any attempt to produce the same level of manual skills required of the tradesman. Conversely, the tradesman should not be subjected to excessive academically based theoretical training which is not likely to be used.

The requirements briefly stated, are that the professional, the technician, and the tradesman should each receive an appropriate mix of academic, practical, and administrative training to equip them to carry out the duties of their respective positions effectively and efficiently in a cooperative manner.

30.4.2   Extra-Mural (Academic) Training

As mentioned earlier (in Section 30.4.1) the academic training programs must be so devised that the professional, the technician, and the tradesman, receive and develop skills appropriate to their positions in the overall organization. Some allowance for an overlap situation must be made so that in the right circumstances a tradesman may extend his training and qualify for appointment as a technician. Similar remarks apply in the case of a technician who may have the capability to achieve professional standard.

The possession of appropriate professional ability and qualifications, does not necessarily indicate that a person can properly manage situations involving control or measuring equipment, although they are certainly very useful attributes for the purpose. A training and understanding of the needs of measuring systems must be coupled with a full appreciation of the means and resources required to develop the full potential of existing measurement systems. Additionally, the integration of new items into existing systems, or the phasing out of superfluous, unsatisfactory or obsolete items must be coupled with an understanding of the necessary financial and accounting controls which may, or will be, imposed by the directors of the company. Additionally the human factors such as the effects on the staff including the need for any retraining, requires very careful consideration.

Thus far the principal requirements of the manager's academic training have been set out. The training requirements academically will also include comprehensive training in mathematics, physics, and chemistry, as well as thorough and rigorous training in at least one language, the aim being to ensure the production of high quality reports as a matter of course, and not as the result of persistent rewriting and recomposing of such reports with the consequent wastage of valuable time. Also, where necessary, training would replace physics and chemistry with biochemistry, biology, or metallurgy, as may be appropriate.

30.4.3 In House (On-the-Job) Training

This aspect of training is of fundamental importance, so much so that the academic training (which should be integrated with on-the-job training) should, for both the technicians and trades staff, assume a secondary place, that is, academic training should have prime importance only for the measurement professional person. Training in the necessary manual skills required to service, maintain, and calibrate measuring equipment is best received on the job, although it is necessary to combine this with the special short courses commonly run by manufacturers, especially in the early days of the marketing effort when the manufacturer is attempting to establish a secure market for these new items. Very commonly these are of such nature that there are no new basic skills required, mainly the necessity to learn that the procedures set out by the manufacturer should be followed fairly rigorously for best results.

This all points to the need to work on and emphasize the development of the fundamental skills and manual abilities of the individual. For the tradesman, the understanding of and ability to carry out such tasks as light gauge sheet metal work in both ferrous and non-ferrous metals, and the reasons for the use of each; the ability to achieve high reliability in effecting soldered joints for both mechanical and electronic purposes; and understanding of light mechanisms and their adjustment, for example, cord drives, slide wire potentiometers, and relay and printing or inking mechanisms. (That list is not intended to be comprehensive but to show that the competent instrument tradesman should rate with, or even above, the best toolmakers in terms of status and salary.)

The technician, whilst needing to understand and possibly carry out similar work to the tradesman also needs to have a clear understanding of his role in checking the work of the tradesman, carrying out a calibration or checking a calibration carried out by others; being able to check systems and to determine any faults therein; and later to restore such systems to working order, if necessary effecting any required recalibrations and ensuring that all relevant records are kept in clear and concise form.

30.4.4 Information

At first glance this heading may seem almost frivolous, but on a little reflection it is rather obvious that for training to be given there must be information as a basis. However, the tremendous range and diversity of items available for use in instrumentation and control systems require the person controlling the training to be highly selective in regard to the information to be assembled into an effective training program.

In attempting to delineate information there are a number of factors to be considered. The training can be given the form of a short course dealing with the specific items making up a control or instrumentation system. Each item is then treated in great detail and the trainees made very familiar with all necessary information to either resolve a system problem, or repair and calibrate an individual instrument, dependent on whether the short training course has been prepared for and aimed at the professional, the technician or the tradesman.

Alternatively, the training program may be a broader, more diverse, and more searching one spread over a longer time period. It could take in the broad general principles applicable to all the instruments in a generic type such as, say, recorders or controllers, and then progress to dealing with the features common to the various types. Should further training in such matters be needed then probably a reversion to the short course situation outlined earlier is probably the best approach.

The information used in any training must be verified to ensure that it is accurate and conveys the essential details in a clear and concise manner. There must always be relevance to the subject matter which is clear and direct, that is, the information must be readily absorbed and understood at the relevant level, be that trade, technical or professional.

In many cases the information to be presented will be novel and will, therefore, require to be repeated a number of times during the training period. Repetition of information should not be avoided and may be necessary particularly in a training course of longer duration. However, repetition should not be done as padding or in an attempt to brainwash the trainees, but should be judiciously spaced and given its rightful places in the training program. Refresher courses, in particular, require careful consideration in this regard to ensure that a proper balance of delivery and reinforcement by repetition is retained.

30.4.5 Skills Required

The officer effecting the training program should have had training and experience as a training officer, which means the aquisition of skill and ability in the training process. Amongst the requisites are the use of visual training aids such as films and slides, overhead projectors, and what may be broadly termed as blackboard techniques which may mean white panels and felt-tipped pens, or the use of light quality white paper with felt-tipped pens. The visual processes are the most important process in the communication of information and should be accorded a corresponding place in the training officer's work.

Aural methods rank second to visual methods in training and communication and deserve good attention. The spoken word should be clear in presentation, as natural as can be achieved, it should not be unnecessarily repetitive or padded but should not lack detail sufficient to make the subject clearly understood (refer to Section 30.4.4).

Various other means, of much lesser importance, such as set reading may be employed in training but the visual, the audio, and the combination of the two represent the most powerful and effective means currently available for training purposes.

The demonstration of the actual instruments or measuring equipment will also require mechanical skills as a minimum with possibly also some electronic and pneumatic or hydraulic knowledge to supplement the mechanical or manipulatory skill. More modern equipment requires, in addition to these, at least some programming ability for minicomputer or microprocessor based equipment. This in turn will require some knowledge of a computer language which at the present time would probably be FORTRAN or BASIC. There has been a quite rapid progression from binary to hexadecimal to the two high-level languages mentioned, and although there are other high-level languages extant FORTRAN and BASIC will probably dominate this area for some time to come. This is not to say that the newer high-level languages should not be studied. In fact the reverse is the case so that a ready adaptation of working language can be effected quickly should this become necessary.

30.4.6 Cost Effectiveness

In considering the training of personnel, the results to be obtained must be set down, and the means to achieve those results examined to ensure that the most effective training for the lowest possible cost is obtained. This may mean that various methods may be adopted at differing times to suit a variety of circumstances, for example, a decision to say change a particular secondary-standard instrument to one of different form and using principles different to those used previously may require a pressure-cooker response to train the staff to the new requirements. Normally such a situation will be sufficiently well planned that the training program will start even before the actual item is available.

Also, it should be apparent that a training course which can be so devised that all grades of staff join in the one training program will be much more cost effective as well as engendering greater *esprit de corps*. Nevertheless, this should not be taken to the stage where: (a) either the professional thinks that his time is being wasted on unnecessarily low levels of activity; or (b) the tradesman thinks that someone is trying to give the professionals an easy ride while he (the tradesman) is finding the going pretty heavy.

In an area so complex as measurement, calibration, and control instrumentation, the training requirements are extensive and hence tend to be expensive, but it must be remembered that not putting appropriate training programs into effect can be even more expensive because of errors brought about by a lack of understanding of the principles of operation of the equipment being used.

30.4.7   Use of Set Procedures and Standards

From the preceding sections, it becomes apparent that there is a requirement to seek out, define, and refine procedures to set levels or standards. Preferably all staff will have a common understanding of the problems being faced, not only in the overall operations of the plant or laboratory, but will clearly grasp the problems faced by their individual position or section of the work.

Properly motivated, and with a positive working example displayed by every senior member of the staff, all members will realize the value of this approach to the problems confronting the organization.

The essential requirement from the management is that this attitude be encouraged and displayed at every opportunity. In this regard the use of incentives should be of good value. The essential requirement from the staff is a total pride not only in their work, but in their working area, their laboratory or plant, and in the encouragement of their workmates to use and follow set procedures and to use recognized standards; to question work results to determine whether in fact the set procedures may need revision, or the recognized standard should be modernized either by recalibration or replacement.

Free discussion right across all work-levels is necessary to achieve and sustain such a working environment, it being incumbent on the management to heed the views of the lowliest staff member, as it is incumbent on the lowliest staff member to carry out the duties required of them by the management. There is no single human infallible repository of knowledge, and collective knowledge and experience of all staff members will be required to establish set procedures, determine the standards to be used, and to cooperate in producing the desired output which is a measurement in which the user can have utter and complete confidence.

30.5 RECALIBRATION

30.5.1 Necessity

The need for calibration and recalibration of measuring instruments and systems of measuring instruments is long established. Wherever and whenever it is necessary to determine the value of an object in terms comprehensible to the human mind, that is, to perform a measurement or measurements; then there is a need to know that the measurements, whenever taken, will have some common reference or datum. This is reinforced in those cases where a common item may be made in a number of geographically separated places, or where various parts of a final assembly may be made in separate plants and assembled in a different place again. For example, the punched wad should fit neatly in the screw-cap for the bottle; the screw-cap in its turn fitting well to the screw-thread moulded into the glass or plastic bottle top or rolled into the metal can.

Considerable emphasis in the past has been placed on the measurement of length and angle (the term calibration deriving from the word calibre). This has occurred to the extent that many consider these facets of measurement as the full extent of metrology to the exclusion of other measurement knowledge. The problems of physical and chemical metrology are equally as important as that of length and angle, as would also be others such as biological and radiological metrology. True metrology is essential to all activities and should be accorded a corresponding importance.

30.5.2 Importance in Quality Control

Quality control must, in essence, be predicated on the repeatability and reproducibility of measurements and these in turn are underpinned by a calibration system which relates the work-face measurements to some basic datum or reference point (preferably the National Standard). The instruments used at the work-face are not taken to the National Standard for calibration as this would obviously be wasteful and inappropriate.

Preferably the work-face instruments would be calibrated against a locally held substandard instrument at fairly frequent intervals. The substandard instrument would then be calibrated by either internal cross-comparison with other secondary-standards, or by return to the laboratory where the National Standard is established and maintained. Calibration or verification of the secondary-standards would be effected at intervals which could range from, say, two years to five years depending on the instrument, its construction and reliability, and the extent of its usage. Return of the secondary-standard instrument for verification purposes to the laboratory where the National Standard is maintained, should not be delayed beyond five years, in order to sustain a good level of quality control. (Such periods can only be given as examples for experience with the particular leads to knowledge of intervals to be used—some military control equipment is recalibrated each 10 days.)

Recognition of the calibration aspects of quality control is increasing rapidly as shown by the increasing work being done by various National Standards Institutes, such as the International Electrotechnical Commission (IEC) and the International Standards Organisation (ISO), to set out clear statements of the requirements they consider essential to satisfy the needs inherent in a good quality control system. These bodies are now delineating in some detail the calibration system requirements as well as other quality control and management aspects. (Chapter 3 discusses these issues.)

30.5.3   Integration with Production Operations

The proper integration of calibration and recalibration with the production operations of a manufacturing plant is of great importance, requiring close attention and detailed planning to ensure that not only is the product output rate sustained at its planned or optimum rate, but that there is little or no risk of there being a rise in the rate of reject parts caused by the measuring devices being out of calibration.

It is important to remember that if it is necessary to recall and rework parts found not to fit together properly, or to have been incorrectly heat-treated, or any one of many other fault conditions; the net result may be a total loss of profit on the manufacturing operation. This in itself could result in the closure of the plant and its attendant loss of jobs: a terrible price to pay for not giving proper attention to ensuring that instrument calibrations are effected in due time.

Preferably a set of records of the calibration dates for each of the measuring instruments used in a plant should be maintained and regularly reviewed to ensure that each of the instruments is not only calibrated within its due calibration period but also that a back-up instrument is available to supply the necessary coverage should there be any delay, for any reason, in returning the instrument to its normal service. This would also cover any emergency condition which may arise within the process or the plant.

30.5.4   Methods of Managing

The remarks of the above section indicate the necessity for detailed management to be exerted in this field, and also underline the penalties which can be incurred through failure to recognize the need for, and to properly plan and manage, this part of the manufacturing operations.

Essentially there is no one set formula by which such a system will best be managed, but the organization must be set up to meet those important criteria set out in Section 30.5.3. This will mean that the size and nature of the manufacturing operations will be the major determinants of the requirements for instruments, and the calibration requirements of those instruments will then be determined in accordance with the plant performance needs and the reliability and performance of the instruments chosen for the various tasks.

From this it can be seen that the selection of an instrument is a matter of major importance. Purchase of an inadequate or unreliable instrument will only lead to substantial calibration costs, or the discarding of the instrument in favour of one which will give superior performance (refer to Section 30.2 on replacements).

One concomitant of this situation is that any lack of proper attention to record keeping (refer to Section 30.5.5) will lead to confusion, and may produce results varying right across the spectrum from the rejection of satisfactory items to the acceptance of unsatisfactory items.

Various methods of attack on these problems have been tried, all expensive—either in initial outlays, or in maintenance costs, or calibration costs. Primarily the situation requires that value judgements be made ahead of all the relevant information being available. The basic instruments having been set up as part of the overall plant system, it is then necessary to obtain spare instruments, or spare parts for the instruments, so that production interruptions are minimized.

As an alternative, stocks of instruments and spare parts can be purchased and stored in anticipation of a requirement. A pool of instruments has the attractions of providing the possibility of rapid replacement in the event of failure; of slight cost-saving resulting from purchase in bulk; and of having an instrument available if a process variation is to be tried and established. There is also the advantage of a tendency to standardize or rationalize instrument purchases, thus reducing the demands on staff for the extent of their knowledge, for the amount of training they will require, and for the library of instruction books and manuals required to be stored.

An instrument pool has the disadvantage that there are numbers of instruments lying idle and that relatively extensive storage space (preferably air-conditioned, or at least dust reduced) is required to house the instruments. To these matters must be added the requirement to record movements of instruments from place to place and to ensure that the state of calibration of a particular instrument is known at any given time.

30.5.5 Records and Record-Keeping

The full value of an instrument in performing its function of effectively making measurements which are reliable, rests on its being properly serviced and maintained, and regularly calibrated. In order to assure that this situation is properly managed and controlled it is essential that adequate records of these events are kept, and that the calibration record or certificate in particular is kept in duplicate form, one copy of which should accompany the instrument at all times.

As will be realized, the keeping of a full set of records as delineated requires a considerable volume of paperwork and attention to detail. As a rule the demands are relatively so great that all record details are not kept. Rather, some details, in particular servicing details, are say ticked off on a summary sheet. This means that some details which may be important could be left unrecorded and hence either be totally neglected, leading to possible failures, or at the least to faulty performance, in each case causing the instrument to lose calibration. Alternatively, the work detail may be carried out unnecessarily, which is in itself wasteful, but can also lead to problems and difficulties including loss of calibration.

Detailed examination of the use of automatic data processing (ADP) in the recording and analysis of all these details should be carried out (refer to Section 30.5.9). By the proper choice of programs to analyse the collected data, detailed procedures for routine servicing and routine maintenance can be set up, with regular print-out forms providing effective reminders of the maintenance and servicing to be carried out in the set period of the program. (Refer to Sections 30.6 and 30.7 for further information.)

30.5.6   Reduction Methods (Records Analysis)

In order that the demands for proper and effective recording of details, as set out in Section 30.5.5, can be minimized, the data gathered must be analysed to determine proper courses of action to be carried out. Whilst computers can be used to sort and collate data, and cause a printer to make hard copy of such processed data, care must be taken to ensure that the data gathered are the data required; that the data entered into the computer are verified as being correct; and that information required on a once-only basis is kept that way and not repeated unnecessarily.

Review of the records to ensure that minimal, but adequate and effective information only is retained, is necessary on a regular, planned, basis. Servicing records (refer to Section 30.7.4) should be reviewed to ensure that an item requiring repeated service is either cured of its problem, or replaced with an item which will not present problems to the user. Similarly maintenance records should be reviewed and similar actions taken where appropriate.

Calibration presents a different aspect as the only available change in these circumstances is the time-scale or interval between calibrations. Costs can be minimized by proper review of the performance of calibrated instruments and by setting the time interval between calibrations to provide the optimum combination of maximum usage time between calibrations and minimum out of service time whilst the instrument is checked and recalibrated.

Considerable information is available on calibration in the *Newsletter* of the National Conference of Standards Laboratories (NCSL) of the USA. The extensive committee work of that organization is also very valuable to standards personnel. Membership is open to any laboratory in the world.

30.5.7 Traceability Concepts

Considerations of traceability are relatively new to the calibration field. Here the term is defined as being the ability to trace, through written records, the calibration of an instrument by means of intermediate standards to the ultimate reference point, which is the National Standard of measurement, whatever it may be. Traceability provides insurance to an enterprise. Experience gained using a calibrated instrument can be completely lost if it is damaged. Traceability enables those measurements to be re-established in the event of such loss of a measurement reference point.

It is apparent that the establishment of a National Standard is, or should be, the precursor to the establishment of a network of intermediate standards or instruments which can be used to calibrate the working instruments being used in day-to-day working activities in production plants; for maintaining regular working checks in laboratories; and for use by service or maintenance technicians as part of their working apparatus.

Throughout this process, the keeping of records plays a key role (refer to Section 30.5.5), as it is through the continuity of the calibration checks carried out and the cross-comparison activities which result that product standards can be raised. The improvement of measurement should be a constant goal of all of these activities. The establishment and maintenance of the traceability of the calibration of measuring instruments to the National Standard of measurement is the most important single factor in the control of the quality of a product.

30.5.8 Cost Control

With a process which can be so expensive as calibration, cost control is an obvious requirement, and it is essential that all available methods to minimize calibration costs be employed. However, it is stressed that this is a truism which applies throughout all measuring and manufacturing activities with equal force. Care must always be taken that the cost cutting processes are rational, as poor quality assessments and control methods will be as disastrous as their absence.

Proper and effective interpretation of the records (Section 30.5.5) may show that the periods between calibrations can be extended resulting in a lower cost per annum on the particular instrument. However, the process being controlled must be examined in conjunction with the calibration frequency, as a possible rise in rejection rates may cost more than the savings obtained by extending the calibration intervals. Similarly, it may be found that reducing the calibration interval may also reduce the reject rate from the process. The relevant cost savings must be compared to determine wherein lies the proper and most effective action.

At this point the reader should return to study Sections 30.1 and 30.2 of this chapter as a recall of the basic principles to be followed in working out the requirements to be met where a new system is to be set up. Similarly, when a measuring system is at the obsolete stage, or is becoming obsolescent, the general principles are the same except that there may be some value to be recovered from the sale or re-use of parts of the displaced system.

30.5.9   Application of Automatic Data Processing

As set out previously in this section there are several distinct roles which can be played by the proper and effective use of ADP. The use of programmed routine servicing, programmed routine maintenance and programmed calibration are proper studies where there is a sufficiently large group of measuring instruments to be managed. If the measuring system is small then a human computer is normally sufficiently reliable provided the required data is collected, and written out clearly, and stored in a place where it can be checked and analysed as may be necessary.

The ADP system, whether it uses a micro-, mini-, or mainframe computer, has to be properly designed. The computer must be suitable to the task in all its aspects, must have ample data storage or memory capacity, must be capable of processing the data by means of several different programs so that the information required is that which appears on a print-out and it must be capable of being readily reprogrammed to accommodate any change which might occur in the policies of the company, the needs of the system, or the training and capabilities of the staff (both computing and its instruments).

Engaging experts to advise on the ADP system is one way of helping to set up a system which will be effective in supporting the manager of the measuring system. However, it is an inescapable fact that the manager himself must not only instruct the experts in the system requirements, but must study those requirements in depth, and work with the experts to produce a simple but fully effective working tool ideally suited to his needs, or at least, as close to the ideal as can reasonably be achieved without excessive costs.

Given that all of the foregoing factors receive due and proper consideration, the development of the ADP system must be allowed about 12 months to ensure that the programs developed are effective and meet all needs, that the operator(s) understand the programs and processes and, importantly, can recognize whether the print-out is usable or not, that the inevitable bugs in the system have been located and removed, and that all users have developed confidence in the system. These are problems which tend to be common to all ADP systems and which should always be part of the managers' planning of such activities.

30.6 MAINTENANCE

30.6.1 Requirements

There are very few items or equipments connected with manufacturing, and hence measurement, processes which do not require maintenance in some form. In some cases the maintenance is a frequent and regular, but minor requirement, and the reader is referred to Section 30.7 on servicing, where the maintenance requirement falls into such category.

For most measuring equipment the maintenance requirement is usually irregular if it is only applied when failure or breakdown occurs (corrective maintenance), as measuring equipment is generally, but not always, treated more carefully than the other parts of a manufacturing plant. However, there are various ways of meeting the maintenance requirement and these are discussed in later subsections.

As applied to measuring equipment particularly, the quality of maintenance must be of the highest order. It is pointless to carry out maintenance on measuring equipment, and then find in the endeavour to perform a complete calibration that the maintenance work is faulty, is lacking in some feature, or has been incorrectly performed. This would mean a complete waste of the calibration work and of any of the maintenance work which had to be repeated.

Maintenance work on measuring systems should be under the control of, and carried out by, persons of considerable experience, to ensure that the problems mentioned, and hence costs, are minimized. This generally means that some sort of system will be introduced into the maintenance operations in which the methods and actions will be codified and recorded as found most appropriate.

30.6.2 Systems

In order that maintenance does not degenerate into something more like a fire brigade situation, or be left until a catastrophic event occurs which may result in a completely unnecessary plant shutdown, action to plan for preventive maintenance, or the introduction of routine maintenance programs is recommended. These represent means whereby trouble is met head-on and dealt with before it does develop into a serious matter.

Whether the routine maintenance program is based on a chronological pattern set and repeated by the calendar or on the completion of a set number of operational hours, is a decision to be taken by the management of the organization and will depend on the relevant circumstances. The consideration of these circumstances is further set out in the succeeding subsections.

30.6.3 Programming

As with recalibration, the programming of maintenance operations is dependent on the setting-up of reasonably comprehensive records, which will then permit various analytical processes to be applied to those records enabling the derivation of the most effective program (or set of programs) to deal with the overall maintenance problems of the measuring system.

The program or programs so set up or derived, should be subject to periodic review to ensure that the maintenance effort required is minimized, and that any program redundancies are eliminated. If ADP is being used then the print-out format should be so constructed that the information is in the form of working instructions, that is, the printed page can be handed to a technician or tradesman with very little further instruction being required for the work to be undertaken. The print-out should, therefore, define the instrument or device, the work to be effected, previous maintenance history, calibration (and procedure if necessary), and the date or time-frame within which the work is to be performed. The latter should enable the measuring device to be withdrawn and maintained with little or no disruption to production or operational schedules.

30.6.4 Holding Pools

The holding of spare instruments in a pool as a means of minimizing the shut-down of plant has advantages as well as disadvantages. These need to be considered, with each factor being given an appropriate weighting according to the circumstances of the particular plant. For example, where the plant is relatively large, the holding of a fairly small number of carefully selected measuring or recording devices can be cheap insurance in the event of a malfunction or breakdown. In a small plant this may not be economically justifiable. However, it may be possible to arrange with a supplier that particular instruments will be readily available on demand, thereby eliminating the necessity to create an appropriate storage space for instruments, which would be necessary if the organization were to maintain its own holding pool of spare instruments.

In considering whether or not to establish a holding pool of instruments the following important factors must be considered (they are not necessarily ranked in order of importance):

(a) The cost of the instruments in the total system versus the estimated cost of the required spare instruments plus the cost of their storage (would need to be calculated as a cost per annum);

(b) The necessity to maintain the pool instruments in a state of readiness as regards calibration (refer to Section 30.5). This would be additional to the costs in (a) above.

(c) The requirement to list the instruments held, their calibration records and their state of readiness. By this is meant the coverage available in the event of a second failure occurring before the first instrument is ready to be returned to service.

(d) The cost to be levied by a supplier in guaranteeing the availability of spare instruments.

(e) Whether the replacement is of the same make, model and type, or whether it is a substitute requiring some adaptive work when being installed.

In (a) above, the cost of storage has been set down as a cost factor. Preferably all measuring equipment should, as a minimum, be stored in a clean environment at a reasonably stable temperature and relative humidity to minimize the calibration drifts which might occur during such storage. The provision of such storage conditions is clearly more expensive than the mere provision of shelving without further protection. However, this matter is again one where management must include assessment of the relative costs as a basis for the final decisions on storage conditions.

30.6.5 Spare Parts

Similar remarks to those made in connection with holding pools can be made regarding spare parts. In both cases appropriate storage facilities must be provided, accurate records set up and maintained, analysis of consumption and hence replacement rates and costs carried out, and where the operation is large enough, the setting up of an effective ADP system will be necessary. It is generally more likely that the maker or agent will have stocks of spare parts readily available (but may not when they are needed!). In such circumstances it may be found desirable to hold small quantities of regularly used items.

If such items as charts, oils, and inks are considered as spare parts then there will be an absolute need to hold stocks of these items as the purchase of small quantities is usually not possible, or else the single-order items are very expensive.

The multiplicity of items which can be considered under this heading is very great when an 'all-disciplines' approach is taken, for instance, chemical glassware, electronic parts, or wires of special compositions for pyrometry or chemical analysis—the list is endless. Storage arrangements required will vary to suit the particular needs and must be given full consideration.

30.6.6   Routine Maintenance

As mentioned in Section 30.6.1 a system or routine should be introduced into the maintenance aspects of any measuring system. It is far better to detect trouble approaching, rather than to find a problem, and not know how far back in time its ramifications extend. For process plant this may also require the diversion of output items into a holding store for a period so that they can be put right before being sold. This may mean, as a consequence, some loss of sales dependent upon the availability of stocks to tide over such a situation.

Hence it can be seen that the setting up of routine maintenance procedures can be of considerable value even in the operations of a small plant. It is also of value in that it gives a greater degree of control in forward planning and estimating of costs. This permits close control of financial and budgetting operations, revealing relative costs within the overall plant operations, and indicating areas where costs can be more closely scrutinized and probably more closely controlled to improve overall cost efficiency.

30.6.7   On-Demand Maintenance

The argument for on demand or corrective maintenance is often advanced, being supported by the argument that programmed or preventive maintenance incurs a cost which may not be necessary. This is generally the attitude adopted by a somewhat shortsighted management policy of cutting costs regardless of the on-going effects of such cost-cutting.

However, on-demand maintenance is effected after the problem has arisen, and may not prevent catastrophic failures occurring which may result in costs considerably greater than those which would have been incurred under a system of programmed maintenance.

Hence the selection of programmed or 'on-demand' maintenance is a management decision taken after due and proper consideration of all the relevant circumstances. Programmed maintenance offers greater insurance against catastrophic failures with their usually very great attendant costs, but with some additional cost being incurred. There is also a distinct risk of bringing about failures that may not have otherwise occurred. On-demand maintenance offers generally lower maintenance costs, but carries with it a greater attendant risk of catastrophic failure with considerable costs (often unbudgetted) to be incurred in restoring the plant to proper working order.

30.6.8 Control of Maintenance Costs

The preceding subsections have all, in some way, dealt with questions relevant to the costs of maintaining existing measurement systems in proper and effective working order. In the opinion of the writer a program of routine maintenance actions for each item affords the best prospect for cost control of all facets of plant operation. The measuring equipment is the means whereby the quality of the product is controlled and, therefore, has the key role in plant operations and in ensuring output of items of the specified quality. However, it has a further value in that proper analysis of the records will indicate at an early stage where plant deterioration is permitting a drift in product quality. This two-fold contribution has great significance to the overall plant operation, as mentioned earlier, and contributes more to cost control than would appear at first view.

None of the foregoing can, in any way, absolve management of its responsibility to exercise due care in the control and oversight of all operations including the examination of costs, wherever incurred, and of acting in such manner as to obtain the greatest value for each dollar spent.

30.7 SERVICING

30.7.1 Definition

The servicing of measuring equipment is defined herein as being 'those minor activities carried out on a regular or repetitive basis, and which are essential to the continuing satisfactory operation of the equipment'.

A number of examples which will clarify the definition to a greater degree include:

(a) regular replacement of charts on recorders;
(b) checking of inking mechanisms for recorders;
(c) cleaning of glass electrodes for pH measurements;
(d) cleaning of potentiometer slide-wires.

As can be seen these actions are essential to the continuing satisfactory operation of the measuring instruments concerned and would require regular or repeated application. Although these activities can be grouped with maintenance, the necessity for separate calibration as an activity, distinct from maintenance also dictates (to some extent) the separation of the activities, defined herein, from maintenance.

30.7.2 Programs

One consequence of the definition given in Section 30.7.1 above is that programs are essential parts of the activity. For example, recorder charts will be programmed for replacement on a daily, weekly or other fixed time interval basis. This is necessary in the case of circular charts to avoid the super-imposing of recorded information which could result in much confusion, particularly if a part of the recording shows an out-of-specification condi-tion, and no determination can be made as to whether the first or subse-quent pass was that showing the erroneous control situation. This is an argument often advanced against circular charts and in favour of strip-charts of the roll or fan-fold type. However, here again care is required, as information will be lost if the chart is allowed to run out. It is also apparent that at the time a chart is changed, the pen, ink or inking system, or the actual device used to mark the chart, should be checked to see that it is clean, in correct order, and functioning satisfactorily. From there it is a short step to the effecting of a quick check of the functioning of the entire recorder, including such items as the lubrication of moving parts where necessary and a general clean of the instrument.

Besides establishing the point that programs are essential in the servicing of measurement systems, the foregoing leads on to the consideration of routines or codes of practice.

30.7.3 Routine Practices

A further consequence of the definition of servicing is that the repetitive nature of the tasks involved makes the establishment of routine practices almost a matter of course. This, in its turn, is a useful adjunct to the training of junior staff, the compilation of appropriate records, and, for the alert technician, will indicate the possible approach of trouble.

The development and refinement of such routine practices therefore serves a very useful purpose in the overall management of a measuring system. It helps to promote pride in maintaining a clean, well-functioning system; it emphasizes the value and importance of the system itself; it simplifies several of the routine management needs; and it encourages the staff members involved to keep their system in top working order.

30.7.4 Records

It is hard to overestimate the value of a comprehensive set of records of the instruments which comprise a measuring system. The value of records in terms of calibration has already been set out in Section 30.5.5, but the further requirements to meet servicing activities must also be satisfied. Such records would include the details of the materials required, for example, ink pads, oils, electrodes; the procedures detailing each particular servicing activity; the items to be serviced at that particular time; and the routine tests to be effected to ensure that the instrument is functioning correctly before leaving it.

These records could be very comprehensive and hence costly and time consuming to compile. This may be relevant where the item being measured has to perform a vital function, that is, one error or accident is either one too many, or has very costly consequences. Normally management would have to decide the limits and patterns required in the records system. The costs of maintaining the records must be balanced against the effects on production, the cost of the finished goods, and the possible costs of recall or replacement of faulty or unsatisfactory goods.

30.7.5 Stock Holdings

This is an important question which must be based on a bipartisan decision between technical staff and financial management. The technical staff must decide rationally which items are essential to the stock list, and management must decide the amount of money which they can afford to tie up in such stocks. Technical staff have to appreciate that stock holdings represent to some extent idle money, and management has to appreciate that the technical staff must have adequate back-up supplies readily available. Both really have the common goal of maximizing production and productivity; the holding of stocks of spares (including spare parts for repair purposes) is a major part of these functions and must be universally recognized as such.

Some flexibility is necessary in this activity as there will be the need to change the stock holdings in the light of working experience, and to allow for the changes of consumption rates which will occur if there is a change in the production process, such as a simplification or an expansion of the product range.

30.7.6 Cost Control

Servicing can be a run-away situation unless a tight rein is kept on both activities and supplies. There is a fairly common tendency for staff to book time for maintenance or servicing as a cover for other activities, or to provide a way out if cost control pressures are being exerted in other areas of the plant operations.

Nonetheless, it is possible to keep a reasonable level of cost control in the servicing area without the requirement for excessive supervision or intensive scrutiny and examination of records. However, it does require that the manager or engineer responsible for the measurement equipment has a good working knowledge of that measuring equipment, and can therefore quickly assess the situation and determine whether excessive time and/or supplies are being consumed in servicing the measurement systems.

There is little point in attempting to oversight such areas via the cost accounting system only. Management must make use of the data which cost accounting can provide to decide the time for replacement of a piece of equipment, that is when its actual costs of service are becoming excessive and the maintenance program is having little or no effect on those costs.

30.7.7   Use of Bulk Purchasing

Bulk purchasing is in many senses a parallel activity, or an adjunct, to the instrument holding pool (refer to Section 30.6.4), and spare parts requirements (refer to Sections 30.6.5 and 30.7.5). Normally its purpose is to reduce the cost per unit, and this is usually done by purchasing specified quantities of an item, and as a consequence, being able to purchase at a much lower cost than the single item cost. This is particularly so in the case of oils, greases, recorder charts, ink pads, pens, and all of the other consumables associated with the measuring instrument or system.

Incumbent on such a system, however, is the need to ensure that appropriate storage space is provided, and that the items can be readily identified and obtained or supplied for use. Recorder charts present a most interesting aspect of this activity because of the wide ranges of shapes, types, sizes, scales, parameters, and graduations which can be used to make up the various combinations needed by modern recorders in modern measuring systems. Other obvious needs include safe storage for oils; storage for a wide variety of electronic or pneumatic components and the detailed records required of all of these activities. Possibly, computers may ultimately be used to reduce the tremendous numbers of combinations of items presently extant, but this is still some years away. In the meantime computers should be used to eliminate as much of the routine, detailed work, as is possible.

The use of bulk purchasing needs to be controlled and properly managed so that the best balance of economies covering storage needs, usage rates, number of different items, and record-keeping against the amount of idle funds, which are really a proper form of investment, is maintained.

30.7.8   Use of Automatic Data Processing

As indicated in Section 30.7.7 and other parts of this chapter, the introduction and use of ADP is an inevitable part of the future activities required in servicing and maintaining a measurement system. Servicing, in particular, presents an area where considerable benefits could be realized. This refers especially to the wide-ranging nature and variety of the activities covered, the material supplies needed, and the coordination of the whole into an effectively managed, and in relative terms, profitable activity.

An ADP system for servicing activities would require substantial memory capacity because of the large number of individually described items which would have to be stored in a data bank. The programming function requirements are not nearly so extensive as would be the case for the calibration ADP system (see Section 30.5.9). However, it would again be desirable that the output format should delineate the instrument, the service activity required, the availability of the supplies required for the task(s), and the desired time-frame within which the servicing would be effected.

Feedback into the ADP system could be effected by using the same document, which would be appropriately marked with such quantities as supplies used and date of servicing. This would complete the circle of information flow. An appropriate summary print-out of all servicing operations could be programmed as an on demand management feature, enabling the manager responsible for servicing to maintain proper control and oversight of the activity.

As in all ADP systems a thorough investigation to determine the features required of the system, the means necessary to provide those features in satisfactory format, and the most cost effective means of providing those features is necessary. Although this may require relatively lengthy examination and evaluation processes, failure to do so is inevitably more wasteful and costly.

Handbook of Measurement Science, Volume 2
Edited by P. H. Sydenham
© 1983 John Wiley & Sons Ltd

Chapter

31

P. H. SYDENHAM (Calibration), T. P. FLANAGAN and E. K.
LASKARIS (Evaluation), and J. A. GILMOUR (Accreditation)

Calibration, evaluation, and accreditation

Editorial introduction

The purposes and underlying methodology of measurements have been detailed elsewhere in this Handbook, notably in Chapters 1, 2, 3, 15, 30.

In essence a measurement seeks to quantify a measurand by comparing it with a defined standard for that quantity expressing the result in terms of the declared unit of that measurand.

This seemingly simple process can be broken down into several aspects each of which is essential for an adequate measurement to occur. They are: the definition of the physical unit and the standards and the means for maintaining them; means for using physical standards for comparison with the unknown magnitude to be quantified; and means to ascertain that the measuring equipment will reliably perform the measurements for some required time into the future.

These requirements involve the practices of *calibration, instrument evaluation*, and *accreditation*. This chapter provides greater detail on these topics than is covered in Chapter 3, where a brief overview of their relationships is presented.

Section 31.1 aims to provide familiarization with the essential elements of calibration and where information about the necessary practices can be obtained.

Section 31.2 deals with, what is still a very new approach, formal evaluation of an instrument's performance. This can provide evidence of an instrument system's ability to provide a valid measurement under a wide range of realistic conditions.

Section 31.3 discusses accreditation, that procedure whereby measurements and tests are provided with a level of guarantee by accrediting suitable laboratories as being able to provide stated classes of measurement to stated levels of competence.

31.1 CALIBRATION

31.1.1 Introductory Comment

Definition

Calibration is defined in AS 1514, Part 1 1980 as follows:

'All the operations for the purpose of determining the values of the errors of measuring instruments, material measures, and measurement standards (and, if necessary, to determine other metrological properties, including influence quantities)'.

It deprecates the use of the term *calibration* to describe such actions as gauging, adjustment, and scale graduation. This definition is selected because AS 1514 is, at the time of writing, the most recently approved standard on terms to be used in metrology.

In mid-1980 the International Electrotechnical Commission (IEC) released the first draft of an international list of terms under the title 'International vocabulary of basic and general terms in metrology'. (It is to be expected that the final form of this draft would be approved in late 1982.) The definition contained there is shorter

'The ensemble of operations for the purpose of assigning values to the errors of a measuring instrument (and in some cases to determine metrological properties)'.

Until this draft is approved the international definition stands as the dual language version in (OIML) PD 6461:1971. It is very similar to that in the draft.

Value of calibration

Measuring instruments are used to provide quantification of physical parameters using a suitable process of measurement. Calibration is the process of providing the measurement instrument with a complementary statement about the errors that the instrument will have.

Although a measuring instrument can be used to assign quantity to a measurand, there can be no guarantee that the actual numbers assigned are adequately accurate. For example, consider measuring, for legal purposes, the speed of a motor car using radar speed measuring equipment. If the instrument is not calibrated against some other form of agreed standard for velocity then it is not objectively possible to assert that the speed of the vehicle is truly that read from the output. Where this is the case it is impossible to legally enforce a fine because no one can be certain, within a reasonable degree of certainty, that the value indicated is correct.

This situation applies with most measuring equipments but there are a few exceptions. These occur when the instrument readings are inherently correct because the principle used, and its method of implementation, can be regarded as always giving results for which the errors are within the required levels of uncertainty. As an example the laser interferometry system manufactured by Hewlett-Packard for length dimension measurements has been approved, to a given level of uncertainty, by the National Bureau of Standards (NBS) of the USA as an instrument that does not need to be traceably calibrated to the agreed SI base standard for the metre (interferometry using krypton radiation but see also Section 24.3).

Realization of the need for traceable calibration is not new. In its legal manifestation it can be traced at least as far back as Babylonian times. A statue of Gudea, ruler of Lagash in the third millenium BC, exists today that has a length scale carved into a tablet held on the knees of the ruler's image. Commercial practices, presumably the dubious ones, required some form of legal process to ensure the measuring instrument—a beam balance, volume measure or length ruler—was performing accurately and had not been modified to measure in favour of one of the two parties—*diverse practice* as the Bible called it.

Many recorded instances exist that showed that early traders were legally required to compare their instrument scales and masses against declared standard.

There also exists evidence of early use of standardization and calibration procedures for less pragmatic purposes. The Chinese, at least as early as 164 AD, maintained jade and terra-cotta templates for calibrating the length of shadows cast from sundials. Further information about such matters can be found in and through (Sydenham, 1979).

It is clear to most that calibration is needed for such everyday devices as petrol pumps, shop scales, and liquid measures. It is, however, far less evident that technological systems' managers always realize the value of maintaining proper calibration of the measuring parts of the plant that they manage. It is for this reason that this discussion begins with why calibration is so vital to modern sophisticated technological activity. The following case study illustrates the need for calibration.

Nickel cobalt magnets are formed by a metals' casting process followed by extensive normalizing and other cyclic treatments. Whether the final product conforms to the desired technical magnetic specification cannot be known until the magnet is completely manufactured, a process that can take about one week to complete from the first melt stage. At least twenty critical stages are involved over that time, each of which must be set to quite exacting limits. The value of the parameters to be used with any particular shape are established primarily from experience of past production runs. They build up a body of knowledge of the critical settings that will produce a satisfactory product.

Calibration is vital in such a case for several reasons. The first is that the experience gained, in terms of measured data on finished samples at each of the stages, is used to tune the process efficiency and to optimize the performance of the final magnetic flux level. Variations in the error levels of the measurements, for such purposes, need to be well established as being either small enough to ignore or well enough characterized to be allowed for in subsequent use of the data. The second reason for maintaining the instruments in traceable calibration relates to the likely circumstance that any of the instruments used to obtain the data, which ultimately is used to direct the operators when a process is ready to proceed, may fail due to damage, or because of internal failure reasons. An uncalibrated instrument, once repaired in such circumstances, can only be regarded as a unique item having its own unknown transduction coefficient. When repaired or replaced there can be no easily established degree of uncertainty about the data set produced after that time. In contrast a traceably calibrated instrument can be immediately adopted, the user being in the happy and confident position of knowing that the absolute values of measurements lie within the same band of uncertainty as did the former unit.

It may be possible to reconstruct the calibration from samples retained and the data recorded in the past but this is a most unsatisfactory way to proceed when a properly devised and maintained calibration mechanism could have been adopted in the first place.

The above two reasons are worthwhile within the plant. A third reason for maintaining calibrations results when the product is offered for sale. In this example such magnets rarely sell in large numbers to other than highly informed users requiring high performance and tight specification of magnetic flux levels and dimensional shape, both in the short and long term. If traceable calibration is practised by both the manufacturer and the customer than each will have confidence that the other's measurements will agree for measurements on the same unit. This being the case enables product acceptance to be made on one set of measurements. This feature of calibration applies particularly to international trade and requires a system of accreditation of calibrating agencies.

A further reason, one allied to the above, is that traceable calibration ensures that component parts manufactured in isolation from each other will assemble together as designed. It also ensures that the final product will perform as intended, for instance, that an aircraft will arrive at an absolute place within the circle of error of location that the navigational guidance system is designed to achieve. It is no good arguing that the system is precise; it must also be adequately accurate.

In the above case study, and the infinite number of others that are similar in their needs, calibration is but part of the sequence that provides an assessment of the instruments error. Calibration is meaningless unless there also exists a base system of standards to which to trace the calibration. Thus standards maintenance and a system of traceability is vital to calibration effectiveness. To slavishly recalibrate a system of factory gauge blocks to a standardized set held in the factory's standards room without ensuring that the tool-room standard used is periodically compared with the next upward level of standard, toward the base unit apparatus, only retains traceable calibration within the factory itself.

The costs of calibration can be very significant. Perhaps it is because it appears to produce nothing tangible, except employment of staff who are not directly productive, that management allows, or never really begins to improve, poor calibration conduct. The associated costs must be seen as important for the same reasons as those requiring similar costs for inspection and quality control, in general. It must be very obvious that in the above case study of the magnets a week's production may have to be returned to the melting pot because of such factors as the pyrometer used to decide when the temperature of the melt is ready for it to be poured—a very critical parameter—is repaired and given inadvertently a different transduction coefficient to that existing before it was worked upon. Mitchell (1981), Seed and Pawlik (1974), and Sydenham (1979) provide information on the value of measurements.

All technological devices made by man do not possess infinitely stable characteristics. Wear of mechanism, damage through abuse, failure through chemical and other corrosion, influence of abnormal environmental parameters can each cause the instrument's transduction coefficient to change. Some instruments, no matter how carefully stored, change with time. As an example, some military quidance units need recalibration every 10 days whilst in store. In brief a golden rule to adopt is that *an instrument out of calibration must be considered to have failed.* To quote a part of NATA (1976):

'An undetected change in a laboratory standard may easily initiate a chain of errors that will propagate through a plant and cause losses in time and material exceeding manyfold the cost of the standard and the calibration'.

Literature

It appears that as yet no textbook has been published on the particular subject of calibration as a system function. It is, therefore, not possible to refer the reader to a singular source that can be used to guide those interested in adopting or improving calibration practice. This section assists such a need by referring those interested to some of the many published papers on the subject. In this category of published material the reader may encompass difficulty in procurement of papers for most of them are contained in in-house documentation and in less than well known publications which may not be immediately available in libraries.

Possibly the most extensive source of English language information, that is openly available, is that contained in the quarterly *Newsletter* and annual conference papers of the over twenty year old National Conference of Standards Laboratories (NCSL) that has its secretariat in the National Bureau of Standards, Boulder, USA. This organization is comprised mainly of some two-thirds of the nominal five hundred standards laboratories existing in the USA. Membership is open to any interested person or group in the world. This organization has many expert committees in operation and can assist with training aids and advice. Various issues of *NCSL Newsletter* contain useful bibliographies on the general topic of calibration systems management.

This section now provides consideration of the salient aspects of calibration and its practice addressing, in turn, suitable staff, laboratory practice, kinds of calibration, calibration interval, and the relationship between this and the quality of the instrument, automatic calibration and, finally, means to assure that calibration chains maintain the standards of excellence needed.

31.1.2 Calibration Personnel

The relatively small number of standards and testing laboratories deal with a diverse set of measurand variables, which are spread across very large geographical regions. These factors provide the situation where specialized training of personnel for staff of calibration facilities cannot be efficiently organized in the same manner as can general engineering training.

In general, staff for this activity have been provided with the required background of theory and experience with the calibration laboratory. The resultant apprentice–master relationship, the means by which a person gains the necessary background for calibration work, usually is implemented with little, if any, interaction with other laboratories. Despite many serious attempts to provide organized formal education for this category of staff, for example, the not inconsiderable efforts of the Education Committee of the NCSL and private enterprise ventures, the prevailing situation is still that an employer will usually need to provide training as an in-house activity.

Before training needs are to be met it is first quite critical that a suitable person be located. The following comments from King-Jones (1977) describe the situation well:

'Many calibration and measurement facilities . . . which are accepted as satisfactory in meeting current demands have dedicated and experienced personnel. All too frequently the key personnel is one man with inadequate or no back-up. Such facilities carry the inherent risk of being suddenly de-activated, and the expertise which makes them viable, lost.

The practice of calibration is an exacting and painstaking task. It requires a particular set of human characteristics, education, training and patience. It is a dedicated task. It is a career. No equipment, however expensive or sophisticated will compensate for the lack of these human characteristics and skills. Hence an adequate career structure is essential to attract and sustain suitable talents'.

The author also makes a very significant point when he states that a nation cannot import calibration as it must be performed locally.

Two major constraints on improvement occurring in the quality of oncoming staff are also mentioned. The first is that many see the tasks as *not fit for professionals* and the second that *it is costly and difficult to see visible return.*

Given that suitable persons do exist it is agreed that the implementation of high-quality educational programs aimed to train new personnel and to broaden the scope and level of those already committed to this work are needed. The difficulties are to find formulae for implementing this need in the face of lack of interest by much of top management because of the lack of visible return. Making a case will be aided by consulting Mitchell (1981).

NATA prepared the following statement for delivery at the 1978 NCSL annual conference held in Washington, DC; it comments upon the benefits of formalized training:

'The testing skills associated with most of NATA's fields of testing are taught at both the professional level appropriate to laboratory management and at the technician level appropriate to the routine tester. Length and mass measurement stand out as skills which are not covered rigorously at any level of tertiary education, although some aspects are taught in a number of different courses.

Traditionally, the higher echelons of staff associated with engineering metrology (length measurement) have been drawn from mechanical engineering graduates and diplomates. Technicians are usually drawn from the toolmaking trade. While such training may provide a useful background for a career in measurement, it does not provide very much training in measurement itself. Most of this must be learned in the laboratory under the guidance of a suitably expert superior.

In the area of physical metrology (mass measurement) training opportunities are even more restricted. Except for the highest levels, staff receive only job-training.

Establishment and acceptance of formal courses in metrology in the major centres of engineering would assist the Association in a number of ways. Advantages would include:

(i)   the standard and uniformity of measurement could be expected to improve,

   (ii)   employers would have a ready basis for judging potential employees,

   (iii) professional engineers could extend their practical abilities to better manage measurement facilities,

   (iv) assessment of the adequacy of staff training would be facilitated. Currently, practical tests are the most effective method of judging competence.

Very few technicians are required to measure both length and mass. Formal courses should be so structured that it is possible for students to study, and gain qualification, in only one section.'

Staff of a calibration facility need the following characteristics, few of which are deliberately developed in teaching institutions:

(a)   Ability to accept and use responsibility to conduct the work with care and integrity.

(b)   Possession of a strong in-built motivation to succeed in the tasks set, to constantly improve his or her ability at the task, to improve laboratory practice.

(c)   Possession of adequate experience before being given the next level of responsibility.

(d)   A personality suited to the task.

(e)   Possession of a well developed appreciation of the value of the plant used in calibration and have a positive attitude for its proper care.

(f)   Ability to manage others and implant the need for them to give sound and constant service.

The qualities needed of staff have been defined in documents of the British Calibration Service (BCS). These state that the head of the laboratory must be of adequate calibre, status, and experience to ensure compliance with all the requirements of the BCS. He should possess professional or equivalent qualifications appropriate to the measurement field in general and the work of the laboratory in particular. He must be responsible for all operational aspects of the laboratory, including the equipment and staff. He must be competent to assess and evaluate the work of his laboratory, and should be able to give advice and guidance on measurement problems presented to the laboratory.

Of staff in general the BCS asserts that the prime responsibility for the correctness of measurements lies with the person making them. Each member of the laboratory staff must therefore possess understanding of and experience in measurement principles and practice appropriate to his level and area of responsibility.

Staff of limited but adequate qualifications and experience may be employed on calibration work of a routine nature provided that they have received appropriate training, are adequately supervised, and follow fully documented procedures.

Accreditation of personnel may be required in the near future, because of the increased degree of regulatory control occurring in such industries as nuclear power generation and drug manufacture.

Specific courses for training calibration facility workers have been designed and delivered in several countries. The UNIDO organization has set up workshops on this topic in Chile, Japan, and Ethiopia. The US Army course at the Redstone Arsenal Centre is well known. In 1977 the George Washington University ran a course entitled 'Operating techniques for standards and calibrating laboratories', for which detail is available in the October 1977, *NCSL Newsletter*.

A list of courses available in the US was compiled in 1978 by the Education Comittee (NCSL, 1978) of the NCSL. In the same year the Committee conducted a survey of US training needs. These reports, plus other material of relevance, were presented at the 1979 NCSL conference held at the Boulder Campus of the NBS; Werner (1980) addresses development of training.

One paper of that conference (Taff, 1980) was concerned with the minimum requirements for calibration technicians. Its author, of the Tennessee Valley Authority TVA, who maintain a particularly extensive calibration laboratory, states there that they need knowledge, wisdom, personal integrity, skills, and manual dexterity. He also provided details of the TVA training program.

31.1.3   Calibration Laboratory Practice

Calibrations do not happen without help; they are caused to happen by allocation of staff to tasks on instruments and processes. Management of these factors is variously described as laboratory practice, calibration management, calibration systems management, and laboratory management. The bibliographies published in several issues of the *NCSL Newsletter* are relevant.

Calibration systems management was covered in six workshops reported in *NCSL Newsletter*, October 1977. Particular interest was shown in establishing calibration intervals, in work and performance standards, in data taking and its retention, and in selling calibration systems to top management. The *NCSL* committee for calibration systems management sees its role, among other things, to be in research and dissemination about management methods. Past activity was concerned with sharing of experiences but in more recent times this has changed to tackling the problems arising out of the proliferation of legislation and regulations in this area.

The following calibration centres are described in the *NCSL Newsletter* issues as indicated: NASA Marshall Space Flight Centre, Huntsville, Alabama, November 1973; Autonetics Division of Rockwell International, Southern California, March 1974; Eli Whitney Laboratory of the Bendix Corporation, Dayton, Ohio, October 1974; Lockheed Measurement Standards Laboratories, Sunnyvale, California, December, 1977; Rockwell International, Avionics and Missiles Group Metrology Standards Laboratory, and a mobile laboratory, July, 1978; Teledyne Systems Company, Northridge, California, March 1979; AVCO Measurement Standards Laboratories, Wilmington, Massachusetts, December 1980; and the TRW Defence and Space Systems Group at Redondo Beach, California, June, 1981.

Mobile calibration cans are reviewed in Daneman (1978). Each of the above reviews includes extensive photographs of laboratories and their equipment.

The BCS has issued notes on laboratory accommodation and services (BCS, 1969). The following aspects are detailed:

(a)   compatibility with the need;
(b)   definition of what constitutes the area;
(c)   use of other areas for calibration purposes;
(d)   use of the laboratory for other purposes;
(e)   location, construction, furniture;
(f)   space, office for head;
(g)   storage for apparatus, special furniture needs;
(h)   environmental conditions (ambient temperature, relative humidity, monitoring equipment, dust levels, noxious fumes, magnetic fields levels, electromagnetic radiation, and illumination levels)
(i)   services, stores;
(j)   housekeeping and safety.

In recent years several standards and calibration laboratories have been established in developing countries. A consultant in this area has published a review of planning of such laboratories (Daneman, 1977).

31.1.4   Kinds of Calibration

Physical features to be calibrated

There exist many physical features of a measuring instrument that may need calibration. Chapters 16, 17 address characterization. Often the manufacturer's original test certificate will indicate which parameters will need periodic calibration. The following list provides a checklist of some of the many possible error sources that may need quantifying:

(1) zero setting;
(2) full scale value;
(3) error of non-linearity (four methods are described in AS 1514, Part 1);
(4) gain or transduction coefficient;
(5) effect of external influence parameters (temperature, relative humidity, atmospheric pressure, magnetic fields, vibration, r.f. interference, ambient illumination level, and more);
(6) power supply variations and noise on supply;
(7) response to pulsations on measurand;
(8) frequency response;
(9) step response and other inputs;
(10) maximum safe voltage levels;
(11) leakage resistance to ground;
(12) level of operation of a protective device;
(13) general inspection of instrument integrity;
(14) battery voltages;
(15) transmission and reflectance levels in optical systems.

Before a new instrument is commissioned it is first necessary to establish a suitable calibration plan. The instrument should be calibrated before use and proper records created. The group who are responsible will be required to maintain a record system. Often a history card is used, on which is recorded all required technical details about the instrument including calibration information.

Where automatic testing equipment ATE is used the system will generally produce a calibration printout which can act as the certified calibration document. Clearly, the data displayed is selected to suit the need.

It is often convenient to hold a copy of the test certificate with the instrument but such cases the original should still be filed elsewhere in a secure location. In the event of severe damage to the instrument, or its complete loss, the certificate will then be most useful to help phase in a replacement.

Labels

Many different reasons exist for requiring one or more of the many physical parameters listed above to be calibrated. A guide of the types of calibration *labels* that are in use was published in the *NCSL Newsletter* of July 1978, this being as the summary of a study organized by one of the Regional Groups of the NCSL membership. The following list is extracted from that report:

(1) standards calibration;
(2) calibration: manufacturer's specifications;
(3) calibration: limited specifications;
(4) calibration: facility specification;
(5) calibration: limited facility specification;
(6) not usable for acceptance testing;
(7) cross-check;
(8) operational check;
(9) calibration not required;
(10) preventive maintenance;
(11) calibrate before use;
(12) inactive or do not use;
(13) reject;
(14) reference or primary standard;
(15) transfer standard;
(16) working standard.

The list was compiled from government, military, and private users. It was thought by that meeting that, in many cases, simplification of labeling systems might be possible.

31.1.5 Calibration Interval

Associated with every calibration is the cost to make a new calibration and the cost resulting because of not having done it when it needed to be done. The first cost is relatively easy to decide. It depends upon the instrument complexity, ease of adjustment, degree of automation, calibration procedure, ability of staff, and less obvious factors such as cost of loss of production whilst it is being calibrated, loan or spares holdings needed to reduce such loss of production, transport costs, and costs of the associated paperwork. Clearly, the cost to make a calibration far exceeds the staff time of those involved.

Costing the effect of using an instrument that might be (or might not be!) out of calibration is the intangible factor. Overcalibration is often considered to be—but see Gebhart (1980)—the safe path but in the interests of preserving maximum financial efficiency as well as, perhaps, maximum instrument integrity it is usually necessary to seek a suitable calibration interval along with guiding rules which will decide when, in the future, the instrument should be calibrated.

Obviously the interval will depend upon the equipment's characteristics and the purpose of the instrument. In nuclear power stations the torque wrenches used to tighten critical nuts, such as those that hold on the reactor vessel top, are recalibrated after a dozen or so pulls: for each nut and its associated pull there is an individual record of the circumstance. At the other end of the scale the multimeter of the servicing electrician may not need to be calibrated after the initial build unless the instrument becomes damaged.

The top priority problem faced by calibration facilities is establishing suitable calibration intervals to use. At present it does not appear that there will ever be developed a system that guarantees minimum cost of calibration and zero loss through not having recalibrated an instrument in time. It is necessary to build up information about an instrument as its life passes, making intelligent informed estimates of need that are gained from this experience and by study of what others have done beforehand. The essential state to attempt to reach is an adequate state of satisfaction that a responsible decision has been made.

To assist those without such experience the following information is extracted from various issues of the *NCSL Newsletter.*

A suitable paper that describes how a very large organization handles the problem—the US Army—is provided in Westmoreland (1980).

The bibliography of the June 1977 issue extends beyond *Newsletter* contents and is specifically concerned with calibration interval topics.

During early 1978 the NCSL Calibration Systems Management Committee conducted a US survey on this subject. Their findings are published in the July 1978 issue. It contains the following outline of how a typical member laboratory characterizes its calibration interval system:

'1. Initial intervals are established based upon the recommendation of others, usually the manufacturer of the instrument.
2. All instruments with the same function (e.g. DMM's, scopes, etc.) are calibrated on the same interval.
3. Intervals are adjusted (lengthened or shortened) periodically (about twice a year) based on analysis of data for the manufacturer and model number group of instruments.
4. Major test systems are either calibrated as a system, or individual instruments within the system are removed and calibrated separately.
5. A maximum interval (between 1 and 2 years) is imposed, but there is no minimum interval.
6. The interval system is intended to provide either at least 85% or at least 95% in tolerance of instruments at recalibration'.

It is left to the reader to consult the detail of the report. Calibration intervals are generally initially based upon one of two procedures these being *recall after a given calendar time period* or, less favoured, on the basis of *instrument usage time*. These are discussed in Greb (1973). Once established the period is varied in accordance with experience. The impact of the *smart* instrument generations containing self-calibration of a kind will add a new dimension to the task.

The recall periods of a large number of commercial instrument products were surveyed and reported in the May 1976 issue, the topic being followed up in Gebhardt (1980).

It is not easy to provide any simple rule to adopt. Recall periods vary from 1 to 36 months, varying considerably from user to user for the same product.

It might be considered that the better the *quality* of an instrument the longer can be the calibration interval. Quality in this case is defined to be a factor of recalibration interval. An extensive study (Greb, 1976a, 1976b) provides detailed comment on whether such simplistic assumptions are correct. The results of the study are expressed in terms of the in-tolerance rates (ITR) that occur.

On the quality of precision instrumentation, in the more general sense of the term, the reader is referred to Moss (1978). There is summarized the experience of the Aerospace and Guidance Metrology Centre, US of the acceptance of new equipment from industry. It is stated there that a considerable amount of incoming equipment was rejected—21% over one period involving 62 contracts. Of those failing to meet specifications, 43% were adjustable to specification but many were either condemned or returned for other, less serious, reasons. The author states that some improvement was evident at the time of study but that 1 in 6 complex facilities did not comply and 1 in 10 needed action beyond adjustment. These findings would probably not surprise many purchasers of instrumentation!

31.1.6 Automated Test Equipment

Where the number of instruments and degree of variation and type arises and the number of calibrations is high, a case may exist where automation of many of the test processes becomes economically viable. Remembering that calibration and testing work is an overhead cost it is usually very necessary to produce a cost-benefit analysis which can clearly show that there is an economic case to support introduction of automated test equipment (ATE). A case study (Pearson, 1980) provides one group's experience of getting started. Also relevant is the committee report included in the October 1977 issue of the *NCSL Newsletter*. That study provides parameters of the situation that might be met, those enabling costing and justification to be established.

The NCSL Calibration Laboratory Automation Committee have been very active. In 1979 they compiled a directory of some 120 users' systems! Some use commercially available systems and others have developed their own plant for this purpose. Semi-automatic systems are also included in the survey. It was published in the June 1979 issue of the *NCSL Newsletter*.

Another activity of that committee is the generation of a listing of desk-top calculator programs used in automation. At the time of writing the most recent listing was that published in the June 1981 issue. Programs are available only to stated rules of participation by users.

In use the ATE system operator, who may be relatively unskilled in many instances, typically calls up computer-based file information about the instrument on a visual display terminal. The display states the questions to be answered by the operator and indicates the actions that the operator may need to take—such as connecting the appropriate terminals of the ATE plant to the instrument. The system then cycles the instrument through the required tests by applying programmed sources in turn and measuring the corresponding outputs. When connection changes are needed the operator is requested to take the necessary action. Upon completion the operator removes the instrument, attaching whatever record information is needed to it and checking that the print-out of the test report is complete. One ATE system manufacturer offers 125 different test routines in the software instructions package.

Compared with manual methods these systems can perform a much faster calibration, can retain more consistent test quality and are capable of automatically providing management data. Manufacturers of ATE plant generally provide design assistance to potential customers because ATE systems are generally sold on a one to one basis. Needs will vary from one case to the other requiring customizing to some degree.

Initially ATE was introduced as fixed plant to which instruments were taken for calibration. With the introduction of the *smart* instrument a considerable amount of calibration is now being done within the instrument itself. This method enables more frequent checking. It must, however, be remembered that the calibration is only traceable if the internal calibration is compared with the traceable standard at periodic intervals.

Calibration need not always be checked by comparison with the same quantity. In many cases a reliable indication of system integrity can be obtained by indirect testing. For example, piezo-electric accelerometers can be checked for satisfactory operation using purely electrical tests that do not require the sensor to be in a state of acceleration. Such techniques, however, do not replace traceable calibration but only provide more convenient interperiod checks of performance.

31.1.7   Measurement Assurance Programs

These are programs of surveillance that provide the calibration laboratory with tools for evaluating and improving the quality of the measurements being performed in the laboratory. They are based on comparison with similar activities in other laboratories. Synonomous names with measurement assurance program (MAP) are *audit, sample audit,* and *round robin.* The *NCSL Newsletter* of May 1976 describes a MAP as follows:

> 'MAP is a departure from traditional metrology. It treats measurements as a continuous process subject to control and rigorous analysis and it leads to the assignment of uncertainties based on documented, long-term performance rather than on error assignment by estimation and occasional experimental verification. The MAP approach will be especially important to industries and other activities where documented proof of measurement uncertainties and close tolerance quality control are essential. The program is easily modified to serve less stringent requirements and still provide the vital monitoring presently missing in so many measurement activities'.

MAP's are conducted between laboratories having similar capabilities and test each group's ability to perform a set of stated measurement tasks. The results are then intercompared in a manner that retains confidentiality of participants' returns, each laboratory being in a position to assess its own performance relative to the group's overall performance.

The concept is not new; standard specification organizations have often conducted round robins to establish the real level of performance. One example was the circulation of several large-sized mechanical components to several laboratories in the UK in the early 1960's. The results of the measurements were used to assist revision of the standard for mechanical limits and fits tolerancing of large size parts.

The National Association of Testing Authoratories (NATA), Australia, has conducted a MAP for the Australian Petroleum Laboratories for several years. It created one for the dairy industries in 1977. The value of MAP's is discussed in this passage from *NATA Newsletter,* June 1977:

> 'Interlaboratory test programmes can establish confidence on the part of participating laboratories in their own ability to achieve reliable results for specific tests. They can identify testing problems not otherwise evident to laboratory management. They can provide a statistical basis for determining repeatability and reproducibility figures for the standard methods used. And by no means last, they reinforce NATA's confidence in the testing ability of registered laboratories'.

In 1977 the NBS issued a paper on measurement assurance (Cameron, 1977). Another review paper is the report of a conference workshop published in the *NCSL Newsletter* of December 1977.

MAP's have been conducted and reported on gage blocks, voltage standards, regional MAP's, mass (*NCSL Newsletter*, December 1978); on gage blocks in more detail (a series of documents of the NBS which are summarized in *NCSL Newsletter*, May 1976) and on medical devices (Hooten, 1980). A film on the related topic, confidence in measurement, is also available (CIO, 1974).

REFERENCES

BCS (1969), 'General criteria for laboratory approval', *BSC Document No.* 0302, British Calibration Service, London.

Cameron, J. M. (1977), 'Measurement assurance', *NBSIR* 77-1240, National Bureau of Standards, Washington, DC (Reported in (1978), *NCSL Newsletter*, **18,** No. 1, 25–37).

CIO (1974), *Confidence in Measurement*, 16 mm colour film, 16 min, Central Office of Information, Department of Trade and Industry, UK.

Daneman, H. (1977), 'Laboratory planning in developing countries', *NCSL Newsletter*, **17,** No. 4, 27–33.

Daneman, H. (1978), 'Mobile calibration vans', *NCSL Newsletter*, **18,** No. 2, 43–5.

Gebhardt, C. (1980), 'Recall period pilot program', *NCSL Newsletter*, **20,** No. 1, 14–18.

Greb, D. J. (1973), 'Calibration on basis of instrument usage', *NCSL Newsletter*, **13,** No. 3, 26–7.

Greb, D. J. (1976a), 'Calibration intervals and instrument quality—introduction, *NCSL Newsletter*, **16,** No. 3, 26–32.

Greb, D. J. (1976b) 'Calibration intervals and instrument quality—parts II, III', *NCSL Newsletter*, **16,** No. 4, 29–35.

Hooten, W. F. (1980), 'Measurement assurance and medical devices', *NCSL Newsletter*, **20,** No. 4, 36–7.

King-Jones, W. (1977), *A Study of Standards Calibration and Measurement in Australia*, Department of Productivity, Melbourne.

Mitchell, J. D. (1981), 'Getting top management support for metrology', *NCSL Newsletter*, **21,** No. 2, 9–11.

Moss, C. (1978), 'Quality of precision measuring equipment', *NCSL Newsletter*, **18,** No. 1, 23–4.

NATA (1976), 'Why calibrate?', *NATA Newsletter*, **9,** 1.

NCSL (1978), 'Metrology course register', *NCSL Newsletter*, **18,** No. 3, 45–7.

Pearson, T. A. (1980), *NCSL Newsletter*, **20,** No. 1, 19–20.

Seed, J. R. and Pawlick, P. V. (1974), 'The value of making measurements', *NCSL Newsletter*, **14,** No. 1, 34–43.

Sydenham, P. H. (1979), *Measuring Instruments: Tools of Knowledge and Control*, Peter Peregrinus, London.

Taff, H. A. (1980), 'Minimum requirements for calibration technicians', *NCSL Newsletter*, **20,** No. 1, 21–2.

Werner, H. B. (1980), 'Development of an educational program in metrology', *NCSL Newsletter*, **20,** No. 1, 29–32.

Westmoreland, F. G. (1980), 'Intervals and calibration system management', *NCSL Newsletter*, **20,** No. 1, 10–13.

31.2 INSTRUMENT EVALUATION

31.2.1 Introduction

The most widely used definition of instrument evaluation among the major cooperative instrument evaluation groups is 'the investigation and testing of the performance of instruments and related equipment in accordance with an agreed program based on the manufacturer's specification, relevant national and international standards and specific user requirements'. The agreed program will usually include performance evaluation under reference conditions and simulated environmental conditions (new usually termed influence conditions) together with an assessment of standards of construction and adequacy of documentation such as operating instructions and maintenance handbooks.

The process can most succinctly be described as a comprehensive technical audit. The concept of instrument evaluation has been developed over some 18 years to meet the needs of users and manufacturers of industrial instrumentation and process control equipment. It is aimed at producing the maximum possible amount of information on the performance of instruments and related equipment bearing in mind the needs of the ultimate user and the rights of the manufacturer to fair and consistent assessment of his product.

31.2.2 Need for Instrument Evaluation

With the increasing post-war use of sophisticated measurement and control equipment in process plants (which themselves were increasing in size, complexity and cost) came the realization of the increasing financial consequences of failure. Such consequences do not only apply in the event of complete failure of an instrument which, because of a fail-safe system philosophy, causes a costly plant shutdown. They also, and more commonly, apply in cases where instrument performance gradually deteriorates in an undetected manner. Such deterioration can cause a shift in the performance of an automated high-throughput plant away from a state of optimum performance as defined by a required cost and quality target. It is relatively straightforward, though costly, to provide instrument redundancy so that fail-safe plant shutdown only occurs if, say, two out of three instruments fail simultaneously in a catastrophic sense. It is less simple to arrange that such redundancy will compensate for the gradual deterioration of the measurement integrity of the instruments or their susceptibility to the variations of a particular environmental parameter such as temperature, humidity or vibration. This financial consequence of instrumentation failing to perform to its required accuracy has led to increasing insistence by user industries of comprehensive evaluation of instrumentation and control equipment as a pre-requisite to purchasing decisions.

The earliest practitioners of equipment evaluation were the defence establishments. Military personnel have always been conscious of the exacting environments in which their equipment (and themselves) have to operate, and they were the first to insist on equally exacting type approval tests

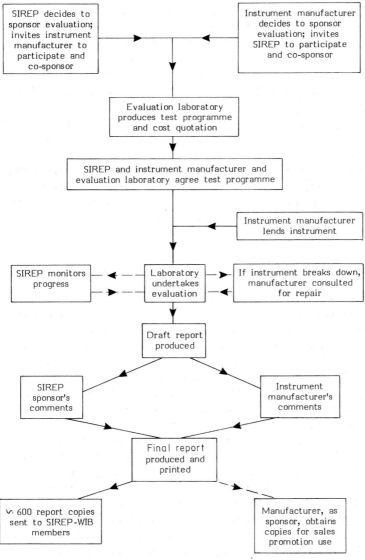

Figure 31.1   SIREP evaluation procedure

for such equipment. This gave rise to a series of detailed test specifications and standards which have subsequently formed the basis of many industrial standards. The defence procurement authorities also realised that, while type approval tests could ensure that a design was capable of meeting its specification, it would not ensure that all models from the production line would perform satisfactorily. This led to the defence services taking an active role in guiding the quality assurance procedures used by their suppliers, with the aim of raising the proportion of production models which met the performance achieved in the type approval tests.

The same trend can be seen in industry. The industrial users of instruments are conscious of the cost penalties of instrumentation failing to perform to specification. The major companies have conducted or commissioned instrument evaluations for years as a basis for purchasing decisions. Again, the major companies have insisted on vetting the quality assurance procedures of their suppliers. While this causes some dissatisfaction among suppliers who object to a multiplicity of inspection authorities, it is probable that manufacturers and major users in the UK, through their trade associations, will cooperate in a common scheme for rating and vetting the quality assurance procedures of control and instrumentation equipment manufacturers.

While the needs of manufacturers for instrument evaluation are less obvious they nevertheless exists. Particularly in relation to new designs, it is important to identify a serious design weakness before a significant number have been delivered. The direct cost penalty of retrieving and modifying such instruments is obvious enough. The penalty in terms of lost customer confidence and future sales is more serious, yet there is a surprising proportion of cases where an independent comprehensive evaluation reveals design weaknesses which have not previously been identified by the manufacturer. The results of evaluation (see below) indicate that a high proportion of evaluations lead to necessary design modifications, and many manufacturers now commission an independent evaluation prior to a major launch of a new instrument.

A more detailed account is given later in this chapter of the different phases of evaluation, and Figure 31.1 shows the main activities involved in a typical SIREP evaluation project.

31.2.3   History of Cooperative Instrument Evaluation

By cooperative instrument evaluation is meant the sponsorship of evaluation programs by groups of companies on the basis of cost-sharing and results-sharing. Although instrument evaluation has been practiced individually for many years by the major instrument using companies in the chemical, oil, steel, and other industries as a means of guiding purchasing decisions, cooperative instrument evaluation began in the late 1950's. In 1958 and 1959 Sira Institute (then the British Scientific Instrument Research Association) commenced instrument evaluation work under contract separately to the Central Electricity Generating Board and the United Kingdom Atomic Energy Authority. It was known that other major users of instruments had their own individual programs of instrument evaluation, and it was suggested that such users would obtain benefits from joining forces to sponsor evaluations cooperatively and share the results. With a great deal of help from the UKAEA a proposal for such a cooperative group project was put to a number of instrument user companies, and in 1961 a meeting was held at Sira, chaired by the then Director, Dr John Thomson, at which the cooperative instrument evaluation scheme was formalized. The founder members comprised CEGB, UKAEA, ICI, Distillers Co. (now BP Chemicals), Ministry of Defence, and Shell Research. The group which eventually gave itself the acronym SIREP (Sira Instrument Evaluation Panel) had its inaugural meeting on 6 December 1961. It has since grown to a 1980 membership of 30 organizations. The membership includes North American and European companies including the group membership of a Swedish consortium SIP (Stiftelsen for Instrumentprovning) which itself comprises 22 members. SIREP evaluations are carried out largely by Sira Institute.

In 1963 a similar group was formed in The Netherlands. This group is WIB (working-party on instrument behaviour) whose membership is principally Dutch companies but includes other European and North American companies. Although WIB use mainly the laboratories of TNO (Toegepast Natuurwetenscappelijk Onderzoek) in Holland for evaluations, they also have evaluations carried out in the laboratories of LCIE (Laboratoire Central des Industries Electriques) in France, KEMA (Keuring van Electrotechnische Materialen) in Holland, and Sira Institute in the UK. In more recent years a French group, Association des Exploitants D'Equipments de Regulation Automatique (EXERA), has also been formed.

Although the adminstrations of SIREP and WIB differ in detail (SIREP is currently administered by Sira Institute, and WIB is administered by an appointed manager) the general objectives are the same, and over the years a close collaboration has grown between the two organizations. Evaluation procedures and report format have been standardized and there is continuing technical and administrative contact through a liaison committee.

31.2.4 Finance and Communication

Taking SIREP as an example of a typical cooperative evaluation group, each member of SIREP pays an agreed annual sum to fund evaluations, the amount being agreed by the SIREP members at their annual general meeting. The 1980 figure is £3600. This is used to sponsor evaluations of the member's own choice, either alone or on a cost-sharing basis. If he knows that an instrument of interest to him is already being sponsored by another member, he can look for an alternative evaluation on which to use his money. From this it will be appreciated that it is essential that each member shall know at any time exactly what evaluation projects have already been authorized, or alternatively what potential projects are of interest to other members seeking co-sponsorship.

To deal with this, a project list and a shopping list are issued periodically. The project list shows the state of all authorized projects from inception to issue of final report. Each equipment has an entry indicating manufacturers, country of origin, model description, date proposed, delivery date expected, actual delivery date, date evaluation started, date draft report expected, date draft report actually issued, date of final report, laboratory and/or report reference number, and sponsor. The shipping list includes those instruments which have been proposed by members looking for co-sponsors, and is an early indication to the laboratories of changing priorities with regard to types of instrumentation for which performance data are required.

31.2.5 Specifications and Standards

Although the basic reference point in evaluation is the manufacturer's specification, it is often a very incomplete statement of the instrument performance, particularly in relation to influence conditions which depart significantly from laboratory reference conditions. In such circumstances it is necessary to look for other guides to performance which are acceptable both to manufacturer and user. Whether or not the specification is complete, the actual methods used for evaluation need to be acceptable to manufacturer and user and not subject to significant modification when applied to different instruments of the same kind.

For these reasons a great deal of attention has been given to specifications and testing procedures by national and international standards bodies. A standard specification is a technical specification or other document which has been evolved by cooperation between various bodies and represents a consensus of the parties concerned. Such a standard may set out the quantities to be specified, the boundary values of accuracy classes, an explanation of terminology used, and an outline of methods of measuring the quantities. In the absence of a sufficiently detailed manufacturer's specification, an evaluation will be based on a national or international standard specification. The manufacturer concerned thus has the reassurance that his instrument performance will be compared with a specification evolved by consensus, that competitive instruments will be subject to the same comparison, and that the test methods used for evaluation will be the same as used for other instruments.

31.2.6 The procedures of Instrument Evaluation

A number of steps can be identified in a typical instrument evaluation program, consisting of: agreement of test program; tests under reference conditions; tests under simulated environmental (influence) conditions; assessments of construction standards and literature quality; draft report; final report. Table 31.1 gives an indication of the range of equipment dealt with in current evaluation programs, and Figure 31.2 summarizes the main tests carried out during an evaluation.

The test program

The test program is a key document. It is prepared by the evaluator, who is guided by the manufacturer's specification, standard specifications, and standards relating to methods of evaluation and calibration. In particular his experience guides him to seek performance information not revealed by the manufacturer's specification. The test program is submitted to the sponsor and the manufacturer (if not the sponsor) and the evaluation only proceeds when the three parties agree the program. Sometimes the agreement to a program is almost a formality. In other cases the methods to be used to establish basic accuracy and environmental effects can be the subject of controversy and lengthy discussions may be necessary to reach agreement. For example, evaluation and calibration procedures for temperature or pressure transmitters are widely documented and accepted, but this is not the case for instruments for chemical analysis such as water quality or process gas analysers. In such cases, evaluation work is often preceded by a

Table 31.1 Some examples of equipment evaluated

Pressure, temperature, level, and flow switches, transmitters, and indicators
Chemical composition analysers, hygrometers
Mass flow computers
General process and other controllers
Valve positioners, I/P converters
Control valves
Liquid and gas density meters
Flashpoint, pourpoint, and other analysers
Gas analysers for O_2, NO_x, SO_2, CO_2, and other gases
Sulphur-in-oil analysers, water quality analysers
Flammable gas monitors/alarms
Safety monitors
Signal processors
Recorders
Switches and alarms
Microprocessor-based instrumentation

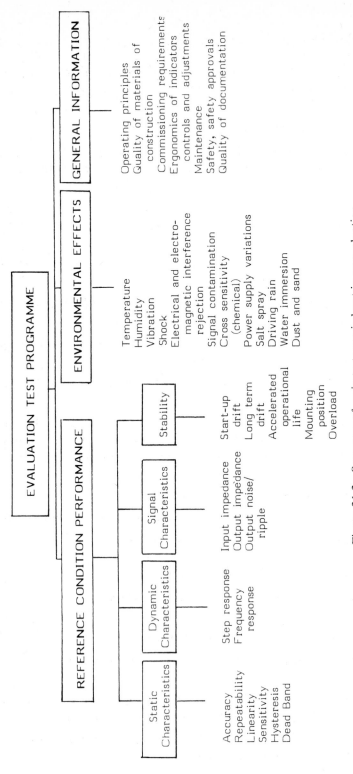

Figure 31.2 Summary of main tests carried out in an evaluation

significant amount of research and development in order to arrive at mutually acceptable methods of evaluation.

Tests under reference conditions

Basic accuracy. Average error, repeatability error, linearity error determined from at least 5 readings at up to 11 points on all ranges for rising and falling inputs separately.

Warm-up time. After the instrument has been switched off for 24 hours, the level of the output signal 5 min, 10 min, 30 min, and 1 hour after switching on, for up to 7 hours.

Short term drift. With a constant mid-span input, the maximum variation in output over a 24 hour period.

Response time. The time for the output to attain 63% of the final reading after a step change of input.

Frequency response. The frequency response measured with a 10% span peak to peak input centred on 50% span.

Dead zone. The smallest change in input which results in a detectable change in input.

Long term drift. With a constant mid-span input, the maximum variation in output, measured once daily, over a 30 day period or longer. (This is usually the penultimate test, followed by a final check on accuracy.)

Tests under simulated environmental conditions

Climatic tests. Specified tests to determine effects of exposing instrument to

(a)   dry heat (+55 °C maximum or higher if applicable);
(b)   low temperature (−10 °C or as applicable);
(c)   damp heat (+40 °C, 95% relative humidity);

in comparison to operation under reference conditions.

Power supply variations. Effect on output of changes in voltage and frequency (or changes in pressure for pneumatic instruments).

Power supply interruptions. Effect on output of random interruptions of 10, 40, 100, 500 ms.

Overload. Effect on output (after a short resting period) of exceeding maximum specified input by a given amount for a given time.

Series mode interference. Effect on output of supply frequency signal injected in series with input (if applicable).

Common mode interference. Effect on output of applying 240 V (or 100 V) r.m.s. supply frequency signal between each input lead and earthed case (if applicable).

Flash test and insulation resistance. 2 kV mains frequency supply applied for 1 min between each power terminal and earth followed by measurement of insulation resistance at 500 V d.c. between same pairs of terminals.

Output terminal isolation. Effects on output of connecting each output terminal in turn to earth and instrument case.

Mounting position. Effect on output of 10° tilt in attitude about two mutually perpendicular horizontal axes.

Magnetic field interference. Effect on output of mains frequency magnetic field of 400 A/m.

Load variation. Effect on the accuracy of varying the output load within the specified limits.

Cross sensitivity (for chemical analysers and monitors). Effect on output of various concentrations of chemically interfering species in the sample.

Sample temperature compensation (for chemical analysers). Effect on output of different sample temperatures of, say, 5 °C and 35 °C compared to 20 °C reference.

Vibration. Effect on output of vibrating the instrument along three mutually perpendicular axes at specified peak accelerations over a specified frequency range. Resonating components are detected with phase-locked stroboscope.

Drop tests. 6 drops through 25 mm onto each of three mutually perpendicular faces, followed by inspection for damage.

Driving rain. Detection of water ingress with test conducted according to BS 2011, Part 20g (8).

Dust test. Detection of penetration of dust into instrument interior.

Salt spray. Determination of any corrosion effects.

Assessment of construction standards and literature quality

Standards of construction and finish. Overall design quality, robustness, quality of components, ease to operation.

Packaging. Adequacy and method of delivery.

Maintenance. Maintenance requirements, component accessibility.

Safety.   Clarity of labelling, accessibility to live terminals, whether certified for use in potentially explosive atmospheres.

Documentation.   Quality, comprehensiveness and language(s) of documentation, adequacy of installation, operating and servicing instructions, inclusion of circuit diagrams, fault-finding routines.

The evaluation report

A standardized format is used for reporting. The principle followed is to present findings in a concise and easy-to-refer-to manner. Major findings are presented immediately after a short introduction and contain enough information to enable a reader who does not wish to immerse himself in details to assimilate quickly the salient facts. A very important aspect of the report is a section devoted to the manufacturer's comments. This contains verbatim comments, additional information, and criticism (occasionally) by the manufacturer on the findings.

This section is followed by tabular and graphical results and the report is concluded with a description of the test methods and references.

31.2.7   Results of Instrument Evaluation

An analysis made in 1979 of the results of a sample of 700 evaluation reports, covering equipment supplied from 12 different countries, shows the proportion of equipment deemed unsatisfactory in various categories. The categories and the relevant proportions are as follows:

1	Unsatisfactory in condition as received	25%
2	Outside specification under reference conditions	27%
3	Outside specification under simulated environmental (influence) conditions	64%
4	Breakdown during evaluation	24%
5	Inadequate handbook/manual/documentation	38%

In addition, 29% of the equipments evaluated in the sample were the subject of modification to design or manufacturing method as a direct result of evaluation.

Category 1 records the proportion of instruments which in some way had a fault on receipt, sometimes trivial such as missing a fuse, sometimes serious such as completely uncalibrated or missing one or more components. In categories 2 and 3, the manufacturer's specification was used wherever possible. Where performance was not specified by the manufacturer it was judged in relation to a specification based on current practice in instruments of a similar kind. In the case of category 3 the performance was judged in relation to environmental influence conditions of high temperature, low temperature and humidity; supply variations of voltage, frequency and pneumatic pressure; other influence conditions as agreed in advance with the manufacturer. Category 4 records the proportion of instrument breakdown which was usually traceable to a specific component fault.

Category 5 records the results of an assessment of the documentation accompanying the instrument. The main documents are the operating handbook and the maintenance instructions, and the evaluator forms an opinion of the adequacy of these as a guide to the purchaser.

It was indicated above that 29% of the total instruments in the sample were subjected to modification of design or manufacturing method as a direct consequence of the evaluation. The criterion for inclusion in this figure was that the manufacturer had made a statement in his published comments to the effect that such a change would be made.

While a number of conclusions can be drawn from these results, category 1 results probably cause more concern to users than the other categories. Other than failure occurring during finished goods storage or transit, the unpalatable conclusion is that some instruments do not receive sufficient final testing for an obvious malfunction to be detected. There was no significant evidence of physical damage which could be correlated with the instruments found to be faulty. Instruments shipped from abroad and generally travelling greater distances do not show a higher proportion of faults as received. The SIREP Instrument Evaluation Panel has drawn the attention of UK manufacturers and user trade associations to the figures, but as yet there is no significant evidence to show that high 'unsatisfactory as received' figures apply to any one country of origin more than others. A number of users have reported similar figures from their own records. Manufacturers have pointed out that such a high proportion of returns would not be commercially realistic. Users have commented that actual returns are lower because faults are often rectified by the user.

While in categories 2 and 3 no attempt has been made to assess the proportions of instruments which could be regarded as serious or less serious departures from specified performance, the figures suggest a tendency to optimism in specifications. Some individual manufacturers have indicated that this is so with regard to certain instruments and state that it is often due to the commercial pressures caused by competitors making performance claims known to be exaggerated. This problem is made more difficult by variations in the way in which performance specifications of similar kinds of instruments are described. For example, it is sometimes unclear whether a specified performance relates only to reference conditions or takes some account of influence conditions. A significant improvement could be effected in this situation if the methods of specifying performance of instruments of a similar kind could be standardized. Some work along these lines is being pursued in national and international standards bodies, but a vast effort would be required to cover the principal areas of industrial instruments.

Category 4, dealing with failures leading to instrument breakdown, cannot be used to draw conclusions about component reliability. The number and types used in instruments vary widely, and deductions of component reliability are better made from controlled tests on particular categories of components. It may be possible however to ascribe part of the high figure in category 3 to insufficient use by manufacturers of 'burn-in' procedures, in which instruments are allowed to run in their operating state for a sufficient time to enable unreliable components to reveal themselves by their well-known tendency to fail early in their life.

Category 5 results indicate that insufficient attention is being given to the preparation of information which ensures that users get maximum effectiveness from the instrument after delivery. The subject of documentation for instruments is covered by British Standard 4462:1969 'Guide for the preparation of technical sales literature for measuring instruments and process control equipment' and it is recommended that manufacturers re-appraise their documentation in the light of this standard.

A high proportion of instruments are modified as a direct result of evaluation. The figure of 29% emphasizes both the benefit of and the need for instrument evaluation. It is a direct measure of action to improve performance resulting from evaluation, and it indicates the positive way in which instrument manufacturers respond to the results of impartial evaluation.

31.2.8 Quality Control

Even with adverse performance figures such as those given above, it is often said that the results of an evaluation lose some validity in that they are based on tests on a model which has been hand-picked and is therefore unrepresentative. The figures suggest that such hand-picking does not occur, and the authors' knowledge of organization within major instrument companies suggests that it is unlikely to occur. However, even if such selection did take place it does not invalidate the conclusions of an evaluation. If the instrument performs entirely satisfactorily in the evaluation it is evidence that the design is capable of meeting the claimed specification. Such a single evaluation cannot, however, give a reliable measure of the probability of a high proportion of the production models meeting the specification. A single comprehensive evaluation is therefore capable of verifying that the design is such that the instrument can meet its specification but it is not a substitute for the quality assurance procedures necessary to ensure that, as far as possible, every model actually does meet its specification.

31.2.9   Uses and Benefits of Cooperative Evaluation

The factors which users consider when justifying membership of the evaluation groups are as follows.

Purchasing decisions

Instrument evaluation reports cannot be the only information input for purchasing decisions, but most members regard them as an important factor. They at least help to narrow a choice to two or three, the final choice being decided by other considerations. In particular, evaluation can prevent disastrous purchasing decisions.

Value for maney

Many companies used to evaluate all instruments themselves. Nowadays, they only deal with equipment of a specialist nature to themselves. One member said that, after distributing evaluation reports to each factory, the cost per factory per instrument evaluated was less than £10. It depends critically on each member getting effective distribution of the reports within his organization.

Approvals

Some companies operate a formal or informal 'approvals scheme' where purchasing is restricted to an approved list, and a satisfactory evaluation report is the basis of approval.

Sponsorship

Apart from the foregoing benefits, a member's contribution can be used by him to sponsor or co-sponsor instrument evaluations of particular interest to his company. It is the sum total of such evaluations which represent the collective benefit eventually derived by all members.

Feedback to manufacturers

Experience with the methods of carrying out evaluation projects, including the procedures of consultation with manufacturers and reporting of difficulties during evaluation, have led manufacturers to realize that equipment faults developing during evaluation are not used to condemn the instrument, but are considered in the context of the overall performance. Manufacturers thus have confidence that their product will be treated fairly, and the reporting of faults in design and construction methods is usually followed by improvements to design or manufacturing technique. The evaluation laboratories now find that a significant proportion of their work is directly sponsored by manufacturers seeking independent verification of instrument performance.

Conclusions

Cooperative instrument evaluation has become a significant international activity, involving a number of sponsoring groups who cooperate in the exchange of information on instrument performance. Reports are circulated widely throughout the companies in membership of the groups. A number of specialized instrument evaluation laboratories exist, equipped with costly evaluation facilities and staffed with engineers and scientists skilled in the methodology of evaluation and the procedures of measurement and interpretation. Instrument manufacturers cooperate willingly in evaluation projects and where necessary use the results to improve their products. A significant number of instrument manufacturers now sponsor evaluations directly.

The objective of instrument evaluation is better instruments, and the cooperating user groups can justifiably claim that steady progress is being made towards this objective.

31.3 ACCREDITATION

31.3.1 Introduction

The International Organization for Standardization (ISO) defines laboratory accreditation as 'formal recognition that a testing laboratory is competent to carry out specific tests or specific types of tests'.

Such an accreditation is formally granted only after the accrediting organization is satisfied that the particular laboratory meets all pre-defined criteria. Current accreditation procedures provide for:

(a) initial assessment of all aspects of laboratory management and operation by panels of expert assessors;
(b) reassessment at prescribed intervals;
(c) proficiency testing or other form of objective audit programs (where possible) on a regular basis.

Use of the expression *laboratory accreditation* is very recent but the concept dates from the 19th century when laboratory test results were first used as a basis for acceptance of goods or services. Even then the buyer sought assurance that the testing laboratory was competent.

The history of laboratory accreditation began with individual customers, such as defence services, requiring such assurances and it developed on a very *ad hoc* basis with the proliferation of special purpose systems to suit the needs of individuals or narrow groups. Some laboratories had to maintain separate approvals by a number of different systems.

In recent years, however, it has been recognized that there is a great degree of common ground between existing systems and that there is room for considerable rationalization at the national level. This has led to the development of comprehensive systems for the accreditation of testing laboratories to serve the needs of most sectors of a community whether they be for statutory purposes, purchasing, production, manufacturing, health services, product certification, or other purposes.

Need for accreditation

The following examples illustrate the range of individuals and organizations within a community who need access to laboratories of demonstrated competence.

Government departments and agencies
(a)   tests required by regulation or ordinance;
(b)   forensic work;
(c)   investigative work for projects of national importance.

Public or private purchasing authorities.   To ensure compliance of goods and services with specification.

Producers and manufacturers
(a)   quality control;
(b)   product development;
(c)   failure investigation;
(d)   advertising data.

Health services.   Laboratories form an integral part of modern public and private health services.

Consumers and consumer organizations.   Challenges to manufacturers on technical grounds require test data.

Certifying bodies.   Acquisition of test data is an essential element in any certifying operation.

General contracting parties.   Contract arrangements may require access to either in-house or third party laboratories.

In all of these examples it is important that laboratories be competent in the area of testing for which they issue test reports. It should also be recognized that the competence of a laboratory is independent of its ownership and that government or institutional laboratories are not necessarily more reliable than those operated by manufacturers or commercial interests. Regardless of its purpose or ownership, the principal concern of users of a laboratory is that it performs competently and reliably. It is in this regard that accreditation systems serve their most vital function, for the user, without needing to make his own evaluation, may reasonably have greater confidence in the competence and reliability of a laboratory that has been authoritatively accredited than in one which has not been so accredited.

The community, or significant sections of it, gains a direct benefit from the fact that accredited laboratories will produce test results that are more likely to be found useful, reliable, and achieved efficiently. There is, therefore, general interest in ensuring that laboratory services at all levels and for all purposes are adequate for their tasks. Although some of the needs identified above are obviously of more public significance than others, the performance of most testing laboratories is of concern not only to the laboratory proprietor but also to the public at large. Hence the widespread interest in various laboratory accreditation systems which provide assurance that laboratories are examined by expert assessors and are subject to a degree of technical audit.

The introduction of formal laboratory accreditation systems in a number of countries is also a recognition of the need to rationalize and harmonize the many existing special purpose and *ad hoc* schemes, which often compete or overlap. The recognition that the criteria against which the laboratories are to be judged is applicable to various areas of science or technology has enabled such comprehensive programs to be launched.

31.3.2 Objectives

Existing and proposed laboratory accreditation systems throughout the world are designed to meet particular national needs which may differ from country to country but the main objectives of all such systems are included amongst the following:

(a) Ensure validity of test data—to meet whatever need.

(b) Promote acceptance of test data by users of laboratory services—test data produced by one laboratory may be accepted at another location without recourse to further tests.

(c) Facilitate trade and commerce, both national and international—test results obtained at point of manufacture can be accepted at point of receipt of the goods without recourse to further tests.

(d) Make more efficient utilization of testing facilities and resources within a country by the coordination of existing testing capabilities—identifies centres of competence irrespective of geography or ownership by publicizing location and availability of expensive and sophisticated test facilities.

(e) Give credibility to a greater number of laboratories—prevent unnecessary duplication and overloading.

(f) Give additional status to competent laboratories.

(g) Promote good testing practices—some accreditation agencies are also engaged in educational activities.

(h) Improve test methods—the accreditation procedures provide a feedback to standards-producing bodies on the adequacy of test methods used in laboratories.

(i) Provide technical and other relevant information on accredited laboratories

31.3.3 Existing National Systems

Existing accreditation systems take many forms and serve many purposes. For the purposes of this discussion, national systems are taken to be those that operate comprehensive laboratory accreditation systems throughout a particular country. They are usually, but not necessarily, operated by national government agencies or instrumentalities. A few examples of such systems are described below. These are followed by some examples of special purpose systems.

Australia

The world's oldest national comprehensive system is operated by National Association of Testing Authorities (NATA), Australia (Gilmour, 1978). Formed in 1947 by a decision of the federal government, NATA is an independent organization which is supported by governments (state and federal) and industry and is recognized as the national system. NATA is incorporated as a non-profit-making company and is governed by a council consisting of representatives of the federal government, each state government, professional institutions, and individuals elected from the accredited laboratories.

Laboratories are accredited in all areas of science and engineering subdivided into nine fields of testing, and accreditation is available for tests done in the field as well as in more formal laboratory accommodation. Accreditation is granted to laboratories which can demonstrate compliance with the Association's criteria for competence. Accredited laboratories are not obliged to offer facilities for general use.

The criteria are defined by committees of experts which are established for each field of testing. Operating within common guidelines, each committee considers what is required by way of:

staff	qualifications and relevant experience
equipment	availability, calibration, and maintenance
laboratory practices	sampling, test methods, operations, records, quality control, reporting
accommodation	suitability, safety

Laboratories are assessed by panels of assessors, the members of which are drawn from government establishments, academic institutions, research organizations, consultants, and industrial and commercial enterprises. The assessment of a laboratory is therefore truly a *peer review.*

Accredited laboratories are audited by means of regular peer assessment as well as participation in objective proficiency testing progarms run by the Association and other cooperating bodies.

The Association employs professional and clerical staff to provide essential technical and secretarial support to council, criteria committees, and assessors.

NATA is funded by fees paid by accredited laboratories and from a substantial grant from the federal government. The grant currently (1981) provides approximately 65% of total revenue. The system is designed to provide a truly national service and therefore there is no financial penalty for laboratories situated in remote locations. It must be appreciated that Australia is a country approximately the size of mainland USA and, therefore, there is a significant number of laboratories located over a considerable distance from the main centres of population on the eastern seaboard. These remote laboratories cost much more to service but are subsidized from general funds.

Denmark

The Danish National Testing Board (STP) was established by statute in 1973 to provide a national laboratory accreditation system. The system uses the term *authorized* rather than *accredited* and there is an emphasis on *official* requirements and the need to satisfy regulations and public needs.

Laboratories may be accredited in any field of engineering or science provided there is a need for such an accredited or authorized laboratory. Broadly, testing is divided into industry groups such as building and construction, electrical engineering, and chemical industries. For each group there is a criteria committee which defines the detailed requirements that must be met by laboratories prior to accreditation.

Assessment of laboratories is undertaken by panels of independent experts drawn from all sectors of the community. The Board is prepared to engage experts from elsewhere in Europe if necessary.

Authorized laboratories are monitored by the Board by regular reassessment and by surveillance of test reports issued by the laboratories. Authorized laboratories are obliged to accept commissions for the performance of authorized tests.

The operations of the Board and its supporting professional and clerical staff are funded by some government support and fees paid by accredited laboratories. It is intended that the system become entirely self-supporting within a few years.

New Zealand

New Zealand established its Testing Laboratory Registration Council (TELARC) late in 1972 (Gilmour, 1978). It is incorporated as a statutory board responsible to the Minister of Science. Members of the council represent government, industry, and the interests of the wider community. Criteria for accreditation are defined by expert committees and cover all fields of testing including testing for medical purposes.

Assessors are drawn from all sectors of the community and from Australia. The procedures for the appraisal of a laboratory are modelled on Australian practice and there is considerable interchange of staff and assessors between the systems of the two countries as the need and opportunity arises.

The organization is funded by fees from laboratories and a substantial government grant.

United States of America

The three countries mentioned so far all have in common the feature of small homogeneous populations—a feature which has facilitated the development of national comprehensive accreditation systems. The USA, on the other hand is a large country with a large and well dispersed population and, is, therefore, very much more fragmented politically and administratively. A single national system for accreditation of testing laboratories has been very much more difficult to develop. A survey of existing systems in the USA reveals the existence of approximately 80 special purpose systems and it was only in the late 1970's that an attempt was made to develop a coordinated approach to the problem by establishing a comprehensive national system known as the national voluntary laboratory accreditation program (NVLAP).

NVLAP was formally established in 1976 as a program within the Department of Commerce (DOC). The program is managed by staff of the Office of Productivity, Technology and Innovation within DOC in collaboration with the National Bureau of Standards (NBS).

Accreditation is available to laboratories for testing of specified products within a defined laboratory accreditation program (LAP). Before such a program is developed, however, there must be a public finding of need for the program. Each LAP for a specific product category will, however, provide for accreditation for virtually all the standard tests performed on that product irrespective of the range of scientific or engineering disciplines involved.

Assessors in the NVLAP system are drawn from NBS staff or paid consultants. They are not true peers in that they are not drawn from equivalent laboratories to those under assessment.

NVLAP is funded by government grant and fees. It is intended to become self-supporting in the longer term.

General observations

The four national systems described above range from an incorporated company to a division of the national public service. All, however, have the support of their respective national governments.

Despite some fairly superficial operational differences and variations in administration, all the organizations have developed, more or less independently, criteria and standards of operation that are remarkably similar to each other. The brief descriptions above are intended to highlight the differences rather than the similarities between them. At this time the organizations are moving towards the same goals, using essentially the same tools and in considerable harmony with each other.

31.3.4 Special Purpose Systems

There are literally hundreds of *laboratory approval* schemes in operation throughout the world serving a wide variety of special interests from local building authorities, trade associations, manufacturers to government departments and statutory authorities.

The standard of their operation also varies enormously from a simple registration without having any requirement for compliance with any technical criteria to fully fledged accreditation systems with rigorous assessment and auditing procedures.

This section briefly describes a few of these to illustrate the differences in approach.

New Zealand

Dairy products produced in New Zealand for export are subject to testing and inspection by nominated laboratories operated under the Dairy Industry

Agency Certification Program of the Department of Agriculture. The scheme is limited to testing of milk products for export only and is separate from other laboratory approval schemes for other agricultural materials.

Laboratories are inspected by departmental staff for compliance with some general criteria but the main emphasis of the program is on an extensive proficiency testing program run by the department's National Dairy Laboratory.

United Kingdom

The UK pioneered a special purpose accreditation program known as the British Calibration Service (BCS) (Thurnell, 1979) which is devoted to accreditation of calibration laboratories and which is the model for a number of similar schemes being established throughout Europe.

BCS provides a network of laboratories within the UK which calibrate all types of measuring instruments for scientific and industrial purposes. Laboratories are accredited for specific measurements to specified least uncertainties of measurement.

Its general statements of criteria and its methods of operation are very similar to the more comprehensive national laboratory accreditation programs referred to earlier. BCS operates a very elaborate proficiency testing program and collaborates with its European counterparts in this regard.

United States

There are a large number of special purpose systems operating within the US instituted by federal, state, and local governments and by private organizations. The following are just a few to illustrate the diversity of approach.

Department of defense

Defense Electronics Supply Center (DESC). This program, established in 1962, developed from an earlier one in operation from the early 1940's. It is an integral part of a program designed to qualify specific electronic components against specific military standards. Approval is therefore available to laboratories performing tests associated with controlling the quality of electronic components supplied to the Department of Defense and the scheme is very much related to the total quality assurance program operated by the Department.

Laboratory assessments are conducted by engineering employees of

DESC against criteria generally in agreement with the specifications ASTM E548.

Ohio State Board of Building Standards

This program was established in 1965 to provide a register of approved agencies for the mandatory testing of specified products used in the State of Ohio. The accreditation is based essentially on a review of documentation which is used to provide an assurance of the capability of the laboratories.

United States Potters' Association

This Association runs a laboratory approval program to perform lead and cadmium release tests on ceramics and laboratories are approved to Food and Drug Administration testing procedures and specifications of ASTM. The assessment of each laboratory is purely by checking known samples. At the end of 1979 fifteen laboratories had been approved under this scheme.

31.3.5   New National Systems

NATA was formed in 1947, TELARC in 1972, the Danish National Testing Board in 1973, and NVLAP in 1976. Since that time there has been a remarkable upsurge of interest in national comprehensive laboratory accreditation programs. This interest has been shown in countries ranging in size from the United Kingdom and France to Trinidad and Tobago and Jamaica. Many other countries have also shown interest in developing separate national programs and this interest will be discussed in a later section.

In this section the emerging programs of France, Jamaica, and the United Kingdom are outlined.

France

The Reseau National d'Essais (RNE) (National Network for Testing) was established in December 1979 and the detailed criteria and administration of the system are currently under development.

It is intended that RNE operate a fully national comprehensive accreditation program which will involve government and private testing laboratories operating in all fields of testing except for calibration activities which are under the control of a similar separate organization which specializes in that area of activity.

The program as it is currently envisaged will have some organizational differences from the existing national systems but in essence it seems that the criteria and procedures will be in harmony with existing systems.

Jamaica

Standards in Jamaica are the responsibility of the Jamaica Bureau of Standards which was established in 1968. JBS is currently establishing a laboratory accreditation program which it is anticipated will be operational in 1981. Again the proposed criteria and procedures are similar to other national systems that operate in much larger and more industrially developed countries. The way in which a small country like Jamaica will cope with gaps in basic calibration facilities and lack of an adequate pool of expert assessors within its own national boundaries will be interesting and will really determine the viability of its accreditation system. It is an ambitious program which is designed to provide this small nation with a network of testing laboratories which have demonstrated capability to meet national and international needs and commitments.

United Kingdom

The British Calibration Service was established in 1966 and it was not until June 1980 that the National Testing Laboratory Accreditation Scheme (NATLAS) was founded to complement the existing limited scheme. Both BCS and NATLAS are operated by agencies of the Department of Industry through the National Physical Laboratory (NPL).

There are a number of special purpose laboratory accreditation systems existing in the United Kingdom and it is anticipated that NATLAS will absorb some if not all of these.

NATLAS is proposing to establish criteria for, and accredit laboratories in, all conventional fields of testing except biological testing at this stage. It may extend into other areas as time goes on.

31.3.6 International Laboratory Accreditation Conference (ILAC)

By the mid 1970's a number of isolated but interrelated technical activities were emerging:

(a) Plans were under way to establish international certification arrangements particularly in relation to electronic components, but also for other materials and products.
(b) The General Agreement on Tariffs and Trade (GATT) was well along the way to developing codes of conduct for international trade, particularly in relation to developing a treaty which would restrict the de-

velopment of technical barriers to trade through standards, testing, or certification.

(c)　Organization for Economic Cooperation and Development (OECD) began developing a code of good laboratory practice for the testing of hazardous chemicals.

(d)　A number of international organizations, such as ISO and the United Nations Economic Commission for Europe (UN/ECE) were interested in the need for acceptance of test results from various countries.

All these activities and developments had something in common—the need for some form of recognition or accreditation of testing laboratories in various countries throughout the world (Gilmour, 1981).

The GATT standards code for instance, imposes an obligation on signatories to the code to accept tests done in foreign countries. Obviously the only way in which this can work is if the receivers of test data have some assurance that the laboratories in the foreign country are competent. It is unreasonable to expect acceptance of test data from just any laboratory, foreign or local. One avenue for obtaining this necessary assurance of competence is the utilization of a laboratory accreditation system in the foreign country.

Most of these developments referred to above relate to international acceptance of test results for trading purposes and as a result of these influences a conference was convened in Copenhagen in October 1977 of representatives of known national laboratory accreditation systems and of others who it was thought would possibly be interested in the subject of laboratory accreditation.

This first conference, now known as ILAC 77, was attended by 18 national delegations (ILAC, 1977) which included countries as diverse as Australia, Brazil, Denmark, France, Israel, Japan, and the United States, as well as a number of international organizations such as the UN/ECE, ISO, and International Union of Testing and Research Laboratories for Materials and Structures (RILEM).

The conference was really a meeting at which various national delegations presented statements of activities in their own countries, but it did lead to the appointment of three task forces which were given the task of studying:

(a)　Legal problems associated with international acceptance of test results.

(b)　Publication of a directory of laboratory accreditation systems.

(c)　With ISO, the development of criteria for accreditation of laboratories.

The conferences have continued each year since 1977 with growing attendances, (ILAC, 1978, 1979, 1980). Conferences have attracted delegates from Eastern Europe as well as from the central secretariats of the European Economic Community (EEC) and GATT. The central, but not exclusive, theme of the conferences is facilitating acceptance of test results internationally and while considerable progress has already been made, widespread international arrangements are some years away. An international directory of testing arrangements and testing laboratory accreditation systems has been published (ILAC, 1981).

What is emerging, therefore, are national measurement and testing systems based on hierarchies of laboratories with demonstrated *traceability* of measurement from accredited testing laboratories back through accredited calibration services to national standards laboratories and hence to international standards of measurement. The formal and informal relationships between international standards and national standards laboratories has existed for many years. The formal second and third echelons of laboratories are only now being developed through the accreditation process.

REFERENCES

Gilmour, J. A. (1978). 'Laboratory accreditation in Australia and New Zealand', *NCSL Newsletter*, **19**, No. 1, 15–22.
Gilmour, J. A. (1981). 'Laboratory accreditation—international progress', *American Society for Quality Control, Annual Conference, Milwaukee.*
ILAC (1977). *Proc. Int. Conf. on Recognition of National Programs for Testing Laboratories*, Danish National Testing Board, Copenhagen.
ILAC (1978). *Proc. Int. Laboratory Accreditation Conf. 1978*, US Department of Commerce, Washington, DC.
ILAC (1979). *Proc. Int. Laboratory Accreditation Conf. 1979*, National Association of Testing Authorities, Sydney.
ILAC (1980). *Proc. Int. Laboratory Accreditation Conf. 1980*, Laboratoire National d'Essais, Paris.
Thurnell, D. P. (1979). 'The British Calibration Service', *NCSL Newsletter*, **19**, No. 1, 23–4.

Handbook of Measurement Science, Volume 2
Edited by P. H. Sydenham
© 1983 John Wiley & Sons Ltd

Chapter

32 P. H. SYDENHAM

Sources of Information on Measurement

Editorial introduction

The fragmented nature of the literature reporting measurement knowledge and the diversity of measuring instrumentation application makes retrieval of literature very difficult and often inefficient.

This chapter provides means to enter the literature along organized lines, doing this with an understanding of the problems, lists of terms, book classification code numbers, relevant journals, and sources of relevant bibliographies. It also provides an introduction to the use of computerized searching methodology.

32.1 INTRODUCTORY FOUNDATION

32.1.1 The Need for Information

Each new encounter with a need for measurement will more than likely present new facets not previously encountered or the need for detail on a technique of which one has less than adequate experience.

Selection, design, adoption or modification of already reported methodology may be the need. This will require the user to become familiar with the specific circumstance in order to be part of the gradually maturing design process through which specific decisions emerge out of the wide generality at first available. Sound final decisions are based on the availability of reliable and relevant information.

Due to the fragmented and widely applied nature of the basics of measurement systems, as will be discussed later in this chapter, arriving at a final set of decisions about hardware often requires considerably more thought and study than might at first be considered necessary. Too often practical action concerning instrumentation is begun, and irrevocable commitment made, before a sound basis for proceeding has been formulated.

This particularly occurs when instrument users are more interested in obtaining results about a subject than they are in the all-important processes by which they obtain them.

Many kinds of questions can arise

(a) Does a suitable measurement system exist; has it been reported?
(b) Where can a suitable system be purchased?
(c) Has a given principle been applied to the application of interest?
(d) What performance characteristics are reasonable to expect?
(e) How is a given measurement system calibrated and standardized?
(f) How can a given kind of instrument be made more cheaply?
(g) How novel is the measurement system that a group has just developed?

There already exists, in print, a vast quantity of information on all aspects of measurement systems—this Handbook is a collection of part of such information.

The practical reality is that there is no straightforward and simple procedure whereby questions can be answered easily and cheaply.

This chapter is provided to assist efficient location of appropriate information. It will also assist authors to decide where best, and in what form, to publish their contributions so that others can retrieve them with the least difficulty. An abridged version is available in Sydenham (1982). The booklet by Melton (1978) is also relevant.

32.1.2 Nature of the Information in the Literature

Study of the dominant features of the discipline of measurement systems provides broad insight into the kinds of problems that exist in this particular discipline. (The following sections provide a bases for seeking out the diffused information.)

Measurement systems are widely used. They are to be found in use in just about all fields of endeavour. Practitioners in each field are not generally aware that they are often using the same basic fundamental approaches and principles as are also in use in other fields. Consequently they often describe existing methodology and hardware in different linguistic terms. The result is that the same information is available in many different guises. Figure 32.1 is a diagrammatic representation of this feature.

The number of possible applications of measurement fundamentals far exceeds the principles involved. To date, however, the information is more usually considered in respect to its application, not its underlying principles. This gives the appearance that there exists a vastly complex set of rules and methods, each of which does not readily fit the framework of a discipline. This Handbook aims to standardize the set of fundamentals into recognizable subsets.

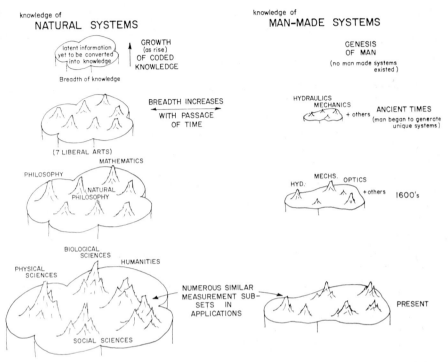

Figure 32.1   Diagrammatic representation of the place of measurement system information in the total body of knowledge (From Sydenham (1979) reproduced by permission of Peter Peregrinus Ltd)

In use, most measurement systems are a *tool* in a much larger system. For example, a thermocouple thermometer is a small part of a whole blast furnace. In respect of capital cost, energy consumption, materials deployment and indeed, almost every respect, a measurement system is only a small feature of a much larger whole. (In the process industry measurement equipment costs 1–10% of the total.) Their value, of course, is in providing vital information that is used to control the whole sytem.

The capabilities, relevant experience, and training of many measurement system practitioners is generally not specifically, or highly developed, in respect of measurement fundamentals. Because of this the information reported can be of low quality and veiled in different terminology than that to which a reader is accustomed. A suspicious attitude is worth cultivating when reading reports on instrumentation. It is considerably more difficult to design and commision apparatus, that is reliable and reproducible over its life, than it is to report it!

A most relevant feature, relating to the literature of measurement systems, is the lack of a widely established, and universally appreciated, teaching and training discipline. The courses that do exist are, with few exceptions, too short to produce an expert in measurement systems. The result is that there are no established places in the information storage system of the world to report information, nor an accepted standardized jargon for describing it. This in turn, does little to engender classification of the fundamentals; emergence of a worthwhile universal training discipline is yet to become reality.

32.2 SOURCES OF INFORMATION

Information is published in many forms. Each has its particular kind of value.

It begins with people who convert the latent information about existence into a recordable form that suits the storage classification then in vogue. Figure 32.2 depicts how information passes around a circuit in which that useful to measurement finds its way back to be the basis of the design of more measuring instruments that, in turn, aid the flow of more information (Sydenham, 1979).

Figure 32.3 provides an appreciation of the flow of information though storage systems. Scholarship and experimental observation lead to latent information being made available: the first stage is termed primary information. This may occur as internal reports within organizations, in the personal records of workers, or in the published literature.

Within a few months of being established (weeks in some cases) primary information is published in the primary journals where it is openly available through library services. Material published in primary journals is usually

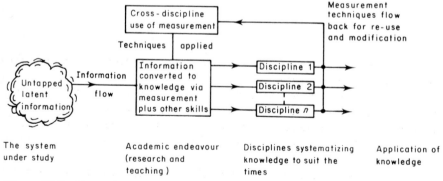

Figure 32.2   Extraction of latent information leads to new measuring instruments. (From Sydenham (1979) reproduced by permission of Peter Peregrinus Ltd)

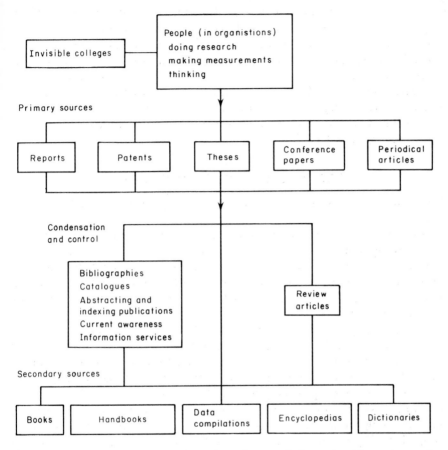

Figure 32.3 Flow of information through recorded, retrievable systems of documentation. (From Melton (1978) reproduced by permission of The Institute of Physics)

subject to expert refereeing that helps maintain but cannot guarantee, the quality of reported material.

Primary information is also available in student theses (dissertations), patent specifications, and in some conference proceedings.

There are no absolute rules to regulate what is published where. Primary information often also appears in part in the next class, the secondary outlets.

Primary information is, for a variety of reasons, not always presented in the form that is required. It may need to be collected together to form a work on a common topic, or be re-expressed in a more acceptable form for others to make use of it. In this class are secondary journals, reviews, books, encyclopedias, dictionaries, video-tapes, audio-tapes, and films.

A major kind of published secondary source is that comprising the reference material that aids retrieval of primary material—bibliographies, abstracts, catalogues, information services. (Section 32.3 discusses information retrieval.)

Reviews and bibliographies are formed from secondary information but their construction, classification, and presentation can present a primary viewpoint often being original in nature.

The rate at which information is published is almost incomprehensible. By 1900 the Royal Society in London had to abandon its attempts to catalogue all of the scientific publications from 1800 to 1900—the Author Index alone was 19 volumes in length! In 1974 there were being published an estimated 80,000 regular scientific and technical journals in the total of 150,000 of all kinds. Abstracting periodicals then numbered 1500, in a total of 3500. An estimate of that time suggested there were 300,000 books published in a year with over 3000 million books in print.

The diverse nature of measurement means that information needed could rest almost anywhere in the vast system of documentation. Well devised retrieval is essential to glean information in an efficient manner.

32.3 SEARCHING FOR MATERIAL

32.3.1 Use of Standard Library Catalogue and Reference Facilities

Figure 32.4 is a flowchart showing how to retrieve information from a library. The first step is to clarify the information sought. General reading around the subject helps: encyclopedias, technical dictionaries and subject reviews in books, articles and in trade literature can help define, more exactly, what might be the key words or library classification codes associated with the topic. Later sections are provided in this chapter to assist in this regard.

Refinement of the originally conceived need continues as a search progresses because the act of searching raises the knowledge level of the searcher.

Consider first that a book is sought. With a background defined it is then time to consult the library catalogues to establish which specific books, or which classification location might be worth consulting. Usually Library catalogues, prior to some date in the 1970–80 period, will be in card form. Later acquisitions are generally on microfiche. Both will usually need to be consulted. In either case they contain a subject, a title, by author order, index. Subject indexes tend to be more general than the specific topics of

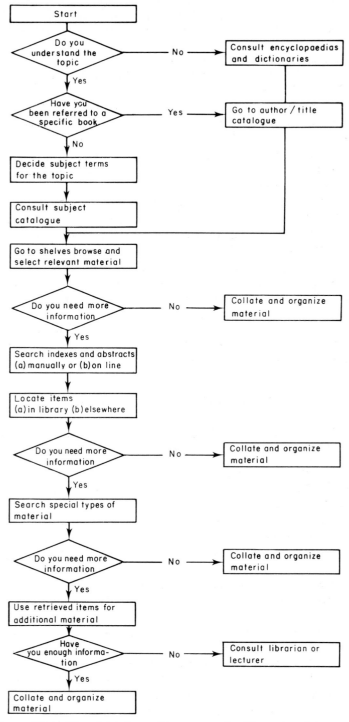

Figure 32.4 Flowchart for a library search. (Reproduced by permission of M. Thompson)

measurements would often require for direct indexing in this way. Section 32.4 shows (see Table 32.5), in the selected list of Dewey Decimal Classification codes, that material can lie in many locations. There is generally no single location in which measurement books of all kinds will be solely located as, say, are books on graph theory in mathematics. Where a title and author is known, the search is straightforward.

If a search of the library catalogue does not yield a satisfactory title it is then usual to use other information sources to locate titles which can then be requested by the interlibrary loan (ILL) service.

Some sources for locating titles are the general material mentioned above, the several forms of books in print volumes, and use of catalogues of library holdings for major libraries (such as microfiche available for purchase from the British Library on the holdings of the Science Reference Library). The above sources are often not obviously available in a library, in which case it may be necessary to request librarian assistance.

Locating book titles on measurement topics—due to the fragmented nature of the practice of measurement—is particularly difficult. The bibliography (Sydenham, 1980a), described in more detail in Section 33.6, is a prime source for this topic. The chapters of this Handbook will also provide an entry into the appropriate literature.

Much of the information relevant to instrument design retains its pertinence with time. Unfortunately many technical libraries, due to pressure of new acquisitions and lack of space, have to adopt the policy of discarding or storing material predating around the 1950 point. Material prior to that date is consequently hard to procure or learn about. The extensive bibliographies contained in Sydenham (1980b) and Sydenham (1979) cover mechanical design of instruments and historical development of measurement systems, respectively. Entry into the literature of measurement in the empirical sciences is assisted by consulting the bibliography contained in Van Brakel (1977) and the longer term bibliography project of that author. The collected papers (Jones, 1982) also provide succinct reviews of many aspects of relevance. Library holdings also contain bibliographies on broad subjects. Some of these are listed in Melton (1978).

The location of literature published in periodical form requires a different approach to that used for books. For this form the identification of material can be much more detailed but the stock to retrieve it from will be greater in quantity of articles. The information will be more up-to-date (that is, at the time of publication).

The usual path here is again to first decide the subject which then leads to use of the suitable abstract and index periodicals. Abstract journals contain an abstract describing the content of an article; indexing journals do not.

These sources can lead to primary and secondary material of all forms. They sometimes only lead to a precis of a paper presented at a conference which was never published in full.

The process of tracing references in this way requires careful documentation of each step. It will gradually expand the searcher's knowledge of the topic but will also often lead to many disappointing papers, in that they are not as relevant as at first thought.

The above procedures lead to the full bibliographic references for many articles. These articles must then be obtained in their full versions from the library system. The library serials catalogue will lead to the location of the appropriate journals, within the sequence of which the year, volume, perhaps number, and page numbers define the article's position.

Having located useful articles it is important to record the full bibliography reference ready for possible later citation: i.e. author(s), date, title, journal name, volume, number, page numbers.

If the serial is not held locally the article can usually be obtained through the ILL network. This, however, does require time, ranging from a few days to several weeks.

A comparatively recent publishing innovation was the publication of the *Science Citation Index*. In this reference work it is possible, knowing an author on a topic, to quickly establish who has cited that author's publications in subsequent works (and the paper citing it). This can lead to other work in the same field. Articles, books, and the like that are cited by a known author are also traceable in this way.

Abstracting, index-journals, and library microfiche indexes take time to compile, print, and publish. To keep right up to date there exist several sources of airmail, weekly release, current awareness publications. These describe articles just in print, or even about to be printed. The most relevant to measurement are *Current Contents* and *Current Papers*.

As journals are received libraries often put them on display for a short period before transferring them into the main holdings locations. Periodic browsing of these is also an effective way to keep up-to-date.

The abstracting networks of information will enable location of a vast quantity of literature but there is much they do not pick up.

Theses (dissertations), patent documents, conference proceedings, standards of specification, and the many government publications, each have their own means of identifying locations of material. These must be searched individually as no overall system exists as yet (and possibly never will—each minute a new standard of specification is published in the world, patents flow continually, and so on).

32.3.2 Personal Experience

Despite the vast resources of librarianship available to the individual, it is often the case that the expert on a topic can be more helpful in information gathering. This is because individuals collect information on specific topics, doing this from numerous forms of source—library material, manufacturers' catalogues, reprints from colleagues, handbooks, personal notes, and so on.

To make effective use of such people it is necessary to present a well devised case which will assist them to rapidly produce a workable interface with the enquirer.

32.3.3   Computer Aided Information Searching

Abstracting and index periodicals moved to the use of computer storage, sorting, and printing control from the late 1960's onward. It was a logical step to then also offer a service in which computer programs could search in the stored abstracting material for material, printing out the details of the bibliographic references.

The 1970's decade saw immense growth in this form of service. At first the bases were used to produce lists according to a predetermined search profile, the read-out being supplied at a later time by the off-line procedure.

The need to adjust the search strategy as the user learned more about the subject led to on-line methodology. In this a quite simple search profile is predetermined (usually with help of a trained on-line operator) using the documentation relevant to the data base chosen to be searched. The operator, in close collaboration with the searcher, then goes on-line adjusting the input requests to suit the situation.

The steps of a typical on-line search are as follows:

(a)   Establish about five keywords using the relevant thesaurus and the specifications of the problem.

(b)   Decide how the words might be combined in the *and* and *or* manner. For example the terms might be micro-electronics, management, environment, measurement, review. Management by itself would be listed in the abstract of a vast number of articles (10,000's) but combined as 'management *and* micro-electronics' (that is, both words appearing in the abstract) reduces the number considerably. Still further, incorporation of *and* environment will eliminate management of a micro-electronics activity.

(c)   Go on-line (which only takes minutes at the most), entering identification code as required. Await instructions about choice of base.

(d)   Choose base required.

(e)   Key in all terms and await reply about number of entries existing with at least one of those terms in the abstract. Usually it will be in the 10,000's and the return is almost immediate, in which case more selective searching is needed.

(f)   Decide combinations to use to reduce number of titles.

(g)   When the numbers seem reasonable to study (less than 100, as a guide), call-up brief details, such as titles only. This will provide appreciation of how well the selection is working.

(h) Of these titles select those worthy of calling up abstracts and either print abstracts whilst on-line (this can become an expense) or instruct the system to print out the list at the host computer and mail the hardcopy, which may delay return by several days.

The above list can only be a guide. Considerable variety exists for each base and in the methodology possible. Direct experience is necessary to become efficient in the use of on-line systems.

Major internationally available bases are located in the USA, UK, and the European continent. Gradually various other countries are supplying bases to the world network. They are accessed from other countries by satellite communication links.

Computer data bases are of limited value in locating book literature. They are now also tending to gradually take in more of the non-journal literature. They do not contain full text. Except for GIDEP (see Section 32.5) none is specifically concerned with measurement and instrumentation as a major facet. Many may need to be used for a thorough search on an instrument topic.

The scenario envisaged for EDP based systems is that a user can *couple* (with a small terminal unit) to any international telephone line via the telephone handset and go on-line at any time of the day, in any geographical location. This concept is yet to find full acceptance, a stumbling block being that each base has its own search requirements which are outlined in quite extensive guides to their use. Training is needed in their use if costs of operation are to be kept within reasonable limits.

32.4 TERMINOLOGY AND CLASSIFICATION

32.4.1 Terminology

Information retrieval by browsing through original material makes use of many recognition features—words, illustrations, mathematical formulae—but is invariably too slow for searching the bulk of the literature.

Realistic retrieval procedures are almost exclusively based on the principle of tracing linguistic terms used by authors in their titles, abstracts, and index keys. Consequently the choice of terms written into abstracts and used in searching is of paramount importance.

The diversity and uncoordinated multiplicity of the sources of generation (and regeneration!) of instrument concepts has brought with it an highly varied and inconsistent nomenclature. It is only in recent times that standardization of terms has begun to be accepted; they are not, as yet, widely used nor broadly developed.

Chapter 3 (of Volume 1) addresses the problems of terminology with Table 3.1 listing relevant standards documents on nomenclature of metrology.

The wide and often incorrect use of synonymous terms makes literature retrieval difficult. Apparently obvious terms often lead nowhere— metrology, measurement, and instrumentation in particular. The name instrument-science, for example, might be indexed using at least ten alternatives such as, measurement systems, measurement physics or measurement-science.

Terms describing the process of measurement exist in great abundance. Table 32.1 lists many of these.

Measurement, in specific disciplines often has its own name. Table 32.2 lists some of these.

The fundamental term metrology also has at least two different meanings.

Use of data bases, in particular, requires careful choice of terms in the search profile. In many only terms given in a standard thesaurus are allowable inputs. Table 32.3 lists a selection of the approximately 650 (of a

Table 32.1   Terms describing the process of measurement

accounting	effect	prospection
accredit	evaluation	quality
adjustment	expectation	rate making
administer	experiment	recognize
a measure of	exploration	resolution
analyse	features	sensing
appraisal	group	standardize
appreciation	incidence	style
arbitrate	indication	taxonomy
assay	influence	tests
assessment	information	value
choice	inspection	weighing up
classification	interpretation	worth
classify	judge	
composition of	logistic	
criticize	management	
criticism	material constants	
decide	measurement	
decision	mensuration	
design	metricate	
detection	monitoring	
determining	number	
diagnosis	performance	
distribution	precision	
document	properties	

Table 32.2   Some names of disciplines involved with measurement

Accounting	Measurement systems engineering
Anthropometrics	Measurement technology
Archaeometry	Mensuration
Assay	Metallography
Automatic testing	Metrology
Biometry	Nuclear engineering
Cybernetics	Pathology
Decision theory	Posology
Diagnostics	Psychrometrics
Econometrics	Quality control
Epidemiology	Quantity surveying
Ergonomics	Radiometry
Faunistics	Scientific instrumentation
Geophysics	Seismology
Instrument physics	Sensitometry
Instrument-science	Sensory physiology
Instrument technology	Stoichiometry
Management	Symptomatology
Man–machine-systems	Taxonomy
Measurement engineering	Teleology
Measurement physics	Time and motion study
Measurement-science	Value engineering

total 9000) terms that are contained in the thesaurus of the data base INSPEC.

32.4.2   Classification Location

Author's and reader's classification

Consider first, classification from the author's and reader's viewpoint. It is found that numerous possibilities exist for deciding the key list of index terms.

In general authors contributing to measurement align more with their field of application or discipline than with sources publishing fundamentals of instrument-science. Often they see the instrument content as insignificant and not worth abstracting. Many facets of a report might be chosen as keys. The reality is that the report, although given several descriptors can only be published in one journal. As an example consider an hypothetical report containing, as part only the

calibration of a novel, laser-based, alignment system used, by the XYZ Company, to test the military load-carrying capacity of an historic concrete bridge whilst the bridge is conveying a mobile nuclear reactor.

Table 32.3   Some INSPEC terms related to measurement activity (Reproduced by permission of The Institution of Electrical Engineers).

Approximation theory	Non-electric sensing devices
Artificial intelligence	Nuclear instrumentation
Automatic testing	Oceanographic techniques
Character recognition	Optical systems design
Characteristics measurement	Optical instruments
Computerized instrumentation	Parameter estimation
Cybernetics	Patient diagnostics
Data acquisition	Performance index
Data processing	Physical instrumentation control
Data reduction and analysis	Physiological models
Digital instrumentation	Predictor-corrector methods
Display instrumentation	Probes
Dynamic testing	Production testing
Education	Program testing
Electric variables measurement	Quality control
Environmental testing	Quantization
Error analysis	Radiation monitoring
Error detection	Recorders
Fundamental law tests	Relay protection
Geophysical equipment	Research and development management
Height measurement	Reviews (see also published lists
History	of bibliographies with author
Impulse testing	indexes)
Information science	Robots
Instrumentation	Seismology
Instruments (53 kinds fall into this)	Self-organising systems
Laboratory apparatus and techniques	Sensitivity analysis
Logic testing	Sensory aids
Machine testing	Signal processing
Magnetic variable measurement	Social and behavioural sciences
Man-machine systems	Spatial variables measurement
Materials testing	Standardization
Measurement (29 specific variables	Student laboratory apparatus
are listed: all appear elsewhere).	Switchgear testing
This is is also used instead of	Telemetering systems
term metrology	Testing
Measurement standards	Thesauri
Measurement systems	Transducers
Measurement theory	Units (measurement)
Mechanical testing	Value engineering
Medical diagnostic equipment	Water pollution and detection
Meteorological instruments	
Noise measurement	
Nomenclature and symbols	

Table 32.4 Classes and example journals in which an instrument might be reported

Example
Calibration of a novel, laser-based, alignment system used, by the XYZ Company, to test the military load-carrying capacity of an historic concrete bridge whilst the bridge is conveying a mobile nuclear reactor.

Class basis chosen	One example of typical journal
Physical principle used (interferometer)	*Applied Optics*
Discipline based on a device used (laser)	*Laser Focus*
Contemporary nature (novel and timely)	*New Scientist*
Military relevance	Confidential report of military organization
Discipline of use (surveying)	*Photogrammetric Engineering*
Parameter of measurement (measurand is related to strain)	*Journal of Strain Analysis*
Measurement principle used (angle measuring interferometer)	*Journal of Physics E: Scientific Instruments*
Constructional material used (concrete)	*Magazine of Concrete Research*
Standardization (calibration)	*Newsletter of National Conference of Standards Laboratories*
Design implications	*Journal of Elasticity*
Testing	*International Journal of Non-destructive Testing*
Instrumentation detail	*Instrumentation and Control Systems*
Engineering heritage	*Transactions of the Newcomen Society*
Trade journal	*XYZ Affairs* (hypothetical name)

Plus other aspects such as nuclear systems, transportation, power generation, environmental issues, civil engineering and many more.

Table 32.4 gives many likely key features chosen to classify the report. The report is only to be published in one journal—which one would you choose? If the instrument content is minor it may not be abstracted as a contribution to instrument-science at all! Authors should give attention to the need to prepare several papers on a major topic, each on the key aspects.

Library classification

Now consider classification from the viewpoint of library locations. The location of book literature is identified in a library by the classification code number—a numbering method. Retrieval is, therefore, (when a title or author is not known) based on establishing suitable code numbers, not by use of a set of key terms as is the case in computer retrieval of journal literature abstract information.

Several different schemes of library locational classification are in use. The predominant systems are:

(a)   Dewey Decimal Classification;
(b)   Library of Congress (LC);
(c)   Universal Decimal Classification (UDC).

Other schemes are used—local subject code allotment and even author order only in some cases. (Subject retrieval is virtually impossible in this last case.)

Each of the established systems has its prime catalogue—such as the *Dewey Decimal Classification and Relative Index* (the 19th edition is the most recent). The UDC system is fundamentally supported by an extensive set of British Standards Institution standards documents which enable the greatest possible detail in subject identification. In practice, however, some libraries using the UDC system make use of a much abridged version of this. UDC and Dewey are somewhat alike but not identical. The LC system is quite different in code allocations. Equivalence between LC and Dewey codes is published in a conversion book.

The diverse and ubiquitous nature of measurement often makes classification of such material a difficult decision.

At the point of publication the publisher may assign a code number to a book to make it ready for shelving. Alternatively, the local librarian (who seldom specializes in instrumentation) decides where a book, or periodical, should be placed. It is possible for a book to be placed in one of many alternatives—it is not unheard of for a second edition of a book to be located differently to the first!

The lack of structure of the knowledge of measurement-science is reflected in library holdings: material will be found diffused throughout many classes. An impression of the situation is obtained by study of the selected list of subjects and code numbers given in Table 32.5. (The Dewey Classification contains over 25,000 named code numbers of which some 650 are instrument related.) For most effective searching the full classification would need consulting—a matter of many days of work! Libraries generally maintain an alphabetical subject index but this will seldom provide the detail sought. The wide choice of instrument topics requires searching based on detailed study of the handbook of the classification system.

It is appropriate here to explain the development of the Dewey Decimal System for this has a strong bearing on current day problems in its use.

The Dewey system was devised in the late 19th century. Its construction

Table 32.5   Sample list from the likely 650 Dewey Decimal codes that relate to measurement: for full information consult Dewey Decimal Classification. (Reproduced from *Dewey Decimal Classification*, Edition 19 (1979) by permisssion of Forest Press Division, Lake Placid Education Foundation, owner of copyright)

Dewey Decimal code	Classification volume entry
001.422	Statistical method (formerly 311.2)
001.51	Communication theory
001.53	Cybernetics
001.533	Self-organizing systems
001.6	Data processing
003	Systems
016	Bibliographies and catalogs of specific disciplines and subjects. *Note:* Other codes used such as ---.0 of notation
119	Number and quantity
120	Knowledge, cause, purpose, man
121	Epistemology
152	Physiological and experimental psychology
169	Analogy
330	Economics
339.3	Measures of national income
364.12	Criminal investigation (Detection)
368.011	Rates and rate making (Insurance)
371.26	Educational tests and measurements
378.16	Educational measurement and student placement
384.1	Telegraphy
384.9	Visual signaling
389	Metrology and standardization
501	Philosophy and theory
502.8	Techniques, apparatus, equipment materials
511.43	Error theory
512.7	Number theory
515.42	Theory of measure and integration
519.52	Theory of sampling
519.86	Quality control and other statistical adjustments
522.2–522.5	Instruments (astronomy)
526.3	Geodetic surveying
529.75	Time systems and standards
530.16	Measurement theory
530.7	Instrumentation (for measurement, control, recording)
530.8	Physical units, dimensions, constants
531.382	Elastic constants and their measurement
532.053	Flow properties and their measurement
534.4	Measurements, analysis and synthesis of sound
535.33	Optical instruments (see also specific functions class) Microscopes formerly 578.1

Table 32.5 (cont'd)

Dewey Decimal Code	Classification volume entry
536.5	Temperature, (536.51–536.54 Measurement)
539.77	Detection and measurement of particles and radioactivity
542.3	Measuring apparatus (laboratory-chemistry)
543.07	Instrumentation (chemistry)
544	Qualitative analysis
545	Quantitative analysis
548.1	Geometrical crystallography
551.4	Geomorphology (.4607 is Deep sea surveys and exploration)
573.6	Anthropometry
574.028	(no key title given but biological instrumentation here see 610.28)
578.4	Use of microscopes
591.18	Movements senses, control mechanisms (Zoology)
610.28	Medical instrumentation (see also 574.028)
611.8	Nervous system and sense organs
612.8	Neurophysiology and sensory physiology (extensive subgrouping)
614.7	Environmental sanitation and comfort (614.71–614.77 cover pollution but no specific class for measurements)
615.19018	Assay methods (pharmacy)
616.07	Pathology (includes diagnoses and prognoses)
617.05	Surgery by instrument and technique
617.752	Optical work
617.89	Audiology
617.9178	Surgical instruments, apparatus, material
620.0042	Engineering analysis, synthesis, design
620.0044	Testing and measurement (engineering)
620.1127	Non-destructive testing of materials (subgroups on radiographic, tracer, ultrasonic, magnetic methods)
620.32	Measurements (of mechanical vibration)
620.72	Systems analysis, synthesis, design
621.313	Generating machinery and converters (includes maintenance and testing)
621.37	Electrical testing and measurement
621.372	Units and standards of measurement
621.373	Recording meters
621.374	Measurement of electric quantities (various subgroups exist)
621.379	Measurement of non-electrical quantities
621.381043	Measurements (electronic)
621.38137	Testing measurements standardization (Classifier instructed to place measurements in group for application)

Table 32.5 (cont'd)

Dewey Decimal Code	Classification volume entry
621.381548	Testing and measuring devices and their use (Electronic)
621.383	Specific instruments and apparatus of wire telegraphy
621.38417	Measurements and standardization (radio)
621.3843	Specific instruments and apparatus of radio telegraphy
621.3887	Measurements and standardization (television)
621.3895	Underwater devices (subgroups include hydrophics and Sonar)
621.4831	Reactor physics (includes testing of physical phenomena occurring within reactors)
621.4835	Operation, control, safety measures
621.756	Inspection technology (in factory operations engineering)
621.902	Machine tools (including numerical control and other automation techniques)
621.994	Measuring tools
622.15	Geophysical exploration (several subgroups)
622.8	Mine health and safety
623.71	Intelligence and reconnaissance topography (military engineering)
623.819	Design tests (nautical)
623.863	Nautical instruments
624.1513	Soil mechanics (includes tests)
624.17720– 624.1779	Specific structural elements (includes strength tests)
625.794	Traffic control equipment
627.81	General principles (Dams and reservoirs) (includes surveying)
628.177	Measurement of consumption (sanitary engineering)
629.045	Navigation (celestial in 527)
629.1345– 629.1346	Aircraft tests, measurements, standards, maintenance, repair (see subgroups)
629.135	Aircraft instrumentation and systems (many subgroups)
629.273	Panel instrumentation (vehicles)
629.4775	Control of temperature, humidity, air supply and pressure
634.9285	Mensuration (forestry)
637.127	Quality and purity determinations (cows milk)
640.73	Consumer education (includes guides to quality and value of products and services)
657.48	Analytical (Financial accounting) (Measurement of profitability, of financial strength, of income, of liquidity, of flow of funds)

Table 32.5 (cont'd)

Dewey Decimal Code	Classification volume entry
658.28	Equipment for safety and comfort (includes noise control)
658.516	Standardization (equipment, procedures, in general management)
658.562	Standards and specifications (form, size, dimensions, quality, materials, performance, including standardization formerly in 658.16)
658.568	Inspection (includes statistical methods of quality control)
658.834	Consumer research (studies of consumer preferences, attitudes, motivations, behaviour)
660.283	Process equipment (chemical instruments, apparatus, machinery)
662.622	Properties, tests, analysis (of coal)
664.07	Tests, analysis, quality controls (food— for texture, taste, odour, colour, contaminants)
665.0288	Tests analyses, quality controls (oils, fats waxes, greases)
666.13	Tests, analyses, quality controls (ceramics)
669.95	Metallography
674.13	Properties and their tests (lumber)
675.29	Properties, tests, quality controls (leather, fur)
676.121	Properties, tests, quality controls (Pulp)
676.27	Properties, tests, quality controls (paper and paper products)
681	Precision instruments and other devices (several groups)
697.9315	Psychrometrics
697.9322	Temperature controls
771.37	Focusing and exposure apparatus
774	Holography
781.91	Musical instruments
788.971	Mechanical and electrical reproducers
913.0310285	Interpretation of remains (includes dating, use of data processing in ancient studies)

did not cater for the subsequent explosion of technical knowledge. For this reason subjects that often began as a quite reasonable, and small, subset of knowledge have had to be allocated a relatively extended code length (deep into the system). For this reason already long code numbers had to be greatly extended to cope with, what are now, major subsets of knowledge. For example, the 620–629 series was devoted to engineering and allied operations requiring electrical measurements (621.381043). This contrasts with a 19th century awareness to time measurement that allocated that topic the code 529.78. Thus length of the code number is not an indicator of depth of specialism.

Many aspects of modern measurement are not placed, by the Dewey Decimal Classification, where they could now be expected to be in a library. For example telegraphy (384.1) is placed in the Social Science 300's group. Holography (774) appears in the Arts 700's group along with games theory at 795.01.

In some, but not all, groups there is often a specific code for relevant instrumentation for that application. An example is flow properties and their measurement (532.053). This does not appear to be a standard practice— many principles or application areas do not possess such a measurement aspect code.

Table 32.6 Some journals relevant to measurement science

Applied Optics
Australian Journal of Instrumentation and Control
British Communications and Electronics
Control
Control Engineering
Electronic Engineering
Engineers Digest
Instrument Practice
ISA Journal
Journal of Applied Measurements
Journal of Optical Society of America
Journal of Physics E: Scientific Instruments
Laser Focus
Measurement and Control
Metrologia
Microtechnic
Nature
NCSL Newsletter
New Scientist
Nuclear Instruments and Physics
Optica Acta
Opto-electronics
Physics Today
Proc. IEEE-Control Section
Review of Scientific Instrumentation
Science
Sensor Review
Sensors and Actuators
Strain
Trans, IEEE, Instrumentation and Measurement (*Vol. IM*)
Trans. IEEE, Ultrasonic (*Vol. U*)

Occasionally, to add yet more difficulty, a new volume of the index is published. This requires some changes to codes (deletion and placement elswhere, new code numbers) but usually this produces only minor changes in an established libraries positionings as it generally only affects books catalogued after that event.

The problems of using the catalogue to locate books require library users to also use other methods of location. If an author, a title, and a publisher can be established, retrieval is then greatly simplified. The IMEKO bibliography—see Section 32.6—is a key aid. Table 32.6 lists journals of major relevance to measurement.

32.5 INFORMATION DATA BASES OF RELEVANCE TO MEASUREMENT

This subsection provides detail of the computer based information retrieval systems, generalized use of which is described in Section 32.3.3.

As with text literature, journal sources on measurement are also spread widely throughout the whole of the periodicals holdings. For this reason no one base can be recommended as entirely adequate. The following bases are listed in order of their likely usefulness (as seen by the author). The first is covered in some detail, as an example.

INSPEC is a base covering (since 1969) physics, electrical and electronic engineering, computers and control. It is operated by the Institution of Electrical Engineers (London). It selects (Figure 32.5) abstract detail from the entire contents of about 500 journals with partial selection from another 2300 journals. INSPEC publishes a regular list of the journals abstracted. It also publishes an annual *INSPEC Thesaurus* of terms. This contains over 9000 terms of which at least 650 are relevant to a search for an instrument-related subject. Table 32.3 lists a sample of these. As can be seen they are diffused throughout the whole set. Considerable time is needed to properly choose suitable key terms.

An example of print-out of an INSPEC data search entry follows:

0050465 A78032803
REPAIR AND CHECKING OF FLOW AND VOLUME MEASURE-
MENT DEVICES STALLOCH, G.: SCHULZ, E.
MSR (MESS, STEUERN REGELN) (GERMANY) VOL. 20. NO. 11
629–9 NOV. 1977 TO Coden: MSRGAN
Treatment: GENERAL, REVIEW JOURNAL PAPER
Languages: GERMAN
FOR PREVIOUS PART SEE IBID., NO. 7, P. 394. A TUTORIAL
ACCOUNT IS GIVEN OF LABORATORY PRACTICES. A RE-
VIEW OF EQUIPMENT MANUFACTURED IN E. GERMANY IS
PRESENTED (7 Refs)

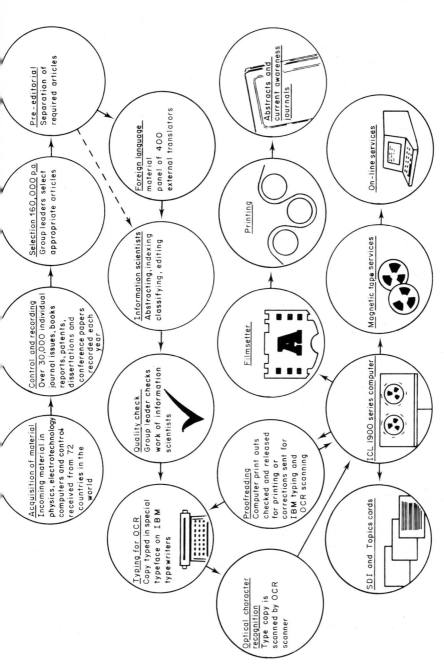

Figure 32.5 The INSPEC production system. (Reproduced by permission of the Electrical Engineers)

Descriptors: FLOWMETERS; LABORATORY APPARATUS AND TECHNIQUES; TESTING; VOLUME MEASUREMENT
Identifiers: CHECKING; VOLUME MEASUREMENT DEVICES; LABORATORY PRACTICES; FLOWMETER; REPAIR; EAST GERMAN DEVICES; LABORATORY APPARATUS
Section Class Codes: A0660V, A4780, A0670D, A0630C, A0130R

Study of this shows the parameters used, and provided, in an INSPEC search. The length of an output item can vary from a few lines to several pages. It is possible to retrieve less information than the whole above by appropriate input instructions: this is useful in early stages of search.

COMPENDEX corresponds to the periodical *Engineering Index Monthly*. It covers (since 1970) a wide spectrum of engineering, including electrical, electronics, chemical, and industrial. It also takes in management, mathematics, physics, and instruments. About 1500 serial publications and 900 monographs are used to feed the base. It grows at over 80,000 entries a year.

GIDEP is the only currently existing file that predominantly contains metrology material. It also has other subfiles on system reliability, failures, and experience. It is a US file and is normally only available to groups who subscribe to it with a sufficient quantity of input material.

The following files are less used to seek instrumentation information but nevertheless, could be useful. The dynamic pattern of base development is, however, bringing about name changes as groups combine or change direction. Detail will need to be checked at the time of a search.

AUSINET is an Australian national interest content base. It is listed here as an example of such. It contains material of national interest, such as government department reports. It may also have other-country files available and also provide miscellaneous services to local people (such as personal file keeping).

BIOSIS, BIO 6973 is an entire life sciences file corresponding to *Biological Abstracts* (BA) and *Bio Research Index* (BioI). It includes (from 1969) biomedical material and is sourced from over 8000 serials.

CHEMCON, CHEM 7071 contains an analytical chemistry section. Correspondence is with *Chemical Abstracts Condensates*. It covers from 1970 in part, all from 1972.

CDI, the *Comprehensive Dissertation Index*, contains all dissertations at academic doctoral level granted in the USA and in over 125 other-country institutions. It covers from 1861 and grows at over 300 citations a month.

CLAIMS is a multifaceted base filing data on US patents.

CPI provides (since 1973) a *Conference Papers Index* and contains over 800,000 entries of papers given at over 1000 major conferences.

ELECOMPS is a base of the European Space Agency. It covers passive and active electronic components from 1975.

ENVIROLINE contains citations on all areas of environmental studies. It corresponds to the journal *Environment Abstracts. ENVIROBIB* is in the same field.

GEOARCHIVE: geoscience material is abstracted here. Its contents begin in 1969 and it indexes material from over 5000 serials and the books from over 1000 publishers. The area of geophysics is here relevant to instrumentation.

ISMEC includes (from 1973) mechanical, production engineering, and engineering management. A specific topic area is measurement and control.

NTIS corresponds to the US *Weekly Government Abstracts* and semi-monthly *Government Reports Announcements.* It leads to titles in research and development activities published by over 200 US federal agencies.

SCI SEARCH files life sciences.

WPI, the *World Patent Index*, files (partially, since 1963) patent specifications issued by the major industrial nations. It corresponds to *Central Patents Index* and *World Patent Index.* It grows at over 250,000 new inventions per year.

32.6 THE IMEKO BOOK BIBLIOGRAPHY STUDY

The practical problems of establishing the existence of book material on measurement science and technology were recognized, in 1978, by the Higher Education Committee (TC-1) of the International Measurement Confederation (IMEKO).

It was decided that the Committee should establish a list of works for free distribution to member nations of IMEKO and any other interested user. The bibliography was completed in 1980 (Sydenham, 1980a).

Initially it was thought that the list would be quite short and that key articles in the serial literature could also be included. Entries were to be for works in the four official IMEKO languages: English, French, German, and Russian.

It was soon established that computer searching was not productive for book location—too few are filed that way, for too recent a time period. The best approach was found to be to select entries from a decade collection of publishers' catalogues and promotional mailings, from committee members' personal book collections, and from visits to the catalogues of libraries that were closely related to measurement interests.

Individual contributor's lists were found to rarely overlap each other, yet each person was convinced he was in touch with the bulk of related book literature. The final list includes over 800 titles. Books related to, but not directly concerned with, measurement and instrumentation (such as control theory, systems design), were not included.

Sections of the (by author only) listing, are on:

(a) fundamental concepts (theory of measurement for instance);
(b) units, standards, calibration;
(c) measurement uncertainty, results processing, errors;
(d) instrumentation practice (in ten groups);
(e) instrument system design, construction, and evaluation;
(f) education and training;
(g) miscellaneous (mainly historical).

An entry is included only once in the listing. A preferable manner to make use of this bibliography is for it to be placed on a local computer file for no cross-referencing is provided in the printed form—see Section 32.7.

The problems of deciding what are significant papers in the journal literature and the vast quantity available were realized when the first 18 instrument-related terms (of a possible 650 from the INSPEC thesaurus) were combined as *and* with review and tested on a short file length of INSPEC. It produced over 800 abstracts, most of which appeared valid entries for a bibliography. Clearly the user must go direct to the bases for citations, sorting them on-line to yield those of relevance.

32.7 SPECIALIZED MEASUREMENT-SCIENCE FILES

32.7.1 Personal Files

Browsing, current awareness services, and other sources such as reprint requests and conference attendance, provide the individual with a steady influx of quite particular information. For this to be useful it, in turn, requires ordered storage.

Traditional personal methods involve card indexes including those that can be manually sorted by indexing holes—called aperture cards—using a needle.

When the number of entries reaches a few hundred (which can occur rapidly as the IMEKO bibliography experience proved) it becomes worthwhile to enter the references into a personal computer or into available centralized services that can accommodate private files. At this level computer based options require only elementary sort routines because the problem of locating a few references in a thousand is far easier than in a few million entries. It is quite realistic to sort hardcopy, or VDU display, of a few tens of abstracts to decide which is relevant. Systems with a VDU display can also allow users to browse through the abstracts contained in order to locate those references that are known to be in the file but for which the key terms used are inappropriate for their recovery.

32.7.2 Need for a Major Base

The difficulty of sorting the 1500+ primary book title entries for the IMEKO bibliography proved the need for an information data base devoted to measurement-science and technology entries.

The IMEKO bibliography could begin the base but abstracts would be needed for each entry. Addition of relevant periodical literature would present great problems due to its diffuse and diverse nature.

Nevertheless, it appears that the now emerging maturity of measurement-systems studies into a recognizable and teachable systematic science of great importance to most of endeavour creates the case for establishment of an internationally available file.

REFERENCES

Jones, B. E. (Ed.) (1982). *Instrument Science and Technology*, Vol. 1, Adam Hilger, Bristol.

Melton, L. R. A. (1978). *An Introductory Guide to Information Sources in Physics*, Adam Hilger, Bristol.

Sydenham, P. H. (1979). *Measuring Instruments: Tools of Knowledge and Control*, Peter Peregrinus, London.

Sydenham, P. H. (Ed.) (1980a). 'A working list of books published on measurement science and technology in the physical sciences', International Measurement Confederation IMEKO (Delft: Applied Physics Dept, Technische Hogeschool Delft).

Sydenham, P. H. (1980b). 'Mechanical design of instruments', *Measurement and Control*, **13**, 365–72 and subsequent issues.

Sydenham, P. H. (1982). 'The literature of instrument science and technology', *J. Phys. E: Sci. Instrum.* **15**, 487–91 (To be reprinted in B. E. Jones (Ed), *Instrument Science and Technology*, Vol. 2, Adam Hilger, Bristol).

Van Brakel, J. (1977). *Meten in de Empirsche Wetenschappen* (*Measurement in the Empirical Sciences*'), Work Group for Basics in Physics, University of Utrecht, Utrecht.

Index